T0344968

Quantum Mechanics for Nanostructures

The properties of new nanoscale materials, their fabrication and applications, as well as the operational principles of nanodevices and systems, are solely determined by quantum-mechanical laws and principles. This textbook introduces engineers to quantum mechanics and the world of nanostructures, enabling them to apply the theories to numerous nanostructure problems.

The book covers the fundamentals of quantum mechanics, including uncertainty relations, the Schrödinger equation, perturbation theory, and tunneling. These are then applied to a quantum dot, the smallest artificial atom, and compared with the case of hydrogen, the smallest atom in nature. Nanoscale objects with higher dimensionality, such as quantum wires and quantum wells, are introduced, as well as nanoscale materials and nanodevices. Numerous examples throughout the text help students to understand the material.

VLADIMIR V. MITIN is SUNY Distinguished Professor at the Department of Electrical Engineering and Adjunct Professor of Physics at the University at Buffalo, The State University of New York. He is the author of eight textbooks and monographs and more than 490 professional publications and presentations.

DMITRY I. SEMENTSOV is Professor of Physics at Ulyanovsk State University, Russia. He is the author of more than 420 papers in peer-reviewed journals.

NIZAMI Z. VAGIDOV is Research Assistant Professor of Electrical Engineering at the University at Buffalo, The State University of New York. He is the author of about 90 professional publications in the fields of solid-state electronics, nanoelectronics, and nanotechnology.

Quantum Mechanics for Nanostructures

Vladimir V. Mitin
University at Buffalo, The State University of New York

Dmitry I. Sementsov
Ulyanovsk State University

Nizami Z. Vagidov
University at Buffalo, The State University of New York

CAMBRIDGE
UNIVERSITY PRESS

University Printing House, Cambridge CB2 8BS, United Kingdom

One Liberty Plaza, 20th Floor, New York, NY 10006, USA

477 Williamstown Road, Port Melbourne, VIC 3207, Australia

314-321, 3rd Floor, Plot 3, Splendor Forum, Jasola District Centre, New Delhi-110025, India

79 Anson Road, #06-04/06, Singapore 079906

Cambridge University Press is part of the University of Cambridge.

It furthers the University's mission by disseminating knowledge in the pursuit of education, learning and research at the highest international levels of excellence.

www.cambridge.org
Information on this title: www.cambridge.org/9780521763660

First published 2010
Reprinted 2015

A catalogue record for this publication is available from the British Library

Library of Congress Cataloging in Publication data
Mitin, V. V. (Vladimir Vasil'evich)
Quantum mechanics for nanostructures / Vladimir V. Mitin, Dmitry I. Sementsov, Nizami Z. Vagidov.
p. cm.
Includes index.
ISBN 978-0-521-76366-0 (hardback)
1. Quantum theory. 2. Nanostructured materials. I. Sementsov, Dmitry I. II. Vagidov, Nizami Z. III. Title.
TA357.M58 2010
620′.5 – dc22 2010000700

ISBN 978-0-521-76366-0 Hardback

Contents

Preface

Nanoelectronics is a field of fundamental and applied science, which is rapidly progressing as a natural development of microelectronics towards nanoscale electronics. The modern technical possibilities of science have reached such a level that it is possible to manipulate single molecules, atoms, and even electrons. These objects are the building blocks of nanoelectronics, which deals with the processes taking place in regions of size comparable to atomic dimensions. However, the physical laws which govern electron behavior in nanoobjects significantly differ from the laws of classical physics which define the operation of a large number of complex electronic devices, such as, for example, cathode-ray tubes and accelerators of charged particles. The laws that govern electron behavior in nanoobjects, being of quantum-mechanical origin, very often seem to be very strange from a common-sense viewpoint. The quantum-mechanical description of electron (or other microparticle) behavior is based on the idea of the *wave–particle duality of matter*. The wave properties of the electron, which play a significant role in its motion in small regions, require a new approach in the description of the electron's dynamic state on the nanoscale. Quantum mechanics has developed a fundamentally new probabilistic method of description of particle motion taking into account its wave properties. This type of description is based on the notion of a wavefunction, which is not always compatible with the notion of a particle's trajectory. This makes electron behavior harder to understand.

The main objects of research in nanoelectronics are quantum-dimensional structures such as *quantum wells*, *quantum wires*, and *quantum dots*, where electron motion is limited in one, two, and three directions, respectively. The size of these quantum-mechanical objects is comparable to the *electron de Broglie wavelength*. In such structures electronic properties become different from those of bulk materials: new so-called low-dimensional effects become apparent. Quantum-mechanical laws govern various processes and define a significant modification of the energy spectrum, which is the main characteristic of an electronic system. The energy spectrum which characterizes the electron motion in the limited region becomes discrete. The structures with such an energy spectrum are the basis for the development of new types of nanoelectronic devices.

The physics of quantum-dimensional structures is currently developing rapidly and is beginning to form a separate field with quantum mechanics as its basis. Only a small number of undergraduate engineering students take quantum-mechanics courses. However, there are only a few textbooks that are simple enough to understand for a wide range of engineering students, who would like to learn theoretical methods of analysis of the electronic properties of low-dimensional structures. While writing the current textbook we pursued two main goals: to present the main low-dimensional structures clearly from the physical point of view and to teach the reader the basics of quantum-mechanical analysis of the properties of such structures. Therefore, the experimental and theoretical material which will help the reader to understand the quantum-mechanical concepts applied to *nanostructures* is presented. Special attention is paid to the physical interpretation of quantum-mechanical notions. Theoretical material as well as the mathematical apparatus of quantum mechanics necessary for carrying out quantum-mechanical calculations independently is presented.

The book is written in such a way that it can be used by students who have studied classical physics to a sufficient extent as well as by students who have not had such an opportunity. The book consists of eight chapters and three appendices. The appendix material contains the main aspects of classical physics (particle dynamics, oscillations and waves in crystals, and electromagnetic fields and waves) which students can use while studying quantum mechanics.

In Chapter 1 we give a review of milestones in the development of *nanotechnology* and *nanoscience*. The main types of nanostructures are described and it is substantiated why it is necessary to use quantum physics for the description of their properties.

In Chapter 2 the main experimental facts which required the introduction of such unusual (for classical physics) notions as *wave–particle duality and uncertainty relationships*, among others, are described. The main notions and principles of the quantum-mechanical description are introduced. The Schrödinger equation – the main equation of non-relativistic quantum mechanics – is discussed in detail and its validity for the description of nanostructures is presented.

In Chapter 3 the solutions of the stationary Schrödinger equation are obtained for several important cases of one-dimensional motion. The main peculiarities of free electron motion as well as confined electron behavior are discussed. The main advantage of these solutions is in explanation and quantitative definition of the discrete energy levels of an electron when it moves in potential wells of various profiles.

In Chapter 4 the peculiarities of electron motion for structures wherein electron motion is confined in two and three dimensions are considered. It is shown that the discrete electron energy levels are characteristic for electron motion in potential wells of particular dimensionalities, in contrast to the continuous energy spectrum of a free electron. The structure's dimensionality and potential profile define the positioning of energy levels in the discrete energy spectrum.

The calculation of electron quantum states in various types of nanostructures generally encounters big mathematical difficulties. Therefore, approximate methods become very important for finding solutions of the Schrödinger equation. We consider in Chapter 5 several important and widely used approximate methods for calculation of electron wavefunctions, energy states, and transition probabilities between quantum states.

Chapter 6 is dedicated to finding wavefunctions, the geometry of electron clouds corresponding to them, and energy spectra of the simplest atoms and molecules using approximate methods.

When the size of the potential well is several times larger than the distance between atoms in a crystal, a fundamental reconstruction of the energy spectrum, which leads to a change in the physical properties of nanostructures, takes place. In Chapter 7 the main peculiarities of the electron energy spectrum in low-dimensional quantum structures (quantum wells, wires, and dots) as well as in periodic structures (superlattices) consisting of these low-dimensional nanostructures are considered.

In the last chapter – Chapter 8 – we consider the main methods of fabrication and characterization of nanostructures as well as their prospective applications in modern nanoelectronics.

Practically all chapters and appendices contain a large number of detailed examples and homework problems, which the authors hope will help students to acquire a deeper understanding of the material presented.

The authors have many professional colleagues and friends from different countries who must be acknowledged. Without their contributions and sacrifices this work would not have been completed. Special thanks go to the Division of Undergraduate Education of the National Science Foundation for the partial support of this work through its Course, Curriculum and Laboratory Improvement Program (Program Director Lance Z. Perez). The authors would like especially to thank Professor Athos Petrou for his editorial efforts in a critical reading of this book and for many valuable comments and suggestions. The authors also would like to thank undergraduate student Brian McSkimming for his thorough reading of the manuscript and helpful comments. We would like to thank undergraduate student Jonathan Bell for his help in preparation of figures.

Vladimir Mitin acknowledges the support and active encouragement of the faculty of the Department of Electrical Engineering and the Dean of the School of Engineering and Applied Sciences, Harvey G. Stenger Jr., as well as the members of the Center on Hybrid Nanodevices and Systems at the University at Buffalo, The State University of New York. He is also grateful to his family and friends for their strong support and encouragement, as well as for their understanding and forgiveness that he did not devote enough time to them while working on the book, and especially to his mother, grandson Anthony, and granddaughter Christina whom he missed the most.

Dmitry Sementsov thanks Tatiana Sementsova for her encouragement and help during the work on the manuscript.

Nizami Vagidov thanks his wife Saadat, his sons Garun, Timur, and Chingiz, his sisters Rukijat and Aishat, his brother Aligadji and their extended families for their constant support. Last but not least, he would like to thank his dissertation advisor, Professor Zinovi Gribnikov, for his encouragement and help.

Notation

Symbols

A – amplitude
A_{wf} – work function
a – lattice constant
a – acceleration
$\mathbf{a}_1$, $\mathbf{a}_2$, $\mathbf{a}_3$ – basis vectors
$\mathbf{B}$ – magnetic flux density
$\mathbf{C}$ – wrapping vector
C – capacitance
c – speed of light in vacuum
D – superlattice period
$\mathbf{D}$ – electric displacement
$\mathbf{d}$ – translation vector
E – energy of a particle
E_{c} – bottom of conduction band
E_{g} – bandgap
E_{i} – ionization energy
E_{v} – bottom of valence band
E_{F} – Fermi energy
$\mathbf{E}$ – electric field intensity
e – elementary charge
$\mathbf{e}_r$ – unit vector directed along radius vector $\mathbf{r}$
$\mathbf{e}_x$, $\mathbf{e}_y$, $\mathbf{e}_z$ – unit coordinate vectors
$\mathbf{F}_{\mathrm{gr}}$ – gravitational force
$\mathbf{F}_{\mathrm{L}}$ – Lorentz force
$\mathbf{F}_{\mathrm{m}}$ – magnetic force
$\mathbf{F}_{\mathrm{e}}$ – electric force
g – acceleration due to gravity; density of states
$\mathbf{H}$ – magnetic field intensity
H_n – Hermite polynomials
$\hat{\mathcal{H}}$ – Hamiltonian operator
h – Planck's constant
$\hbar$ – reduced Planck constant

I – current
I_T – tunneling current
i, **j**, **k** – unit coordinate vectors
K – kinetic energy; superlattice wavenumber
k – spring constant; wavenumber
k – wavevector
k_B – Boltzmann's constant
$k_\mathrm{e} = 1/(4\pi\epsilon_0)$ – coefficient in SI system
l – orbital quantum number
L – angular momentum
L_x, L_y, L_z – dimensions of a sample
m – orbital magnetic quantum number
m^* – effective mass of an electron
m_0 – mass of particle at rest
m_e – electron mass in vacuum
m_s – magnetic quantum number
N – number of states
N_A – Avogadro constant
n – principal quantum number; concentration
P – Poynting vector
$\mathcal{P}$ – pressure
P – probability
p – momentum
q – wavevector
Q – charge
q – charge of a particle
R – universal gas constant
r – magnitude of radius vector
r_1 – first Bohr radius
r – coordinate vector
R_∞ – Rydberg's constant
$\mathbf{R}_\mathrm{c}$ – radius vector of center of mass
S – spin
S – cross-section
t – time
T – time period; ambient temperature
$\mathcal{T}_\mathrm{d}$ – translation operator
U – potential energy; applied voltage
U_0 – height of potential barrier
U_G – gate voltage
u – displacement of atoms from their equilibrium positions
V – volume
v – velocity

$v_{\rm gr}$ – group velocity
$v_{\rm ph}$ – phase velocity
$V_{\rm c}$ – velocity of center of mass
W – work done by a force
$X(r)$ – radial function
x, y, z – spatial coordinates
Y_{ml} – spherical harmonics
α – angle; Madelung constant
β – force constant; $b/(2m)$
γ – gyromagnetic ratio
δ – logarithmic decrement of damping
$\delta(x)$ – Dirac's delta-function
ϵ – dielectric constant of the medium; relative deformation
ϵ_0 – permittivity of free space
ε – energy
ϕ – electrostatic field potential
φ – azimuthal angle; chiral angle; phase difference
λ – wavelength; parameter in characteristic equation
$\lambda_{\rm Br}$ – de Broglie wavelength
μ – magnetic permeability; magnetic moment
μ_l – orbital magnetic moment
$\mu_{\rm B}$ – Bohr magneton
$\nabla^2 = \frac{\partial^2}{\partial x^2} + \frac{\partial^2}{\partial y^2} + \frac{\partial^2}{\partial z^2}$ – Laplace operator
ψ – stationary wavefunction
Ψ – time-dependent wavefunction
Ω – angular velocity of a particle
ω – frequency
$\omega_{\rm e}$ – frequency of electron oscillations
$\omega_{\rm q}$ – frequency of harmonic oscillator
ρ – three-dimensional density
$\boldsymbol{\tau}$ – torque
θ – polar angle

Chapter 1
The nanoworld and quantum physics

1.1 A review of milestones in nanoscience and nanotechnology

It is extremely difficult to write the history of nanotechnology for two reasons. First, because of the vagueness of the term "nanotechnology." For example, nanotechnology is very often not a technology in the strictest sense of the term. Second, people have always experimented with nanotechnology even without knowing about it. Ironically enough, we can say that the medieval alchemists were the founding fathers of nanoscience and nanotechnology. They were the first researchers who tried to obtain gold from other metals. The ancient Greek philosopher Democritus also can be considered as a father of modern nanotechnology, since he was the first to use the name "atom" to characterize the smallest particle of matter. The red and ruby-red opalescent glasses of ancient Egypt and Rome, and the stained glasses of medieval Europe, can be considered as the first materials obtained using nanotechnology. An exhibition at the British Museum includes the Lycurgus cup made by the ancient Romans. The glass walls of the cup contain nanoparticles of gold and silver, which change the color of the glass from dark red to light gold when the cup is exposed to light. In 1661 the Irish chemist Robert Boyle for the first time stated that everything in the world consists of "corpuscules" – the tiniest particles, which in different combinations form all the varied materials and objects that exist.

In modern history the first practical breakthrough in nanotechnology was made by the American inventor George Eastman, who in 1884 fabricated the first roll film for a camera. This film contained a photosensitive layer of silver bromide nanoparticles. In 1931 the German physicists Max Knoll and Ernst Ruska developed an electron microscope, which for the first time allowed one to study nanoobjects.

The development of modern optical, microelectronic, material science, chemical, biological, and other technologies, which took into account quantum-dimensional effects, and, subsequently, the development of the main concepts and methods for the formation and control of nanoparticles has accelerated at an explosive rate. This development was based on the achievements and discoveries made by researchers in diverse fields of science.

The notion of "nanotechnology" was introduced for the first time by Richard Feynman in 1959 in his famous Caltech lecture "There's plenty of room at the bottom: an invitation to enter a new field of physics." Richard Feynman imagined the world of the nanoscale where the fundamental laws of quantum physics define the behavior of a single atom and control the formation of different structures from individual atoms. This vision of the great scientist ushered in the modern era of nanotechnology. The main achievements of this era are the following.

In 1952, L. V. Radushkevich and V. M. Lukyanovich published the first clear images of 50-nm-diameter carbon nanotubes. Carbon nanotubes were rediscovered many times after that.

In 1966, Robert Young suggested the use of piezomotors for positioning; these are currently used to move the tip in scanning-tunneling microscopes (STMs) and atomic-force microscopes (AFMs) with an accuracy of $10^{-2}-10^{-3}$ nm.

In 1968, Alfred Cho and John Arthur developed the theoretical foundations of nanotechnology for the processing of surfaces.

In 1974, the Japanese physicist Norio Taniguchi in his report "On the basic concept of nanotechnology" coined the term "nanotechnology," which he suggested using to name all the processes which take place in objects of size less than 1 μm.

In 1981, Gerd Binnig and Heinrich Röhrer developed their first STM, which enabled them to see individual atoms.

In 1985, Robert Curl, Harold Kroto, and Richard Smalley discovered fullerene – a molecule that resembles a soccer ball and contains 60 carbon atoms. This discovery accelerated the development of the fabrication technology of other carbon nanomaterials such as carbon nanotubes.

In 1986, the atomic-force microscope was introduced by Gerd Binnig, Calvin Quate, and Christoph Gerber. The same year the book *Engines of Creation*, by Eric Drexler which has been called the Bible of nanoscience, was published. Eric Drexler described in his book molecular self-replicating robots, which can assemble molecules, decompose molecules, record in a nanocomputer's memory programs for self-replication, and realize these programs. The predictions for a 20-year period made in this book are incredibly becoming reality. Also in 1986, the American physicist Arthur Ashkin invented *optical tweezers* – the device for manipulation of microobjects and nanoobjects with the help of a focussed laser beam.

In 1987, the French physicist Jean-Marie Lehn introduced the notions of "self-organization" and "self-assembly."

In 1990, Donald Eigler showed that it is possible to develop a molecular automaton. With the help of STM he wrote on one of the crystallographic edges of nickel the name of his company "IBM" using 35 individual xenon atoms.

Further studies showed that it is possible to fix atoms to the surfaces of other materials. Submolecular assembly became a reality from this moment on.

In 1991, the first artificial metamaterial, which was called by its creator, the American physicist Eli Yablonovich, "photonic crystal," was produced.

In 1998, the Dutch physicist Cees Dekker fabricated the first field-effect nanotransistor, which was based on a carbon nanotube. The technology for fabrication of nanotubes of length larger than 300 nm was developed.

In 1999, the American physicists Mark Reed and James Tour formulated the principles of the manipulation of a single molecule as well as chains of molecules.

In 2000, the principles of nanotomography, i.e., the creation of three-dimensional images of the inner structure of matter with a resolution of 100 nm, were developed.

In 2001, IBM researchers developed the first examples of logical circuits constructed on the basis of carbon-nanotube field-effect transistors.

In 2002, Cees Dekker created the first bionanostructure – a synthesis of a carbon nanotube and a DNA molecule.

In 2003, an international team of researchers deciphered the sequence of the human genome.

In 2004, British and Russian scientists obtained the first samples of graphene – a single layer of graphite, which has a two-dimensional hexagonal lattice.

In 2001–2005 a team of American scientists deciphered the mechanism of the replication of genetic information by cells.

In 2007 an international group of physicists from the USA, Germany, and Holland developed a scanning-electron microscope with subatomic resolution of 0.05 nm. The same year a group of American scientists developed the technology of scanning nanolithography with a resolution of 12 nm and a recording speed of more than 1 mm s^{-1}.

At present it is commonly accepted that Nobel laureate Richard Feynman in his lecture "There's plenty of room at the bottom" was the first to relate nanostructures and nanotechnology. In his lecture Feynman suggested that in the future it will be possible to move individual atoms with the help of devices of the same size. Using such devices, macroobjects can be assembled atom by atom, making the fabrication process cheaper by several orders of magnitude. It will be enough to supply these nanorobots with the necessary amount of molecules and write a program for the fabrication of the required product. In his lecture Feynman also mentioned the prospects of nanochemistry for the synthesis of new materials. As soon as physicists create these devices, which will be able to operate with individual atoms, most of the traditional methods of chemical synthesis will be replaced by the methods of atomic assembly. The development of such a technology at the atomic scale will help to solve many problems of chemistry and biology. One can only wonder how the great scientist envisioned the enormous potential of nanotechnology.

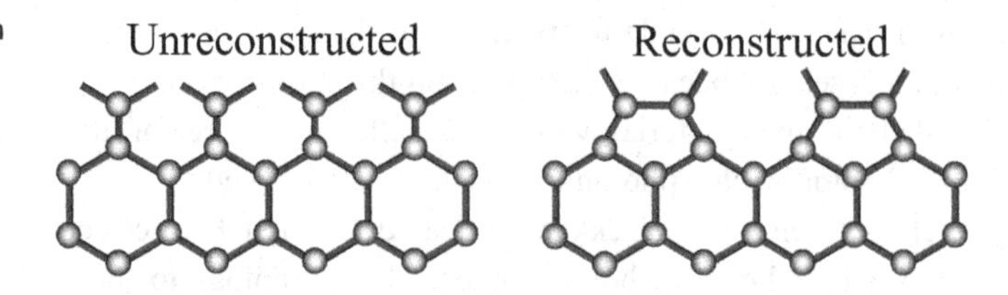

Figure 1.1 Reconstruction of Si surface.

1.2 Nanostructures and quantum physics

The prefix "nano" means one billionth part of something. Therefore, from the formal point of view nanostructures can be any objects with size (at least in one of the directions) of the order of 100 nm or less. Thus, nanostructures are objects whose sizes range from individual atoms (the size of an atom is about 0.1 nm) to large clusters consisting of up to 10^8 atoms or molecules. The transition of material structures from macroscale to nanoscale results in sharp changes of their properties. These changes are due to two reasons. The first reason is the increase of the proportion of surface atoms in the structure. The surface of the material can be considered as a special state of matter. The higher the proportion of atoms on the surface, the stronger are effects connected with the surface of a specimen. The ratio of the number of atoms located within a thin near-surface layer ($\sim$1 nm) to the total number of atoms in a specimen increases with decreasing volume of the specimen. Also the surface atoms are under conditions, which are very different from the conditions for the bulk atoms, because they are bound to the neighboring atoms in a different way. Atoms in the surface layer have some of their chemical bonds broken and therefore they are free to make new bonds. This results in a tendency of those electrons which do not form a pair to form a bond either with atoms of some other type that the surface adsorbs or with atoms of the same type.

If the surface is clean and smooth and there are no other atoms then the surface atoms establish bonds with each other. In the simplest case neighboring atoms of a surface layer unite to give so-called *dimers* (or pairs). The atoms of each dimer approach each other and at the same time move away from the other neighboring atoms which have formed dimers. Therefore, the lattice constant of the surface changes. Such a process is called *reconstruction*. As a result of atomic reconstruction a new type of atomic arrangement occurs at the surface (see Fig. 1.1). Also for those atoms at the edges of monatomic terraces and cavities, where the number of neighboring atoms is much smaller than that in the bulk volume, there exist special conditions. For example, the interaction of electrons with the free surface creates specific near-surface energy states. These facts lead us to consider the near-surface layer as a new state of matter.

Less clear is the second group of dimensional effects, which can be explained only by using a quantum-mechanical description. As will be shown further on, this group of effects is related to a significant increase of quantum effects when the size of the region where an electron moves significantly decreases. Therefore, the properties of nanoparticles strongly change compared with the properties of macroparticles of the same material. This happens mostly at characteristic sizes of 10–100 nm. According to quantum mechanics an electron can be presented as a wave, whose physical meaning will be explained in the following chapters of the book. The propagation of an electron wave in nanosize structures and its interaction with the boundary surfaces lead to the effects of energy quantization, interference of incident and reflected waves, and tunneling through potential barriers. Such a wave, which corresponds to a freely moving electron in an ideal crystalline material, can propagate in any direction. The situation radically changes when an electron is confined within a structure, whose size, L, along one of the directions of propagation is limited and is comparable to the electron de Broglie wavelength. In this case the electron cannot propagate in this specific direction and the electron can be described by a standing wave: only an integer number of electron half-wavelengths can fit within the structure of length L. This leads to the existence of non-zero discrete values of energy that an electron can have in this direction, i.e., the electron energy in this direction is no longer continuous but instead its spectrum consists of a set of separate energy levels. As a result, quantum confinement of electron motion increases the electron minimum energy. In the case of nanometer length of L the distance between energy levels exceeds the energy of thermal motion of the electron, which allows control of the electron energy by external fields. If in the two other directions the size of the structure is not limited, the energy of electron motion in these directions is not quantized and the electron may have any energy values. All this leads to the situation when the electric properties of nanosize structures differ from the well-known bulk properties of the materials from which the nanostructures are fabricated.

The self-interaction of electron waves in nanosize structures as well as their interaction with inhomogeneities and interfaces can be accompanied by the phenomenon of interference, which resembles the interference of electromagnetic waves. The distinctive feature of electron waves is that they are charged waves because the electron is a charged particle. This allows one to control the propagation of electron waves in nanostructures by the application of electric and magnetic fields.

The wave nature of microscopic particles, including electrons, is manifested by their ability to penetrate through an obstacle even when the particle's energy is lower than the height of the potential barrier of the corresponding obstacle. This phenomenon is called *tunneling* and it is a purely quantum phenomenon. According to classical mechanics an electron with energy E that encounters an obstacle with the potential barrier $U_0 > E$ on its path will reflect from this obstacle. However, the electron as a wave is transmitted through the obstacle (see Fig. 1.2). Quantum confinement in nanostructures specifically affects the processes of tunneling in them. Thus, the quantization of electron energy in very

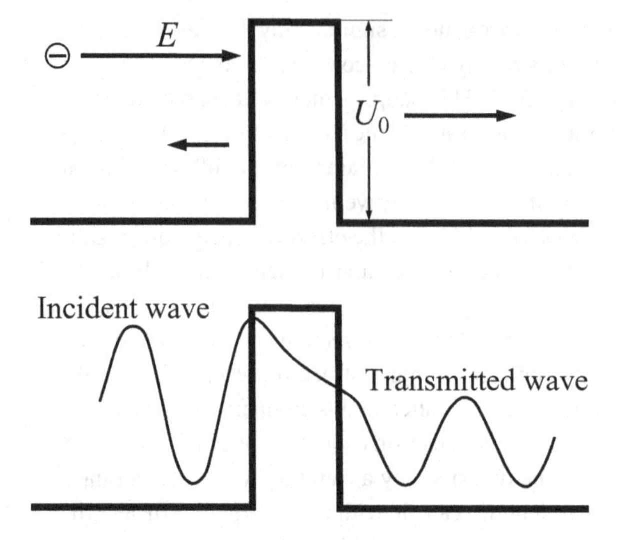

Figure 1.2 An electron with energy E tunneling through the potential barrier $U_0(U_0 > E)$.

thin and periodically arranged potential wells leads to the electron tunneling through these structures only having a certain energy, i.e., tunneling has a *resonance character*. Another such effect is *single-electron tunneling* when a charge is transmitted in an external electric field in portions equal to the charge of a single electron. After each tunneling event the system returns to its initial state. The quantum effects discussed above are widely used in nanoelectronic devices and elements of informational systems, but applications of electron quantum phenomena are not limited to these systems and devices. Currently active research is continuing in this direction.

The development of nanotechnology, which includes molecular-beam epitaxy, modern methods of molecular-beam lithography, diagnostics of nanoobjects, scanning-electron microscopy, scanning-tunneling microscopy, and atomic-force microscopy, is providing fundamentally new tools for the development of the elements of silicon, heterostructure, carbon, and nanomagnetic electronics. Nanotechnology that uses effects of self-organization, and molecular and atomic self-assembly, has become an alternative to the fabrication of macroobjects. The elemental basis of nanoelectronics includes a large number of structures and devices whose operation is based on various physical principles. Considering a variety of prospective directions, special attention must be paid to three of them: (1) the direction related to information technologies, (2) carbon nanotubes, and (3) nanoelectromechanical systems (NEMSs).

When considering any transport process (electric current, thermal conductivity, etc.), we assign the carriers a certain effective mean-free-path length, l. For a characteristic size of the structure $L \gg l$ the scattering of carriers takes place in the bulk of structure and it does not depend on the geometry and the size of the object. If, on the other hand, $L \leq l$, then the situation radically changes and all

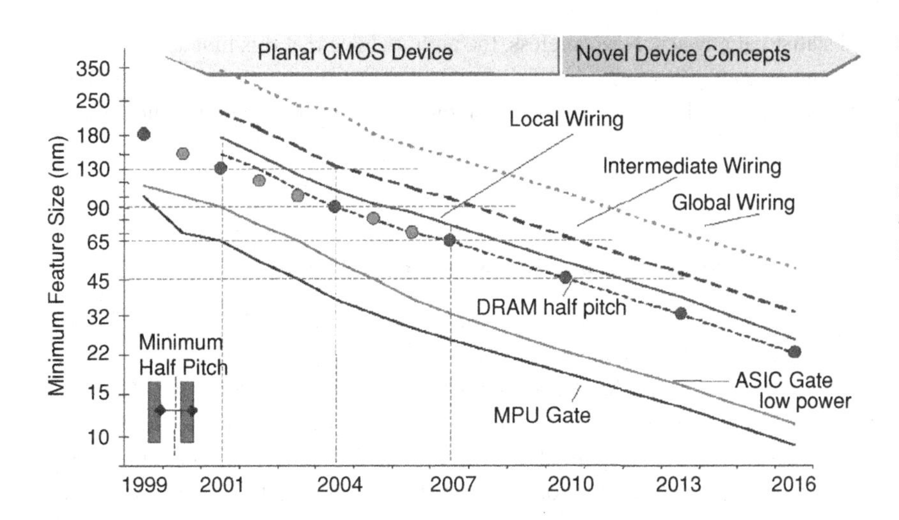

Figure 1.3 Technology nodes and minimum feature sizes from the ITRS 2002 Roadmap: MPU, microprocessing unit; ASIC, application-specific integrated circuit.

transport characteristics depend on the size of a specimen. Some of these effects can be described within the framework of classical physics, but there is a group of size effects that can be understood only on the basis of quantum mechanics.

Any achievements in nanoscience first of all are considered in terms of how they can be applied to the information technologies. Despite increasing difficulties, a very high rate of improvement of all significant parameters of electronics has been maintained during the last few decades. The most revolutionary achievements approach quantum limits when the working elements become a single electron, a single spin, a single quantum of energy, and so on. This may increase the speed of operations to close to 1 THz (10^{12} operations per second) and the writing density to about 10^3 Tbit cm^{-2}, which is significantly higher than the existing values, and energy consumption may be reduced by several orders of magnitudes. Having such a density of writing, we can store on a disk the size of a wristwatch a whole library (see Fig. 1.3 for the trends in miniaturization).

Quantum phenomena are currently widely used in nanoelectronic elements for information systems. However, utilization of the electron quantum properties is not limited to this. It is important to understand that the nanoscale is not just the next step in miniaturization. The behavior of nanostructures, in comparison with individual atoms and molecules, shows important changes, which cannot be explained by the traditional models and theories. The development of these new fields of science undoubtedly will lead to further scientific progress.

This book introduces the reader to the main ideas and laws of quantum mechanics using numerous examples, such as how to calculate energy spectra and other physical characteristics of certain types of nanostructures. In this book the authors had no intention to cover all the aspects of modern quantum physics and nanoelectronics because this task cannot be accomplished without deeper knowledge of subjects such as solid-state physics, the physics of semiconductors, and statistical physics. Nevertheless, the authors hope that this first acquaintance with quantum physics for most of the readers of this book will be useful in their future professional careers and will encourage them to study quantum phenomena at a higher level.

1.3 Layered nanostructures and superlattices

Atoms and molecules until recently were considered the smallest building bricks of matter. As the latest achievements of nanotechnology show, materials can be built not only from single elements but also from whole blocks. Clusters and nanoparticles may serve as such building blocks. Crystalline materials that consist of nanoscale blocks are called *bulk nanocrystalline materials*. These materials may have unique properties. For example, from everyday life we know that, if a material is durable, then it can be simultaneously fragile. The best example of a very durable but fragile material is glass. It turns out that some nanocrystalline materials are especially durable and elastic simultaneously. The unique mechanical properties of nanocrystalline materials in many respects are connected with the existence of an interface between nanoparticles. Such materials have properties that differ from those of the corresponding bulk material.

Below we discuss a class of nanocrystalline materials known as *superlattices*. There are different types of superlattices. Those of one type – heterostructure superlattices – can be grown by alternating layers of two different semiconductor materials, e.g., GaAs and AlGaAs, which have very similar lattice constants. Therefore, heterostructure superlattices can be referred to as *layered structures*. The main elements of the layered structures are two types of layers: (1) the layer of the so-called *narrow-bandgap semiconductor* (GaAs) and (2) the layer of the *wide-bandgap semiconductor* (AlGaAs). These two elements can be used to create another layered structure called a *quantum well*. A thin layer of GaAs between two layers of AlGaAs creates a potential well for an electron, where its motion is restricted. In the next chapters of the book we will consider theoretically the electron motion in such layered structures (see Fig. 1.4).

More generally, superlattices are structures with periodic repetition along one, two or three directions of regions with different values of some physical quantity (dielectric or magnetic permeability, the type and mobility of carriers, the work function, elasticity, and so on). Periodicity along one direction of such a layered structure results in a *one-dimensional superlattice*. If there is a periodicity along two directions, then the superlattice is *two-dimensional*. An example of such a superlattice is a two-dimensional system of quantum wires of a semiconductor formed on the surface of another semiconductor. In such a two-dimensional material the electric properties of material periodically change along two directions. In a *three-dimensional superlattice* the periodicity of physical properties can be observed along three directions.

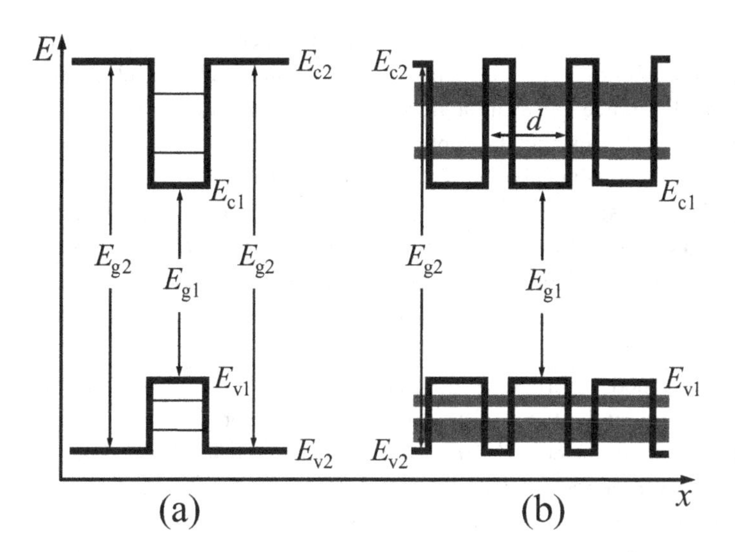

Figure 1.4 The energy bandstructure of a semiconductor quantum well (a) and a type-I heterostructure superlattice (b): E_{g1} and E_{g2} are the bandgaps, E_{c1} and E_{c2} are the bottoms of the conduction bands, and E_{v1} and E_{v2} are the tops of the valence bands of narrow-bandgap and wide-bandgap semiconductors, respectively; d is the period of the heterostructure superlattice.

Semiconductor superlattices, which consist of thin semiconductor layers alternating in one direction, i.e., heterostructure superlattices, have a wide range of applications. The period of such a superlattice usually is much larger than the lattice constant but is smaller than the electron mean free path. Such a structure possesses, in addition to a periodic potential of the crystalline lattice, a potential due to the alternating semiconductor layers. The existence of such a potential significantly changes the energy bandstructure of the semiconductors from which the superlattice is formed. A very important peculiarity of a superlattice is the existence of its own *minibands*. These peculiarities become apparent when studying optical and electric properties of semiconductor superlattices. Since semiconductor superlattices have special physical properties, we can consider superlattices as a new type of semiconductor materials.

The superlattices can be of several types. The most common are *heterostructure* superlattices and *modulation-doped* superlattices. A heterostructure superlattice is a representative of layered nanostructures: it consists of epitaxially grown alternating layers of different semiconductors, which have similar lattice constants. Historically the first heterostructure superlattices were grown for the semiconductor system $GaAs/Al_xGa_{1-x}As$. The success in the growth of such a superlattice was due to the fact that Al has the same valence and ionic radius as Ga, and therefore the incorporation of Al does not cause noticeable distortions of the crystalline structure of the material. At the same time Al may sufficiently modulate the amplitude of the superlattice potential. Depending on the relative position of the semiconductor energy bands, heterostructure superlattices can be divided into two main types: type I and type II. $GaAs/Al_xGa_{1-x}As$ superlattices belong to the first type. The conduction-band minimum and the maximum of the valence band for GaAs are situated inside of the bandgap of $Al_xGa_{1-x}As$ (see Fig. 1.4). Such band alignment leads to a periodic system of quantum wells for current carriers in GaAs, which are separated from each other by potential barriers created by $Al_xGa_{1-x}As$. The depth of the quantum wells for electrons is defined by the difference between the minima of the conduction bands of the two semiconductor materials, and the depth for quantum wells for holes is given by the difference between the maxima of the valence bands.

In the type-II heterostructure superlattices the minimum of the conduction band of one semiconductor is situated in the energy bandgap of the second, and the maximum of the valence band of the second semiconductor lies in the bandgap of the first. A representative of this type of superlattice is the system $In_xGa_{1-x}As/GaSb_{1-y}As_y$.

In modulation-doped superlattices the periodic potential is formed by the alternating layers of n- and p-types of the same semiconductor. These layers can be separated by undoped layers. The most common material for fabrication of modulation-doped superlattices is GaAs.

Besides heterostructure and modulation-doped superlattices, other types of superlattice are possible: they differ by the way in which the modulation potential is created. In spin superlattices, the semiconductor material is doped with magnetic impurities. A periodic potential occurs in such superlattices when an external magnetic field is applied. A superlattice potential can be created also by periodic deformation of a specimen in the field of a powerful ultrasound wave and an electromagnetic standing wave. Superlattices have a wide range of applications in diverse semiconductor devices. The most striking example is their use in cascade semiconductor lasers.

1.4 Nanoparticles and nanoclusters

Nanoparticles are atomic or molecular structures, whose size is equal to 100 nm or less. Such nanoobjects consist of 10^8 or fewer atoms (or molecules). Their properties differ from the properties of bulk materials consisting of the same atoms (or molecules). Nanoparticles whose size is equal to 10 nm or less that contain up to 10^3 atoms are called *nanoclusters* or, simply, *clusters*. Numerous studies have shown that for a given material there exist clusters having only a certain number of particles. This means that clusters consisting of these numbers of particles are the most stable ones. The corresponding numbers are called *magic numbers*. The set of magic numbers shows how clusters (from the smallest to the biggest) are formed from individual particles. An example of this structure of stable clusters is the *closest packing* of identical spheres.

The first magic number is 13, which corresponds to the packing when the internal sphere is surrounded by 12 spheres of the same radius. If subsequent shells of identical spheres are also filled, then their total number corresponds to the following magic numbers: 55, 147, 309, 567, and so on. The number of particles, N_n, in the nth shell can be calculated using the following formula:

$$N_n = 10n^2 + 2. \tag{1.1}$$

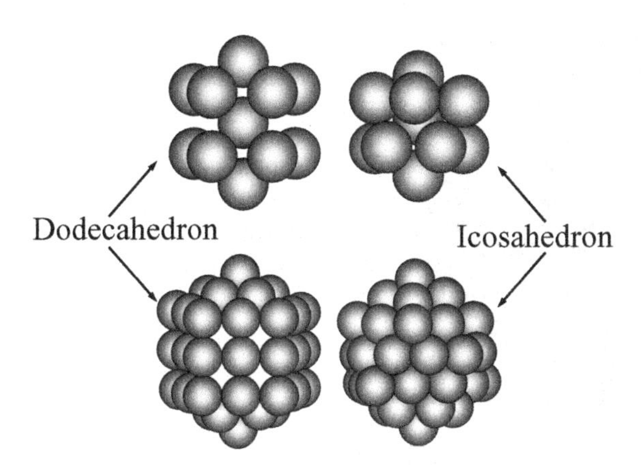

Figure 1.5 Clusters of 13 and 55 nanoparticles.

Thus, in the first shell there are 12 spheres ($N_1 = 12$) around a single sphere and, therefore, the magic number is equal to 13. In the second shell there are 42 spheres ($N_2 = 42$) and, by adding this number to 13, we obtain the next magic number which is equal to 55. In this case the cluster has the form of an *icosahedron* and this structure is the most stable. In some cases clusters have the form of a *dodecahedron*. Both types of packing of spheres are shown in Fig. 1.5. Such clusters are formed mostly during growth under vacuum conditions from liquid or gaseous phases.

Many physical properties of nanoparticles or nanoclusters differ from the properties of a bulk material consisting of the same type of atoms. Thus, nanoparticles and nanoclusters have a crystalline structure slightly different from the bulk crystal, which is due to the significant influence of the surface of the specimen. For example, gold nanoclusters of size 3–5 nm crystallize not having a face-centered cubic lattice as bulk gold does, but as an icosahedral structure. Let us estimate the number of surface atoms in a cluster that consists of N atoms and whose form is close to spherical. The volume of such a cluster is

$$V \approx \frac{4\pi}{3} R^3 = V_0 N, \tag{1.2}$$

where R is the radius of a sphere and V_0 is the volume that *corresponds* to an individual atom. Note, it is not the volume of the atom itself! Let us assume that the volume V_0 can be presented as a sphere of radius a. Therefore,

$$V_0 = \frac{4\pi}{3} a^3. \tag{1.3}$$

For the structure's closest packing the parameter a is almost equal to the radius of an atom. Then, from Eqs. (1.2) and (1.3), we find that the size of a cluster and the radius of an individual atom are related as

$$R = a N^{1/3}. \tag{1.4}$$

For most clusters the size of constituent atoms is close to $a \approx 0.1$ nm. From Eq. (1.4) it follows that a cluster that consists of 10^3 atoms has a size of about $R \approx 1$ nm. For a cluster that consists of molecules this size is significantly larger.

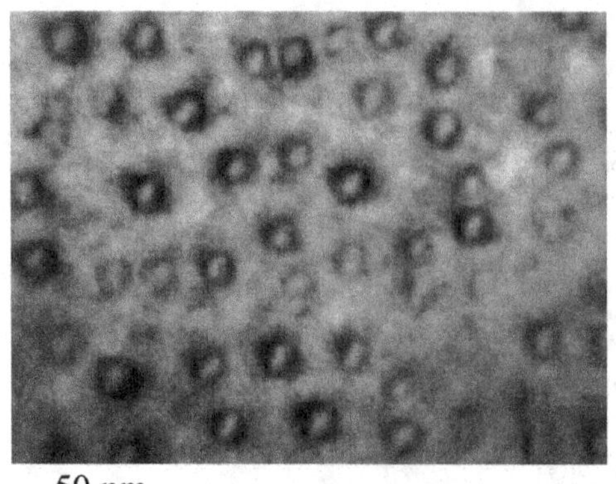

Figure 1.6 A scanning electron microscope image of InAs quantum dots in a GaAs matrix.

The surface area, S, is a very important characteristic of a cluster. Let us estimate S for a cluster with a spherical surface:

$$S \approx 4\pi R^2 = 4\pi a^2 N^{2/3}. \tag{1.5}$$

The number of atoms on the surface of a cluster, N_S, is connected with the surface area, S, as

$$S \approx S_0 N_S = 4\pi a^2 N_S, \tag{1.6}$$

where $S_0 = 4\pi a^2$ is the area that corresponds to an individual atom on the surface of a cluster. Let us find the ratio of the number of atoms, N_S, on the surface of a cluster to the number of atoms, N, in a cluster using Eqs. (1.2) and (1.6):

$$\frac{N_S}{N} = \frac{SV_0}{S_0 V} = \frac{3V_0}{RS_0} = \frac{a}{R} = \frac{1}{N^{1/3}}. \tag{1.7}$$

As we see from the above expressions, the proportion of atoms on the surface of a cluster rapidly decreases with increasing cluster size. Noticeable influence of the surface occurs at a cluster size less than 100 nm.

The formation of energy bands of crystals is manifested by the quantum-dimensional effects, which become apparent when the size of the region of electron motion is comparable to the electron de Broglie wavelength in the material. In metals this wavelength is about 0.5 nm, whereas in semiconductors this wavelength can be up to 1 μm. Therefore, quantum-dimensional effects can be observed at much larger sizes of semiconductor nanoparticles than for metal nanoparticles. In semiconductor clusters called quantum dots electrons are confined in all three directions. Such individual quantum dots or their arrays are frequently created in the matrix of some other semiconductor material. In this case quantum dots are regularly positioned as "islands" of one semiconductor on the surface of the other semiconductor. Figure 1.6 shows the example of an InAs quantum-dot array grown on the surface of GaAs.

The discreteness of electron energy in an individual quantum dot allows one to call it an *artificial atom*. On the basis of quantum dots highly-efficient miniature sources of light can be developed. By varying the size and composition of quantum dots new light-emitting diodes (LEDs) of different colors may be fabricated. Owing to the tunneling effect, electron transport along a chain of quantum dots is possible. This can be used for the development of numerous electronic devices.

In semiconductor devices of micrometer size it is necessary to control (including turn on and turn off) the current, which corresponds to the flux of hundreds of thousands of electrons. With the help of quantum dots we can control the motion of single electrons, which opens new possibilities for the further miniaturization of semiconductor devices and further decrease of power consumption.

Example 1.1. Estimate the size of a spherical nanocluster of water consisting of 100 atoms. Estimate the area and diameter of monomolecular water film that is formed when the nanocluster is spread over the surface of a sample.

Reasoning. The number of molecules, N, in a water cluster of volume V and mass m is defined by the following equation:

$$N = \frac{m}{\mu} N_A, \tag{1.8}$$

where $N_A = 6.02 \times 10^{23}\ \text{mol}^{-1}$ is the Avogadro constant and μ is the molar mass of a molecule. The volume of one molecule of water, V_0, can be estimated according to the following formula:

$$V_0 = \frac{V}{N} = \frac{V\mu}{mN_A} = \frac{\mu}{\rho N_A} = \frac{4}{3}\pi r_0^3, \tag{1.9}$$

where $\mu = 18\ \text{g mol}^{-1}$ is the molar mass of water and $\rho = 1\ \text{g cm}^{-3}$ is the density of water. From Eq. (1.9) we can find the radius of a molecule, r_0:

$$r_0 = \left(\frac{3\mu}{4\pi\rho N_A}\right)^{1/3} \approx 0.2\ \text{nm}. \tag{1.10}$$

Let us write the relationship that connects the cluster's volume, V, and radius, R, with the radius of a water molecule, r_0:

$$V = \frac{4}{3}\pi R^3 = N\frac{4}{3}\pi r_0^3. \tag{1.11}$$

From Eq. (1.11) we find the cluster's radius, R:

$$R = N^{1/3} r_0 \approx 0.9\ \text{nm}. \tag{1.12}$$

Since the diametral cross-section of a molecule is defined as

$$S_0 = \pi r_0^2, \tag{1.13}$$

the area of the water spot is defined by the following expression:

$$S = \pi R_S^2 = N\pi r_0^2 \approx 12.6\ \text{nm}^2, \tag{1.14}$$

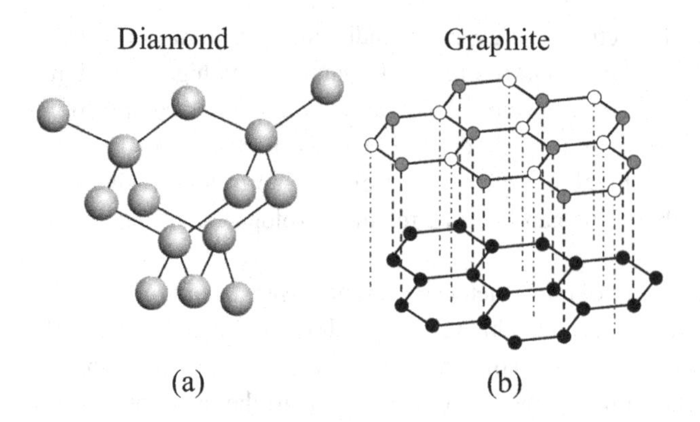

Figure 1.7 The crystalline structures of diamond (a) and graphite (b).

where R_S is the radius of the water spot which occurred after spreading of the water cluster. From the last expression we find R_S:

$$R_S = N^{1/2} r_0 \approx 2 \text{ nm}. \tag{1.15}$$

1.5 Carbon-based nanomaterials

Carbon (C) is an element of group IV of the Periodic Table of the elements. Therefore, it has four valence electrons. It is widespread and it is the basis of living matter, but there is only 0.19% of carbon in the Earth's crust. The ability of atoms of the same chemical element to combine in different spatial configurations is called *allotropy*. Carbon possesses this property in full measure: there are several allotropic forms of carbon. First of all are diamond and graphite, which are shown in Fig. 1.7. In the diamond crystal each carbon atom is in the site of a tetrahedral lattice with average distance between atoms equal to 0.154 nm. Four valence electrons of each carbon atom form four strong C—C bonds. It is difficult to break them since there are no conduction electrons: diamond crystal is a dielectric. For the same reason diamond has exceptional hardness and a high melting temperature ($T_{mt} = 3277$ K).

Another allotropic form of carbon is graphite, which has exceptionally different physical properties from those of diamond. Graphite is a soft black substance consisting of easily flaked layers, which are called *graphene sheets*. Within the plane of a graphene sheet carbon atoms have strong covalent bonds with each other. These bonds form the lattice consisting of regular *hexagons*. However, in contrast to the case of diamond, in graphite only *three electrons* participate in establishing bonds. The fourth electron of each carbon atom does not participate in the formation of interatomic bonds and therefore it is free. This makes graphite such a good conductor. There is only a weak attraction between graphite layers due to van der Waals forces. The weakness of the attraction between graphite layers results in their easy sliding with respect to each other. Therefore, they can be easily flaked apart.

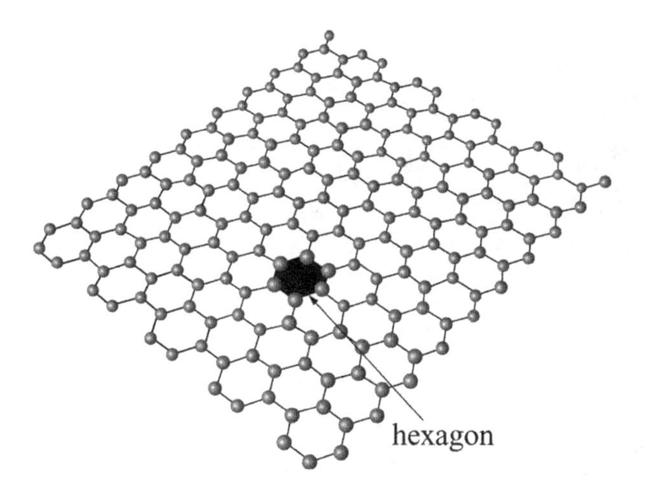

Figure 1.8 A schematic model of graphene in the form of a two-dimensional hexagonal lattice.

1.5.1 Graphene

Although graphite had been known for a long time, researchers managed to obtain single graphite layers and study them only in 2004. The material, which is one gigantic two-dimensional plane carbon molecule of monatomic width, was called *graphene* (see Fig. 1.8). Thus, graphene is a *two-dimensional crystal*, which consists of a single layer of carbon atoms composed in a *hexagonal lattice*. The properties of graphene turned out to be amazing. It is well known that graphite is a *semimetal*, i.e., it does not have a *bandgap*. The *bandstructure* of graphene also does not have a bandgap. At the points of intersection of *valence band* and *conduction band* the energy spectrum, $E(\mathbf{k})$, of electrons and holes has a linear dependence. A similar spectrum is possessed by *photons*, whose mass at rest is equal to zero. Therefore, it is said that the *effective mass* of electrons and holes in graphene near the intersection point is equal to zero. However, let us note that, despite the fact that photons and massless carriers in graphene have similarities, there are significant differences between them that make carriers in graphene unique. First of all, electrons and holes are *fermions* and they possess charge. There are no analogs for these massless charged fermions among known elementary particles. Second, graphene possesses unusual physical and chemical properties. Since the effective mass of electrons in graphene tends to zero, they have a high mobility, 100 times larger than the mobility of electrons and holes in crystalline silicon – the most widely used material in nanoelectronics. This explains the gigantic thermal conductivity and good electrical conductivity of graphene. These properties, together with the transparency and outstanding mechanical properties, make graphene a prospective material for nanotechnology. Using graphene as a basis, a new class of materials with extreme consumer properties can be developed. However, the most interesting effects are the electronic properties of graphene since their application opens new possibilities for

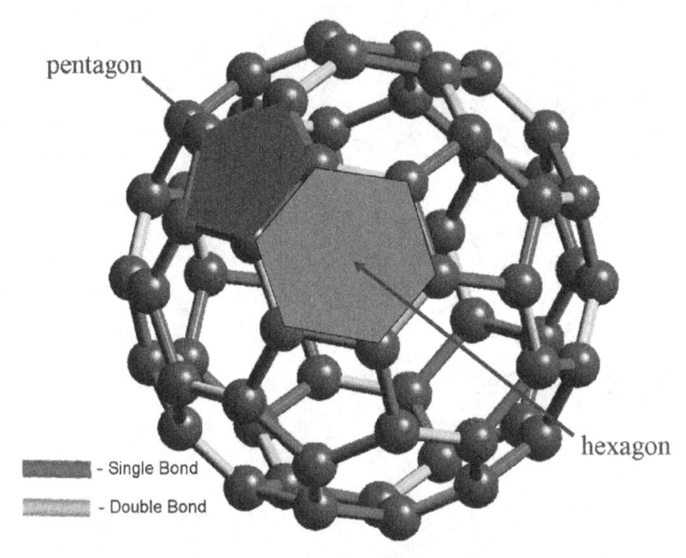

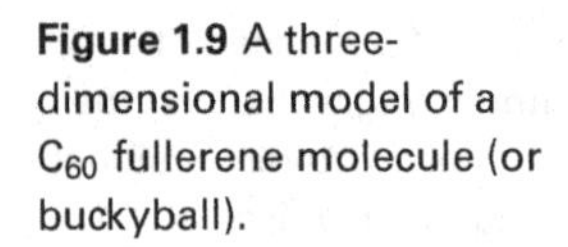
Figure 1.9 A three-dimensional model of a C_{60} fullerene molecule (or buckyball).

the development of elements for nanoelectronics. Thus, depending on the type of substrate and geometrical dimensions, the electronic properties of graphene may be either metallic or semiconducting.

1.5.2 Fullerenes

In 1990, one more crystalline modification of carbon called *fullerite* was discovered. Fullerite has as a structural unit not a carbon atom, as in the case of graphene or a carbon nanotube, but a molecule of *fullerene*. Fullerenes are a new class of carbon material, whose molecules have the form of skeleton structures reminding one of a soccer ball. In such molecules carbon atoms are at the vertices of regular pentagons and hexagons, which are placed on the surface of a sphere or a spheroid (see Fig. 1.9). Different fullerene molecules can consist of 28, 32, 50, 60, 70, 76, and so on carbon atoms. Although there are various geometrical forms of fullerenes, quantum-mechanical calculations of stable carbon structures show that their formation obeys certain rules. Stable carbon clusters have the form of polyhedra. The outer electron shell of an individual carbon atom provides stable bonds, which result in the formation of carbon pentagons or hexagons. The most stable fullerene molecule is C_{60}. The skeleton of the C_{60} molecule consists of 12 regular pentagons and 20 inequilateral hexagons. Each hexagon has three pentagons and three hexagons as its neighbors, whereas pentagons have as neighbors only hexagons. Such a structure provides the C_{60} molecule with an extraordinary stability. During the formation of closed geometrical figures pentagons provide the bending of a graphene sheet. The length of C—C bonds depends on the boundary on which they are located. The length of this bond at

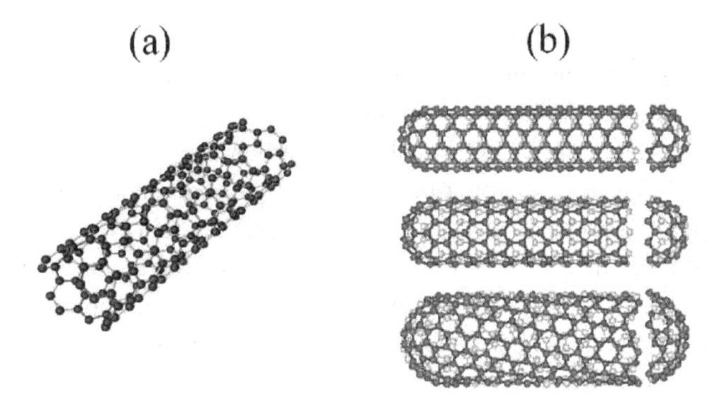

Figure 1.10 Three-dimensional models of carbon nanotubes: (a) open and (b) closed nanotubes.

the hexagon–hexagon boundary is equal to 0.132 nm and it is a double covalent bond. The length of the bond on the hexagon–pentagon boundary is equal to 0.144 nm and it is a single covalent bond. Three electrons of each carbon atom participate in the formation of the fullerene structure. The fourth electron has a free chemical bond. Therefore, these molecules possess the important property of adsorbing atoms of other materials (for example, atoms of hydrogen or fluorine).

In the fullerite crystalline structure, C_{60} fullerene molecules are attracted to each other by the weak van der Waals forces. In the face-centered cubic lattice the centers of C_{60} molecules are at a distance of 1 nm from each other. In the unit cell of fullerite 26% of the volume between spherical molecules of fullerene is hollow. Atoms of alkali elements can be easily placed in this empty space. The C_{60} crystal is a dielectric, but, when it is doped with atoms of alkali elements, it becomes a conductor. Thus, doping of this crystal with potassium forms a K_3C_{60} compound. In this compound the potassium is in an ionized state and each C_{60} molecule acquires an additional three electrons weakly connected with the molecule moving around the crystal, which makes the compound a conductor. On decreasing the temperature to $T_{cr} = 18$ K, a doped K_3C_{60} fullerite undergoes a transition into a superconducting state. The record temperature for the superconducting transition for $CHBr_3C_{60}$ of $T_{cr} = 117$ K has been demonstrated recently by researchers from IBM. In another crystal, $CHCl_3C_{60}$, the critical temperature was $T_{cr} = 80$ K. The measured lattice parameters of crystalline samples of the above-mentioned fullerites are 14.45 Å and 14.28 Å, respectively.

1.5.3 Carbon nanotubes

After the discovery of fullerenes it was established that graphene sheets can, under certain conditions, roll up into tubes. These objects were called *carbon nanotubes*. Carbon nanotubes are hollow elongated cylindrical structures of diameter from one to several tens of nanometers. The ideal carbon nanotube is rolled up into a cylindrical graphene sheet. There are different forms of carbon nanotubes. Carbon nanotubes can be classified as single-walled or multiwalled, chiral and non-chiral, long and short, and so on. Carbon nanotubes can be open and closed, as shown in Fig. 1.10. Nanotubes are unusually strong with respect to stretching and flexing. Under high mechanical stress carbon nanotubes cannot be torn or broken. They just reconstruct their structure. Carbon nanotubes possess important properties for practical applications: they can sustain electric high-density currents, change their properties when they adsorb other atoms or molecules, emit electrons from their ends at low temperatures, and so on. Carbon is not the only material for the growth of nanotubes. So far nanotubes of boron nitride, boron and silicon carbides, and silicon oxide have been grown.

The authors recommend the following textbooks on quantum mechanics for further reading.

R. P. Feynman, R. B. Leighton, and M. Sands, *The Feynman Lectures on Physics*, Vol. III (Reading, MA, Addison-Wesley, 1977).

L. Landau and E. Lifshitz, *Quantum Mechanics* (Oxford, Pergamon Press, 1968).

D. Bohm, *Quantum Theory* (New York, Prentice Hall, 1951).

P. A. M. Dirac, *The Principles of Quantum Mechanics* (Oxford, Oxford University Press, 1958).

L. I. Shiff, *Quantum Mechanics* (New York, McGraw-Hill, 1968).

D. K. Ferry, *Quantum Mechanics* (London, IOP Publishing, 1995).

I. I. Gold'man and V. D. Krivchenkov, *Problems in Quantum Mechanics* (New York, Dover Publications, 1993).

G. L. Squires, *Problems in Quantum Mechanics* (Cambridge, Cambridge University Press, 1995).

Chapter 2
Wave–particle duality and its manifestation in radiation and particle behavior

2.1 Blackbody radiation and photon gas

2.1.1 Thermal radiation

Heated bodies emit electromagnetic radiation over a wide frequency range because a part of their internal energy transforms into radiation energy. This type of radiation is called *thermal radiation* and it is caused by the supply of energy to the radiating body. If we place several bodies, heated to different temperatures, in a closed cavity with perfectly reflecting walls, then after a certain period of time the whole system will transfer to a state of thermal equilibrium with all bodies having the same temperature T. The notion of temperature, T, is introduced to characterize the amount of average kinetic energy, E_{kin}, of thermal motion of a body's particles. Energy in the Système International (SI) units is measured in joules (J) and temperature in degrees Kelvin (K). The temperature can be introduced as

$$T = \frac{2}{3}\frac{E_{\text{kin}}}{k_{\text{B}}}, \tag{2.1}$$

where $k_{\text{B}} = 1.38 \times 10^{-23}\,\text{J}\,\text{K}^{-1}$ is known as *Boltzmann's constant.* When the temperature of all bodies in the system becomes the same, then the energy emitted by the body becomes equal to the energy absorbed by the same body, and such a state of the system does not change with time. At the same time the radiation emitted by the heated bodies and cavity is in thermodynamical equilibrium, and therefore it is called *equilibrium radiation* (it is also known as *blackbody radiation*). Numerous experimental studies of this radiation have shown that its spectrum is continuous, i.e., the frequency interval of this radiation spans the entire electromagnetic spectrum. The distribution of energy of blackbody radiation significantly depends on the temperature of the radiation-emitting body. With increasing temperature the maximum of this distribution shifts to the region of higher frequencies and at the same time the total energy emitted by a body in the entire spectral range increases. Thus, bodies at room temperature

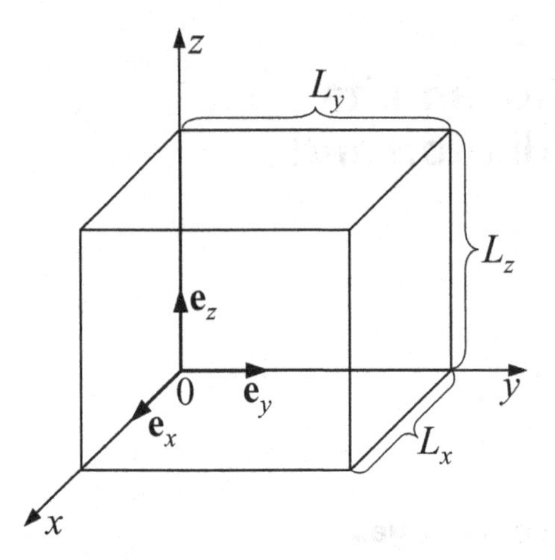

Figure 2.1 A cavity containing thermal radiation.

($T = 300$ K) emit energy with a maximum in the invisible infrared part of the spectrum (the wavelength is about $\lambda \approx 14\,\mu$m). The surface of the Sun, which has the temperature $T \approx 6000$ K, has a maximum of radiation in the region of visible light ($\lambda \approx 0.5\,\mu$m). During a nuclear explosion the temperatures attained are $T \approx 10^7$ K. Therefore, most of the energy of the explosion is carried away by highly penetrating X-ray radiation ($\lambda \approx 10^{-4}\,\mu$m).

The distribution of blackbody radiation of frequency ω is characterized by the spectral density u_ω, which has the dimension [J s m^{-3}]. The magnitude $u_\omega\, d\omega$ defines the energy of the radiation per unit volume in the interval of frequencies from ω to $\omega + d\omega$. The spectral density, $u_\omega(T)$, of blackbody radiation is a universal function of frequency and temperature, and it does not depend on the nature of the radiation-emitting body. Since this function describes the distribution of energy of an electromagnetic field, for a long time it was believed that it could be calculated on the basis of classical concepts. However, all the attempts to do this did not succeed.

2.1.2 The number of states and density of states

Let us consider a cavity with ideally reflecting walls, which has the form of a rectangular parallelepiped with edge lengths L_α, where $\alpha = x, y$, and z. In such a cavity the blackbody radiation as an electromagnetic field can exist only in the form of a superposition of standing electromagnetic waves, which have nodes at the walls of the cavity. Let us define the number of such standing waves which can be established in the cavity (see Fig. 2.1). The conditions for standing-wave formation along each of the axes are

$$L_x = n_x \frac{\lambda_x}{2}, \qquad L_y = n_y \frac{\lambda_y}{2}, \qquad L_z = n_z \frac{\lambda_z}{2}, \tag{2.2}$$

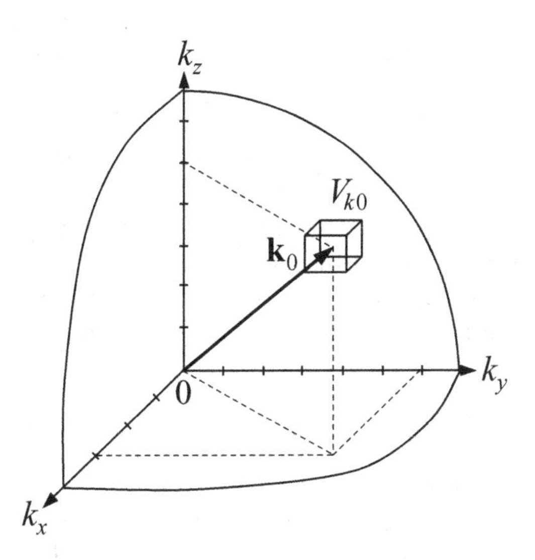

Figure 2.2 The region of *k*-space which corresponds to standing electromagnetic waves with positive components of the wavevector, **k**.

where n_α are positive integer numbers, i.e., each length L_α between reflecting walls has to contain an integer number of half-waves with the wavelength λ_α along the corresponding direction. For the standing wave which is established in an arbitrary direction, its wavevector can be presented as

$$\mathbf{k} = k_x \mathbf{e}_x + k_y \mathbf{e}_y + k_z \mathbf{e}_z. \tag{2.3}$$

Taking into account that

$$k_\alpha = \frac{2\pi}{\lambda_\alpha} \tag{2.4}$$

(see Eq. (B.58) from Appendix B), the projections of a wavevector **k** onto each of the coordinate axes take the form

$$k_\alpha = \frac{\pi n_\alpha}{L_\alpha}.$$

As a result the expression for the wavevector of an arbitrarily directed standing wave can be written in the following form:

$$\mathbf{k} = \pi \left(\frac{n_x}{L_x} \mathbf{e}_x + \frac{n_y}{L_y} \mathbf{e}_y + \frac{n_z}{L_z} \mathbf{e}_z \right). \tag{2.5}$$

In wavevector space (*k*-space) (see Fig. 2.2) the point with projections k_x, k_y, and k_z is defined by three numbers (n_x, n_y, and n_z), and corresponds to a standing wave. As a result we can assign for each standing wave in *k*-space a volume, V_{k0}:

$$V_{k0} = \frac{\pi^3}{L_x \times L_y \times L_z}. \tag{2.6}$$

This volume is often referred to as a *unit* or *primitive cell*. The wavenumber of such a standing wave, $k = |\mathbf{k}|$, is defined as

$$k = \sqrt{k_x^2 + k_y^2 + k_z^2} = \pi \sqrt{\frac{n_x^2}{L_x^2} + \frac{n_y^2}{L_y^2} + \frac{n_z^2}{L_z^2}}. \tag{2.7}$$

Blackbody radiation in a rectangular cavity can be considered as a sum of standing electromagnetic waves with different wavelengths and frequencies. Their values are defined by the set of numbers n_α and L_α. Let us define the number of standing waves in the cavity with wavenumbers less than the given value of k. For this purpose let us select in k-space a sphere of radius k with volume V_k:

$$V_k = \frac{4\pi k^3}{3}. \tag{2.8}$$

Each point inside this sphere, (k_x, k_y, k_z), or more precisely each primitive cell, corresponds to two independent standing waves with fixed frequencies and with orthogonal polarizations. For positive numbers n_α all three projections of the wavevector, k_α, are positive, i.e., they are within the first octant of the space of wavenumbers. The number of standing waves, Z_k, that correspond to $1/8$ of a sphere, which is the above-mentioned octant, can be defined as $2/8$ of the ratio of V_k and V_{k0} (Eqs. (2.8) and (2.6)). The factor of 2 is due to the two polarizations of waves (more details about waves and their polarizations are discussed in Appendix B):

$$Z_k = \frac{2}{8}\frac{4}{3}\frac{\pi k^3}{\left(\pi^3/L_x L_y L_z\right)} = \frac{L_x L_y L_z}{3\pi^2} k^3. \tag{2.9}$$

Taking into account the relation between the wavenumber and frequency, $\omega = ck$, we can find the number of standing waves, Z_ω, corresponding to the entire frequency interval from 0 to ω:

$$Z_\omega = \frac{V}{3\pi^2 c^3}\omega^3, \tag{2.10}$$

where $V = L_x L_y L_z$ is the volume of the cavity. We can find the number of standing waves corresponding to an infinitesimal interval of frequencies from ω to $\omega + \mathrm{d}\omega$ by differentiating Eq. (2.10):

$$\mathrm{d}Z_\omega = V \frac{\omega^2}{\pi^2 c^3} \mathrm{d}\omega. \tag{2.11}$$

Let us introduce the *density of states*, N_ω, i.e., the number of standing waves corresponding to a unit volume of the cavity and to a frequency interval $\mathrm{d}\omega$:

$$N_\omega = \frac{1}{V}\frac{\mathrm{d}Z_\omega}{\mathrm{d}\omega} = \frac{\omega^2}{\pi^2 c^3}. \tag{2.12}$$

Taking into account Eq. (2.12), we can write the expression for the *spectral density of blackbody radiation*, u_ω, defined as

$$u_\omega(T) = N_\omega \langle \varepsilon_\omega \rangle = \frac{\omega^2}{\pi^2 c^3} \langle \varepsilon_\omega \rangle, \tag{2.13}$$

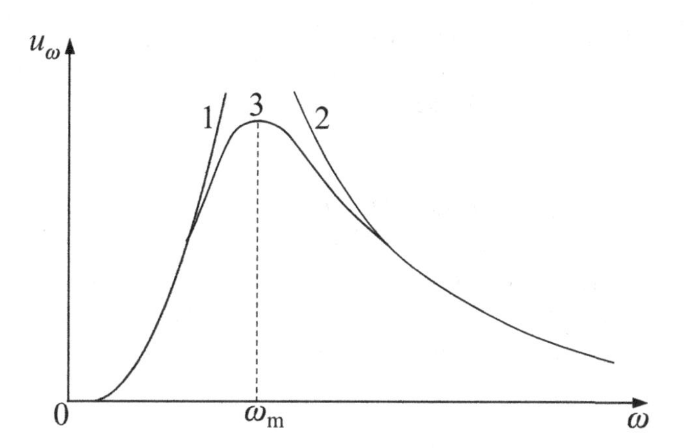

Figure 2.3 The dependences of the spectral density of blackbody radiation on frequency. Curve 1 corresponds to the Rayleigh–Jeans law, curve 2 to Wien's radiation law, and curve 3 to experimental data.

where $\langle\varepsilon_\omega\rangle$ is the average energy, which corresponds to one standing wave with frequency ω and generally does not depend on temperature and frequency.

2.1.3 The classical limit. The Rayleigh–Jeans law

Let us define the density of energy of an ensemble of standing waves, which exist inside of a cavity at a temperature T. From the energy point of view each standing wave can be considered as a harmonic oscillator with one degree of freedom. According to the classical point of view for a standing wave the average thermal energy is

$$\langle\varepsilon_\omega\rangle = k_B T. \tag{2.14}$$

This is because the same average kinetic and potential energy corresponds to one-dimensional motion of the particle. Therefore, the total thermal energy is equal to

$$\langle\varepsilon_\omega\rangle = 2\frac{k_B T}{2}. \tag{2.15}$$

The classical expression for the average thermal energy of each oscillator does not depend on the oscillation's frequency. Therefore for blackbody thermal radiation at temperature T we obtain the following expression for the spectral density, u_ω:

$$u_\omega(T) = \frac{\omega^2}{\pi^2 c^3} k_B T, \tag{2.16}$$

which is called the *Rayleigh–Jeans law*. This fits the experimental data sufficiently well in the region of low frequencies (waves with long wavelengths) (see Fig. 2.3, curve 1). However, with increasing frequency the discrepancy between the experimental data and expression (2.16) increases significantly (see Fig. 2.3, curve 3). The failure of the classical approach to describe correctly the experimental results is seen especially clearly when you calculate the total energy of blackbody radiation, i.e., the energy of standing waves in the cavity in the entire frequency interval:

$$u(T) = \int_0^\infty u_\omega(T)\mathrm{d}\omega = \frac{k_\mathrm{B}T}{\pi^2 c^3}\int_0^\infty \omega^2\,\mathrm{d}\omega. \tag{2.17}$$

The integral of Eq. (2.17) diverges and $u(T) \to \infty$, which contradicts the simple fact that the total energy radiated by a heated body is finite. The divergence of $u(T)$ to infinity is known in physics as the *ultraviolet catastrophe.*

2.1.4 Energy quanta and Planck's equation

The main difference between the classical approach to the blackbody radiation problem and the quantum approach (which was suggested by Max Planck in 1900) is in the calculation of the average energy of thermal radiation. According to the hypothesis suggested by Planck, the radiation is not emitted and absorbed continuously by a heated body, but rather is emitted and absorbed in finite portions of energy with the minimal portion for a given frequency defined as

$$\varepsilon = h\nu \tag{2.18}$$

or

$$\varepsilon = \hbar\omega, \tag{2.19}$$

where $h = 6.62 \times 10^{-34}$ [J s], which was later called *Planck's constant.* Very often instead of Eq. (2.18) the equation (2.19) is used with the constant, $\hbar$, which is called the *reduced Planck constant*:

$$\hbar = \frac{h}{2\pi} = 1.05 \times 10^{-34}\ \mathrm{J\,s}. \tag{2.20}$$

Planck postulated that the possible energy states which can have wave oscillators (standing waves) in the cavity are given by the equation

$$\varepsilon_n = n\hbar\omega, \tag{2.21}$$

where n is an arbitrary integer number known as the *quantum number*. All wave oscillators at a given temperature are in different energy states.

According to classical concepts, the probability that in the state of thermodynamical equilibrium at temperature T the wave oscillator will have energy ε_n is defined by the expression

$$P_n = C\mathrm{e}^{-\varepsilon_n/(k_\mathrm{B}T)}. \tag{2.22}$$

The constant C is determined from the condition that the total probability must be equal to unity, i.e.,

$$\sum_{n=0}^{\infty} P_n = C \sum_{n=0}^{\infty} \mathrm{e}^{-\varepsilon_n/(k_B T)} = 1, \tag{2.23}$$

where we obtain

$$C = \left(\sum_{n=0}^{\infty} \mathrm{e}^{-\varepsilon_n/(k_B T)} \right)^{-1}. \tag{2.24}$$

Taking into account the above-mentioned expressions, the average energy of a wave oscillator has to be calculated as

$$\langle \varepsilon_\omega \rangle = \sum_{n=0}^{\infty} P_n \varepsilon_n = \frac{\sum_{n=0}^{\infty} n\hbar\omega \mathrm{e}^{-n\hbar\omega/(k_B T)}}{\sum_{n=0}^{\infty} \mathrm{e}^{-n\hbar\omega/(k_B T)}}. \tag{2.25}$$

After carrying out simple calculations (see Example 2.1), we obtain for the average energy one of the most important relationships in physics:

$$\langle \varepsilon_\omega \rangle = \frac{\hbar\omega}{\mathrm{e}^{\hbar\omega/(k_B T)} - 1}. \tag{2.26}$$

On substituting this expression into Eq. (2.13) we obtain the well-known *Planck formula for the spectral density of blackbody radiation*:

$$u_\omega(T) = \frac{\hbar\omega^3}{\pi^2 c^3} \frac{1}{\mathrm{e}^{\hbar\omega/(k_B T)} - 1}. \tag{2.27}$$

Equation (2.27) is in complete agreement with experimental data over the entire spectral range of blackbody radiation (see Fig. 2.3, curve 3). In the limiting case of low frequencies ($\hbar\omega \ll k_B T$, $\mathrm{e}^{\hbar\omega/(k_B T)} \approx 1 + \hbar\omega/(k_B T)$) this formula approaches the Rayleigh–Jeans law: $u_\omega \sim \omega^2$ (see Eq. (2.16)), which was obtained on the basis of classical concepts. In the region of high frequencies ($\hbar\omega \gg k_B T$) expression (2.27) approaches the experimentally established radiation law of Wilhelm Wien, according to which the spectral density of energy, $u_\omega(T)$, exponentially decreases with increasing radiation frequency (see Fig. 2.3, curve 2).

Example 2.1. Using the expression for the average energy of a wave oscillator, Eq. (2.25), derive Eq. (2.26) and Planck's formula, Eq. (2.27).
Reasoning. Let us write Eq. (2.25) in the following form:

$$\langle \varepsilon_\omega \rangle = \hbar\omega \frac{\sum_{n=0}^{\infty} n \mathrm{e}^{-n\gamma}}{\sum_{n=0}^{\infty} \mathrm{e}^{-n\gamma}}, \tag{2.28}$$

where we introduced

$$\gamma = \frac{\hbar\omega}{k_B T}. \tag{2.29}$$

The sum in the denominator of Eq. (2.28) represents a geometric progression:

$$F(\gamma) = \sum_{n=0}^{\infty} \mathrm{e}^{-n\gamma} = \frac{1}{1-\mathrm{e}^{-\gamma}}. \tag{2.30}$$

After differentiating two different expressions for the function $F(\gamma)$ in Eq. (2.30) and equating them, we will obtain for the numerator in Eq. (2.28)

$$\frac{\mathrm{d}F}{\mathrm{d}\gamma} = \frac{\mathrm{d}}{\mathrm{d}\gamma}\left(\sum_{n=0}^{\infty} \mathrm{e}^{-n\gamma}\right) = -\sum_{n=0}^{\infty} n\mathrm{e}^{-n\gamma}, \tag{2.31}$$

$$\frac{\mathrm{d}F}{\mathrm{d}\gamma} = \frac{\mathrm{d}}{\mathrm{d}\gamma}\left(\frac{1}{1-\mathrm{e}^{-\gamma}}\right) = -\frac{\mathrm{e}^{-\gamma}}{(1-\mathrm{e}^{-\gamma})^2}, \tag{2.32}$$

$$\sum_{n=0}^{\infty} n\mathrm{e}^{-n\gamma} = \frac{\mathrm{e}^{-\gamma}}{(1-\mathrm{e}^{-\gamma})^2}. \tag{2.33}$$

After the substitution of Eqs. (2.30) and (2.33) into the initial expression (2.28), we obtain for the average energy of a wave oscillator

$$\langle\varepsilon_\omega\rangle = \hbar\omega\frac{\mathrm{e}^{-\gamma}}{1-\mathrm{e}^{-\gamma}} = \frac{\hbar\omega}{\mathrm{e}^{\gamma}-1} = \frac{\hbar\omega}{\mathrm{e}^{\hbar\omega/(k_\mathrm{B}T)}-1}, \tag{2.34}$$

which coincides with Eq. (2.26). On combining Eqs. (2.13) and (2.34) we obtain Planck's formula (2.27).

2.1.5 Photon gas

Using Planck's ideas, Einstein suggested that the quantum properties of electromagnetic radiation (light) become apparent not only in the emission and absorption of radiation by materials, but also during the propagation of electromagnetic radiation. According to his hypothesis, radiation can be imagined in the form of a gas, which consists of particles called *photons*. The photons possess energy defined as

$$\varepsilon_\mathrm{ph} = \hbar\omega = \frac{2\pi\hbar c}{\lambda}. \tag{2.35}$$

The photons move in vacuum with the speed of light, c, and they cannot stay at rest at any time. If a photon "stops" after some inelastic collision with a surface or after a collision with another particle, it simply disappears, giving its energy to the object with which it collided.

For a photon, as for any other real particle, we can define a mass. A photon's mass at rest must be equal to zero because a photon has speed equal to c and only particles with mass equal to zero can move at the speed of light. Using the relation between a relativistic particle's energy and its mass, m_ph, we can write

$$m_\mathrm{ph}c^2 = \hbar\omega, \tag{2.36}$$

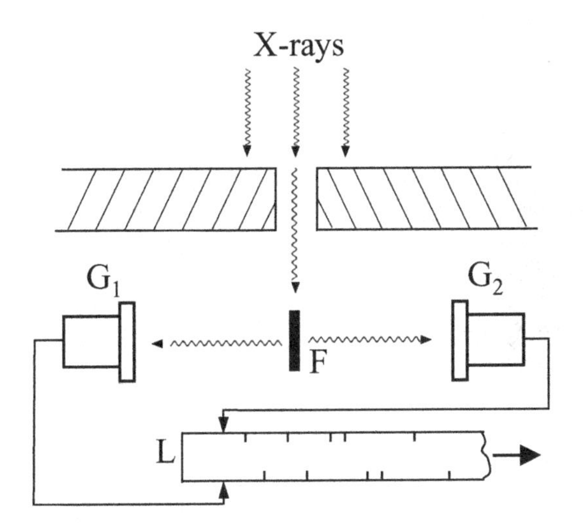

Figure 2.4 The scheme of Bethe's experiment for photon registration. By F we denote metallic foil, G_1 and G_2 are detectors of photons, and L is a registration device.

or

$$m_{\mathrm{ph}} = \frac{\hbar\omega}{c^2}. \tag{2.37}$$

The photon, which has mass and speed, must also have a momentum, the magnitude of which is defined as

$$p_{\mathrm{ph}} = m_{\mathrm{ph}}c = \frac{\hbar\omega}{c} = \frac{2\pi\hbar}{\lambda}. \tag{2.38}$$

For a photon considered as a particle and for electromagnetic radiation, whose quantum is a photon, the directions of propagation coincide and can be defined by a wavevector $\mathbf{k}$, whose modulus is equal to $k = |\mathbf{k}| = 2\pi/\lambda$, i.e.,

$$\mathbf{p}_{\mathrm{ph}} = \hbar\mathbf{k} \tag{2.39}$$

(compare this with Eqs. (2.38) and (2.4)). The first direct proof of the existence of a photon as a particle was provided by the experiment carried out by Hans Bethe: a metallic foil, F, was exposed to weak X-rays, and it became itself the source of secondary radiation (see Fig. 2.4). If the radiation were propagating in the form of spherical waves, then two independent counters, G_1 and G_2, placed at the opposite sides of the metallic foil, F, would simultaneously detect the arrival of waves of secondary radiation. However, when one of the counters detected a signal the other did not show that the signal had arrived, although the number of detections of each counter was practically the same. This can be explained only by the assumption that the radiation from the foil propagated in the form of separate quanta, which were detected by one or the other counter but not by both.

Thus, blackbody radiation can be represented in the form of a photon gas, which fills up the cavity. The particles of this gas – photons – propagate with equal probability along all directions, just like molecules at thermal equilibrium.

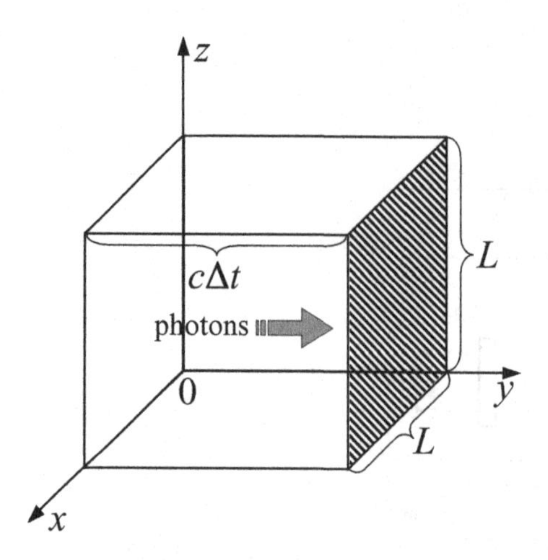

Figure 2.5 The number of incident photons along one of the possible six directions in a parallelepiped.

However, the photon gas cannot be described as a classical ideal gas. The particles of such a gas do not have a velocity distribution: at any temperature they have the same speed, equal to c. Their energy distribution cannot be described by the classical Maxwell–Boltzmann distribution since the number of these particles significantly depends on temperature. If we write the expression for the average energy of a wave oscillator, Eq. (2.26), as

$$\langle \varepsilon \rangle = \hbar\omega \langle n(T) \rangle , \tag{2.40}$$

then the quantity $\langle n \rangle$ will define the average quantum number n, which depends on temperature and frequency as

$$\langle n \rangle = \frac{1}{e^{\hbar\omega/(k_B T)} - 1} . \tag{2.41}$$

This quantity can be interpreted as the average number of photons in the cavity with energy $\varepsilon = \hbar\omega$ at blackbody radiation temperature T. This quantity defines exactly the distribution of photons over the energy states.

Let us derive the state equation of photon gas, which relates its thermodynamical parameters – pressure, temperature, and volume. Let us select in a rectangular cavity only the photons which move towards the cavity's wall along one of the coordinate axes (see Fig. 2.5). The number of photons with frequency ω incident on the wall during time Δt is given by the expression

$$\Delta N_{\text{inc}} = \frac{n_\omega}{6} L^2 c\, \Delta t, \tag{2.42}$$

where n_ω is the total number of photons of the corresponding frequency in the unit volume. Note that the number of possible directions in the cavity is equal to six. During the emission and absorption of the photon by the wall the momentum $\hbar\omega/c$ is transferred to the wall. During the reflection of the photon by the wall the doubled momentum $2\hbar\omega/c$ is transferred to the wall. Let us suppose that the number of photons absorbed by the wall during time Δt is equal to ΔN_{abs}, and the number of reflected photons to $\Delta N_{\rm ref}$. It is clear that the total number of reflected and absorbed photons is equal to the number of photons incident on the wall, i.e.,

$$\Delta N_{\rm inc} = \Delta N_{\rm abs} + \Delta N_{\rm ref}. \tag{2.43}$$

In thermodynamical equilibrium the number of photons absorbed and emitted by the wall must be the same, i.e., $\Delta N_{\rm abs} = \Delta N_{\rm em}$. At the same time the internal energy of a wall and the energy of radiation do not change with time.

Let us write the momentum transferred to the wall during the time Δt:

$$\Delta p = (\Delta N_{\rm abs} + 2\,\Delta N_{\rm ref} + \Delta N_{\rm em})\frac{\hbar\omega}{c} = 2\,\Delta N_{\rm inc}\,\frac{\hbar\omega}{c} = \frac{1}{3}n_\omega \hbar\omega L^2\,\Delta t. \tag{2.44}$$

Knowing the momentum transferred by the photons with frequency ω, we can find the pressure, $\mathcal{P}_\omega(T)$, experienced by the wall:

$$\mathcal{P}_\omega(T) = \frac{F}{L^2} = \frac{1}{L^2}\frac{\Delta p}{\Delta t} = \frac{1}{3}n_\omega\hbar\omega = \frac{u_\omega(T)}{3}. \tag{2.45}$$

The total pressure of the photon gas, $\mathcal{P}(T)$, is the sum of individual pressures, $\mathcal{P}_\omega(T)$, of photons of all frequencies:

$$\mathcal{P}(T) = \int_0^\infty \mathcal{P}_\omega(T)\mathrm{d}\omega = \frac{1}{3}\int_0^\infty u_\omega(T)\mathrm{d}\omega. \tag{2.46}$$

By substituting into the integrand in Eq. (2.46) Planck's expression for the spectral density (2.27), we obtain

$$\mathcal{P}(T) = \frac{u(T)}{3} = \frac{k_{\rm B}^4 T^4}{3\pi^2 c^3 \hbar^3} I = \frac{4\sigma}{3c}T^4, \tag{2.47}$$

where we introduced $u(T) = \int_0^\infty u_\omega(T)\mathrm{d}\omega$, the volume energy density of photon gas, the integral I,

$$I = \int_0^\infty \frac{y^3\,\mathrm{d}y}{\mathrm{e}^y - 1} = \frac{\pi^4}{15}, \tag{2.48}$$

and the Stefan–Boltzmann constant σ:

$$\sigma = \frac{\pi^2 k_{\rm B}^4}{60 c^2 \hbar^3} = 5.67 \times 10^{-8}\ \mathrm{W\,m^{-2}\,K^{-4}}. \tag{2.49}$$

It follows from Eq. (2.47) that the photon-gas pressure does not depend on the volume filled by the photon gas. In contrast to an ideal gas of molecules, whose pressure linearly depends on the temperature, the pressure of a photon gas is proportional to T^4, which is observed in many physical phenomena. Thus, the gravitational squeezing of stars because of their enormous mass (compared with the mass of the Earth) is prevented by the pressure of the photon gas. As a result of the thermonuclear reactions taking place inside of the stars their temperature is $T \sim 10^8$ K. At these temperatures the pressure of the photon gas is about $\mathcal{P} \sim 10^{16}$ Pa (in the center of the Earth $\mathcal{P} \sim 10^{11}$ Pa). After having

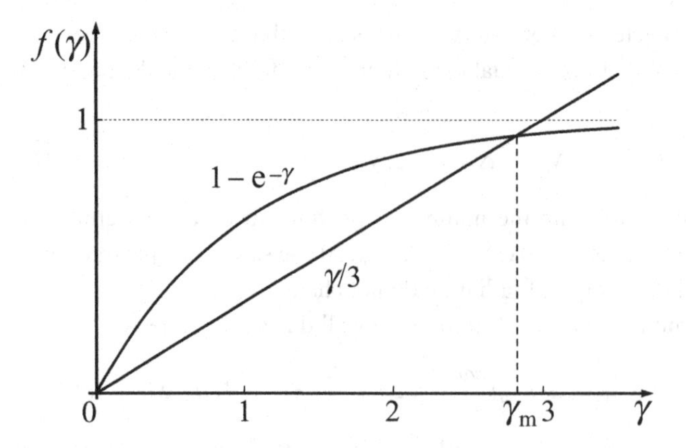

Figure 2.6 Graphical solution of the transcendental equation (2.53).

burned its nuclear fuel, the temperature inside a star drastically decreases and the gravitational collapse of the star begins.

Example 2.2. Using Planck's formula for the spectral density of blackbody radiation, find how the frequency at which the heated body emits maximum energy depends on the temperature of the body.
Reasoning. Let us rewrite Eq. (2.27) in the following form:

$$u_\omega(T) \equiv u_\omega(\gamma) = A\frac{\gamma^3}{\mathrm{e}^\gamma - 1}, \tag{2.50}$$

where

$$A = \frac{k_\mathrm{B}^3 T^3}{\pi^2 \hbar^2 c^3} \tag{2.51}$$

and

$$\gamma = \frac{\hbar\omega}{k_\mathrm{B} T}.$$

The maximum of the emitted energy corresponds to the maximum of $u_\omega(T)$. Let us find the value of γ that corresponds to the maximum of the function $u_\omega(\gamma)$. To find it, we write the first derivative, $\mathrm{d}u_\omega/\mathrm{d}\gamma$, of the function u_ω with respect to γ and equate it to zero. Then, we obtain the following equation:

$$3\mathrm{e}^\gamma - 3 - \gamma\mathrm{e}^\gamma = 0. \tag{2.52}$$

It is convenient to rewrite the last equation as

$$1 - \mathrm{e}^{-\gamma} = \frac{\gamma}{3}. \tag{2.53}$$

The graphical solution of this transcendental equation is shown in Fig. 2.6. The intersection point of the functions $1 - \mathrm{e}^{-\gamma}$ and $\gamma/3$ gives the

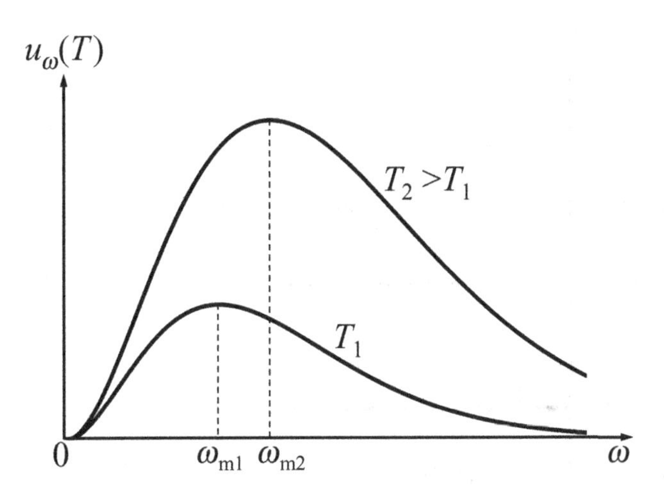

Figure 2.7 Wien's displacement law: the dependences of the spectral density of blackbody radiation on frequency for two different temperatures T_1 and T_2; ω_{m1} is the maximum of the spectral density corresponding to temperature T_1 and ω_{m2} is the maximum of the spectral density corresponding to temperature T_2 ($T_2 > T_1$).

root $\gamma = \gamma_m$:

$$\gamma_m = \frac{\hbar\omega_m}{k_B T} \approx 2.82. \tag{2.54}$$

From this equation it follows that the frequency which corresponds to the maximum of the spectral density of blackbody radiation and the temperature are related as

$$\omega_m = bT, \tag{2.55}$$

where b is

$$b = 2.82\frac{k_B}{\hbar}. \tag{2.56}$$

Thus, from Eq. (2.55) it follows that the maximum of the function $u_\omega(T)$ shifts into the region of higher frequencies with increasing temperature. This statement is called *Wien's displacement law* (see Fig. 2.7).

2.2 The quantum character of the interaction of radiation with matter

2.2.1 The photoelectric effect

The emission of electrons by the surface of a body exposed to light is called the *external photoelectric effect*. This effect was discovered in 1887 by Heinrich Hertz, who observed the emission of electrons by the surface of different materials under the influence of light. The external photoelectric effect is especially easily observed in metals. The reason for it is as follows. The distinctive feature of metals is the existence of almost-free electrons, i.e., valence electrons, which are separated from their atoms by the crystal's internal electric field. Inside of the metal these electrons move freely, but they cannot leave the metal since they

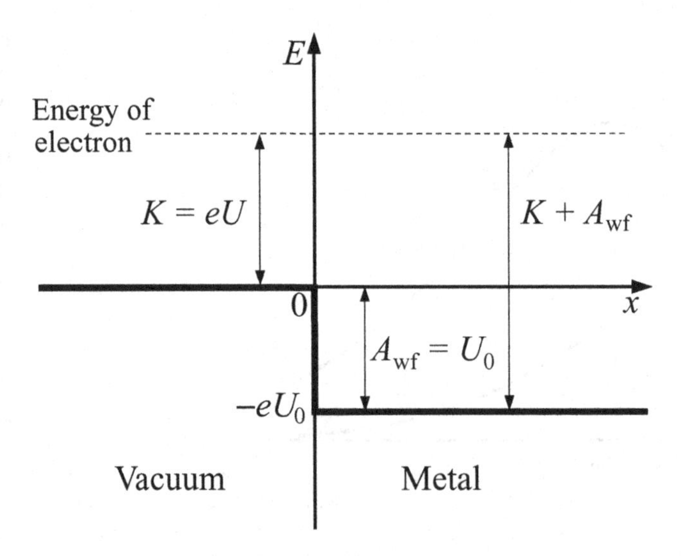

Figure 2.8 The interface between metal and vacuum: A_{wf} is the work function of an electron, and $K = eU$ is the kinetic energy of an electron that escapes from the metal.

do not have enough energy to overcome the potential barrier (see Fig. 2.8). The magnitude of the energy barrier is defined by the *work function*, A_{wf}, which is several electron-volts (1 eV $= 1.6 \times 10^{-19}$ J) (see Fig. 2.8).

From the classical point of view the escape of electrons from the metal to the surrounding vacuum is possible under the influence of an electromagnetic wave. However, all attempts to explain the main characteristic features of the photoelectric effect – the absence of a dependence of the energy of photoelectrons on the intensity of the light source, the absence of a time delay, and the existence of a threshold value of frequency, below which the photoelectric effect cannot be observed – on the basis of the wave nature of light have been unsuccessful.

All the questions that the photoelectric effect raised were resolved by Albert Einstein in 1905 by utilizing the hypothesis of the existence of light quanta – photons. After the absorption of a photon by an individual electron the photon's energy, $\varepsilon = \hbar\omega$, is transferred completely to the electron. If the energy of the photon is sufficient to overcome the potential barrier which keeps the electron in the metal, the energy that is left after escape from the metal is transformed into kinetic energy, K. The absorption of a photon and release of an electron from the metal happen practically simultaneously, which explains the absence of a time delay.

The equation that relates the above-mentioned quantities was introduced by Einstein in the following form:

$$\hbar\omega = A_{\mathrm{wf}} + \frac{m_e v_{\max}^2}{2}, \tag{2.57}$$

where $v_{\max}$ is the highest velocity of photoelectrons. From this equation it follows that the highest energy of photoelectrons depends on the frequency of incident light and does not depend on the intensity. If $\hbar\omega < A_{\mathrm{wf}}$, then the photoelectric effect is not observed. The threshold frequency, ω_{th}, which is equal to

$$\omega_{th} = \frac{A_{wf}}{\hbar}, \tag{2.58}$$

is called the *cut-off frequency of the photoelectric effect.*

For practical applications of the photoelectric effect an important quantitative characteristic called the *external quantum efficiency* (or *quantum efficiency*), η, is used. It defines the ratio of the number of emitted electrons to the number of incident photons. Near the cut-off frequency of the photoelectric effect for most metals the quantum efficiency is about $\eta \sim 10^{-4}$ (electrons per photon). The small efficiency in this frequency range is due to the fact that only those electrons which are close to the surface of the metal possess sufficient energy to leave the metal. At the same time the predominant portion of the radiation incident on the metal is reflected. With increasing photon energy, i.e., increasing frequency of the radiation, the quantum efficiency increases and η is about 10^{-2} for the photons with energy close to 1 eV. For X-rays with the energy of photons $\varepsilon_{ph} \sim 10^3$ eV the quantum efficiency is about 10^{-1} electrons per photon.

Example 2.3. Find the work function of an electron for an unknown metal, if after exposure of its surface to light with wavelength $\lambda = 0.35$ μm the photocurrent that occurs in the circuit of a photoelement disappears if the negative voltage $U_d = -1.2$ V is applied.
Reasoning. If we apply a negative (decelerating) voltage, U_d, to the circuit with the photoelement,

$$eU_d = \frac{m_e v_{max}^2}{2}, \tag{2.59}$$

the photocurrent disappears. In this case Einstein's equation for the photoelectric effect can be rewritten in the form

$$\hbar\omega = A_{wf} + eU_d. \tag{2.60}$$

Taking into account that $\omega = 2\pi c/\lambda$, we obtain for the work function of the unknown metal

$$A_{wf} = \hbar\frac{2\pi c}{\lambda} - eU_d \approx 3.84 \times 10^{-19}\,\text{J} = 2.4\,\text{eV}. \tag{2.61}$$

This unknown metal may be lithium, whose work function is equal to $A_{wf} =$ 2.39 eV.

2.2.2 X-ray bremsstrahlung

If the energy of a photon incident on a metal significantly exceeds the work function of a metal, i.e., $\hbar\omega \gg A_{wf}$, then A_{wf} can be neglected in Einstein's equation (2.57) for the photoelectric effect. Thus, the equation can be simplified

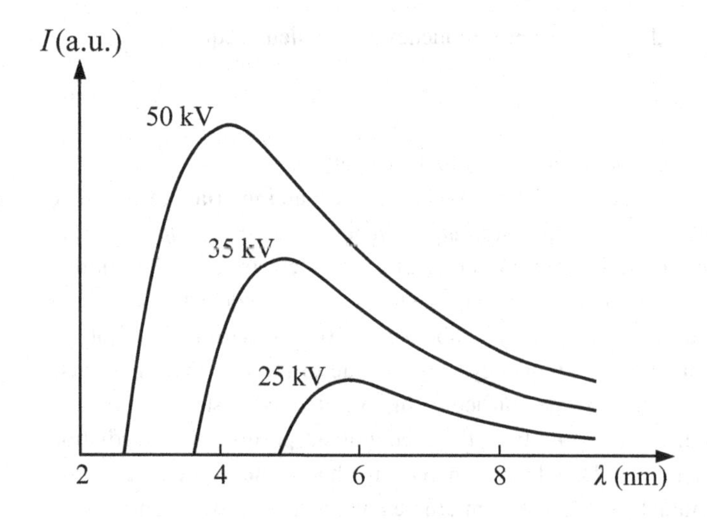

Figure 2.9 The dependences of the distribution of the intensity, I, of X-ray bremsstrahlung on the wavelength λ at accelerating potentials U equal to 50, 35, and 25 kV.

to

$$\hbar\omega = \frac{m_e v_{\max}^2}{2}. \tag{2.62}$$

In the X-ray range, where the energy of a quantum is $\hbar\omega \sim 10^2 - 10^3$ eV, this equation is valid with a high degree of precision since the work function is only about $A_{\rm wf} \sim 1$ eV.

Let us consider the process which is the reverse of the photoelectric effect, i.e., let us consider the transformation of the kinetic energy of electrons incident on a metal into the energy of the photons that are created after a sudden deceleration of electrons in the metal. If we take into account the fact that the electrons, in order to gain the necessary kinetic energy, are accelerated by a corresponding potential drop U, then Eq. (2.62) can be rewritten in the form

$$eU = \hbar\omega. \tag{2.63}$$

The process of acceleration of electrons takes place in the X-ray tube, where the given potential drop is created between the cathode and the anode (anti-cathode). After colliding with the anode the electrons abruptly begin to decelerate and, because of this negative acceleration, the electrons emit electromagnetic waves in the X-ray range. This radiation, which is called *X-ray bremsstrahlung*, has a continuous electromagnetic spectrum. Figure 2.9 shows experimental dependences of distributions of the radiation intensity on the wavelength, obtained for a tungsten anode at several values of the accelerating potential drop U. The characteristic feature of these curves is the existence of a short-wavelength cut-off frequency. From the particle point of view this fact can be simply explained. Since the radiation occurs because of the energy lost by an electron during its deceleration, the energy of the photon cannot be greater than eU. Therefore, there is a threshold frequency

$$\omega_{\max} = \frac{eU}{\hbar}, \tag{2.64}$$

which is the highest for the given accelerating potential drop. Correspondingly, the minimal wavelength in the spectrum of bremsstrahlung is defined as

$$\lambda_{\min} = \frac{2\pi\hbar c}{eU}. \tag{2.65}$$

The existence of a short-wavelength cut-off frequency in the X-ray spectrum is one of the most important quantum features of X-ray radiation. According to the classical wave theory such a threshold in the radiation spectrum must not exist.

2.2.3 The Compton effect

In 1923, while studying scattering of hard X-ray radiation on samples consisting of light atoms (graphite, paraffin, etc.), Arthur Compton discovered that the wavelength λ' of the radiation scattered at an angle θ is greater than the wavelength λ of the incident radiation. The wavelength of the scattered radiation depends on θ according to the equation

$$\lambda'(\theta) = \lambda + \Lambda(1 - \cos\theta), \tag{2.66}$$

where $\Lambda = 2.42 \times 10^{-12}$ m is a constant called the *Compton wavelength of the electron*, which can be found experimentally.

Classical theory was unable to explain the appearance of the observed wavelength shift of the radiation scattered at some angle. According to classical concepts the scattering can be considered as a process during which an electron undergoes forced motion under the influence of the electric field of an incident wave. At the same time the electron behaves as the source of secondary (scattered) waves with the frequency of the incident radiation. The results of the experiment become clear if the scattered radiation is considered as the consequence of an elastic scattering of the photon on the free electron. For the light atoms of the materials on which the experiments were carried out, the binding energy of valence electrons with their atoms is smaller than the energy of the X-ray photon. Therefore, these electrons can be considered as free electrons. In such a case the formula (2.66) is a consequence of energy and momentum conservation laws for the elastic collision of the photon (with energy $\hbar\omega = 2\pi\hbar c/\lambda$ and momentum $\hbar\mathbf{k}$) and the electron at rest with energy $m_e c^2$ and zero momentum:

$$\frac{2\pi\hbar c}{\lambda} + m_e c^2 = \frac{2\pi\hbar c}{\lambda'} + m(v)c^2, \tag{2.67}$$

$$\hbar\mathbf{k} = \hbar\mathbf{k}' + m(v)\mathbf{v}, \tag{2.68}$$

where wavenumbers k and k' are equal to $k = 2\pi/\lambda$ and $k' = 2\pi/\lambda'$, respectively. For Eqs. (2.67) and (2.68) we used relativistic expressions for the energy and momentum of the electron because as a result of collision with an X-ray photon the freed electron may become relativistic, i.e., it can acquire a velocity v that is close to the speed of light, c. In these equations the mass of the electron depends on its velocity in accordance with the relativistic equation

$$m(v) = \frac{m_e}{\sqrt{1 - v^2/c^2}}. \tag{2.69}$$

Self-consistent solution of Eqs. (2.67)–(2.69) leads to the expression (2.66), where

$$\Lambda = \frac{2\pi\hbar}{m_e c} = 2.42 \times 10^{-12}\ \text{m}. \tag{2.70}$$

By analyzing his results Compton proved that the photon is a particle that is characterized not only by its energy but also by its momentum. This allows application of the laws of conservation of energy and momentum to the processes of interaction of photons with other particles.

Example 2.4. Show that the absorption of a photon in the case of its inelastic collision with a free electron is not possible because it is prohibited by the laws of conservation of energy and momentum.
Reasoning. Let us suggest that before collision with a photon the electron was at rest. In this case the laws of conservation of energy and momentum can be written in the form

$$\hbar\omega + E_0 = E(p), \tag{2.71}$$

$$\frac{\hbar\omega}{c} = p, \tag{2.72}$$

where $E_0 = m_e c^2$ is the energy of the electron at rest and the energy of the moving electron is related to the momentum as

$$E(p) = c\sqrt{m_e^2 c^2 + p^2}. \tag{2.73}$$

On squaring the left-hand and right-hand sides of Eqs. (2.71) and (2.72) we obtain

$$\hbar^2\omega^2 + 2m_e c^2\hbar\omega = c^2 p^2, \tag{2.74}$$

$$\hbar^2\omega^2 = c^2 p^2. \tag{2.75}$$

At $m_e \neq 0$ these two equalities contradict each other. From the above we can draw the conclusion that an inelastic collision of a photon and a free electron is impossible, i.e., a free electron cannot absorb a photon. Such a process can take place in the case of the existence of a third particle, which will be able to take part of the energy and/or momentum of the photon. In Compton's experiment, which we discussed previously, the third particle is the scattered photon.

2.2.4 Wave–particle duality

As a result of thorough experimental and theoretical analysis of the properties of light its dual nature was established. In some processes light behaves as a wave (interference, diffraction, etc.), whereas in others it behaves as a flux of particles (blackbody radiation, the photoelectric effect, etc.). The equations

$$\varepsilon_{\text{ph}} = \hbar\omega, \tag{2.76}$$

$$\mathbf{p}_{\text{ph}} = \hbar\mathbf{k} \tag{2.77}$$

relate the particle and wave properties of light. The left-hand sides of these equations (with ε_{ph} and $\mathbf{p}_{\text{ph}}$) characterize the photon as a particle and the right-hand sides (with ω and $\mathbf{k}$) characterize the photon as a wave.

There is an important trend in the observations of the dual nature of light. For long-wavelength radiation (e.g., infrared radiation) its quantum properties are not so obvious and mainly its wave properties are detected. However, on going to the shorter wavelengths the quantum properties of light become more apparent. Wave and quantum properties of light are connected, and they supplement each other. The quantum properties of light become apparent by virtue of the fact that the energy, momentum, and mass of radiation are concentrated in particles – photons. The probability of finding photons at particular points of space is defined by the amplitude of the light wave, i.e., by the wave properties of light.

The wave properties are inherent not only to large ensembles of photons, but also to each individual photon. This is evident because it is not possible to specify the location of a photon and the direction which it will have after a collision with an obstacle. We can talk only about the probability of finding an individual photon in one place or another. The description of the behavior of such an object on the basis of classical laws is impossible. Nevertheless, experimental facts allow us to state that this duality in light's behavior is a *law of nature*. Light was the first object that allowed physicists to observe and to interpret the wave–particle duality of matter. The further development of physics greatly enhanced the class of such objects. As will be discussed later, other particles, such as the electron, proton, neutron, etc., may have wave properties too, which broadens our knowledge about matter.

Example 2.5. An X-ray photon with frequency $\nu = 6 \times 10^{18}$ Hz scatters on a free electron at angle $\theta = 90^\circ$. Find the frequency of the scattered photon, as well as the momentum, velocity, and energy of the electron after its collision with the photon.

Reasoning. After the collision of a photon with a motionless electron, the wavelength of the scattered photon increases, in accordance with Eq. (2.66), by the magnitude

$$\Delta\lambda = \Lambda(1 - \cos\theta), \tag{2.78}$$

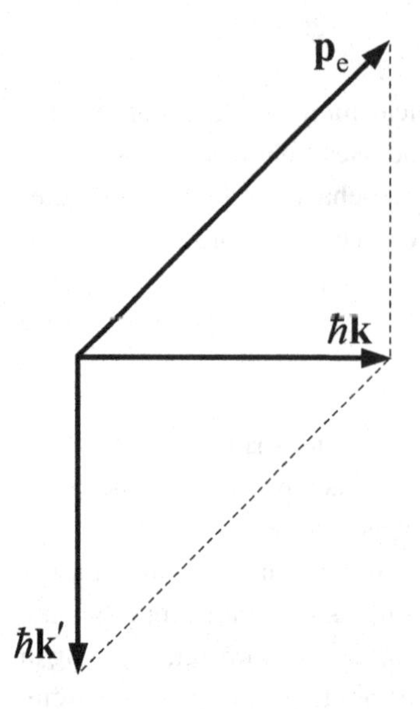

Figure 2.10 Scattering of a photon and an electron.

where Λ is the Compton wavelength of the electron. Taking into account that the scattering angle is $\theta = 90°$, we obtain

$$\Delta\lambda = \lambda' - \lambda = \Lambda. \tag{2.79}$$

Therefore, for the frequencies of the incident, ν, and scattered, ν', photons we have

$$\frac{1}{\nu'} - \frac{1}{\nu} = \frac{\Lambda}{c}. \tag{2.80}$$

Taking into account that $\Lambda = 2.42 \times 10^{-12}$ m and $c = 3 \times 10^{8}\,\mathrm{m\,s^{-1}}$, we find the frequency of the scattered photon:

$$\nu' = \frac{c\nu}{c + \nu\Lambda} = 5.72 \times 10^{10}\,\mathrm{Hz}. \tag{2.81}$$

Let us rewrite the momentum conservation law for the considered collision:

$$\hbar\mathbf{k} = \hbar\mathbf{k}' + \mathbf{p}_e. \tag{2.82}$$

You can see from Fig. 2.10 that the momentum of the electron after collision with the photon is

$$p_e = \sqrt{(\hbar k)^2 + (\hbar k')^2}, \tag{2.83}$$

where for the wavenumber, k, of the incident photon the following equality is valid:

$$k = \frac{\omega}{c} = \frac{2\pi\nu}{c}. \tag{2.84}$$

The same equality can be written for k'. Taking this into consideration, we obtain for the electron's momentum, p_e,

$$p_e = \frac{2\pi\hbar}{c}\sqrt{\nu^2 + \nu'^2} = 1.83 \times 10^{-23}\,\mathrm{kg\,m\,s^{-1}}. \tag{2.85}$$

The velocity of the electron is

$$v_e = \frac{p_e}{m_e} = 2 \times 10^7\,\mathrm{m\,s^{-1}}. \tag{2.86}$$

The energy of the electron can be found from the expression

$$E = \frac{p_e^2}{2m_e} = 1.84 \times 10^{-16}\,\mathrm{J} = 1.15 \times 10^3\,\mathrm{eV}. \tag{2.87}$$

2.3 Wave properties of particles

2.3.1 De Broglie's hypothesis

The understanding of the dual nature of electromagnetic radiation that was established in the first quarter of the twentieth century allowed scientists to explain many phenomena on the basis of the particle picture, according to which electromagnetic radiation is a flux of photons – the particles which have zero mass at rest. The success of such a description of electromagnetic radiation evoked interest in wave–particle concepts and led to its extrapolation to particles that have a finite mass at rest. In 1924 Louis de Broglie suggested the idea of the universality of Eq. (2.38), which related the momentum of a photon to its wavelength. Any particle that has momentum p corresponds to a wave for which the following equality is valid:

$$\lambda_{\mathrm{Br}} = \frac{2\pi\hbar}{p} = \frac{h}{p}. \tag{2.88}$$

This physical quantity λ_{Br} is called the *de Broglie wavelength of a particle.* The second main equality that relates the parameters of the particle and wave is the relation between the energy of the particle and the de Broglie frequency:

$$\omega_{\mathrm{Br}} = \frac{E}{\hbar}. \tag{2.89}$$

A free particle of energy E and momentum p corresponds to a plane wave with frequency ω_{Br} and wavevector

$$\mathbf{k} = \frac{\mathbf{p}}{\hbar}. \tag{2.90}$$

This wave can be described by the scalar function

$$\Psi(t, \mathbf{r}) = C\mathrm{e}^{-\mathrm{i}(\omega t - \mathbf{k}\cdot\mathbf{r})} = C\mathrm{e}^{-\mathrm{i}(Et - \mathbf{p}\cdot\mathbf{r})/\hbar}. \tag{2.91}$$

The de Broglie waves are not electromagnetic waves. We will discuss later their particular nature and the physical meaning of the function Ψ.

The frequency and wavenumber of the de Broglie wave are related to the phase and group velocities as follows. For the phase velocity we obtain

$$v_{\rm ph} = \frac{\omega_{\rm Br}}{k} = \frac{E}{p}. \tag{2.92}$$

For the group velocity, taking into account that for a free particle $E = p^2/(2m)$, we obtain

$$v_{\rm gr} = \frac{{\rm d}\omega_{\rm Br}}{{\rm d}k} = \frac{{\rm d}E}{{\rm d}p} = \frac{\rm d}{{\rm d}p}\left(\frac{p^2}{2m}\right) = \frac{p}{m} = v. \tag{2.93}$$

Thus, the group velocity of the de Broglie wave coincides with the velocity of a particle. Let us estimate the wavelength of the de Broglie wave of an electron for various cases. For a free electron moving inside of a metal at room temperature with thermal velocity, v_T,

$$\frac{m_{\rm e}v_T^2}{2} = \frac{3}{2}k_{\rm B}T \quad \text{and} \quad v_T = \sqrt{\frac{3k_{\rm B}T}{m_{\rm e}}} \approx 10^5\,{\rm m\,s^{-1}}. \tag{2.94}$$

The de Broglie wavelength according to Eq. (2.88) is equal to $\lambda_{\rm Br} \approx 7 \times 10^{-9}$ m, while the distance between neighboring atoms in a crystal lattice is about $a \approx 3 \times 10^{-10}$ m, $\lambda_{\rm Br} \gg a$. The velocity of an electron rotating around a proton in a hydrogen atom according to Bohr's theory is $v = 2 \times 10^6$ m s^{-1}. Then, $\lambda_{\rm Br} \approx 4 \times 10^{-10}$ m, while the extent of the atom is about 10^{-10} m, i.e., $\lambda_{\rm Br}$ is of the order of the atom's size.

In an electric field with potential difference U an electron accelerated to velocities less than the speed of light in vacuum gains kinetic energy $K = eU$. Taking into account Eq. (2.88), we obtain for the de Broglie wavelength of the electron

$$\lambda_{\rm Br} = \frac{2\pi\hbar}{\sqrt{2m_{\rm e}K}} = \frac{2\pi\hbar}{\sqrt{2m_{\rm e}eU}}. \tag{2.95}$$

At $U = 150$ V the de Broglie wavelength is equal to $\lambda_{\rm Br} \approx 10^{-10}$ m $\equiv 1$ Å.

Let us estimate now $\lambda_{\rm Br}$ for heavy particles, using the same procedure as for the electron (see Eq. (2.94)). For a copper ion located at the site of the metal crystal lattice (the mass of the ion is about $m_{\rm Cu} \approx 10^{-25}$ kg), at room temperature the de Broglie wavelength is $\lambda_{\rm Br} \approx 2 \times 10^{-11}$ m, which is one order of magnitude less than the dimension of the ion and the distance between ions. If we consider a macroscopic object even with a small mass, then its de Broglie wavelength appears to be much less than the dimensions of this object. For example, for a speck of dust with mass $m \approx 10^{-9}$ kg and size 10^{-5} m moving with the velocity $v \approx 10^{-2}$ m s^{-1}, the de Broglie wavelength is $\lambda_{\rm Br} \approx 10^{-22}$ m.

Since there is no important difference between microobjects and macroobjects, the following question arises: when are the wave properties of particles pronounced and when do the particle properties dominate? To answer this question let us use the analogy with optics. The wave nature of light becomes

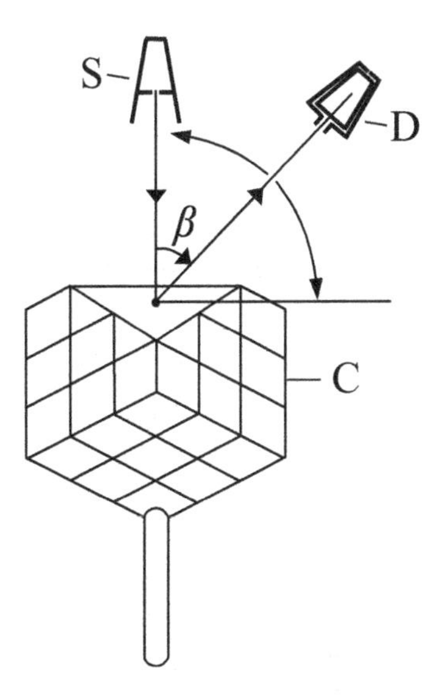

Figure 2.11 A schematic diagram of the Davisson–Germer experiment on the diffraction of electrons by a monocrystal of nickel. The source of electrons is denoted as S, the mobile detector of electrons as D, and the monocrystal of nickel as C.

maximally apparent at wavelengths comparable to the characteristic dimensions of obstacles, i.e., $\lambda \sim L$ (for example the phenomena of interference and diffraction). If $\lambda \ll L$, then the wave properties become negligible and we can use geometrical optics (rectilinear propagation of rays). If we extrapolate this idea to particle waves, i.e., to de Broglie waves, then the wave properties of particles will be more apparent if their dimensions are less than or of the order of the de Broglie wavelength λ_{Br}. If the dimensions of a particle are much larger than λ_{Br}, then the wave properties are weakly pronounced and they can be neglected. The first situation ($L \leq \lambda_{\mathrm{Br}}$) occurs for electrons in metals and in atoms. The second situation ($L \gg \lambda_{\mathrm{Br}}$) occurs for an electron in an electron-beam tube, for a copper ion, and for a speck of dust.

2.3.2 Experimental verification

The de Broglie hypothesis about the existence of wave properties of particles received experimental confirmation in 1927 in the Davisson–Germer experiment, which involved scattering of electrons on a nickel monocrystal. The scheme of the experiment is shown in Fig. 2.11. In the electron gun S there was produced a beam of electrons, whose energy and velocity were defined by the accelerating voltage. A narrow beam of monoenergetic electrons was incident on the surface of a nickel crystal C, and was reflected from it. The crystal could be rotated, changing the angle of incidence of the electron beam. The detector D registered the number of electrons scattered by the crystal in different directions. Intensity measurements were carried out at a fixed angle of incidence of the electron beam by changing the magnitude of the accelerating voltage.

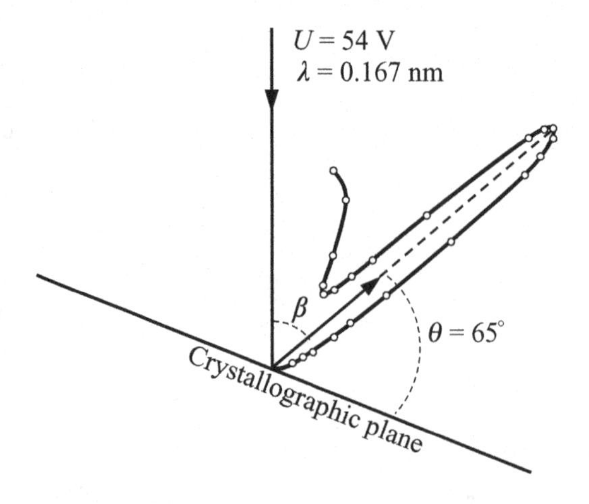

Figure 2.12 The angle distribution of reflected electrons in one of the diffraction maxima.

If electrons behaved as classical particles, they would have had to reflect from the crystal in accordance with the laws of geometrical optics, i.e., the incidence angle would have had to be equal to the reflection angle. However, the experiments showed that the intensities of scattered electrons in different directions differ. There was also alternation of the maxima and minima of the number of electrons scattered at different angles, i.e., in short the phenomenon of electron diffraction was verified. Analogously to the diffraction of X-rays, the theoretical analysis of the diffraction of electrons by a crystal can be carried out using Bragg's law of diffraction Eq. (B.177),

$$2d \sin\theta = n\lambda_{\mathrm{Br}}, \tag{2.96}$$

where, instead of X-rays of wavelength λ, the de Broglie wavelength of the electron, λ_{Br}, is used. Taking into account Eq. (2.95), Eq. (2.96) can be rewritten as

$$2d \sin\theta = n\frac{2\pi\hbar}{\sqrt{2m_e eU}}. \tag{2.97}$$

The angle β between incident and reflected rays is related to the angle θ in Eq. (2.96) via the equation

$$\theta = \frac{180^\circ - \beta}{2}. \tag{2.98}$$

The measurements were carried out at a fixed voltage U and various angles θ, as well as at a fixed angle θ and different voltages U. In Fig. 2.12 the results of measurements that correspond to a fixed voltage $U = 54$ V are shown. The diffraction maximum becomes clearly pronounced at an angle $\theta = 65^\circ$. The results of measurements for the fixed angle $\theta = 65^\circ$ and various values of U are shown in Fig. 2.13. If we solve Eq. (2.97) for the square root of the accelerating voltage we get the equation

$$\sqrt{U} = \frac{\pi\hbar}{d \sin\theta\sqrt{2em_e}}n. \tag{2.99}$$

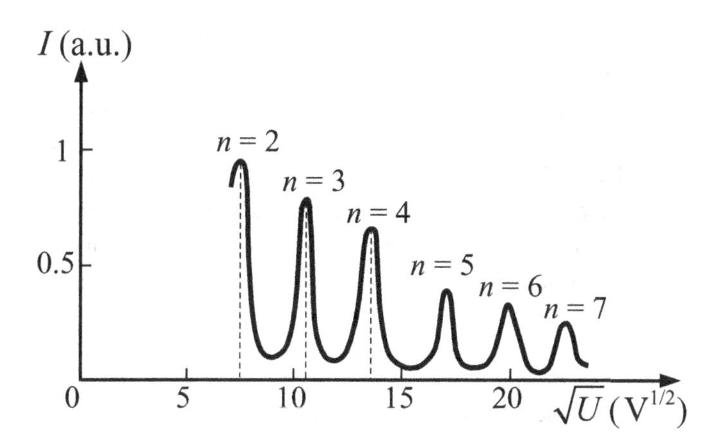

Figure 2.13 The dependence of the intensity of the electron beam on the square root of the accelerating voltage for its diffraction by a monocrystal of nickel.

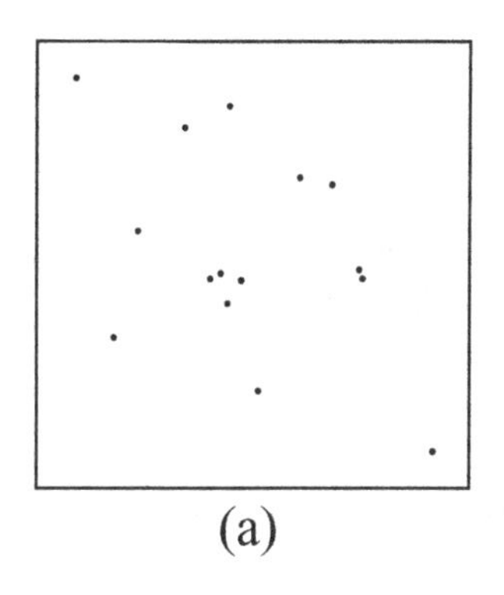

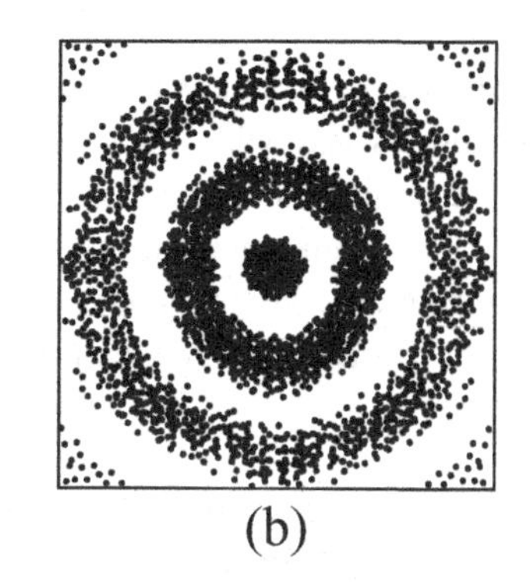

Figure 2.14 The distribution of diffracted electrons over the surface of a photographic plate for (a) a short and (b) a long time of exposure.

Since n is an integer number, the diffraction maxima on the axis $\sqrt{U}$ must be distributed equidistantly, and this is clearly seen from the results of the experiment (see Fig. 2.13).

The de Broglie wavelength (at $U = 54$ V) calculated with the help of Eq. (2.95) is equal to $\lambda_{\text{Br}} = 0.167$ nm. At the same time the corresponding wavelength found from Eq. (2.96) for the lattice parameter $d = 0.215$ nm and $n = 2$ is equal to $\lambda_{\text{Br}} = 0.165$ nm. The results obtained from this experiment convincingly confirmed the de Broglie hypothesis.

Diffraction experiments studying the wave properties of individual electrons are of great interest. The size of the electron beam in these experiments is chosen to be so small that only one electron at a time is incident on the thin metallic film. The electrons transmitted through the film form diffraction patterns in the form of concentric circles on a photographic plate after a sufficiently long exposure time. At short exposure times the points on the photographic plate, which correspond to incident electrons, are randomly distributed (see Fig. 2.14(a)). However, with an increase in the exposure time the distribution of points acquires the characteristic form for the diffraction of X-ray radiation by a polycrystalline sample (see Fig. 2.14(b)).

Note that not only electrons possess wave properties. In 1936 the diffraction of neutrons by a crystal was observed for the first time. Neutrons are particles that are found in atomic nuclei and have mass $m_n = 1.67 \times 10^{-27}$ kg. If the velocity of the incident neutrons and the lattice constant of the crystal are known, then, according to Bragg's law, Eq. (2.96), the de Broglie wavelength of a neutron can be found. For thermal neutrons, the energy is defined as

$$E = \frac{3}{2} k_B T, \tag{2.100}$$

and we obtain for the de Broglie wavelength of the neutron

$$\lambda_{Br} = \frac{2\pi\hbar}{\sqrt{3 m_n k_B T}}. \tag{2.101}$$

At room temperature the de Broglie wavelength of a neutron is $\lambda_{Br} \approx 0.1$ nm. The diffraction of neutrons has a wide range of applications in the study of solid-state structures. Wave properties of atoms and molecules were also discovered. For atomic or molecular beams incident on the surface of a crystal additional diffraction maxima were observed in addition to specular reflection, for which the angle of incidence is equal to the angle of reflection.

2.3.3 The criterion governing the classical and quantum properties of a particle

The dimensionality of Planck's constant, h, is [J s]. The same dimensionality, [J s], can be obtained for the following products of two quantities: (1) energy and time, (2) momentum and coordinate, and (3) angular momentum and angle. The quantity with such a dimensionality, S, is called an *action*. Therefore, Planck's constant, which has the same dimensionality, can be called the *quantum of action*.

If for an object the value of an action is comparable to the value of h, then the behavior of such an object must be governed by the quantum laws. If the value of an action, S, is much larger than Planck's constant, h, then the behavior of this object is described with high precision by classical physics. Let us note that the smallness of an action does not always correspond to the non-applicability of a classical approach. In many cases the classical approach gives a qualitative picture of the behavior of the object studied, which can be refined further with the help of a quantum approach.

Among particles the photon has the smallest possible action. The energy of a photon is equal to $\varepsilon_{ph} = \hbar\omega$ and the characteristic time is equal to the period $T_{ph} = 2\pi/\omega$. Therefore, the value of an action for the photon is

$$S_{ph} = 2\pi\hbar = h, \tag{2.102}$$

i.e., S_{ph} is comparable to the quantum of action, h. It is obvious that photons obey quantum laws.

For comparison let us calculate the action of an ion that is undergoing thermal oscillations in a crystal at room temperature with a frequency $\omega_{ion} \sim 10^8\ s^{-1}$. The average energy of thermal oscillations is $E = k_B T$, and the period of oscillations is equal to $T_{ion} = 2\pi/\omega$. Therefore, the ion's action is

$$S_{ion} = k_B T \frac{2\pi}{\omega} \approx 2.5 \times 10^{-28}\ \mathrm{J\,s}. \tag{2.103}$$

The ratio of the ion's action to the quantum of action, h, in the case being considered is $S_{ion}/h \approx 4 \times 10^5$. Thus, the behavior of an ion in a crystal is described sufficiently well by the laws of classical physics.

Example 2.6. A parallel beam of non-relativistic electrons accelerated by the potential difference U is incident normally at a diaphragm with two narrow slits with the distance between the slits being equal to d. Find the distance between the neighboring maxima in the interference pattern on a screen located at a distance l, which is much greater than the distance between the slits d, i.e., $l \gg d$.
Reasoning. Using the analogy with wave optics (see Fig. B9(a)), we can state that the interference maxima are formed in the place where the path-length difference of rays contains an integer number of de Broglie waves, i.e.,

$$L \approx r_2 - r_1 = m\lambda_{Br}. \tag{2.104}$$

Since the first interference maxima are observed at small angles ($\theta \ll 1$), the path-length difference, L, can be presented as

$$L = d \sin\theta \approx \theta d. \tag{2.105}$$

At the same time the coordinate of the mth interference maximum, x_m, is much smaller than the distance between the slits, d, and that to the screen, l. Therefore, we can consider that $\theta \approx x_m/l$. Thus, for the coordinate x_m we obtain

$$x_m = m\frac{l\lambda_{Br}}{d}. \tag{2.106}$$

The maximum corresponding to zero path-length difference of the de Broglie waves is located at the point $x = 0$. The number of the next interference maximum changes by unity and its coordinate changes by a magnitude

$$\Delta x = x_{m+1} - x_m = \frac{l\lambda_{Br}}{d}, \tag{2.107}$$

which is called the *width of the interference fringe*. On substituting the expression for the de Broglie wavelength (2.95) into Eq. (2.107) we get

$$\Delta x = \frac{2\pi\hbar l}{d\sqrt{2m_e eU}}. \tag{2.108}$$

From the last expression it follows that with increasing accelerating potential difference, U, the distance between the interference maxima of electron waves for two-slit interference decreases.

Example 2.7. Using the expression for the phase velocity of de Broglie waves write the refraction constant of these waves passing through the inner potential of a crystal U_0 and applied potential U.
Reasoning. An electron in a metal is affected by an inner crystalline field, which is created by the positively charged lattice ions. This field changes periodically from ion to ion. The magnitude of the potential of this field averaged over the volume of a crystal is called the *inner crystalline potential*. To extract an electron from a metal it is necessary to spend energy equal to the work function $A_{\text{wf}} = eU_0$ (see Fig. 2.8). As we would do in optics, let us define the refractive index, n, of the de Broglie waves as the ratio of the phase velocities of these waves in vacuum and in a medium, i.e.,

$$n = \frac{v_{\text{vac}}}{v_{\text{med}}} = \frac{\omega k_{\text{med}}}{\omega k_{\text{vac}}} = \frac{\lambda_{\text{Br}}^{\text{vac}}}{\lambda_{\text{Br}}^{\text{med}}}. \tag{2.109}$$

Taking into account Eq. (2.95), according to which the de Broglie wavelength is inversely proportional to the square root of an electron's kinetic energy, we can write

$$n = \sqrt{\frac{K + A_{\text{wf}}}{K}} = \sqrt{1 + \frac{A_{\text{wf}}}{K}}. \tag{2.110}$$

On writing the energy through the potential, U_0, we obtain

$$n = \sqrt{1 + \frac{U_0}{U}}. \tag{2.111}$$

Since the work function of metals tends to be several electron-volts for electrons, which are subjected to the high accelerating potential difference ($U > 100$ V and $U \gg U_0$), the index of refraction is practically equal to unity.

Example 2.8. Find the de Broglie wavelength of an electron that is moving with velocity v, at which the electron mass is equal to double the electron mass at rest.
Reasoning. The relativistic mass of a moving electron is defined by the expression

$$m(v) = \frac{m_{\text{e}}}{\sqrt{1 - v^2/c^2}}. \tag{2.112}$$

Let us introduce the variable $\eta = m_{\text{e}}/m(v)$. Then, the velocity of an electron is equal to

$$\sqrt{1 - v^2/c^2} = \frac{m_{\text{e}}}{m(v)} = \eta, \tag{2.113}$$

$$v = c\sqrt{1 - \eta^2}. \tag{2.114}$$

The de Broglie wavelength of an electron moving with velocity comparable to the speed of light in vacuum is equal to

$$\lambda_{\mathrm{Br}} = \frac{2\pi\hbar}{p} = \frac{2\pi\hbar}{m_e v}\sqrt{1-\frac{v^2}{c^2}} = \Lambda\frac{\eta}{\sqrt{1-\eta^2}}, \tag{2.115}$$

where we introduced the Compton wavelength (Eq. (2.70)): $\Lambda = 2\pi\hbar/(m_e c) = 2.42 \times 10^{-12}$ m. In our case $\eta = 1/2$, and

$$\lambda_{\mathrm{Br}} = \frac{\Lambda}{\sqrt{3}} = 1.4 \times 10^{-12}\,\mathrm{m}. \tag{2.116}$$

Thus, the wave properties of relativistic ($v \leq c$) electrons can be revealed when they are in a volume with linear dimensions of about 10^{-12} m (e.g., in the vicinity of an atomic nucleus).

2.4 The uncertainty relations

In classical physics a complete description of a particle's state is defined by its dynamic parameters (coordinate, momentum, torque, energy, and so on). However, the real behavior of microscopic particles shows that there is a limit in principle to the precision with which the above-mentioned parameters can be measured. The limits of application to the classical description of the behavior of microscopic particles are determined by the uncertainty relations, which were suggested by Werner Heisenberg in 1927.

The most important among them are the following two relations. The first one relates the uncertainty in the values of the particle's coordinate, r_α, to the uncertainty in the corresponding component of momentum, p_α, at the same instant in time:

$$\Delta r_\alpha\, \Delta p_\alpha \geq h, \tag{2.117}$$

where $\alpha = x$, y, and z. Because of the importance of these relations we will rewrite them for each projection on the Cartesian coordinate axes:

$$\Delta x\, \Delta p_x \geq h, \qquad \Delta y\, \Delta p_y \geq h, \qquad \Delta z\, \Delta p_z \geq h. \tag{2.118}$$

The second expression, which relates the uncertainty in a change of the energy of a particle to the uncertainty in time during which this change takes place, is

$$\Delta E\, \Delta t \geq h. \tag{2.119}$$

The impossibility of simultaneous exact measurement of the quantities which are involved in the uncertainty relations is the manifestation of the dual wave–particle nature of microscopic objects. The uncertainty relations (2.117)–(2.119) are fundamental laws of nature and they are confirmed by all available experimental data. Application of the uncertainty relations to specific physical systems allows us not only to establish the limits of application of classical theory, but also to obtain numerical estimates of parameters of systems under study without using complicated mathematical analysis and draw important conclusions about the character of the processes involved.

2.4.1 The uncertainty relations of coordinate and momentum for a particle

In the case of macroscopic objects the applicability of classical concepts is clear. Let us consider for example a flying speck of dust (sufficiently small from an everyday point of view) with mass $m \sim 10^{-7}$ kg and velocity in the x-direction $v_x \sim 0.1\ \mathrm{m\,s^{-1}}$. If we define this velocity by available standard optical methods with a precision $\Delta v_x \approx 10^{-5}\ \mathrm{m\,s^{-1}}$, then the unavoidable uncertainty in measurements of the particle's coordinate according to relation (2.118) is

$$\Delta x \approx \frac{h}{\Delta p_x} = \frac{h}{m\,\Delta v_x} \approx 10^{-21}\ \mathrm{m}. \tag{2.120}$$

The size of a speck of dust with mass 0.1 mg depends on the material composing this speck and is of the order of 10^{-5} m. The uncertainty of measurement of its location is 16 orders of magnitude less than its dimensions, and it lies far beyond our possibility to make measurements of a coordinate with such precision. In this case the behavior of the particle is entirely described by the classical laws of physics.

Let us study the applicability of the different theories for the description of the behavior of an electron. It turns out that there is no definite answer in advance. The answer solely depends on the conditions of the electron's state. For example, in an electron tube with an accelerating potential difference $U \sim 10^4$ V, an electron acquires momentum

$$p = \sqrt{2m_e eU} \approx 5.4 \times 10^{-23}\ \mathrm{kg\,m\,s^{-1}}. \tag{2.121}$$

This momentum is directed along the tube's axis. The diameter of the focussed beam can reach values of $d \approx 10^{-5}$ m. This magnitude defines the uncertainty in the electron's position on the screen. According to the uncertainty relations (2.117) the electron has non-controllable momentum in the transverse direction

$$\Delta p_\perp \approx \frac{h}{d} \approx 10^{-29}\ \mathrm{kg\,m\,s^{-1}}. \tag{2.122}$$

Since $\Delta p_\perp / p \ll 1$ the motion of the electron follows a specific trajectory and can be described by the laws of classical physics.

For the electron in a hydrogen atom the radius of the first Bohr orbit is equal to $r \approx 5 \times 10^{-11}$ m. In accordance with Newton's second law of motion the acceleration, a, of the electron along the circular orbit is defined by the Coulomb force:

$$m_e a = \frac{k_e e^2}{r^2}. \tag{2.123}$$

For a circular orbit,

$$a = \frac{v^2}{r}. \tag{2.124}$$

By combining Eqs. (2.123) and (2.124) we obtain the magnitude of the electron's momentum:

$$p = e\sqrt{k_e \frac{m_e}{r}} \approx 2 \times 10^{-24}\,\text{kg m s}^{-1}. \tag{2.125}$$

Let us use now the uncertainty relations. Since the electron is localized in an atom, the uncertainty of its coordinate is equal to the dimension of an atom, i.e., $\Delta r \approx 2r$. The corresponding uncertainty of momentum is

$$\Delta p \approx \frac{h}{\Delta r} \approx \frac{h}{2r} \approx 6.6 \times 10^{-24}\,\text{kg m s}^{-1}. \tag{2.126}$$

The uncertainty of momentum, Δp, is comparable to the magnitude of the momentum, p. Thus, the laws of classical physics cannot describe the behavior of such a particle and it is necessary to apply the quantum approach instead.

2.4.2 The uncertainty relations of energy and time for a particle

The detailed consideration of the process for the measurement of energy shows that the energy of a system can be measured only with a finite uncertainty, ΔE, i.e., ΔE cannot be equal to zero. It follows from relation (2.119) that the energy conservation principle for a system can be verified by two measurements with accuracy $h/\Delta t$, where Δt is the time interval of measurement. The process of measurement with $\Delta t \to \infty$ is unrealistic; this is why $\Delta E \neq 0$.

Let us apply the energy–time uncertainty relation to a system that can be in a state with energy E for a finite time Δt. After this time the system makes a transition to another energy state. In this case, taking into account relation (2.119), we can write

$$\Delta E \; \Delta t \geq h, \tag{2.127}$$

where ΔE is the unavoidable uncertainty of the energy for that state. This form of the uncertainty relation allows us to estimate the finite width of the energy state of the atomic system, which remains in that state for a finite time Δt.

It is well known that the atomic emission spectral lines are not infinitely narrow – this would correspond to an uncertainty in the radiation of a quantum of energy of $\Delta E = 0$. The spectral lines that are observed in experiments have a so-called *natural linewidth* $\Delta\omega$ (see Fig. 2.15). The width of this spectral line is determined by the spread of the photon's energy values $\Delta E = \hbar\,\Delta\omega$ with respect to the mean value that characterizes the center of the line $E = \hbar\omega$. This width

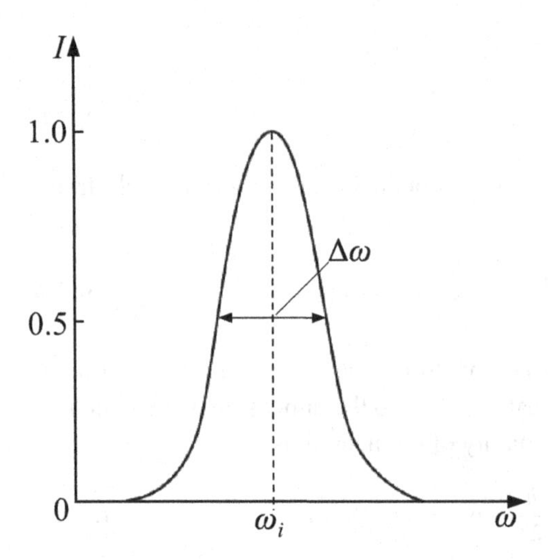

Figure 2.15 The form of an atom's emission spectral line.

is related to the lifetime of an atom in the excited state, Δt, by the uncertainty relation

$$\hbar\,\Delta\omega\,\Delta t \approx h, \qquad \text{or} \qquad \Delta\omega\,\Delta t \approx 2\pi. \tag{2.128}$$

By experimentally measuring the natural linewidth we can find, with the help of relation (2.127), the lifetime of an atom in the corresponding excited state. The natural linewidth which corresponds to the visible range of emission of atoms is of the order of $\Delta\omega \sim 10^8\ \text{s}^{-1}$. Therefore, the atomic lifetime in the excited state is about $\Delta t \sim 10^{-8}$ s. Taking into account that the frequency in the visible range is about $\omega \approx 4 \times 10^{15}\ \text{s}^{-1}$, we can find the relative width of the emission spectral line: $\Delta\omega/\omega \approx 2.5 \times 10^{-8}$. Note that in the ground state, i.e., in the non-excited state, stable (light) atoms can have indefinitely long lifetimes, i.e., in this state the lifetime of an atom $\tau \to \infty$, while the width of the energy level that corresponds to the ground state $\Delta E \to 0$.

The wave properties of particles and the probabilistic character of de Broglie waves indicate that the description of a particle's motion has to be very different from the classical description. In Newton's classical mechanics the motion of a particle under the influence of a force is completely determined by the initial position and velocity of the particle. By solving Newton's equation we can find the trajectory, coordinates, and velocity of the particle at any instant with any degree of accuracy, which is limited only by the accuracy of the measuring devices used. From the uncertainty relations it follows that, in cases in which the uncertainty of a coordinate is comparable to the dimension of the region where the particle is moving, the notion of the trajectory loses its meaning. The simultaneous determination of a particle's position and its velocity cannot be done with an arbitrary degree of accuracy. This is true for the pairs of physical quantities whose product has the dimension of Planck's constant. From this discussion follows the necessity of creating a more complete description of physical phenomena, in which particle and wave properties of the objects under study can be revealed. We can consider the year 1926 as the beginning of the emergence of such a theory, when Erwin Schrödinger discovered the equation for the description of microscopic particle behavior.

Example 2.9. Using the uncertainty relations, we can relate a particle's coordinate and its de Broglie wavelength. Find the condition under which we can neglect quantum effects if the particle occupies a region with linear dimension equal to L.
Reasoning. According to the uncertainty relations (2.118) for the uncertainty of the coordinate we have

$$\Delta x \geq \frac{h}{\Delta p_x}. \tag{2.129}$$

The smallest value of Δx corresponds to the largest value of Δp_x according to the above uncertainty relation. Since Δp_x cannot exceed the total momentum of a particle p,

$$\Delta x \geq h/p. \tag{2.130}$$

Taking into account the relation between the momentum and the de Broglie wavelength (Eq. (2.88)), we can obtain from Eq. (2.130)

$$\Delta x \geq \lambda_{\mathrm{Br}}. \tag{2.131}$$

The notion of a particle's trajectory can be used only when the uncertainty of its coordinate is small compared with the characteristic dimension of the region where it moves, i.e., $\Delta x \ll L$. Therefore, the condition for correctness of the classical description of a particle's behavior is

$$\lambda_{\mathrm{Br}} \ll L. \tag{2.132}$$

When the size of the region, L, becomes of the order of λ_{Br} the quantum effects cannot be neglected, as shown in the example discussed earlier.

Example 2.10. An electron is in a spherical cavity with impenetrable walls (a quantum dot). The dimension (diameter) of this confined region is $d = 3$ nm. Estimate the minimum energy of the electron.
Reasoning. In this case the uncertainty of the electron's coordinate $\Delta r \approx d$ and the uncertainty of the momentum is comparable to the magnitude of the momentum itself: $\Delta p \approx p$. The electron's momentum is related to its energy as

$$p = \sqrt{2m_{\mathrm{e}}E}. \tag{2.133}$$

Taking into account the uncertainty relations, we have

$$\Delta r\, \Delta p = d\sqrt{2m_{\mathrm{e}}E} \geq h. \tag{2.134}$$

From the last inequality for the minimal energy of the electron we obtain

$$E_{\min} = \frac{h^2}{2m_e d^2}, \tag{2.135}$$

which gives $E_{\min} \approx 2.6 \times 10^{-20}$ J ≈ 0.16 eV, i.e., an electron cannot have energy smaller than $E_{\min}$. Because $E_{\min}$ is finite, the electron cannot be in the state of rest that corresponds to $E = 0$ and $p = 0$.

2.5 The world of the nanoscale and the wavefunction

The study of processes and phenomena involving microscopic particles, where the wave properties are taken into account, is carried out in the framework of the so-called *wave mechanics*, which is also frequently called *quantum mechanics*. For the description of a particle's motion well-known classical quantities such as the position vector $\mathbf{r}$, momentum $\mathbf{p}$, energy E, etc. are used. The state of a particle at a given instant is defined by a number of physical quantities, i.e., by a set of variables. The significant peculiarity of the quantum-mechanical description is that this set includes a smaller number of variables than is necessary for a classical description.

The quantum-mechanical description uses three variables. These variables can be either the three projections of $\mathbf{r}(t)$ or the three projections of $\mathbf{p}(t)$, but not both. In classical physics all six variables, $\mathbf{r}(t)$ and $\mathbf{p}(t)$, are required for the description of a particle's state. Usually the coordinates $\mathbf{r}(t)$ are chosen for the quantum-mechanical description as the total set of variables. Then, such a description of a particle's state is called the *coordinate representation.*

According to the uncertainty principle (2.118), in quantum mechanics it is impossible to measure simultaneously a particle's coordinate and its momentum. From the physical point of view this is equivalent to the absence of a trajectory for the particle. Indeed, a trajectory is a series of sequential close positions of a particle, which are defined by the position vectors at close instants in time: $\mathbf{r}(t_0)$, $\mathbf{r}(t_1)$, $\mathbf{r}(t_2)$, ... To define the position of a particle at an instant in time t_k it is necessary to know its position at the previous instant in time, t_{k-1}, and its momentum at that time, $\mathbf{p}(t_{k-1})$, i.e.,

$$\mathbf{r}(t_k) = \mathbf{r}(t_{k-1}) + (t_k - t_{k-1})\frac{\mathbf{p}(t_{k-1})}{m}. \tag{2.136}$$

The impossibility of simultaneous definition of position and momentum leads to the impossibility of indicating the exact position of the particle at any given time. Therefore, the motion of a particle cannot be described by its position vector $\mathbf{r}$ as a function of time. For any microscopic process it makes sense to talk only about the probability of finding a certain value of a physical quantity, i.e., the *quantum-mechanical description is a probabilistic description.*

Another feature of the description on a microscopic scale lies in the fact that some values of physical quantities cannot be measured because these values are *forbidden* by the quantum-mechanical laws. The set of all values that a physical quantity is allowed to have is called the *spectrum*. Depending on the character of this physical quantity and the specific physical conditions, the spectrum can be *discrete*, *continuous*, or *mixed*. A discrete spectrum consists of a set of individual values of the physical quantity, between which intervals of forbidden values occur. A continuous spectrum occurs if the physical quantity can have any values from the given interval. Finding the spectra of physical quantities is one of the main problems of quantum mechanics. Examples of finding the spectra of specific systems will be considered later.

Summarizing the above, it follows that according to the quantum-mechanical description we may talk only about the *probability* of a particle being at a given point at a given instant – namely in an infinitesimal volume $\mathrm{d}V = \mathrm{d}x\,\mathrm{d}y\,\mathrm{d}z$, which includes this point. Accordingly, the so-called *wavefunction* $\Psi(\mathbf{r}, t)$, which depends on time and coordinates, can be introduced. With the help of the wavefunction the above-mentioned probability, P, is defined by the expression

$$\mathrm{d}P = \Psi\Psi^*\,\mathrm{d}V = |\Psi|^2\,\mathrm{d}V. \tag{2.137}$$

In general, the wavefunction Ψ can be a *complex function*, i.e., it consists of real and imaginary parts. Therefore, the wavefunction itself does not have any physical meaning, but the square of its modulus, $|\Psi|^2$, which is defined as a product of the wavefunction, Ψ, and its complex conjugate, Ψ^*, has. According to Eq. (2.137), the physical quantity

$$\rho(\mathbf{r}, t) = \frac{\mathrm{d}P}{\mathrm{d}V} = |\Psi|^2 \tag{2.138}$$

defines the so-called *probability density* (the probability of finding the particle per volume unit). The quantity $\rho(\mathbf{r}, t)$ can be experimentally observed, but the function $\Psi(\mathbf{r}, t)$ cannot. From Eq. (2.137) we can find the finite probability of the particle, described by the wavefunction $\Psi(\mathbf{r}, t)$, being within an arbitrary volume V:

$$P = \int_V \rho(\mathbf{r}, t)\mathrm{d}V = \int_V |\Psi|^2\,\mathrm{d}V. \tag{2.139}$$

Since the probability of finding a particle somewhere must equal unity, its wavefunction, Ψ, is chosen in such a way that it satisfies the so-called *normalization condition*:

$$\int_V |\Psi|^2\,\mathrm{d}V = 1, \tag{2.140}$$

where the integral is taken over the entire space or over the region where the wavefunction, Ψ, is non-zero. A wavefunction that satisfies condition (2.140) is called *normalized*.

In electromagnetism the superposition principle for the wave fields follows from the linearity of Maxwell's equations. The wavefunctions that describe the wave properties of particles satisfy the same principle. The superposition principle for wavefunctions Ψ is one of the main postulates of quantum mechanics and can be formulated as follows: *if a particle can have two different quantum states described by the functions* Ψ_1 *and* Ψ_2, *then a particle can be in the state described by the function*

$$\Psi = C_1\Psi_1 + C_2\Psi_2, \tag{2.141}$$

where C_1 and C_2 are two arbitrary complex numbers. In the general case we can talk about the superposition of any number N of quantum states, in which the particle can be, i.e., we can talk about the existence of states described by a function

$$\Psi = \sum_{n=1}^{N} C_n \Psi_n. \tag{2.142}$$

In this state the quantity $|C_n|^2$ defines the probability that during a measurement we will find a particle in a quantum state described by the wavefunction Ψ_n. Taking into account that all functions, Ψ_n, in Eq. (2.142) are normalized, we obtain the condition for the coefficients C_n:

$$\sum_{n=1}^{N} |C_n|^2 = 1. \tag{2.143}$$

If a particle is in one of its N quantum states it cannot be simultaneously in another quantum state. In other words, if a particle is found in the state described by the function Ψ_n the probability of finding it in the state described by another function Ψ_m (at $m \neq n$) is zero. This property is called the *orthogonality of functions* Ψ_n. The orthogonality condition together with the normalization condition can be written in the form

$$\int_V \Psi_m^*(\mathbf{r}, t)\Psi_n(\mathbf{r}, t)\mathrm{d}V = \delta_{mn} = \begin{cases} 1, & m = n, \\ 0, & m \neq n. \end{cases} \tag{2.144}$$

The wavefunctions that satisfy Eq. (2.144) are called *orthonormalized wavefunctions*.

For a free particle the wavefunction has the form of the de Broglie wave defined by Eq. (2.91). If the wavefunction of a particle is a superposition of wavefunctions of "free particles" then such a particle is no longer a free particle. For example, the wavefunction

$$\Psi(\mathbf{r}, t) = \mathrm{e}^{-\mathrm{i}Et/\hbar}\left(C_1\mathrm{e}^{\mathrm{i}(\mathbf{p}\cdot\mathbf{r})/\hbar} + C_2\mathrm{e}^{-\mathrm{i}(\mathbf{p}\cdot\mathbf{r})/\hbar}\right) \tag{2.145}$$

describes a particle that with probability $|C_1|^2$ moves along the vector $\mathbf{p}$ and with probability $|C_2|^2$ moves in the opposite direction. This particle cannot be considered a free particle.

Example 2.11. Find the uncertainty in the position of a free particle described by a wavefunction

$$\Psi(\mathbf{r}, t) = C e^{i(\mathbf{p}\cdot\mathbf{r} - Et)/\hbar} \tag{2.146}$$

at a given instant in time.

Reasoning. According to Eq. (2.138) the probability density of finding a particle at some point of space, taking into account Eq. (2.146), can be written as

$$\rho(\mathbf{r}, t) = \Psi(\mathbf{r}, t)\Psi^*(\mathbf{r}, t) = |C|^2. \tag{2.147}$$

Since C is a constant, $|C|^2$ is a constant too, which indicates that there is an equal probability of finding the particle at any point of space. Such a conclusion is in accordance with the reasoning based on the uncertainty principle. Because the particle's momentum $\mathbf{p}$ is exactly defined, $\Delta p_x = \Delta p_y = \Delta p_z = 0$. It follows from this that for the uncertainties of a particle's coordinates we have

$$\Delta x \to \infty, \qquad \Delta y \to \infty, \qquad \Delta z \to \infty, \tag{2.148}$$

i.e., a particle is uniformly spread over the entire space. This is not surprising because it corresponds to a classical wave with frequency $\omega = E/\hbar$ and wavevector $\mathbf{k} = \mathbf{p}/\hbar$.

Example 2.12. Find the normalization coefficient for the wavefunction

$$\Psi(x, t) = C e^{i(p_x x - Et)/\hbar - \alpha x^2}. \tag{2.149}$$

Reasoning. Using the normalization condition, Eq. (2.140), we can write

$$\int_{-\infty}^{\infty} |\Psi(x, t)|^2 \, dx = |C|^2 \int_{-\infty}^{\infty} e^{-2\alpha x^2} \, dx = |C|^2 \sqrt{\frac{\pi}{2\alpha}} = 1, \tag{2.150}$$

where we took into account that

$$\int_{-\infty}^{\infty} e^{-2\alpha x^2} \, dx = \sqrt{\frac{\pi}{2\alpha}}. \tag{2.151}$$

Then,

$$|C|^2 = \sqrt{\frac{2\alpha}{\pi}} \qquad \text{and} \qquad C = \left(\frac{2\alpha}{\pi}\right)^{1/4} e^{i\delta}, \tag{2.152}$$

where the phase factor δ is an arbitrary number and can be chosen equal to zero. For $\delta = 0$ the normalization constant, C, becomes real.

Example 2.13. Let the wavefunction have the form

$$\Psi = \frac{C}{r^2} e^{-iEt/\hbar}, \tag{2.153}$$

where $r \geq R$. Find the probability of finding a particle in a spherical shell $R \leq r \leq 2R$.

Reasoning. The volume unit for the spherically symmetric case has the form

$$\mathrm{d}V = 4\pi r^2\,\mathrm{d}r. \tag{2.154}$$

Thus, for the normalization condition we have

$$\int_R^\infty |\Psi(r,t)|^2\,4\pi r^2\,\mathrm{d}r = 4\pi C^2 \int_R^\infty \frac{r^2}{r^4}\,\mathrm{d}r = \frac{4\pi C^2}{R} = 1. \tag{2.155}$$

Therefore, the normalization constant, C, is defined as

$$C = \left(\frac{R}{4\pi}\right)^{1/2}. \tag{2.156}$$

The sought probability, P, is defined by the expression

$$P = \int_R^{2R} |\Psi(r,t)|^2\,4\pi r^2\,\mathrm{d}r = R\int_R^{2R} \frac{\mathrm{d}r}{r^2} = R\left(\frac{1}{R} - \frac{1}{2R}\right) = \frac{1}{2}. \tag{2.157}$$

2.6 The Schrödinger equation

The main equation of non-relativistic quantum mechanics is known as the Schrödinger equation. It allows us to find the wavefunction of a particle or a system of particles in a stationary state as well as its evolution in time. This equation has an *operator form* since in quantum mechanics *each physical quantity corresponds to its own operator.*

2.6.1 The relation between operators and physical quantities

An *operator defines some mathematical manipulation according to which one function is transformed into another function.* The transformation can be (a) just a simple multiplication of a function by a number or by another function on which the operator acts, (b) differentiation of the initial function, (c) a shift of the function in time and/or in coordinate space, and so on. The main operators in quantum mechanics are the so-called *coordinate, momentum, angular-momentum,* and *energy operators.* Further, we will use the *coordinate representation* in which the action of the coordinate operator on the wavefunction is equivalent to its multiplication by the corresponding coordinate:

$$\hat{r}_\alpha \Psi = r_\alpha \Psi \quad \text{or} \quad \hat{\mathbf{r}}\Psi = \mathbf{r}\Psi, \tag{2.158}$$

where $\alpha = x, y,$ and z. The momentum operator in the coordinate representation is the differential operator

$$\hat{p}_\alpha = -\mathrm{i}\hbar\frac{\partial}{\partial r_\alpha} \quad \text{or} \quad \hat{\mathbf{p}} = -\mathrm{i}\hbar\,\nabla, \tag{2.159}$$

$$\vec{\nabla} = \mathbf{e}_x\,\frac{\partial}{\partial x} + \mathbf{e}_y\,\frac{\partial}{\partial y} + \mathbf{e}_z\,\frac{\partial}{\partial z}. \tag{2.160}$$

To find the operator of kinetic energy it is important to know the operator of the square of momentum:

$$\hat{p}^2 = \hat{p}_x^2 + \hat{p}_y^2 + \hat{p}_z^2 = -\hbar^2 \left(\frac{\partial^2}{\partial x^2} + \frac{\partial^2}{\partial y^2} + \frac{\partial^2}{\partial z^2} \right) = -\hbar^2 \nabla^2. \tag{2.161}$$

The operator of angular momentum is introduced taking into account the definition in classical physics of the angular momentum, $\mathbf{L} = \mathbf{r} \times \mathbf{p}$, and after substitution into this expression of operators for the coordinate and momentum. The expressions for the operators of projections of angular momentum $\hat{L}_\alpha$ and the square of angular momentum $\hat{L}^2$ will be introduced in the next chapter.

The operator of the total energy in quantum mechanics is of special importance. This operator is formed by a substitution into the expression for the energy E,

$$E = \frac{p^2}{2m} + U(x, y, z), \tag{2.162}$$

of operators for the momentum and potential energy. Taking into account the form of the operator $\hat{p}^2$ in the coordinate representation and the fact that the operator of the function of coordinates is this function itself, we obtain

$$\hat{H} = \frac{\hat{p}^2}{2m} + \hat{U}(x, y, z) = -\frac{\hbar^2}{2m} \nabla^2 + U(x, y, z), \tag{2.163}$$

where $\nabla^2 = \partial^2/\partial x^2 + \partial^2/\partial y^2 + \partial^2/\partial z^2$. The total energy operator in quantum mechanics is written as $\hat{H}$ and it is called the *Hamilton operator* or the *Hamiltonian*.

To write an operator equation it is necessary to introduce the notion of the *eigenfunction of this operator* Ψ and its *eigenvalues* F in addition to the notion of the *operator* $\hat{F}$ itself. If the operator $\hat{F}$ corresponds to some physical quantity and its eigenfunction describes one of the possible states of the quantum system, where this quantity takes a certain value F_n, then the operator equation has the form

$$\hat{F}\Psi_n = F_n \Psi_n. \tag{2.164}$$

The operator $\hat{F}$ can have several (or infinitely many) eigenfunctions, thus n is an index that corresponds to a certain eigenfunction and thus to a certain state of the system. Some of the eigenvalues F_n may correspond to several eigenfunctions Ψ_{nl}, which differ by the index l. Such eigenvalues F_n are called *degenerate*. The number of eigenfunctions with different indices l corresponding to the given eigenvalue F_n is called the *order of degeneracy of this eigenvalue*.

The operator that corresponds to physical quantities must satisfy the following two conditions. First, this operator must be linear, i.e., the equality

$$\hat{F}(C_1\Psi_1 + C_2\Psi_2) = C_1\hat{F}\Psi_1 + C_2\hat{F}\Psi_2 \tag{2.165}$$

must be valid, which in the simplest case reflects the *superposition principle of quantum states*. Second, a quantum-mechanical operator must be *self-adjoint*, i.e., it must satisfy the integral condition

$$\int_V \Psi_1^*(\hat{F}\Psi_2)\mathrm{d}V = \int_V \Psi_2(\hat{F}\Psi_1)^*\mathrm{d}V. \tag{2.166}$$

Such an operator has real eigenvalues corresponding to the physical quantity described by the operator. The operators introduced earlier, Eqs. (2.158)–(2.161) and Eq. (2.163), are both linear and self-adjoint.

Operators can be added to each other and multiplied. The action of the sum of two operators on a wavefunction can be described as

$$(\hat{F}_1 + \hat{F}_2)\Psi = \hat{F}_1\Psi + \hat{F}_2\Psi. \tag{2.167}$$

The action of the product of two operators on a wavefunction is defined by the following rule:

$$(\hat{F}_1\hat{F}_2)\Psi = \hat{F}_1(\hat{F}_2\Psi) = \hat{F}_1\Psi_2 = \Psi_{12}. \tag{2.168}$$

The product of the same operators in reverse order leads to the following result:

$$(\hat{F}_2\hat{F}_1)\Psi = \hat{F}_2(\hat{F}_1\Psi) = \hat{F}_2\Psi_1 = \Psi_{21}. \tag{2.169}$$

If wavefunction Ψ_{12} is equal to wavefunction Ψ_{21}, then the operators are called *commutative* and the following equality can be written:

$$\hat{F}_1\hat{F}_2 = \hat{F}_2\hat{F}_1. \tag{2.170}$$

If $\Psi_{12} \neq \Psi_{21}$, then the operators are called *non-commutative*. Therefore, we can change the order of multiplication of two operators only if they are commutative. An example of non-commutative operators is a pair of operators of coordinate and its corresponding projection of momentum. Let $\hat{F}_1 = \hat{x} = x$ and $\hat{F}_2 = \hat{p}_x = -\mathrm{i}\hbar\,\partial/\partial x$. Then,

$$(\hat{F}_1\hat{F}_2)\Psi = -\mathrm{i}\hbar\left(x\frac{\partial}{\partial x}\right)\Psi = -\mathrm{i}\hbar x\frac{\partial\Psi}{\partial x}, \tag{2.171}$$

$$(\hat{F}_2\hat{F}_1)\Psi = -\mathrm{i}\hbar\left(\frac{\partial}{\partial x}x\right)\Psi = -\mathrm{i}\hbar\frac{\partial}{\partial x}(x\Psi) = -\mathrm{i}\hbar\left(\Psi + x\frac{\partial\Psi}{\partial x}\right). \tag{2.172}$$

We can see that these two given operators do not commutate. One of the most important postulates of quantum theory is the following statement: *if two different physical quantities f_1 and f_2 cannot be simultaneously measured, then the corresponding operators are non-commutative.*

The operators of kinetic and potential energies are non-commutative. This means that simultaneous measurement of both energies for a quantum system is impossible. Thus, we cannot state that the total energy of the system is the sum of its kinetic and potential energies. This may be valid only for the operators that correspond to these physical quantities: $\hat{H} = \hat{K} + \hat{U}$.

The following pairs of operators are illustrative examples of commutative operators: the operator of kinetic energy $-[\hbar^2/(2m)]\nabla^2$ and the operator of momentum $-i\hbar\,\nabla$; the operator of coordinate $\hat{x} = x$ and the operator of projection of momentum on the axes excluding the x-axis, i.e., excluding $\hat{p}_x$; and the operator of potential energy $\hat{U} = U(\mathbf{r})$ and any of the operators of coordinates $\hat{r}_\alpha = r_\alpha$.

In the quantum-mechanical description of a particle's behavior, which has probabilistic character, the most important characteristic of a physical quantity is its average value in a given quantum state. For a state described by Ψ the average value of the physical quantity F, with the corresponding operator $\hat{F}$, is defined by the expression

$$\langle F \rangle = \int_V \Psi^*(\mathbf{r}, t)\hat{F}\Psi(\mathbf{r}, t)\mathrm{d}V, \qquad (2.173)$$

where the integration is carried out over the entire volume within which the particle moves. In Cartesian coordinates the volume element is equal to $\mathrm{d}V = \mathrm{d}x\,\mathrm{d}y\,\mathrm{d}z$. If the operator $\hat{F}$ is a function of coordinates only, i.e., $\hat{F} = F(\mathbf{r})$, then

$$\langle F \rangle = \int_V F(\mathbf{r})|\Psi(\mathbf{r})|^2\,\mathrm{d}V. \qquad (2.174)$$

Let us consider several examples that illustrate the main properties of quantum-mechanical operators.

Example 2.14. The normalized wavefunction

$$\Psi(r, t) = \frac{1}{\sqrt{2\pi a r}}\mathrm{e}^{-r/a - \mathrm{i}Et/\hbar} \qquad (2.175)$$

is the solution of the Schrödinger equation for a sphere of radius a with the potential $U(\mathbf{r} < a) = 0$, which is surrounded by the potential $U(\mathbf{r} \geq a) = \infty$. Here r is the distance from the center of the sphere. Find the particle's average distance from the center of the sphere.

Reasoning. Let us rewrite Eq. (2.174) for the case being considered:

$$\langle r \rangle = \int r|\Psi(r, t)|^2\,\mathrm{d}V. \qquad (2.176)$$

It is more convenient to carry out the integration in the last expression in spherical coordinates. The corresponding volume, $\mathrm{d}V$, of a spherical layer with radii r and $r + \mathrm{d}r$ will have the form

$$\mathrm{d}V = 4\pi r^2\,\mathrm{d}r. \qquad (2.177)$$

After all substitutions have been done, we obtain

$$\langle r \rangle = \frac{1}{2\pi a}\int r\frac{\mathrm{e}^{-2r/a}}{r^2}4\pi r^2\,\mathrm{d}r = \frac{2}{a}\int_0^\infty r\mathrm{e}^{-2r/a}\,\mathrm{d}r = \frac{2}{a}Y. \qquad (2.178)$$

Here the integral

$$Y = \int_0^\infty r e^{-2r/a}\, dr$$

can be calculated by integration by parts. As a result of integration we obtain $Y = a^2/4$, and for the particle's average distance from the center of the sphere we have $\langle r \rangle = a/2$.

Example 2.15. Determine whether the operator of complex conjugation is linear.
Reasoning. Let us denote the operator of complex conjugation by the symbol $\hat{g}$. Let us apply an operation of complex conjugation to the linear combination of wavefunctions Ψ_1 and Ψ_2:

$$\hat{g}(C_1\Psi_1 + C_2\Psi_2) = C_1^*\Psi_1^* + C_2^*\Psi_2^* = C_1^*\hat{g}\Psi_1 + C_2^*\hat{g}\Psi_2 \neq C_1\hat{g}\Psi_1 + C_2\hat{g}\Psi_2. \tag{2.179}$$

We see that the condition (2.165) is not satisfied. Therefore, the operator $\hat{g}$ is not linear.

Example 2.16. Determine whether the operator of momentum projection $\hat{p}_x = -\hbar\, \partial/\partial x$ is self-adjoint.
Reasoning. To find out whether the operator $\hat{p}_x$ is self-adjoint or not, let us consider the following integral:

$$J_x = \int_{-\infty}^{\infty} \Psi_1^*(x, y, z, t)\hat{p}_x \Psi_2(x, y, z, t) dx = -i\hbar \int_{-\infty}^{\infty} \Psi_1^*(x, y, z, t)\frac{\partial \Psi_2}{\partial x}\, dx. \tag{2.180}$$

Let us carry out the integration by parts, taking into account that at $x \to \pm\infty$ the wavefunction becomes equal to zero:

$$J_x = -i\hbar\Psi_1^*\Psi_2|_{-\infty}^{\infty} + i\hbar \int_{-\infty}^{\infty} \Psi_2 \frac{\partial \Psi_1^*}{\partial x} dx = \int_{-\infty}^{\infty} \Psi_2 \left(-i\hbar \frac{\partial \Psi_1}{\partial x}\right)^* dx$$
$$= \int_{-\infty}^{\infty} \Psi_2 (\hat{p}_x \Psi_1)^*\, dx. \tag{2.181}$$

According to Eq. (2.166) the operator $\hat{p}_x$ is self-adjoint. In the same way we can prove that the operators $\hat{p}_y$ and $\hat{p}_z$ are self-adjoint. Using property (2.167), we can prove that the operator of momentum $\hat{\mathbf{p}}$ is self-adjoint.

2.6.2 The time-dependent Schrödinger equation

Erwin Schrödinger was the first to discover, 1926, the main equation of non-relativistic quantum mechanics, which described the behavior of a particle governed by the laws of quantum mechanics:

$$\hat{H}\Psi(\mathbf{r}, t) = i\hbar\frac{\partial}{\partial t}\Psi(\mathbf{r}, t), \tag{2.182}$$

where $\hat{H}$ is the Hamiltonian of a system and Ψ is the wavefunction that describes this system. Equation (2.182) is called the *Schrödinger equation*.

Let us obtain with the help of the Schrödinger equation several important relationships. For this purpose let us compose the complex conjugate of Eq. (2.182):

$$\hat{H}^*\Psi^* = -i\hbar\frac{\partial\Psi^*}{\partial t}. \tag{2.183}$$

Let us multiply Eq. (2.182) by Ψ^* and Eq. (2.183) by Ψ and subtract the second result from the first:

$$\Psi^*\hat{H}\Psi - \Psi\hat{H}^*\Psi^* = i\hbar\frac{\partial}{\partial t}(\Psi\Psi^*). \tag{2.184}$$

By integrating this equation over the entire space and taking into account the self-adjoint property of the operator $\hat{H}$, we obtain

$$\frac{\partial}{\partial t}\int_V \Psi\Psi^*\,\mathrm{d}V = 0. \tag{2.185}$$

From Eq. (2.185) we get

$$\int_V |\Psi|^2\,\mathrm{d}V = \text{constant}. \tag{2.186}$$

Equation (2.186) shows that the normalization condition of the wavefunction is conserved over time (compare Eqs. (2.186) and (2.140)).

If in Eq. (2.184) we substitute the explicit expression for the Hamiltonian operator, Eq. (2.163), for the particle moving in a potential field, then we will arrive at the equation

$$i\frac{\partial}{\partial t}|\Psi|^2 = \frac{\hbar}{2m}(\Psi\,\nabla^2\Psi^* - \Psi^*\,\nabla^2\Psi). \tag{2.187}$$

In this equation $|\Psi|^2 = \rho$ is the probability density and

$$\Psi\,\nabla^2\Psi^* - \Psi^*\,\nabla^2\Psi = \nabla(\Psi\,\nabla\Psi^* - \Psi^*\,\nabla\Psi). \tag{2.188}$$

Therefore Eq. (2.187) can be rewritten in the form of the continuity equation

$$\frac{\partial\rho}{\partial t} + \nabla\cdot\mathbf{j} = 0, \tag{2.189}$$

where we introduced the probability current density, $\mathbf{j}$,

$$\mathbf{j} = \frac{i\hbar}{2m}(\Psi\,\nabla\Psi^* - \Psi^*\,\nabla\Psi). \tag{2.190}$$

The form of Eq. (2.189) is the same as for the continuity equation obtained in Appendix B (Example B2). The modulus of the vector $\mathbf{j}$ defines the probability per unit area that a particle will cross a surface perpendicular to the vector $\mathbf{j}$ per unit time.

2.6.3 Stationary states

The stationary states in which the physical quantities do not change with time are of special importance. In this case all the dependence of the wavefunction on time is contained in the phase factor, i.e.,

$$\Psi(\mathbf{r}, t) = \psi(\mathbf{r})\mathrm{e}^{-\mathrm{i}Et/\hbar}. \tag{2.191}$$

The exponential factor, which depends on time, changes only the phase of the wavefunction. The probability of finding a system in the corresponding state with energy E does not depend on time, as follows from Eq. (2.191):

$$|\Psi(\mathbf{r}, t)|^2 = |\psi(\mathbf{r})|^2. \tag{2.192}$$

By substituting Eq. (2.191) into Eq. (2.182) we obtain the equation

$$\hat{H}\psi(\mathbf{r}) = E\psi(\mathbf{r}), \tag{2.193}$$

which is called the *stationary Schrödinger equation*. In stationary states the eigenvalues of the operator $\hat{H}$ take certain values E_n. The state with the minimal energy $E_1 = E_{\mathrm{min}}$ is called the *ground state*. Actually this is analogous to the classical state of rest, for which E_{min} is equal to zero. Equation (2.193), after substitution of the Hamiltonian $\hat{H}$, which is defined by Eq. (2.163), takes the form

$$\left[-\frac{\hbar^2}{2m}\nabla^2 + U(\mathbf{r})\right]\psi(\mathbf{r}) = E\psi(\mathbf{r}). \tag{2.194}$$

The solutions, $\psi(\mathbf{r})$, and the corresponding values of the energy, E, are defined by the particular form of the potential energy, $U(\mathbf{r})$. In the next chapter we will consider some examples of quantum-mechanical analysis of a particle's behavior in the potential (quantum) well relevant for the physics of low-dimensional structures.

Example 2.17. Show that in a stationary state the probability density current $\mathbf{j}$ does not depend on time.

Reasoning. Let us substitute the wavefunction of the stationary state, Eq. (2.191), into the expression for the probability current density, Eq. (2.190):

$$\mathbf{j} = \frac{\mathrm{i}\hbar}{2m}\left[\psi(\mathbf{r})\mathrm{e}^{-\mathrm{i}Et/\hbar}\,\vec{\nabla}\psi^*(\mathbf{r})\mathrm{e}^{\mathrm{i}Et/\hbar} - \psi^*(\mathbf{r})\mathrm{e}^{\mathrm{i}Et/\hbar}\,\vec{\nabla}\psi(\mathbf{r})\mathrm{e}^{-\mathrm{i}Et/\hbar}\right]. \tag{2.195}$$

We see that in this expression the time-dependent exponents disappear and

$$\mathbf{j} = \frac{\mathrm{i}\hbar}{2m}\left[\psi(\mathbf{r})\vec{\nabla}\psi^*(\mathbf{r}) - \psi^*(\mathbf{r})\vec{\nabla}\psi(\mathbf{r})\right], \tag{2.196}$$

i.e., $\mathbf{j}$ does not depend on time.

2.7 Summary

1. In the framework of classical physics it was impossible to explain the dependence of the spectral density on the radiation frequency in the spectrum of blackbody radiation.
2. Planck's formula for the spectral density of blackbody radiation fully describes this phenomenon over the entire frequency range. The derivation of this formula is based on the assumption that electromagnetic radiation is emitted by a heated body in the form of single portions of energy (quanta).
3. The quantum properties of electromagnetic radiation become apparent not only during its emission, but also upon its absorption by a substance as well as during the propagation of radiation. Electromagnetic radiation is a flux of photons – particles with mass at rest equal to zero, energy $E_{ph} = \hbar\omega$, and momentum $\mathbf{p}_{ph} = \hbar\mathbf{k}$. These relationships connect the particle and wave properties of radiation.
4. Any particle with momentum p is related to a wave process, whose wavelength is defined by the relationship $\lambda_{Br} = 2\pi\hbar/p$. The wave properties of particles become apparent when the linear size of the region within which the motion of the particles takes place is comparable to the de Broglie wavelength, λ_{Br}.
5. The product of the uncertainties of two observables whose product has the dimensionality of an action, [J s], cannot be less than Planck's constant, h, i.e., it cannot be less than the quantum of action, h. The main uncertainty relationships in quantum physics are $\Delta r_\alpha\, \Delta p_\alpha \geq h$ and $\Delta E\, \Delta t \geq h$.
6. The concept of wave–particle duality applied to a substance and a field allows a complete description of many physical phenomena.

2.8 Problems

Problem 2.1. Using Planck's formula, find the most probable frequency, ω_{prob}, and mean frequency, ω_{mean}, in the blackbody radiation spectrum at $T = 2000$ K. (Answer: $\omega_{prob} \approx 7.4 \times 10^{14}$ s^{-1} and $\omega_{mean} \approx 10^{15}$ s^{-1}.)

Problem 2.2. Monochromatic radiation of wavelength $\lambda = 0.55$ μm is incident on the photosensitive surface of the photoelectric cell and it liberates $N_{ph} = 10^3$ electrons from the cell. The sensitivity of the photoelectric cell, which is defined as $\gamma = N_{ph}e/(N_i\hbar\omega)$, is equal to $\gamma = 10$ mA W^{-1}. Estimate the number of photons, N_i, incident on the surface of the photoelectric cell.

Problem 2.3. An isolated small iron ball of radius $r = 1$ mm is exposed to monochromatic light with wavelength $\lambda = 0.2$ μm. What is the maximum potential, φ_m, of the ball's charging and how many electrons, N_e, will it emit? The work function of iron is equal to $A_{wf} = 4.36$ eV. (Answer: $\varphi_m = 1.85$ V and $N_e \approx 1.3 \times 10^6$.)

Problem 2.4. The cut-off wavelength for a given metal is $\lambda_c = 0.62$ μm. How much larger is the highest kinetic energy, K_{max}, of photoelectrons emitted by the metal surface after its exposure to light of wavelength $\lambda = 0.35$ μm than the mean thermal electron energy, K_T. Assume room-temperature conditions. (Answer: $K_{\max}/K_T \approx 40$.)

Problem 2.5. A laser beam with wavelength $\lambda = 0.35$ μm, power $N = 5$ W, and cross-sectional diameter $d = 200$ μm is perpendicularly incident on a cube face with side length $a = 100$ μm. Find the number of incident photons per unit time, n, and the pressure, P, and force, F, that they create. (Answer: $n \approx 2.8 \times 10^{18}$, $P \approx 1.1$ Pa, and $F \approx 1.1 \times 10^{-8}$ N.)

Problem 2.6. In a homogeneous magnetic field with magnetic flux density $B = 0.1$ T an electron gyrates in a circle with a radius 100 times greater than the electron de Broglie wavelength. Calculate the radius, r, and the electron de Broglie wavelength, λ_{Br}. (Answer: $r \approx 0.8$ μm and $\lambda_{\mathrm{Br}} \approx 8$ nm.)

Problem 2.7. A parallel array of electrons is perpendicularly incident on a narrow slit of width $b = 5$ μm. The electrons are accelerated by a voltage $U = 100$ V. Find the distance, ΔX, between two maxima of the first order in the diffraction picture. The screen is located at a distance $L = 20$ cm from the slit. (Answer: $\Delta X \approx 14.8$ μm.)

Problem 2.8. The wavefunction of a particle moving along the x-axis has the form of Eq. (2.149). Considering all parameters of this function as real parameters, find the dependence on coordinate of the real part of this function, $\mathrm{Re}\{\Psi\}$, and the square of its modulus, $|\Psi|^2$, at time $t = t_0$.

Chapter 3
Layered nanostructures as the simplest systems to study electron behavior in a one-dimensional potential

The description of an electron's behavior is very simple if the corresponding probability current density, $\mathbf{j}$, is constant. In this case, as we already found in Section 2.6.3, the electron wavefunction, $\Psi(\mathbf{r}, t)$, is the product of a wavefunction $\psi(\mathbf{r})$ that is the solution of the stationary Schrödinger equation and an exponential factor, $\exp(iEt/\hbar)$, which depends on time (see Eq. (2.191)). Depending on the configuration of the potential within which the electron motion takes place, the solution of the time-independent Schrödinger equation (Eq. (2.194))

$$\left[-\frac{\hbar^2}{2m_e}\nabla^2 + U(\mathbf{r})\right]\psi(\mathbf{r}) = E\psi(\mathbf{r}) \tag{3.1}$$

defines the allowed energy states and the corresponding wavefunctions. Equation (3.1) is easier to solve than the time-dependent Schrödinger equation (2.182).

In this chapter we will consider several important examples of electron behavior in one-dimensional potential wells and demonstrate that the solutions of the time-independent Schrödinger equation (3.1) explain the main peculiarities of the behavior of confined electrons. The main advantage of the examples considered is that we can find numerical values for electron energy levels (the so-called *electron spectrum*) in potentials of different one-dimensional profiles. The discrete character of the electron spectrum, which was discovered experimentally for various atoms, demonstrates that the classical concepts are irrelevant for the description of microscopic particle behavior on smaller scales. The solutions of the time-independent Schrödinger equation also allow us to explain the probability of electron penetration into the region of the potential barrier as well as into the region behind the barrier, which is a *purely quantum phenomenon*.

As an example of one-dimensional potentials we will use layered nanostructures, which we have briefly considered in Section 1.3, and whose fabrication we will consider in Section 8.1. With the help of a modern technology called molecular-beam epitaxy (MBE) it became possible to fabricate structures with a precision of up to one atomic layer. The solutions of the Schrödinger equation for such structures were available in quantum-mechanical textbooks a long time before the layered structures were actually grown. This set of toy problems from the quantum-mechanical tool box became applicable to real structures and nowadays they are a hot topic of contemporary nanoelectronics. By combining layers of semiconductor materials with different forbidden energy gaps, for example materials such as GaAs and $Al_xGa_{1-x}As$, where x is the fraction of Al in the alloy, we can limit the space within which an electron can freely move. As you see in Fig. 1.4(a), electrons can be confined within the GaAs material, which has a smaller forbidden energy gap, E_g, than $Al_xGa_{1-x}As$. Electrons can move freely only in the y- and z-directions, whereas in the x-direction their motion is restricted. If we allow the percentage of Al in the $Al_xGa_{1-x}As$ alloy, which serves as a barrier for electrons in the GaAs layer, to vary with respect to coordinate, we can create potentials of various profiles, such as those which will be considered in Section 4.3.2.

The simplest form of space in which electron motion is restricted only in one of the directions is the so-called *quantum well*. It can be made in the form of a sandwich: a thin layer of GaAs is placed between two thick layers of AlGaAs alloy (see Fig. 1.4). To analyze the peculiarities of electron motion in such *layered (sandwiched) structures* we will first consider the quantum-mechanical description of free electron motion in a vacuum. After that we will consider electron motion in a quantum well with infinite barriers, which is a simplified model of a quantum well with real barriers. In such a well an electron cannot escape from the well or penetrate under the barrier, but it can freely move in the plane of the well, where its motion is not restricted. Then we will consider the case when one of the barriers is finite and the other is infinite, and after that the quantum well with both barriers being finite, which corresponds to a real quantum well of a layered structure.

In the last sections of Chapter 3 we will consider the case of electron transfer through a barrier (the so-called *electron-tunneling phenomenon*). Such a barrier can easily be fabricated in the form of a layered structure consisting of three layers, whose middle layer creates a potential barrier for electrons instead of a well.

3.1 The motion of a free electron in vacuum

The simplest example that is described by the Schrödinger equation (3.1) is that of a free electron moving in an unbounded region. The potential energy of such an electron does not depend on coordinates, i.e., it is constant, and therefore as a reference point the condition $U(\mathbf{r}) = U_0 = 0$ can be chosen. From classical mechanics it is well known that in such a potential the electron momentum, $\mathbf{p}$, and energy, E, are conserved. Since $U(\mathbf{r}) = 0$ the time-independent Schrödinger equation (3.1) for a free electron with mass m_e has the form

$$-\frac{\hbar^2}{2m_e}\left[\frac{\partial^2}{\partial x^2} + \frac{\partial^2}{\partial y^2} + \frac{\partial^2}{\partial z^2}\right]\psi(x, y, z) = E\psi(x, y, z). \qquad (3.2)$$

Since the electron's motion along each of the coordinate axes can be considered independent, the total wavefunction $\psi(x, y, z)$ can be written as a product of separate wavefunctions $\psi_x(x)$, $\psi_y(y)$, and $\psi_z(z)$ (i.e., we have *separated variables*),

$$\psi(x, y, z) = \psi_x(x)\psi_y(y)\psi_z(z), \tag{3.3}$$

and the total energy E as

$$E = E_x + E_y + E_z. \tag{3.4}$$

On substituting Eqs. (3.3) and (3.4) into the Schrödinger equation (3.2) and dividing it by $\psi_x(x)\psi_y(y)\psi_z(z)$ we obtain the following expression:

$$\frac{1}{\psi_x(x)}\left[\frac{\hbar^2}{2m_e}\frac{\partial^2\psi_x(x)}{\partial x^2} + E_x\psi_x(x)\right] + \frac{1}{\psi_y(y)}\left[\frac{\hbar^2}{2m_e}\frac{\partial^2\psi_y(y)}{\partial y^2} + E_y\psi_y(y)\right] + \frac{1}{\psi_z(z)}\left[\frac{\hbar^2}{2m_e}\frac{\partial^2\psi_z(z)}{\partial z^2} + E_z\psi_z(z)\right] = 0. \tag{3.5}$$

Since each of the three terms in Eq. (3.5) depends only on one of the independent variables x, y, and z, each of these terms should be equated to zero to satisfy Eq. (3.5). Thus, Eq. (3.5) can be rewritten as a system of three independent second-order differential equations:

$$-\frac{\hbar^2}{2m_e}\frac{d^2\psi_x(x)}{dx^2} = E_x\psi_x(x), \tag{3.6}$$

$$-\frac{\hbar^2}{2m_e}\frac{d^2\psi_y(y)}{dy^2} = E_y\psi_y(y), \tag{3.7}$$

$$-\frac{\hbar^2}{2m_e}\frac{d^2\psi_z(z)}{dz^2} = E_z\psi_z(z). \tag{3.8}$$

Note that Eqs. (3.6), (3.7), and (3.8) have the same form, and therefore it suffices to find a solution for one of them, for example the solution of Eq. (3.6) only, and then the solutions of the other two equations can be written by substituting for the variable x the variables y and z, correspondingly.

Since the electron's energy E_α is defined through the wavenumber k_α as

$$E_\alpha = \frac{\hbar^2 k_\alpha^2}{2m_e}, \tag{3.9}$$

where $\alpha = x, y, z$, let us rewrite Eq. (3.6) in the following form:

$$\frac{d^2\psi_x(x)}{dx^2} + k_x^2\psi_x(x) = 0, \tag{3.10}$$

where

$$k_x^2 = \frac{2m_e E_x}{\hbar^2}. \tag{3.11}$$

There are two types of solutions for the one-dimensional equation (3.10). One is for an electron moving in the positive direction along the x-axis, and has the following form:

$$\psi_x(x) = A\mathrm{e}^{\mathrm{i}k_x x}. \tag{3.12}$$

The second is for an electron moving in the negative direction along the x-axis:

$$\psi_x(x) = B\mathrm{e}^{-\mathrm{i}k_x x}. \tag{3.13}$$

The general solution of Eq. (3.10) is the sum of Eqs. (3.12) and (3.13):

$$\psi_x(x) = A\mathrm{e}^{\mathrm{i}k_x x} + B\mathrm{e}^{-\mathrm{i}k_x x}. \tag{3.14}$$

Analogously for the y- and z-directions,

$$\psi_y(y) = D\mathrm{e}^{\mathrm{i}k_y y} + E\mathrm{e}^{-\mathrm{i}k_y y}, \tag{3.15}$$

$$\psi_z(z) = F\mathrm{e}^{\mathrm{i}k_z z} + G\mathrm{e}^{-\mathrm{i}k_z z}. \tag{3.16}$$

Considering an electron moving in the positive direction along the x-axis, we can find the time-dependent wavefunction $\Psi_x(x, t)$ of a free particle in a stationary state if we take into account Eqs. (3.12) and (2.191):

$$\Psi_x(x, t) = \psi_x(x)\mathrm{e}^{-\mathrm{i}E_x t/\hbar} = A\mathrm{e}^{\mathrm{i}(k_x x - E_x t/\hbar)}. \tag{3.17}$$

The solutions in the other two directions have analogous forms and the total wavefunction of an electron moving in an arbitrary direction is

$$\Psi(\mathbf{r}, t) = C\mathrm{e}^{\mathrm{i}(\mathbf{k}\cdot\mathbf{r} - Et/\hbar)}, \tag{3.18}$$

where $\mathbf{k}\cdot\mathbf{r} = k_x x + k_y y + k_z z$ and $E = E_x + E_y + E_z$. The wavefunction (3.18) defines the probability amplitude of finding the particle at any arbitrary point in space as well as at any instant in time. Since this function is complex, only the square of its modulus has physical meaning. It defines the corresponding *probability density*, $\rho(\mathbf{r}, t)$:

$$\rho(\mathbf{r}, t) = |\Psi(\mathbf{r}, t)|^2 = \Psi^*(\mathbf{r}, t)\Psi(\mathbf{r}, t). \tag{3.19}$$

On substituting Eq. (3.18) into Eq. (3.19) we find that the *probability density of a quantum state of a free electron does not depend on time*:

$$\rho(\mathbf{r}, t) = |C|^2. \tag{3.20}$$

Example 3.1. Find the probability current density for a free electron that is in a stationary state with momentum $\mathbf{p}$ and energy E.
Reasoning. Let us choose the positive direction of the x-axis along the direction of the vector $\mathbf{p}$. Then the vector of the probability density flux will have only one component along the x-axis, which, according to Eq. (2.190), has the form

$$j_x = \frac{\mathrm{i}\hbar}{2m_\mathrm{e}}\left(\Psi_x\frac{\partial\Psi_x^*}{\partial x} - \Psi_x^*\frac{\partial\Psi_x}{\partial x}\right). \tag{3.21}$$

Here the wavefunction, $\Psi_x(x, t)$, of the free electron depends only on one spatial coordinate x and is defined by Eq. (3.17). The complex-conjugate function

$\Psi_x^*(x, t)$ is defined as

$$\Psi_x^*(x, t) = A^* e^{-i(k_x x - E_x t/\hbar)}. \tag{3.22}$$

Let us calculate the corresponding derivatives from Eq. (3.21):

$$\frac{\partial \Psi_x}{\partial x} = ik_x \Psi_x, \qquad \frac{\partial \Psi_x^*}{\partial x} = -ik_x \Psi_x^*. \tag{3.23}$$

By substituting these expressions into Eq. (3.21) we obtain

$$j_x = \frac{i\hbar}{2m_e}\left(-ik_x \Psi_x \Psi_x^* - ik_x \Psi_x^* \Psi_x\right) = \frac{\hbar k_x}{m_e}\Psi_x \Psi_x^* = \frac{p_x}{m_e}\Psi_x \Psi_x^*. \tag{3.24}$$

Taking into account Eq. (3.19) for the probability density $\Psi_x \Psi_x^* = |A|^2 = \rho(x)$, we obtain

$$j_x = \frac{p_x}{m_e}\rho(x) = v_e \rho(x), \tag{3.25}$$

where v_e is the electron's velocity. The result obtained for a free electron is consistent with the classical expression for the flux density since it is proportional to the electron's velocity, v_e, and the electron density, ρ.

3.2 An electron in a potential well with infinite barriers

Let us study the energy spectrum of an electron confined in a one-dimensional rectangular quantum well with impenetrable walls. This case corresponds to an electron being in a thin (several nanometers thick or even thinner) semiconductor film. In this case the potential energy of the electron depends only on one coordinate (in the example considered here, on the x-coordinate) and can be written as

$$U(x) = \begin{cases} \infty, & x \leq 0, \\ 0, & 0 < x < L, \\ \infty, & x \geq L. \end{cases} \tag{3.26}$$

The time-independent Schrödinger equation (3.1) for this one-dimensional potential well can be written as

$$\left[-\frac{\hbar^2}{2m_e}\left(\frac{\partial^2}{\partial x^2} + \frac{\partial^2}{\partial y^2} + \frac{\partial^2}{\partial z^2}\right) + U(x)\right]\psi(x, y, z) = E\psi(x, y, z). \tag{3.27}$$

As in the case of a free electron, the wavefunction $\psi(x, y, z)$ can be taken as a product of three separate wavefunctions $\psi_x(x)$, $\psi_y(y)$, and $\psi_z(z)$ (see Eq. (3.3)). On repeating the same procedure as for Eq. (3.2) we find that the Schrödinger equation in the y- and z-directions is the same as for a free electron (see Eqs. (3.7) and (3.8)). Thus, the electron motion in the y- and z-directions remains free and the solutions to the Schrödinger equation, $\psi(y)$ and $\psi(z)$, are defined by

Eqs. (3.15) and (3.16), respectively. The values of the kinetic energy for free motion in the y- and z-directions are defined by Eq. (3.9), i.e.,

$$E_y = \frac{\hbar^2 k_y^2}{2m_e} \tag{3.28}$$

and

$$E_z = \frac{\hbar^2 k_z^2}{2m_e}. \tag{3.29}$$

The Schrödinger equation for the x-coordinate differs from (3.6) since it contains a term with potential energy $U(x)$:

$$\left[-\frac{\hbar^2}{2m_e}\frac{d^2}{dx^2} + U(x)\right]\psi_x(x) = E_x\psi_x(x). \tag{3.30}$$

Equation (3.30) is a one-dimensional differential equation and, to avoid unnecessary use of subscripts, we will henceforth omit the subscript x from the eigenfunction, ψ_x, and eigenvalue, E_x:

$$\left[-\frac{\hbar^2}{2m_e}\frac{d^2}{dx^2} + U(x)\right]\psi(x) = E\psi(x). \tag{3.31}$$

For the region $0 < x < L$ where $U(x) = 0$ and where the electron is confined, the Schrödinger equation (3.31) takes the form

$$\frac{d^2\psi}{dx^2} + k^2\psi = 0, \tag{3.32}$$

where

$$k = \frac{\sqrt{2m_e E}}{\hbar}.$$

Here, for simplicity, we omitted the subscript x from the wavenumber k_x. The solutions to differential equations similar to (3.32) can be sought in the form of a combination either of harmonic functions or of exponential functions as we have already done in Eq. (3.14):

$$\psi(x) = A\sin(kx) + B\cos(kx), \tag{3.33}$$

$$\psi(x) = A'e^{ikx} + B'e^{-ikx}. \tag{3.34}$$

Let us seek solutions for the case of a potential well with infinite walls in the form of Eq. (3.33):

$$\psi(x) = A\sin(kx) + B\cos(kx) = C\sin(kx + \phi). \tag{3.35}$$

Here we introduced two unknown parameters, C and ϕ. To find these parameters we should take into account the boundary conditions imposed by the potential of the well. Since the potential barriers at the boundaries of the well are infinite (see Eq. (3.26)), the electron cannot penetrate into the regions outside of the well.

Thus, we can write

$$\psi(0) = 0, \tag{3.36}$$

$$\psi(L) = 0. \tag{3.37}$$

By substituting the wavefunction (3.35) into Eqs. (3.36) and (3.37) we obtain a system of two equations for three unknowns, C, ϕ, and k:

$$\psi(0) = C \sin\phi = 0 \tag{3.38}$$

and

$$\psi(L) = C \sin(kL + \phi) = 0. \tag{3.39}$$

Since the parameter $C \neq 0$ ($C = 0$ corresponds to the trivial solution $\psi(x) \equiv 0$), from Eq. (3.38) it follows that

$$\phi = 0. \tag{3.40}$$

Equations (3.35) and (3.39) can be rewritten as

$$\psi(x) = C \sin(kx), \tag{3.41}$$

$$\sin(kL) = 0. \tag{3.42}$$

From the last equation we obtain the following allowed values of the wavenumber, k, for the electron's wavefunction in the potential well being considered:

$$k_n = \pm\frac{\pi n}{L}, \quad \text{where } n = 1, 2, 3, \ldots \tag{3.43}$$

Taking into account the relation of this parameter with the electron's energy, we come to the important conclusion that the *electron energy in the potential well cannot change continuously – it can have only certain discrete values of allowed energies* (see Fig. 3.1):

$$E_n = \frac{\hbar^2 k_n^2}{2m_e} = \frac{\pi^2\hbar^2}{2m_e L^2} n^2, \tag{3.44}$$

where $n = 1, 2, 3, \ldots$ The electron energy, E_n, in the potential well is thus *quantized* and the parameter n, which defines the electron energy state, is called the *principal quantum number*. For $n = 1$ the electron's energy is minimal and equal to

$$E_1 = \frac{\pi^2\hbar^2}{2m_e L^2} = \frac{0.3737}{L^2}\ \text{eV}. \tag{3.45}$$

This state, which has the lowest energy, is often called the *ground state*. The energy E_1 is inversely proportional to L^2 and its numerical value in electron-volts for an electron with free electron mass m_e can be calculated with the coefficient 0.3737 when the width of the well, L, is measured in nanometers. The quantum number n cannot be equal to zero because this would mean that the wavenumber $k = k_0 = 0$ and there would be no electron in the well because $\psi(x) = 0$ at $k = 0$

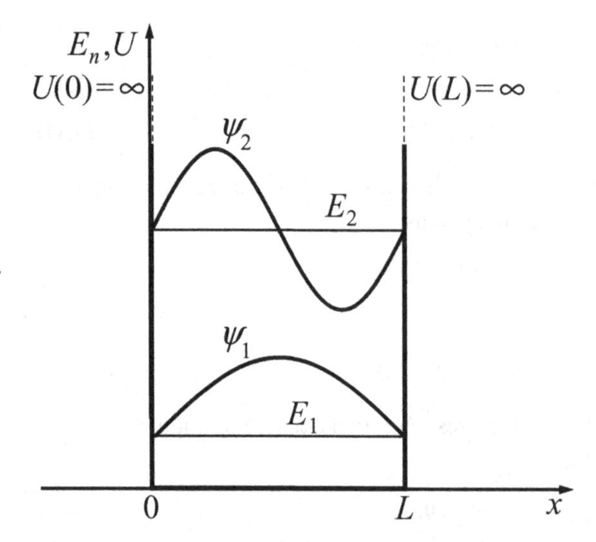

Figure 3.1 Quantization of an electron's energy in a potential well with impenetrable walls of $U(0) = \infty$ and $U(L) = \infty$. The two lowest energy levels, E_1 and E_2, and the corresponding wavefunctions, ψ_1 and ψ_2, are shown.

(see Eq. (3.41)). This leads us to the important conclusion that a *confined electron can never be in a state of rest*. Even in its ground state the electron has non-zero momentum, i.e., from the point of view of classical mechanics the *electron is in a state of constant motion*. This is a direct consequence of Heisenberg's uncertainty principle and wave–particle duality.

For the type of potential well considered here the positions of the energy states are not equidistant. The distance between adjacent energy states increases with increasing quantum number, n, according to the law

$$\Delta E = E_{n+1} - E_n = (2n + 1)E_1. \tag{3.46}$$

If the width of the potential well, L, within which the electron is confined (for example the width of the above-mentioned semiconductor film) increases, then the energy of the ground state E_1 decreases and therefore the distance between the energy states, ΔE, decreases and tends to zero at $L \to \infty$. The energy spectrum transforms from a discrete spectrum into a quasicontinuous or practically continuous spectrum and the electron's behavior becomes classical.

The unknown coefficient in the expression for the wavefunction (3.41) can be found from the normalization condition (2.140):

$$\int_{-\infty}^{\infty} |\psi(x)|^2 \mathrm{d}x = C^2 \int_0^L \sin^2(k_n x)\mathrm{d}x = 1. \tag{3.47}$$

Calculation of this integral gives us

$$\int_0^L \sin^2(k_n x)\mathrm{d}x = \frac{1}{2k_n}\left[-\sin(k_n L)\cos(k_n L) + k_n L\right]. \tag{3.48}$$

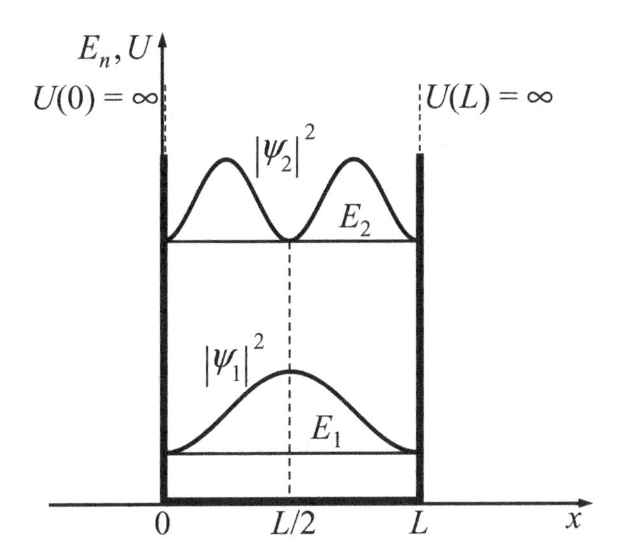

Figure 3.2 The probability density distributions, $|\psi_1(x)|^2$ and $|\psi_2(x)|^2$, for the two lowest energy levels, E_1 and E_2, in a well with impenetrable walls.

Taking into account Eq. (3.42), this integral reduces to

$$\int_0^L \sin^2(k_n x)\mathrm{d}x = \frac{1}{2k_n} k_n L. \tag{3.49}$$

The parameter C is

$$C = \sqrt{\frac{2}{L}}. \tag{3.50}$$

Finally, the wavefunction of the electron in the potential well for the nth energy state can be written as

$$\psi_n(x) = \sqrt{\frac{2}{L}} \sin\left(\frac{n\pi x}{L}\right). \tag{3.51}$$

The graphs of the wavefunctions $\psi_1(x)$ and $\psi_2(x)$ for $n = 1$ and $n = 2$, are shown in Fig. 3.1. The distribution of the probability density, $|\psi_n(x)|^2$, of finding an electron in the region of the potential well whose number of maxima is determined by the quantum number n is shown in Fig. 3.2. We see that in the ground state ($n = 1$) the probability of finding an electron has a maximum at the center of the potential well. In the first excited state ($n = 2$) the probability of finding an electron at the center of the well is equal to zero, whereas the probability of finding an electron has maxima at the points $x = L/4$ and $x = 3L/4$. This behavior of the electron differs drastically from the behavior of a classical particle in a potential well, for which the probability of finding a particle is the same at any point in the well, for any value of the particle's energy (see Appendix A, Section A.2.6).

Note that there are similarities between the mathematical description of electron motion in a potential well with infinite walls and the oscillations of a string with fixed ends. In both cases the boundary conditions are such that the length of the system (in our case it is the width of the potential well or the distance between fixed ends in the case of the string) must be equal to an integer multiple of half of the wavelength, which is described by the wavefunction (in the case of an electron) or by the transverse wave (in the case of a string).

Example 3.2. An electron is located in a rectangular potential well with infinite walls. The width of the well is $L = 10$ nm. Find the energies of stationary states of the electron for which at the well's boundary $x = L$ the derivatives of the wavefunction are negative and equal to $-\tau_n$ (here τ_n are arbitrary positive constants).
Reasoning. The wavefunction, ψ_n, for the nth stationary state in the potential well with infinite walls is given by Eq. (3.51). Let us find its derivative:

$$\frac{d\psi_n}{dx} = \sqrt{\frac{2}{L}}\frac{n\pi}{L}\cos\left(\frac{n\pi x}{L}\right). \tag{3.52}$$

For the point $x = L$, Eq. (3.52) gives the relation between the derivative of the wavefunction and its quantum number, n:

$$\sqrt{\frac{2}{L}}\frac{n\pi}{L}\cos(n\pi) = -\tau_n. \tag{3.53}$$

Since the quantum number n is an integer and positive, and because the left-hand side of Eq. (3.53) must be negative, the quantum number n must be an odd number:

$$\cos(n\pi) = -1, \quad n = 1, 3, 5, \ldots \tag{3.54}$$

On substituting Eq. (3.54) into Eq. (3.53) we obtain the expression for n for a given τ_n:

$$n = \frac{\tau_n L}{\pi}\sqrt{\frac{L}{2}}. \tag{3.55}$$

On substituting Eq. (3.55) into the formula for the energy of the nth state, Eq. (3.44), we obtain

$$E_n = \frac{\pi^2\hbar^2}{2m_e}\left(\frac{n}{L}\right)^2 = \frac{\hbar^2 L\tau_n^2}{4m_e}. \tag{3.56}$$

On substituting the value of $L = 10$ nm into Eq. (3.56) we obtain

$$E_n = 3.1 \times 10^{-47}\tau_n^2 \text{ J} = 1.91 \times 10^{-28}\tau_n^2 \text{ eV}. \tag{3.57}$$

In reality we never deal with the case of given τ_n, and rather relation (3.55) has to be rewritten in the form

$$\tau_n = \frac{n\pi}{L}\sqrt{\frac{2}{L}}, \tag{3.58}$$

i.e., Eq. (3.58) defines the value of the derivative of the wavefunction at the boundary of the potential well and the most important conclusion from Eq. (3.58) is that the derivative τ_n is a quantized quantity. Thus, from Eq. (3.58) we obtain

$$\tau_n \approx 4.4 \times 10^{12} n \ \ \mathrm{m}^{-3/2}, \tag{3.59}$$

where $n = 1, 3, 5, \ldots$ By substituting the first two values of τ_n into Eq. (3.57) we obtain for the magnitudes of the first two odd energy levels $E_1 = 3.7$ meV and $E_3 = 33$ meV. Note that at boundaries with infinite walls the condition of continuity of the derivative of the wavefunction is violated while the continuity of the wavefunction is preserved since each wavefunction becomes zero at the boundaries.

3.3 An electron in a potential well with finite barriers

The technology of fabrication of planar layered structures with given physical properties allows the experimental realization of the confinement of electrons inside potential wells of various profiles. Most important for practical realizations are one-dimensional potential profiles of rectangular wells, (1) with one of the potential barriers being finite and the other infinite, and (2) with both potential barriers finite.

3.3.1 The potential well with one of the barriers finite

Let the dependence of the potential energy of an electron on the coordinate x have the following form:

$$U(x) = \begin{cases} \infty, & x \le 0, \\ 0, & 0 < x < L, \\ U_0, & x \ge L. \end{cases} \tag{3.60}$$

Such a potential profile is shown in Fig. 3.3. The finite value of the potential at the right boundary, $x = L$, considerably changes the solutions of the Schrödinger equation and correspondingly the behavior of an electron. It is now necessary to distinguish between two different types of electron motion in the well. The first type corresponds to $E \le U_0$, and the electron in the potential well can have bound states, which correspond to the discrete set of energy levels. The second type of motion corresponds to $E > U_0$, and the electron does not have bound states and can be found at any point with coordinate $x > 0$. This type of motion corresponds to a continuous energy spectrum. In this section we will consider only the case when $E \le U_0$ and the electron is confined in the potential well.

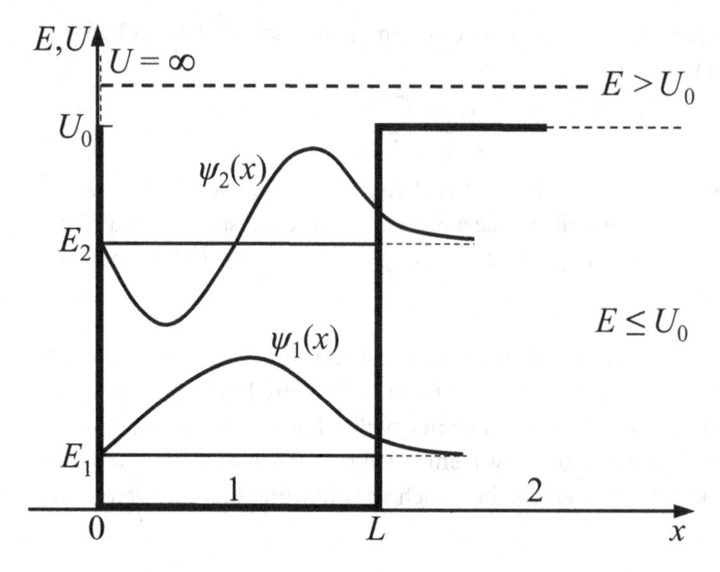

Figure 3.3 The first two energy states, E_1 and E_2, and the corresponding wavefunctions, ψ_1 and ψ_2, for an electron in a potential well with the right-hand-side potential barrier finite.

A classical particle cannot move out of the well because its kinetic energy would be negative when it is outside of the well. On reflecting from the walls of the well the particle would move only inside the well. The behavior of the electron described by the solution of the corresponding Schrödinger equation is fundamentally different from the behavior of a classical particle.

Because the potential profile $U(x)$ is a step-like function, let us divide the electron's accessible region $x > 0$ into regions 1 and 2 with potential energies $U_1 = 0$ and $U_2 = U_0$. In each of these regions we will obtain solutions of the Schrödinger equation. Then, in correspondence to the boundary conditions applied to the wavefunctions, we will join these solutions in such a way that these wavefunctions and their derivatives are continuous at the boundaries.

Let us write the Schrödinger equation for each of these regions, 1 and 2:

$$\frac{\mathrm{d}^2\psi_1(x)}{\mathrm{d}x^2} + k^2\psi_1(x) = 0, \quad 0 \le x \le L, \tag{3.61}$$

$$\frac{\mathrm{d}^2\psi_2(x)}{\mathrm{d}x^2} - \kappa^2\psi_2(x) = 0, \quad x > L. \tag{3.62}$$

Here we introduced for the wavenumbers of the electron wave in the potential well

$$k = \sqrt{\frac{2m_\mathrm{e}E}{\hbar^2}}, \tag{3.63}$$

and outside of the potential well

$$\kappa = \sqrt{\frac{2m_\mathrm{e}(U_0 - E)}{\hbar^2}}. \tag{3.64}$$

The solutions of Eqs. (3.61) and (3.62) can be written as follows:

$$\psi_1(x) = A_1 e^{ikx} + B_1 e^{-ikx}, \tag{3.65}$$

$$\psi_2(x) = A_2 e^{\kappa x} + B_2 e^{-\kappa x}, \tag{3.66}$$

where A_1, A_2, B_1, and B_2 are constants, which can be found from the boundary conditions. In our case the boundary conditions are the continuity of the wavefunction and its derivative at the boundaries of the potential well. From the condition of continuity of the wavefunction at $x = 0$ we obtain the equation for unknowns A_1 and B_1:

$$A_1 + B_1 = 0, \tag{3.67}$$

since in the region $x \leq 0$ the function $\psi(x) = 0$. Taking into account the last equation, Eq. (3.65) transforms into

$$\psi_1(x) = 2iA_1 \sin(kx). \tag{3.68}$$

The wavefunction must be finite everywhere, in region 2 as well. For this reason $A_2 = 0$ because in region 2 the term $A_2 e^{\kappa x}$ increases infinitely with increasing x. At the boundary $x = L$ the following conditions for the continuity of the wavefunction and its derivative must be satisfied:

$$\psi_1(L) = \psi_2(L), \tag{3.69}$$

$$\left.\frac{d\psi_1}{dx}\right|_{x=L} = \left.\frac{d\psi_2}{dx}\right|_{x=L}. \tag{3.70}$$

After substitution of the corresponding expressions for ψ_1 and ψ_2, the boundary conditions (3.69) and (3.70) can be rewritten in the form

$$2iA_1 \sin(kL) = B_2 e^{-\kappa L}, \tag{3.71}$$

$$2ikA_1 \cos(kL) = -\kappa B_2 e^{-\kappa L}. \tag{3.72}$$

Let us divide Eq. (3.71) by Eq. (3.72). As a result we obtain the following equation:

$$\tan(kL) = -\frac{k}{\kappa}, \tag{3.73}$$

which defines the energy spectrum of the electron in the potential well. This is a transcendental equation, whose solution cannot be obtained analytically. For the analysis of such equations very often graphical methods of solution are used, together with numerical methods.

Let us transform Eq. (3.73) in order to carry out the graphical analysis. Taking into account the trigonometric identity

$$\sin\varphi = \pm\frac{\tan\varphi}{\sqrt{1 + \tan^2\varphi}}, \tag{3.74}$$

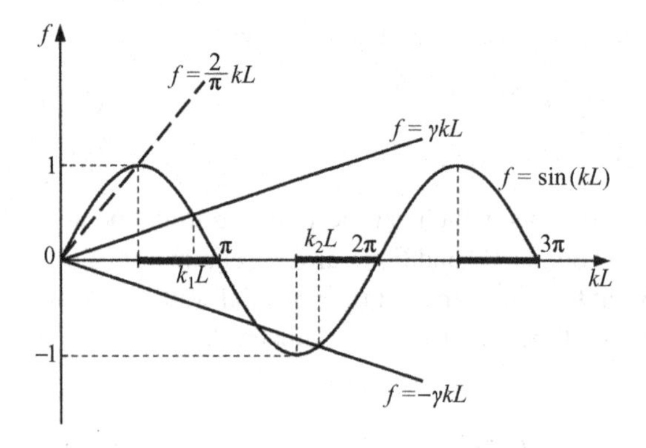

Figure 3.4 Graphical solution of the transcendental equation (3.77).

we can transform Eq. (3.73) into the form

$$\sin(kL) = \pm\frac{k}{\sqrt{k^2 + \kappa^2}}. \tag{3.75}$$

According to Eqs. (3.63) and (3.64), for the terms under the square root in Eq. (3.75) we obtain

$$k^2 + \kappa^2 = \frac{2m_e U_0}{\hbar^2}. \tag{3.76}$$

By substituting Eq. (3.76) into Eq. (3.75) we obtain

$$\sin(kL) = \pm\gamma kL, \tag{3.77}$$

where we introduced the parameter

$$\gamma = \frac{\hbar}{\sqrt{2m_e L^2 U_0}}, \tag{3.78}$$

which defines the slope $f(kL) = \pm\gamma kL$.

Let us plot graphs of the right-hand and left-hand sides of Eq. (3.77) as functions of the parameter kL (see Fig. 3.4). The points of intersection of the sine function with the lines $f = \pm\gamma kL$ define the roots of Eq. (3.77). From all the roots of Eq. (3.77) we separate the roots which are the roots of Eq. (3.73). The chosen roots must satisfy the condition $\tan(kL) < 0$. The tangent function is negative only in the following intervals of values of kL where the roots can exist:

$$\frac{\pi}{2} + n\pi \leq kL \leq \pi + n\pi. \tag{3.79}$$

Thus, we have to choose the roots of Eq. (3.77) which exist in the interval (3.79). In Fig. 3.4 intervals of kL (Eq. (3.79)) are shown by thick lines. Note that from the expressions derived we can identify wells for which the solutions of Eq. (3.73) do not exist. So there would be no roots and no bound energy states of the electron in those potential wells. We show by the dashed line the maximal slope of $f = \gamma kL$ for the parameter $\gamma = 2/\pi$ when the roots for Eq. (3.73) exist. For higher values of the slope, γ, there are no bound states. Thus, the condition for which there is at least one energy state with $E \leq U_0$ reduces to the inequality

$$\gamma \leq 2/\pi. \tag{3.80}$$

Taking into account relation (3.78), inequality (3.80) takes the form

$$\frac{\hbar}{\sqrt{2m_e U_0}} \leq \frac{2L}{\pi}. \tag{3.81}$$

This condition can be rewritten in more convenient form:

$$U_0 L^2 \geq \frac{\pi^2 \hbar^2}{8m_e}, \tag{3.82}$$

where the left-hand side of the inequality is defined by the potential-well parameters only, and the right-hand side by the type of the particle confined in the potential well. In relatively shallow and/or narrow potential wells, there may be no bound states for an electron, i.e., the electron cannot be captured by such a well and the electron has energy $E > U_0$, which is greater than U_0. That corresponds to the case of an unbound particle, which will be considered in Section 3.4.1.

The second energy level in the well can exist at $\gamma \leq 2/(3\pi)$, i.e.,

$$U_0 L^2 \geq \frac{9\pi^2 \hbar^2}{8m_e}. \tag{3.83}$$

As the parameter $U_0 L^2$ increases from the value at which equality is achieved in Eq. (3.80) to the value at which equality is achieved in Eq. (3.83), the first level appears at $E_1 = U_0$ and goes down to $E_1 = 0.3 \times U_0$, and the second level appears at $E_2 = U_0$. Further increase of $U_0 L^2$ brings both levels, E_1 and E_2, down and, at $\gamma \leq 2/(5\pi)$, the third energy level appears. Thus, the spectrum of eigenvalues of the electron energy is discrete.

Let us now write expressions for the wavefunctions in each of the two regions (Eqs. (3.65) and (3.66)), taking into account the relations between the coefficients which we have obtained:

$$\psi_1 = A \sin(kx), \tag{3.84}$$

$$\psi_2 = A \sin(kL) e^{\kappa L} e^{-\kappa x}, \tag{3.85}$$

where we introduced the constant $A = 2iA_1$. From these two expressions it follows that inside the potential well the wavefunction is an oscillatory function, as it was in the case of the potential well with infinite barriers. Outside of the well, as the electron moves away from the potential barrier, its wavefunction decreases exponentially. This means that in a bound state there is a non-zero probability of the electron being outside of the potential well where its energy is $E < U_0$ and therefore the electron's kinetic energy, K, would be negative. This can be explained as follows: in quantum mechanics the total energy, E, cannot be presented as a sum of kinetic and potential energy because, according to Heisenberg's uncertainty principle, kinetic and potential energies (as well as momentum and coordinate) cannot be both exactly measured simultaneously. The uncertainty relations Eq. (2.117) relate the particle's coordinates and its corresponding momentum projections, which define potential (coordinates) and kinetic (momenta) energies. Thus, the situation with $E < U_0$ is possible for a quantized particle, and it is very common in various theoretical and applied problems, as we will see in later chapters. The qualitative dependence of the wavefunctions for the first two energy states on the coordinate x is shown in Fig. 3.3.

Example 3.3. Find the ratio E_1/U_0 for an electron in a well with the potential profile described by Eq. (3.60) when only one bound state exists, which corresponds to the parameter $kL = 3\pi/4$.
Reasoning. The parameter k is defined by Eq. (3.63),

$$k = \sqrt{\frac{2m_e E_1}{\hbar^2}}.$$

Then, from the relation $kL = 3\pi/4$ we obtain for the energy, E_1, the following equation:

$$E_1 = \frac{9\pi^2\hbar^2}{32m_e L^2}. \tag{3.86}$$

The parameter γ, which defines the slope of the function $f = \gamma kL$, as shown in Fig. 3.4, can be found from Eq. (3.77):

$$\gamma = \frac{\sin(kL)}{kL} = \frac{\sqrt{2}/2}{3\pi/4} = \frac{2\sqrt{2}}{3\pi}. \tag{3.87}$$

Using the definition of the parameter γ (Eq. (3.78)), we obtain the expression for the height of the potential barrier U_0:

$$U_0 = \frac{9\pi^2\hbar^2}{16m_e L^2}. \tag{3.88}$$

Taking into account Eq. (3.86), we obtain

$$\frac{E_1}{U_0} = 0.5.$$

Example 3.4. Find the normalization constant A for the wavefunction described by Eqs. (3.84) and (3.85).
Reasoning. Since the electron's behavior in regions 1 and 2 is described by its own wavefunction, the normalization condition must be written in the form

$$\int_0^\infty |\psi|^2\,dx = \int_0^L |\psi_1|^2\,dx + \int_L^\infty |\psi_2|^2\,dx = 1. \tag{3.89}$$

Let us substitute into this condition the expressions for the wavefunctions ψ_1 and ψ_2:

$$|A|^2 \left[\int_0^L \sin^2(kx)\mathrm{d}x + \sin^2(kL)\mathrm{e}^{2\kappa L} \int_L^\infty \mathrm{e}^{-2\kappa x}\,\mathrm{d}x \right] = 1. \tag{3.90}$$

By carrying out the integration we obtain

$$|A|^2 \left(\frac{L}{2} - \frac{\sin(2kL)}{4k} + \frac{\sin^2(kL)}{2\kappa} \right) = 1. \tag{3.91}$$

Let us assume that the constant A is real. Then

$$A = \sqrt{\frac{2}{L - [\sin(2kL)]/(2k) + [\sin^2(kL)]/\kappa}}. \tag{3.92}$$

At $U_0 \to \infty$ the parameters $\kappa \to \infty$ and $kL \to n\pi$. Then

$$A = \sqrt{\frac{2}{L}},$$

which coincides with the normalization constant from Eq. (3.50).

3.3.2 A symmetric potential well with finite potential barriers

The dependence of the electron potential energy in a symmetric potential well with finite potential barriers is defined as

$$U(x) = \begin{cases} U_0, & x \le 0, \\ 0, & 0 < x < L, \\ U_0, & x \ge L. \end{cases} \tag{3.93}$$

Such a potential well is shown in Fig. 3.5. Two cases, $E \le U_0$ and $E > U_0$, as in Section 3.3.1 are possible. Let us consider here one of these cases, in particular the case with $E \le U_0$. By analogy to Eqs. (3.65), (3.66), and (3.68) the solutions of the Schrödinger equation for the three regions 1, 2, and 3 (see Fig. 3.5), taking into account that wavefunctions at $x \to \pm\infty$ must be finite, can be written in the following form:

$$\begin{aligned} &\psi_1(x) = C_1\mathrm{e}^{\kappa x}, \quad x \le 0, \\ &\psi_2(x) = C_2 \sin(kx + \phi), \quad 0 < x < L, \\ &\psi_3(x) = C_3\mathrm{e}^{-\kappa x}, \quad x \ge L. \end{aligned} \tag{3.94}$$

In Eqs. (3.63) and (3.64) we have already introduced the parameters $k = \sqrt{2m_\mathrm{e}E}/\hbar$ and $\kappa = \sqrt{2m_\mathrm{e}(U_0 - E)}/\hbar$. The boundary conditions of continuity of the wavefunctions and their derivatives at points $x = 0$ and $x = L$ give us a system of four equations to find the constants C_1, C_2, C_3, and ϕ as well as the

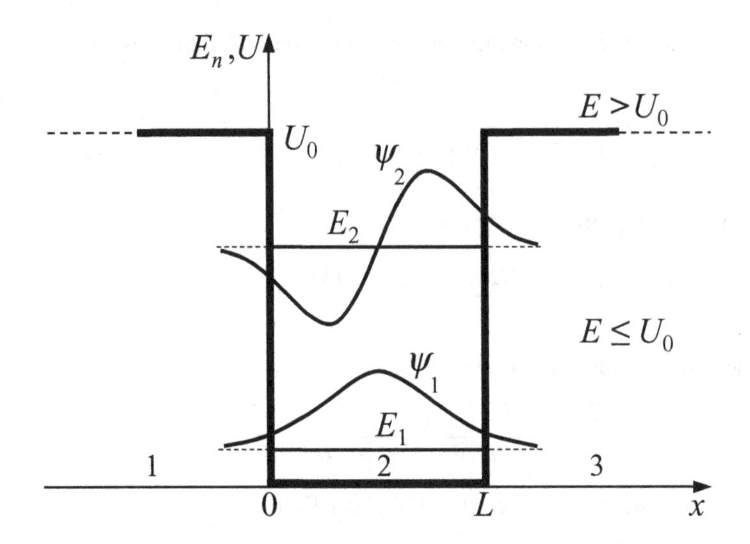

Figure 3.5 A symmetric potential well with potential barriers of finite height. The first two lowest energy levels, E_1 and E_2, and their corresponding wavefunctions, $\psi_1(x)$ and $\psi_2(x)$, are shown.

relationships between the parameters, which replace Eq. (3.73):

$$\tan\phi = \frac{k}{\kappa}, \tag{3.95}$$

$$\tan(kL + \phi) = -\frac{k}{\kappa}. \tag{3.96}$$

These equations can be reduced to

$$\sin\phi = \frac{k}{b}, \tag{3.97}$$

$$\sin(kL + \phi) = -\frac{k}{b}, \tag{3.98}$$

where the parameter $b = \sqrt{2m_e U_0}/\hbar$. If we eliminate from these equations the parameter ϕ we obtain an equation that defines the energy spectrum of an electron in the symmetric potential well with finite barrier heights:

$$kL = n\pi - 2\,\arcsin\left(\frac{k}{b}\right), \tag{3.99}$$

where the quantum number $n = 1, 2, 3, \ldots$ defines the electron energy-level number in the well. Since the positive values of the argument of the arcsine function cannot exceed unity, i.e., $k/b \leq 1$,

$$k_{\max} = b = \frac{\sqrt{2m_e U_0}}{\hbar}. \tag{3.100}$$

The energy of an electron in the potential well is quantized in accordance with Eq. (3.99), i.e., its energy spectrum is discrete. Let us show this graphically.

Figure 3.6 shows the graphs of $f(k) = kL$ and $f_n(k) = n\pi - 2\,\arcsin(k/b)$, which correspond to the left-hand and right-hand sides of Eq. (3.99). The index n of the function $f_n(k)$ coincides with the quantum number n. The points of intersection of the line $f(k)$ with the curves $f_n(k)$ define the roots of Eq. (3.99). We see from Fig. 3.6 that the spectrum of values of k and their related values

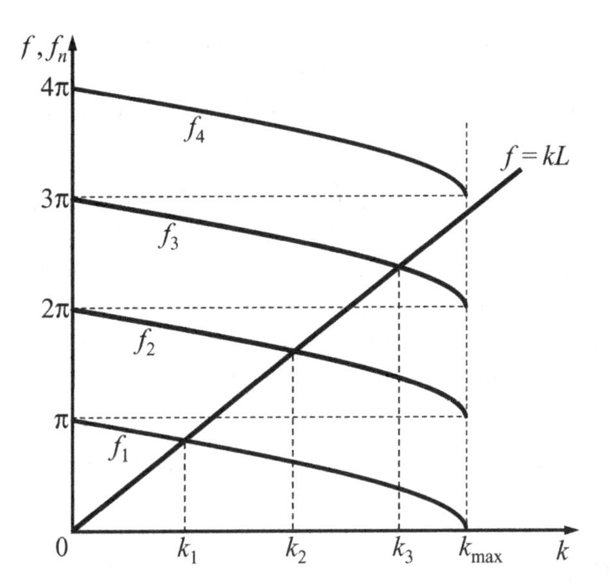

Figure 3.6 Graphical solution of Eq. (3.99). Here, $f_n = n\pi - 2\arcsin(k/b)$.

of electron energy are discrete. For the case shown in Fig. 3.6 there exist only three solutions of Eq. (3.99) and correspondingly three energy levels in the potential well, since the line $f(k) = kL$ and curves $f_n(k)$ have only three points of intersection, $n = 1,\ 2,$ and 3, in the region of allowed values of k: $k \leq k_{\text{max}}$. The larger the width of the well, L, the steeper the line $f(k) = kL$ is, and therefore there are more roots of Eq. (3.99) and correspondingly more energy levels exist in the well.

As $U_0 \to \infty$, the parameter b tends to infinity. Thus, in this limiting case the roots of Eq. (3.99) are $kL = n\pi$ as was calculated earlier for the well with infinite potential barriers (compare with Eq. (3.43)). For the given value of the width of the potential well, L, as the depth, U_0, of the well decreases the magnitude k_{max} also decreases, which leads to a decrease in the number of energy levels in the well. In the case $k_{\text{max}} < \pi/L$, i.e., at

$$U_0 L^2 < \frac{\hbar^2 \pi^2}{2m_{\text{e}}}, \tag{3.101}$$

only one energy level can exist in the well. In contrast to the non-symmetric well with one of the walls being of finite height (Fig. 3.3), for which the bound state of an electron can exist only if condition (3.82) is satisfied, *in the symmetric well of finite depth there is always at least one bound state for an electron.*

To plot the wavefunctions (3.94), let us write the expressions that relate the constants C_1, C_2, and C_3:

$$C_1 = \frac{\hbar k}{\sqrt{2m_{\text{e}} U_0}} C_2, \tag{3.102}$$

$$C_3 = -\text{e}^{\kappa L} \frac{\hbar k}{\sqrt{2m_{\text{e}} U_0}} C_2. \tag{3.103}$$

As has already been mentioned, relations (3.102) and (3.103) were obtained from the conditions of continuity of the wavefunction at the boundaries of the well. The constant C_2 can be found from the condition of normalization, which in our case will have the form

$$\int_{-\infty}^{0} |\psi_1|^2 \, dx + \int_{0}^{L} |\psi_2|^2 \, dx + \int_{L}^{\infty} |\psi_3|^2 \, dx = 1. \tag{3.104}$$

Figure 3.5 shows the first two energy levels, E_1 and E_2, in the potential well with barriers of finite height and corresponding wavefunctions, ψ_1 and ψ_2. We see that for $E \leq U_0$ there is a symmetric probability of finding an electron outside the well and inside the barrier. The higher the energy levels, the higher is this probability.

Example 3.5. Find approximately the energy difference of two levels with energy $E \ll U_0$ for an electron in a symmetric potential well of width L and barrier height U_0.

Reasoning. Since the energy of the electron $E \ll U_0$, the ratio $k/b = \sqrt{E/U_0}$ is a very small number. Therefore, let us substitute the arcsine function by the first term of its expansion in series, $\arcsin(k/b) \approx k/b$; then, from Eq. (3.99), we get

$$kL \approx n\pi - \frac{2k}{b}. \tag{3.105}$$

On solving this equation with respect to the wavenumber, k, we obtain

$$k_n \approx \frac{n\pi}{L + \hbar\sqrt{2/(m_e U_0)}}. \tag{3.106}$$

Let us find from this expression the energy of stationary states of an electron in the potential well. Since $E_n = \hbar^2 k_n^2/(2m_e)$,

$$E_n \approx \frac{\pi^2 \hbar^2 n^2}{2m_e \left(L + \hbar\sqrt{2/(m_e U_0)}\right)^2}. \tag{3.107}$$

The values of k_n and E_n decrease compared with the case of the well with impenetrable barriers (compare Eqs. (3.106) and (3.107) with Eqs. (3.43) and (3.44)). At $U_0 \to \infty$ Eq. (3.107) coincides with the exact expression for the energy levels (3.44). The energy distance between the second and first energy levels is

$$E_2 - E_1 \approx \frac{3\pi^2 \hbar^2}{2m_e \left(L + \hbar\sqrt{2/(m_e U_0)}\right)^2}. \tag{3.108}$$

3.4 Propagation of an electron above the potential well

In the previous section we considered restricted motion of an electron when its energy does not exceed the height of the potential barriers of the quantum well.

The confinement of an electron in the potential well leads to the quantization of its energy, which corresponds to the so-called *quantum-confinement effect* or *quantum-size effect*. Now let us consider the peculiarities of the motion when the energy of an electron, E, is greater than the height of the potential barrier, U_0.

3.4.1 Reflection from an asymmetric well with one infinite potential barrier

Suppose that in the region of the well with one of the barriers having infinite height (Fig. 3.3) an electron is moving along the x-axis in the negative direction. The Schrödinger equation for each of the regions, 1 and 2, has to be written as follows:

$$\frac{\mathrm{d}^2\psi_1(x)}{\mathrm{d}x^2} + k^2\psi_1(x) = 0, \quad k = \sqrt{\frac{2m_\mathrm{e}E}{\hbar^2}}, \tag{3.109}$$

$$\frac{\mathrm{d}^2\psi_2(x)}{\mathrm{d}x^2} + \kappa^2\psi_2(x) = 0, \quad \kappa = \sqrt{\frac{2m_\mathrm{e}(E - U_0)}{\hbar^2}}. \tag{3.110}$$

We see that, in contrast to the case with $E \leq U_0$, only the equation for the second region (Eq. (3.110)) has changed. Equation (3.110) that corresponds to region 2 now has oscillatory solutions as the solutions in region 1. Therefore, the solutions for the above-mentioned equations can be sought in the form

$$\psi_1(x) = A_1\mathrm{e}^{\mathrm{i}kx} + B_1\mathrm{e}^{-\mathrm{i}kx}, \tag{3.111}$$

$$\psi_2(x) = A_2\mathrm{e}^{\mathrm{i}\kappa x} + B_2\mathrm{e}^{-\mathrm{i}\kappa x}. \tag{3.112}$$

Taking into account the boundary condition $\psi_1(0) = 0$, we obtain as before the relation between the constants, $B_1 = -A_1$, and the same expression (3.68) for the wavefunction $\psi_1(x)$. By applying the boundary conditions (3.69) and (3.70) at the point $x = L$, we obtain the expressions for the coefficients A_2 and B_2:

$$A_2 = \mathrm{e}^{-\mathrm{i}\kappa L}\left(\mathrm{i}\sin(kL) + \frac{k}{\kappa}\cos(kL)\right)A_1, \tag{3.113}$$

$$B_2 = \mathrm{e}^{\mathrm{i}\kappa L}\left(\mathrm{i}\sin(kL) - \frac{k}{\kappa}\cos(kL)\right)A_1. \tag{3.114}$$

Since there are no restrictions for the parameters k and κ, the energy E can have any value greater than U_0, i.e., in this case the electron has a continuous energy spectrum.

In each of the regions 1 and 2 the wavefunctions $\psi_1(x)$ and $\psi_2(x)$ are the sums of two de Broglie waves propagating in opposite directions. The wave that approaches the well (the part of the total wavefunction proportional to $\mathrm{e}^{-\mathrm{i}kx}$) undergoes a partial reflection at the boundary $x = L$, and then the reflected wave propagates along the x-axis in the positive direction. The wave that penetrated into region 1 is totally reflected from the boundary $x = 0$. Then, part of this wave penetrates the boundary $x = L$ and also propagates along the x-axis infinitely. The reflection coefficient of the wave incident on the well is defined by the expression

$$R = \frac{|A_2|^2}{|B_2|^2}. \tag{3.115}$$

It follows from expressions (3.113) and (3.114) for the coefficients A_2 and B_2 that the reflection coefficient $R \equiv 1$.

The probability density for all values of $x \geq 0$ is an oscillating function, which has different periods of oscillations λ_1 and λ_2 in regions 1 and 2,

$$\lambda_1 = \frac{\pi\hbar}{\sqrt{2m_e E}}, \tag{3.116}$$

$$\lambda_2 = \frac{\pi\hbar}{\sqrt{2m_e(E - U_0)}}. \tag{3.117}$$

The oscillating character of the probability density is the result of a superposition of the incident and reflected waves. As a result, the wavefunction has the form of a standing wave similar to that for the reflection of an electromagnetic wave from an ideal conductor (see Section B.3.4 in Appendix B).

3.4.2 Reflection from a symmetric well with finite potential barriers

Let us consider now the propagation of an electron over the symmetric potential well with profile defined by Eq. (3.93) (see Fig. 3.5). Suppose that the electron is moving towards the walls of the well in the positive direction along the x-axis and its energy is $E > U_0$. The Schrödinger equation for each of the three regions 1, 2, and 3 can be written as

$$\frac{d^2\psi_1(x)}{dx^2} + \kappa^2\psi_1(x) = 0, \tag{3.118}$$

$$\frac{d^2\psi_2(x)}{dx^2} + k^2\psi_2(x) = 0, \tag{3.119}$$

$$\frac{d^2\psi_3(x)}{dx^2} + \kappa^2\psi_3(x) = 0, \tag{3.120}$$

where the parameters k and κ are defined by the same expressions as in Eqs. (3.109) and (3.110). The solutions for the three regions 1, 2, and 3 can be written as

$$\psi_1(x) = A_1 e^{i\kappa x} + B_1 e^{-i\kappa x}, \quad x \leq 0, \tag{3.121}$$

$$\psi_2(x) = A_2 e^{ikx} + B_2 e^{-ikx}, \quad 0 < x < L, \tag{3.122}$$

$$\psi_3(x) = A_3 e^{i\kappa x} + B_3 e^{-i\kappa x}, \quad x \geq L. \tag{3.123}$$

Let us assume that the electron propagates to the well from the negative part of the x-axis, then in the region $x > L$ only the outgoing wave proportional to e^{ikx} exists. Since there is no electron coming back from $x \to \infty$ into the region of interest, $B_3 = 0$. Using the standard boundary conditions such as (3.69) and (3.70) for the wavefunctions ψ_1, ψ_2, and ψ_3, we arrive at the following expression for the amplitude of the outgoing wave:

$$A_3 = \frac{2\kappa k e^{-i\kappa L}}{2\kappa k \cos(kL) + i(k^2 + \kappa^2)\sin(kL)} A_1. \tag{3.124}$$

According to Eq. (2.190), the probability density currents of the incident wave, $\mathbf{j}_i$, and the wave transmitted through the well, $\mathbf{j}_t$, have the forms

$$\mathbf{j}_i = \mathbf{e}_x \frac{\hbar\kappa}{m_e} |A_1|^2, \tag{3.125}$$

$$\mathbf{j}_t = \mathbf{e}_x \frac{\hbar\kappa}{m_e} |A_3|^2. \tag{3.126}$$

Taking into account these expressions, the transmission coefficient, D, which characterizes the probability of transmission of an electron over the well, can be written as

$$D = \frac{|\mathbf{j}_t|}{|\mathbf{j}_i|} = \frac{|A_3|^2}{|A_1|^2}. \tag{3.127}$$

Taking into account Eq. (3.124) for the transmission coefficient, D, we get the following expression:

$$D = \left[1 + \left(\frac{k^2 - \kappa^2}{2k\kappa}\right)^2 \sin^2(kL)\right]^{-1}. \tag{3.128}$$

Taking into account Eqs. (3.109) and (3.110), Eq. (3.128) can be rewritten in the form

$$D = \left[1 + \frac{U_0^2}{4E(E - U_0)} \sin^2\left(\frac{\sqrt{2m_e E}}{\hbar} L\right)\right]^{-1}. \tag{3.129}$$

The *reflection coefficient*, R, which defines the probability of reflection of the electron from the potential well when it propagates over this well, can be written as

$$R = \frac{|\mathbf{j}_r|}{|\mathbf{j}_i|} = \frac{|B_1|^2}{|A_1|^2} = 1 - D. \tag{3.130}$$

Using Eq. (3.128), we can write the expression for the reflection coefficient, R, as

$$R = \left[1 + \frac{(2k\kappa)^2}{(k^2 - \kappa^2)^2 \sin^2(kL)}\right]^{-1} = \left[1 + \frac{4E(E - U_0)}{U_0^2 \sin^2\left(\sqrt{2m_e E} L/\hbar\right)}\right]^{-1}. \tag{3.131}$$

Thus, the transmission and reflection coefficients for the particle propagating over the potential well strongly depend on the ratio of the energy of the electron, E, and the height of the potential barrier, U_0. When the energy, E, is slightly greater than the potential barrier, U_0, the transmission coefficient $D \ll 1$, and the reflection coefficient is close to unity. As the ratio E/U_0 increases, the transmission coefficient, D, rapidly increases and at $E/U_0 = 2$ its maximal value is $D \approx 0.9$. The electron's reflection from the potential well at $E > U_0$ is a *purely quantum effect, which is caused by the wave properties of the electron.*

With the increase of the electron energy, E, both coefficients, D and R, oscillate. This can be explained by invoking the interference of the de Broglie waves reflected from the potential barrier at the boundaries of the well at $x = 0$ and $x = L$. Since, at $kL = n\pi$, the electron does not undergo reflections, $R = 0$ and $D = 1$. The energy of such an electron has to be equal to

$$E_n = \frac{\pi^2\hbar^2}{2m_e L^2}n^2, \tag{3.132}$$

where n is an integer number from $n = n_{\text{min}}$ to $n \to \infty$, where n_{min} is defined from the condition $E_n \geq U_0$ with E_n found from Eq. (3.132). The absence of electron reflection from the potential well is analogous to the effect of an *antireflecting coating of a lens.* This effect is connected with the destructive interference of waves reflected from the two boundaries of the potential well, which takes place at

$$L = n\frac{\lambda_{\text{Br}}}{2}, \tag{3.133}$$

where $n = 1, 2, 3, \ldots$ and λ_{Br} is the de Broglie wavelength that is defined by Eq. (2.88): $\lambda_{\text{Br}} = 2\pi/k$. The opposite situation, i.e., the minima of the transmission coefficient and, correspondingly, the maxima of the reflection coefficient, occurs if the following condition is satisfied:

$$kL = (2n-1)\frac{\pi}{2}, \tag{3.134}$$

which can be rewritten as

$$L = \frac{2n-1}{4}\lambda_{\text{Br}}. \tag{3.135}$$

Example 3.6. An electron travels over a symmetric rectangular well of width L and barrier height U_0. Find the energy of the electron, E, that corresponds to equal values for the coefficients of transmission, D, and reflection, R. The well should be considered sufficiently narrow, i.e., the condition $kL \ll 1$ must be satisfied.

Reasoning. From the given condition we have that $R = D = 1/2$ and, taking into account Eqs. (3.129) and (3.131), we obtain

$$\sin^2(kL) = 4\frac{E}{U_0}\left(\frac{E}{U_0} - 1\right). \tag{3.136}$$

Taking into account the smallness of kL, we can write $\sin(kL) \approx kL$:

$$k^2L^2 = 4\frac{E}{U_0}\left(\frac{E}{U_0} - 1\right). \tag{3.137}$$

Since $E = \hbar^2k^2/(2m_e)$, on substituting k^2 by $2m_eE/\hbar^2$ in Eq. (3.137) we get the following quadratic equation for the ratio E/U_0:

$$\frac{m_eEL^2}{2\hbar^2} = \left(\frac{E}{U_0}\right)^2 - \frac{E}{U_0}. \tag{3.138}$$

On solving this equation and leaving only the positive root (which is the only one that has physical meaning) we get

$$\frac{E}{U_0} = 1 + \frac{m_eU_0L^2}{2\hbar^2}. \tag{3.139}$$

Therefore, the energy which corresponds to equal values for the reflection and transmission coefficients is defined as

$$E = U_0\left(1 + \frac{m_eU_0L^2}{2\hbar^2}\right). \tag{3.140}$$

3.5 Tunneling: propagation of an electron in the region of a potential barrier

In Section 3.3 we studied the behavior of an electron in a rectangular potential well with one barrier of finite height. This potential barrier has an infinite width, see Fig. 3.3. In this case an electron with energy, E, less than the potential-barrier height, U_0, cannot move very far from the well since the wavefunction decreases exponentially (see Eq. (3.85)), i.e., an electron can be found only in the vicinity of the well in the barrier region. The wavefunctions shown in Fig. 3.3 as well as Eqs. (3.89) and (3.90) clearly demonstrate the finite probability of finding the electron in the barrier region. If the well has potential barriers of finite height and width, then the electron can escape the potential well even when $E < U_0$, i.e., *the electron can penetrate through the potential barrier*. The lower the barrier height, U_0, and the smaller its width, L, the higher is the probability that the electron can escape the well.

To understand the peculiarities of electron transmission (or, as it is often called, *tunneling*) through a barrier of finite width, let us consider a free (at $x \approx -\infty$) electron, which is incident on a so-called *rectangular potential barrier* (see Fig. 3.7). *The potential barrier is a region of space where the potential energy of a particle is greater than in the surrounding area.* For a one-dimensional rectangular potential barrier, as is shown in Fig. 3.7, the potential energy of an

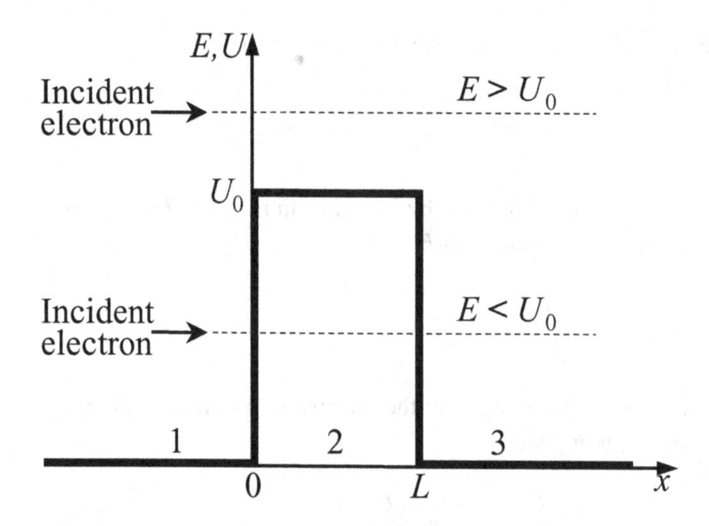

Figure 3.7 Electron transmission through the rectangular potential barrier.

electron in different regions can be defined as follows:

$$U(x) = \begin{cases} 0, & x < 0, \\ U_0, & 0 \le x \le L, \\ 0, & x > L. \end{cases} \tag{3.141}$$

Let us consider further two possible situations, which correspond to under-barrier ($E < U_0$) and over-barrier ($E > U_0$) electron motion.

3.5.1 Electron tunneling under the barrier

Consider an electron with energy $E < U_0$ incident from the negative side of the x-axis on the rectangular potential barrier. The electron can be found on both sides of the barrier. For the three regions 1, 2, and 3 (see Fig. 3.7) the Schrödinger equation can be written as

$$\begin{aligned} &\mathrm{d}^2\psi_1(x)/\mathrm{d}x^2 + k^2\psi_1(x) = 0, \\ &\mathrm{d}^2\psi_2(x)/\mathrm{d}x^2 - \kappa^2\psi_2(x) = 0, \\ &\mathrm{d}^2\psi_3(x)/\mathrm{d}x^2 + k^2\psi_3(x) = 0, \end{aligned} \tag{3.142}$$

where k and κ are defined by Eqs. (3.63) and (3.64): $k = \sqrt{2m_\mathrm{e}E}/\hbar$ and $\kappa = \sqrt{2m_\mathrm{e}(U_0 - E)}/\hbar$. The wavefunctions which are the solutions of the above-mentioned equations can be written as

$$\begin{aligned} \psi_1(x) &= A_1\mathrm{e}^{\mathrm{i}kx} + B_1\mathrm{e}^{-\mathrm{i}kx}, \\ \psi_2(x) &= A_2\mathrm{e}^{\kappa x} + B_2\mathrm{e}^{-\kappa x}, \\ \psi_3(x) &= A_3\mathrm{e}^{\mathrm{i}kx} + B_3\mathrm{e}^{-\mathrm{i}kx}. \end{aligned} \tag{3.143}$$

Since with the chosen direction of the electron's incidence only an outgoing wave can propagate outside of the barrier, $B_3 \equiv 0$ as in the case of Eq. (3.123). From the continuity conditions for the total wavefunction and its derivative at the boundaries of the barrier such as (3.69) and (3.70), we obtain a system of four equations, from which we can find the relation between the coefficients in the solutions (3.143):

$$\begin{aligned} A_1 + B_1 &= A_2 + B_2, \\ ik(A_1 - B_1) &= \kappa(A_2 - B_2), \\ A_2 e^{\kappa L} + B_2 e^{\kappa L} &= A_3 e^{ikL}, \\ \kappa(A_2 e^{\kappa L} - B_2 e^{\kappa L}) &= ik A_3 e^{ikL}. \end{aligned} \tag{3.144}$$

Most interesting for us are the coefficients of reflection and transmission through the barrier, R and D:

$$R = \frac{|B_1|^2}{|A_1|^2}, \tag{3.145}$$

$$D = \frac{|A_3|^2}{|A_1|^2}. \tag{3.146}$$

To find R and D, we determine the coefficients B_1 and A_3 from the system of equations (3.144):

$$B_1 = \frac{(k^2 + \kappa^2)\sinh(\kappa L)}{(k^2 - \kappa^2)\sinh(\kappa L) + 2i\kappa k \cosh(\kappa L)} A_1, \tag{3.147}$$

$$A_3 = \frac{2i\kappa k e^{-ikL}}{(k^2 - \kappa^2)\sinh(\kappa L) + 2i\kappa k \cosh(\kappa L)} A_1, \tag{3.148}$$

where we introduce the hyperbolic sine and cosine,

$$\sinh x = \frac{e^x - e^{-x}}{2}, \qquad \cosh x = \frac{e^x + e^{-x}}{2}.$$

By substituting Eq. (3.147) into Eq. (3.145) we find the expression for the reflection coefficient, R:

$$R = \left[1 + \frac{4k^2\kappa^2}{(k^2 + \kappa^2)^2 \sinh^2(\kappa L)}\right]^{-1}. \tag{3.149}$$

The transmission coefficient can be obtained analogously or from the expression

$$R + D = 1.$$

As a result this coefficient has the form

$$D = \left[1 + \frac{(k^2 + \kappa^2)^2}{4k^2\kappa^2} \sinh^2(\kappa L)\right]^{-1}. \tag{3.150}$$

Using relations (3.109) and (3.110), the coefficients R and D can be rewritten in terms of energy:

$$\frac{k^2\kappa^2}{(k^2 + \kappa^2)^2} = \frac{E(U_0 - E)}{U_0^2} \tag{3.151}$$

and

$$R = \left[1 + \frac{4E(U_0 - E)}{U_0^2 \sinh^2(\kappa L)}\right]^{-1}, \tag{3.152}$$

$$D = \left[1 + \frac{U_0^2}{4E(U_0 - E)} \sinh^2(\kappa L)\right]^{-1}. \tag{3.153}$$

Thus, in the case of a barrier with finite L and U_0, the electron has a finite probability of emerging on the other side of the barrier, which *from the classical-physics point of view is impossible*. This effect was called *tunneling* and the phenomenon of transmission of a particle through the barrier was called the *tunneling phenomenon*. This phenomenon is inherent to any quantum particle regardless of whether the particle has an electric charge.

Suppose that the parameters of the barrier and the energy of the particle satisfy the condition $\kappa L \gg 1$, which can be rewritten in the following form:

$$(U_0 - E)L^2 \gg \frac{\hbar^2}{2m_e}. \tag{3.154}$$

In this case

$$\sinh(\kappa L) \approx \frac{1}{2} e^{kL}, \tag{3.155}$$

then the transmission coefficient can be written as

$$D = D_0 e^{-2L\sqrt{2m_e(U_0 - E)}/\hbar}, \tag{3.156}$$

where

$$D_0 = 16 \frac{E(U_0 - E)}{U_0^2}. \tag{3.157}$$

From these expressions it follows that the transmission coefficient rapidly (exponentially) decreases with increasing barrier width, particle mass, and energy difference $U_0 - E$.

In the case of a sufficiently narrow barrier, for which the condition $\kappa L \ll 1$ or $(U_0 - E)L^2 \ll \hbar^2/(2m_e)$ is valid, the expressions for the reflection and transmission coefficients, R and D, take the forms

$$R = \frac{G}{1 + G}, \tag{3.158}$$

$$D = \frac{1}{1 + G}. \tag{3.159}$$

Here we introduced the parameter

$$G = \frac{m_e}{2\hbar^2 E}(LU_0)^2, \tag{3.160}$$

where the product $L \times U_0$ is often called *"the surface" of the potential barrier*. At finite width, L, and height of the potential barrier $U_0 \to \infty$, or finite U_0 and width $L \to \infty$, the transmission coefficient, D, for passage of the electron

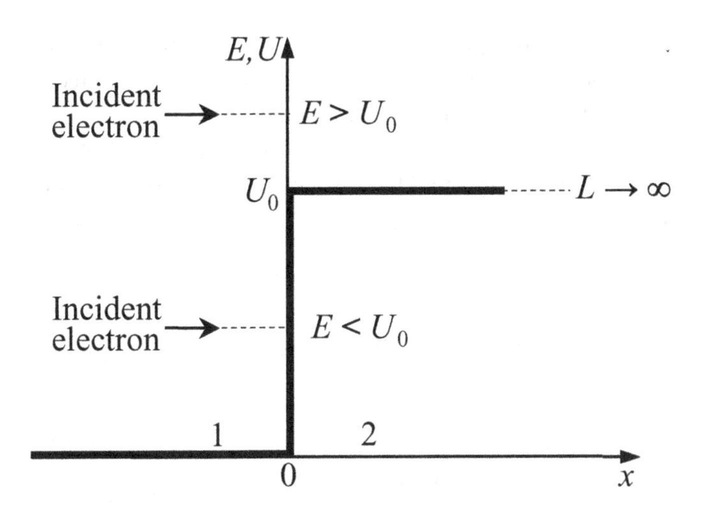

Figure 3.8 Reflection of an electron from a potential step with $U = U_0$.

through the barrier becomes equal to zero, and the reflection coefficient becomes equal to unity: $R = 1$. The case with finite U_0 and width $L \to \infty$ corresponds to the total reflection of the electron from the so-called *potential step* with a finite height of the barrier U_0 (see Fig. 3.8). In this case a phase shift $\Delta\delta$ between the wavefunctions of the incident and reflected waves occurs:

$$B_1 = \sqrt{R}\mathrm{e}^{\mathrm{i}\Delta\delta}A_1. \tag{3.161}$$

Here the phase shift $\Delta\delta$ is defined by the expression

$$\Delta\delta = \arctan\left(\frac{2k\kappa}{k^2 - \kappa^2}\right) = \arctan\left(\frac{2\sqrt{E(U_0 - E)}}{2E - U_0}\right). \tag{3.162}$$

We see that at $E = U_0$ the phase shift is equal to zero, at $E = U_0/2$ the phase shift is equal to $\Delta\delta = \pi/2$, and at $E \ll U_0$ the phase shift is $\Delta\delta \to \pi$. In the case of reflection from an infinitely high barrier ($U_0 \to \infty$) the phase shift between incident and reflected wavefunctions of the electron is equal to $\Delta\delta = \pi$ as in the case of the reflection of an electromagnetic wave from an optically denser medium.

It is interesting to study the probability of finding an electron in the region $x > 0$. In spite of the fact that $E < U_0$ and that the reflection coefficient of an electron from the potential step is equal to unity, the probability of finding an electron in the region $x > 0$ is not equal to zero! The electron wavefunction, $\psi_2(x)$, in this region is equal to

$$\psi_2(x) = B_2\mathrm{e}^{-\kappa x}, \tag{3.163}$$

where

$$B_2 = \frac{2k}{k + \mathrm{i}\kappa}A_1. \tag{3.164}$$

Therefore, $|\psi_2|^2 \neq 0$. The effective penetration depth of the electron under the potential step can be estimated by a quantity $l_{\rm eff} = \kappa^{-1}$. At this distance from the boundary $x = 0$ the probability of finding the electron decreases by a factor of e^2 in comparison with its magnitude at $x = 0$. The probability that the electron penetrates the potential step in this case is defined by the transmission coefficient, D, through the boundary $x = 0$:

$$D = \frac{\kappa}{k}\frac{|B_2|^2}{|A_1|^2}e^{-2kx} = \frac{4k\kappa}{k^2+\kappa^2}e^{-2kx}. \tag{3.165}$$

For the reflection coefficient in this case we obtain the expression

$$R = \frac{|B_1|^2}{|A_1|^2} = \left|\frac{k-i\kappa}{k+i\kappa}\right|^2 = 1. \tag{3.166}$$

Thus, the electron penetrates through the boundary of the potential step into the region $x > 0$, but then it returns back into the region with $x < 0$, reflecting effectively from the potential step with reflection coefficient $R = 1$. Because of such penetration there occurs a phase shift $\Delta\delta$ between the two wavefunctions which correspond to reflected and incident waves.

3.5.2 Electron motion above the barrier

Let us write the general expressions for the electron reflection and transmission coefficients in the case of electron motion with energy $E > U_0$ above the barrier with finite width L (see Fig. 3.7):

$$R = \left[1 + \frac{4E(E-U_0)}{U_0^2\sin^2(\kappa L)}\right]^{-1}, \tag{3.167}$$

$$D = \left[1 + \frac{U_0^2\sin^2(\kappa L)}{4E(E-U_0)}\right]^{-1}, \tag{3.168}$$

where $\kappa = \sqrt{2m_{\rm e}(E-U_0)}/\hbar$. In the case of electron motion above the potential well (Eqs. (3.129) and (3.131)) as well as in the case of motion above the potential barrier (Eqs. (3.152) and (3.153)) the change in electron energy leads to oscillations of both coefficients, R and D. The maxima of the transmission coefficient, D (and minima of reflection coefficient, R), are defined by the condition

$$\kappa_n L = \pi n, \quad n = 1, 2, 3, \ldots \tag{3.169}$$

It follows from the last condition that

$$E_n = U_0 + \frac{\pi^2\hbar^2 n^2}{2m_{\rm e}L^2}. \tag{3.170}$$

The transmission and reflection coefficients are $D = 1$ and $R = 0$, and we are dealing with the so-called *resonant interaction of an electron with the potential barrier*. We note that the quantity $\Delta E_n = E_n - U_0$ coincides with the expression

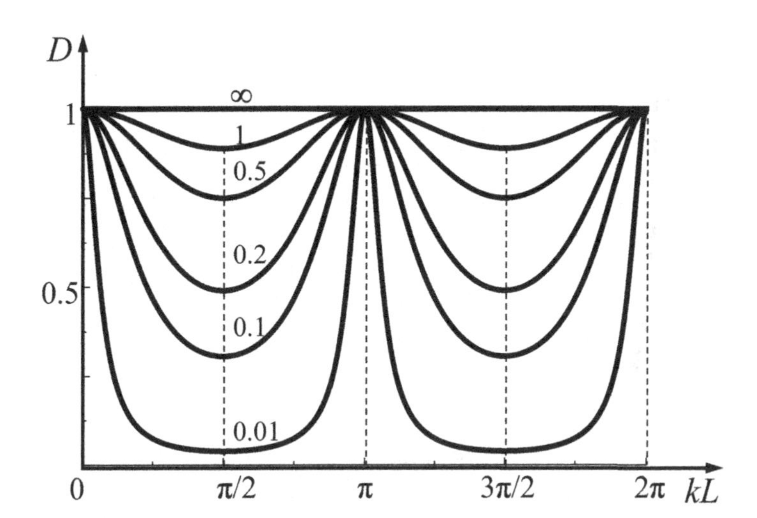

Figure 3.9 The dependence of the electron's transmission coefficient, D, on the magnitude of the potential barrier, U_0, and κL. The six curves correspond to $\Delta E_1/U_0 = \infty$, 1, 0.5, 0.2, 0.1, and 0.01. Here $\Delta E_1 = E_1 - U_0$ and E_1 is the electron's first energy level.

(3.44) for the energy of the nth level of an electron localized inside of the potential well with infinite barriers.

Since the wavenumber is equal to $\kappa = 2\pi/\lambda_{\mathrm{Br}}$, the condition (3.169) means that the width of the potential barrier, L, contains an integer number of de Broglie half-wavelengths, $\lambda_{\mathrm{Br}}/2$. Near values of the wavenumber of

$$\kappa_n = \frac{\pi}{L}\left(n + \frac{1}{2}\right) \tag{3.171}$$

we have maxima of the reflection coefficient and minima of the transmission coefficient. The existence of oscillations in the coefficients R and D is the result of the interference of waves reflected from the potential steps at the boundaries of the rectangular potential barrier. The dependences of the transmission coefficient, D, on κL found for six values of the ratio $\Delta E_1/U_0$, namely 0, 1, 0.5, 0.2, 0.1, and 0.01, where $\Delta E_1 = \pi^2\hbar^2/(2m_e L^2)$, are shown in Fig. 3.9. We see that an increase of the barrier height or an increase of the barrier width results in a decrease of the amplitude of oscillations of the transmission coefficient, D. When the electron energy, E, increases the transmission coefficient, D, quickly increases, tending to unity, whereas the amplitude of oscillations decreases. The locations of the maxima correspond to the resonance values of the energy, E_n (Eq. (3.170)).

The corresponding expressions for the reflection and transmission coefficients at $E > U_0$ for the potential step shown in Fig. 3.8 have the forms

$$R = \left(\frac{k-\kappa}{k+\kappa}\right)^2 = \left(\frac{\sqrt{E} - \sqrt{E-U_0}}{\sqrt{E} + \sqrt{E-U_0}}\right)^2, \tag{3.172}$$

$$D = \frac{4k\kappa}{(k+\kappa)^2} = \frac{4\sqrt{E(E-U_0)}}{(\sqrt{E} + \sqrt{E-U_0})^2}. \tag{3.173}$$

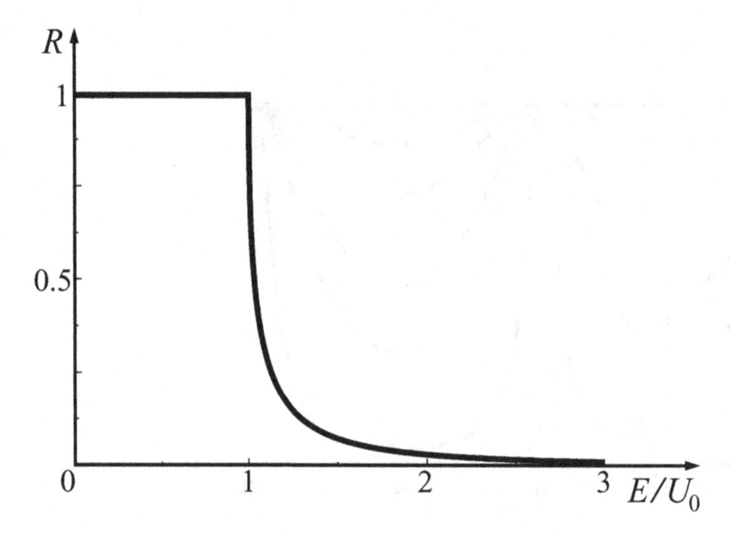

Figure 3.10 The dependence of an electron's reflection coefficient, R, on the barrier height, U_0.

Figure 3.10 shows the dependence of the reflection coefficient, R, on the total energy of an electron, E, for a given height of the potential step, U_0. We see that even when an electron moves above the potential step ($E/U_0 > 1$) the electron still undergoes reflection. At $E = U_0$ reflection stays the same as at $E < U_0$, i.e., $R = 1$.

Example 3.7. An electron traveling from the negative part of the x-axis is incident on a rectangular potential step. The height of the potential barrier U_0 is greater than the total energy of the electron E (see Fig. 3.8). Find the probability of finding the electron "under the barrier" up to the distance a from the boundary $x = 0$.

Reasoning. The probability density of an electron's penetration under the barrier depends on the coordinate x and can be written as

$$|\psi_2(x)|^2 = \frac{4k^2 A_1^2}{(k^2 + \kappa^2)} e^{-2\kappa x}. \tag{3.174}$$

The probability density, $|\psi_2(x)|^2$, of finding the electron at $x > 0$ exponentially decreases with increasing x. If the distance from the barrier at which we can find the electron is equal to a, then the total probability of finding the electron between $x = 0$ and $x = a$ is equal to

$$P = \int_0^a |\psi_2(x)|^2 \, dx = \frac{2k^2 A_1^2}{(k^2 + \kappa^2)\kappa} \left(1 - e^{-2\kappa a}\right). \tag{3.175}$$

Thus, with increasing a the total probability, P, of finding the electron at $0 \le x \le a$ increases, and at $a \to \infty$ this probability tends to its limit:

$$P_{\max} = \frac{2k^2 A_1^2}{(k^2 + \kappa^2)\kappa}. \tag{3.176}$$

Example 3.8. A rectangular potential barrier (Fig. 3.7) has a width $L = 0.15$ nm. Find the barrier height U_0 and the electron energy E for which the electron's total probability of transmission through the barrier is equal to $P = 0.4$ and the coefficient D_0 from Eq. (3.156) is equal to unity.

Reasoning. Considered from the physical point of view, the probability P of electron transmission through the potential barrier is equal to the transmission coefficient D, which is defined by Eq. (3.156):

$$P = D = D_0 \mathrm{e}^{-2L\sqrt{2m_e(U_0-E)}/\hbar}. \tag{3.177}$$

From Eq. (3.157) and the condition $D_0 = 1$ we can find the ratio $\gamma = E/U_0$:

$$16\gamma(1-\gamma) = 1. \tag{3.178}$$

On solving this quadratic equation, we find two possible values of γ:

$$\gamma = \frac{1}{2} \pm \frac{\sqrt{3}}{4}. \tag{3.179}$$

Let us take the logarithm of Eq. (3.177). As a result we obtain

$$\ln P = -\frac{2L}{\hbar}\sqrt{2m_e(U_0 - E)}. \tag{3.180}$$

From the last equation, taking into account $E = \gamma U_0$, we find the expression for U_0:

$$U_0 = \frac{\hbar^2}{1-\gamma}\frac{(\ln P)^2}{8m_e L^2}. \tag{3.181}$$

After the substitution of the given parameters we obtain two values of U_0:

$$U_{01} = 8.51 \times 10^{-19}\,\mathrm{J} = 5.31\,\mathrm{eV},$$

$$U_{02} = 0.61 \times 10^{-19}\,\mathrm{J} = 0.38\,\mathrm{eV}.$$

The corresponding values of the electron's energy are

$$E_1 = 7.94 \times 10^{-19}\,\mathrm{J} = 4.95\,\mathrm{eV},$$

$$E_2 = 0.4 \times 10^{-19}\,\mathrm{J} = 0.25\,\mathrm{eV}.$$

Example 3.9. Find the electron transmission probability through the potential barrier of the following profile:

$$U(x) = \begin{cases} 0, & x < -a, \\ U_0(1 + x/a), & -a \le x \le 0, \\ U_0(1 - x/b), & 0 < x \le b, \\ 0, & x > b, \end{cases} \tag{3.182}$$

where a and b are positive numbers. Assume that the electron is incident on the barrier from the negative part of the x-axis and that the electron's energy is less than U_0 ($E < U_0$).

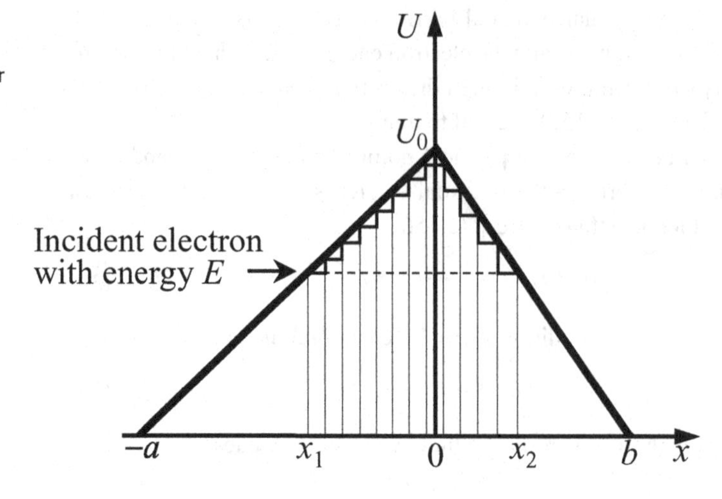

Figure 3.11 An electron's transmission through the triangular potential barrier defined by Eq. (3.182).

Reasoning. Equation (3.156) for the probability of an electron's transmission through the rectangular potential barrier can be generalized for the case of a potential barrier of arbitrary form. If the potential in the barrier region changes slowly, i.e., its change is small at distances comparable to the de Broglie wavelength, the potential of an arbitrary profile can be presented as a sum of a large number of rectangular potentials (see Fig. 3.11). Then for the transmission coefficient the following expression is valid:

$$D = D_0 \mathrm{e}^{-2\sqrt{2m_e}\int_{x_1}^{x_2}\sqrt{U(x)-E}\,\mathrm{d}x/\hbar}, \tag{3.183}$$

where x_1 and x_2 are the points of the electron's penetration under the potential barrier (see Fig. 3.11) and are defined from the solution of the following equations:

$$U_0(1 + x/a) = E, \quad x_1 = -(1 - E/U_0)a,$$
$$U_0(1 - x/b) = E, \quad x_2 = (1 - E/U_0)b. \tag{3.184}$$

Thus, to find D it is necessary to find the following integral:

$$I = \int_{x_1}^{x_2}\sqrt{U(x)-E}\,\mathrm{d}x = \int_{x_1}^{0}\sqrt{U_0 + \frac{U_0}{a}x - E}\,\mathrm{d}x + \int_{0}^{x_2}\sqrt{U_0 - \frac{U_0}{b}x - E}\,\mathrm{d}x. \tag{3.185}$$

In order to evaluate the integrals on the right-hand side of Eq. (3.185) it is necessary to change variables: $t = \sqrt{U_0 + U_0x/a - E}$ in the first integral and $t = \sqrt{U_0 - U_0x/a - E}$ in the second integral. As a result of integration and taking into account the expressions for x_1 and x_2, we obtain

$$I = \frac{2a}{3U_0}(U_0 - E)^{3/2} + \frac{2b}{3U_0}(U_0 - E)^{3/2} = \frac{2}{3}\frac{a+b}{U_0}(U_0 - E)^{3/2}. \tag{3.186}$$

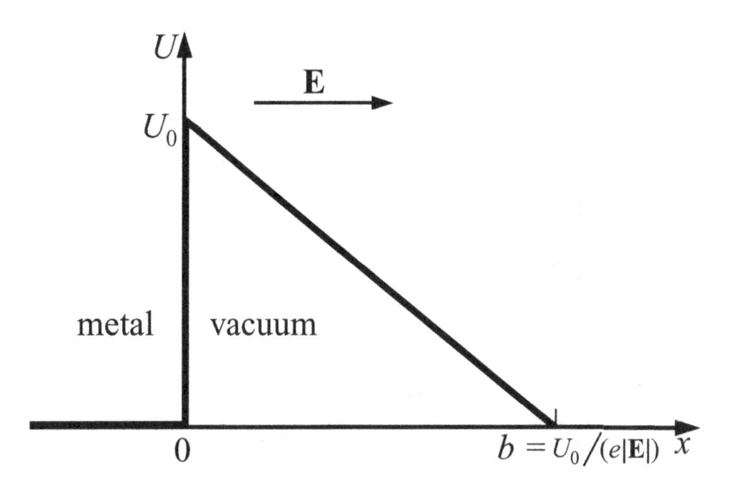

Figure 3.12 The triangular potential barrier at the metal–vacuum interface under an applied electric field, **E**.

On substituting the last expression into Eq. (3.183) we obtain

$$D = D_0 e^{-4\sqrt{2m_e}(a+b)(U_0-E)^{3/2}/(3\hbar U_0)}. \tag{3.187}$$

In practice very often a potential with one of the parameters, a or b, having a value of zero is encountered. For example, at $a = 0$ the potential has the form shown in Fig. 3.12. Such a potential is common for a metal–vacuum interface to which an electric field, **E**, is applied. Without the electric field the interface of the metal with vacuum is a rectangular potential step of finite height, i.e., $U(x) = 0$ at $x < 0$ and $U(x) = U_0$ at $x \geq 0$. If an external electric field **E** is applied along the electron's direction of motion (perpendicular to the interface), then at $x > 0$ the electron's potential energy becomes equal to

$$U(x) = U_0 - e|\mathbf{E}|x, \tag{3.188}$$

and the coordinate for which $U(b) = 0$ is

$$x = b = \frac{U_0}{e|\mathbf{E}|}.$$

For this potential barrier the transmission probability for the electron is defined by the expression

$$D = D_0 e^{-4\sqrt{2m_e}(U_0-E)^{3/2}/(3e\hbar|\mathbf{E}|)}. \tag{3.189}$$

It follows from the last equation that with increasing electric field the probability of an electron's transmission through the barrier rapidly increases. This is because of the decrease of the barrier width.

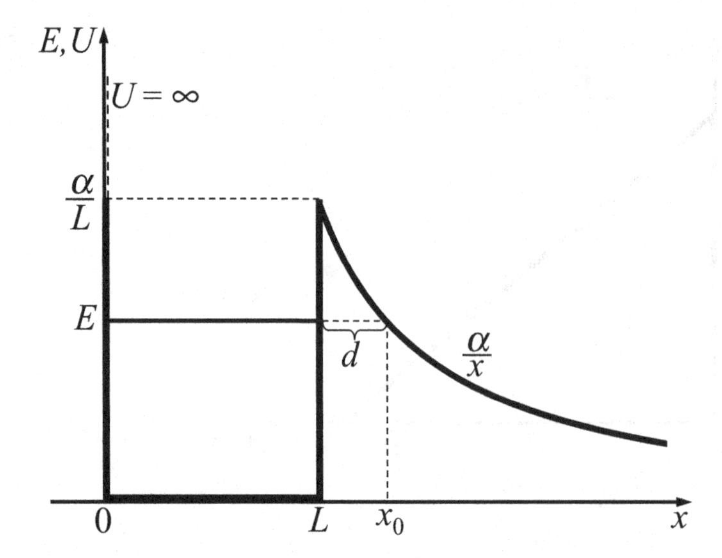

Figure 3.13 A well with the potential profile (3.190).

Example 3.10. Find the dependence on energy of the probability of an electron escaping a potential well through a potential barrier with the following profile:

$$U(x) = \begin{cases} \infty, & x < 0, \\ 0, & 0 \leq x \leq L, \\ \alpha/x, & x > L. \end{cases} \tag{3.190}$$

The graph of the potential well defined by Eq. (3.190) is shown in Fig. 3.13. **Reasoning.** The probability of the electron escaping from the potential well is defined by its transmission coefficient, D. Expression (3.156) was obtained for a barrier of rectangular form. The main difference of the present problem from the potential barrier with profile (3.141) is that the barrier width, d, now depends on the energy, E, and, as follows from Eq. (3.190), it is equal to

$$d(E) = x_0(E) - L = \frac{\alpha}{E} - L, \tag{3.191}$$

where the coordinate x_0 is the solution of the equation $E = \alpha/x$. To calculate the probability of an electron escaping from the well as a result of transmission through the barrier, we have to use expression (3.183):

$$D = D_0 \mathrm{e}^{-2\int_L^{x_0} \sqrt{2m_e(\alpha/x - E)}\mathrm{d}x/\hbar}, \tag{3.192}$$

where the constant D_0 is close to unity and the upper limit of integration is the coordinate x_0. As a result of integration, we obtain the expression

$$D \approx D_0 \mathrm{e}^{-2\alpha\sqrt{2m_e/E}\left[\arccos\sqrt{E/U_0} - \sqrt{(E/U_0)(1-E/U_0)}\right]/\hbar}, \tag{3.193}$$

where $U_0 < \alpha/L$. In the case of a sufficiently high potential barrier, when $U_0 \gg E$, this expression, taking into account that $\arccos(0) = \pi/2$, transforms into the simpler expression

$$D \approx D_0 \mathrm{e}^{-\pi\alpha\sqrt{2m_e/E}/\hbar}. \quad (3.194)$$

From the last expression it follows that with increasing electron energy, E, the transmission coefficient, D, increases. At $E = U_0$, as follows from Eq. (3.193) and from $\arccos(1) = 0$, the transmission coefficient, D, is equal to unity.

3.6 Summary

1. The wavefunction of a free electron can be described by a de Broglie plane wave, whose group velocity coincides with the electron velocity. The energy spectrum of such an electron is continuous.
2. In the case of electron motion in a limited space the main important property of such motion is the discreteness of the energy spectrum, i.e., quantization of electron energy. All quantum states of an electron and the corresponding energy states in the case of one-dimensional motion can be enumerated by one quantum number, n.
3. In a potential well with infinite barriers there is an infinite number of energy levels that correspond to the stationary quantum states of an electron in the well. The energy of corresponding levels is proportional to n^2 and the distance between levels is proportional to $2n + 1$. If we place the center of coordinates at the center of the well, then the wavefunctions of an electron for quantum states with odd numbers of n are symmetric functions of coordinate, and those for quantum states with even numbers of n are antisymmetric functions of coordinate.
4. Higher electron energy corresponds to higher wavenumber. The lowest energy level with $n = 1$ is called the ground state and it corresponds to the minimal electron energy. However, this state is not the state of rest.
5. In confined quantum states the probability density of finding an electron, $|\psi|^2$, has maxima in the region of the potential well and exponentially decreases outside of the well. This means that an electron can be outside of the well with a non-zero probability.
6. In a symmetric potential well with finite barriers there is always at least one energy level, i.e., the energy level E_1, which is lower than the barrier height U_0.
7. An electron, just like any other quantum particle, can tunnel through a potential barrier whose height is greater than the total energy of the electron. The probability of a particle tunneling through a barrier exponentially decreases with increasing width and height of the barrier and with increasing mass of the particle. The total energy of the particle does not change during tunneling through the barrier.

3.7 Problems

Problem 3.1. Using the procedure of calculations of average values of physical quantities, show that for a one-dimensional motion the operator of coordinate is coordinate itself.

Problem 3.2. A particle with mass m is moving in the region $-L/2 \le x \le L/2$. The state of the particle is described by the wavefunction

$$\psi(x) = A \exp\left(\frac{i}{\hbar} px\right), \tag{3.195}$$

where p is the particle's momentum. Find the normalization constant, A, and the average value of the particle's kinetic energy, $\langle K \rangle$.

Problem 3.3. An electron is in a one-dimensional rectangular potential well with barriers of infinite height. Find the well width if the energy difference between the fifth and fourth quantum states, $E_5 - E_4$, is equal to the electron's average thermal energy at room temperature.

Problem 3.4. An electron is in a one-dimensional rectangular potential well with barriers of infinite height. The width of the well is equal to $L = 5$ nm. Find the wavelengths of photons emitted during electronic transitions from the excited states with quantum numbers $m = 2$, λ_{21}, and $n = 3$, λ_{31}, to the ground state with $n = 1$. (Answer: $\lambda_{21} \approx 28.75$ μm and $\lambda_{31} \approx 10.75$ μm.)

Problem 3.5. An electron is in a one-dimensional rectangular potential well with barriers of infinite height. Find the normalized wavefunctions and energy levels if the width of the potential well is equal to L and the wavefunction is symmetric with respect to the coordinate origin placed at the center of the well (see Fig. 3.14).

Problem 3.6. An electron is in a one-dimensional rectangular potential well with barriers of infinite height and with well width equal to L as shown in Fig. 3.14. Find the average values of the coordinate, the momentum projection, and the squares of these magnitudes for an electron that is in quantum states described by symmetric wavefunctions.

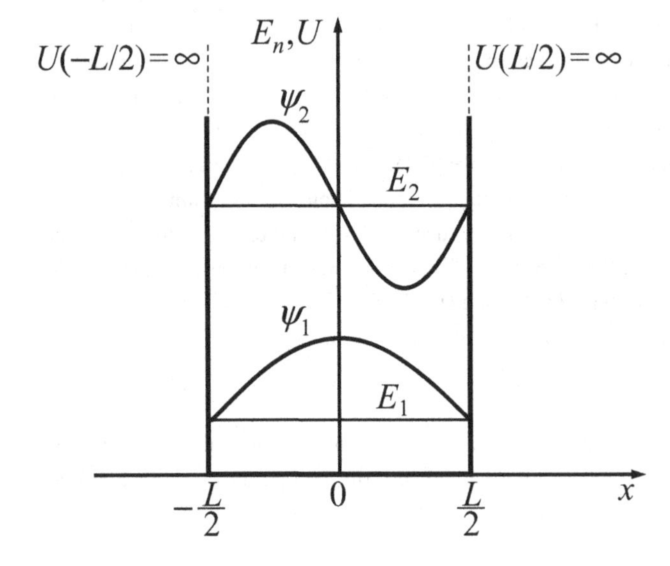

Figure 3.14 A rectangular symmetric potential well with respect to the coordinate origin with barriers of infinite height.

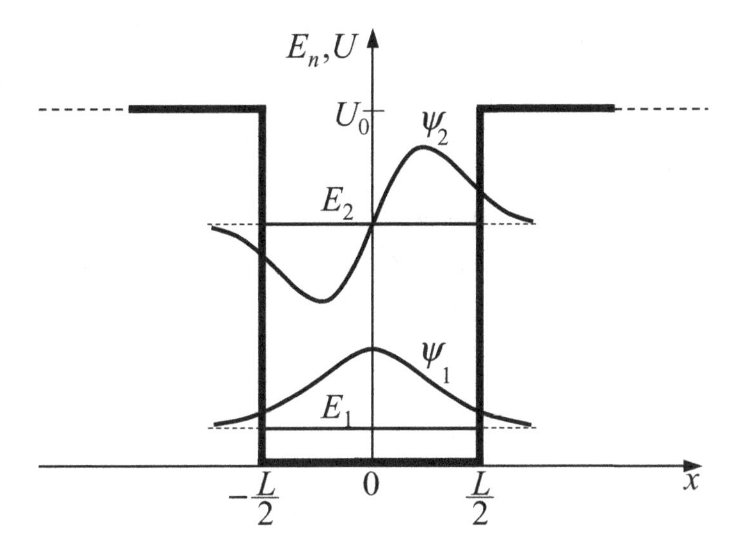

Figure 3.15 A rectangular symmetric potential well with respect to the origin of coordinates with barriers of height $U_0 = 10$ eV.

Problem 3.7. An electron is in a one-dimensional rectangular potential well with width equal to L. One of the barriers has infinite height: $U(x) \to \infty$ at $x < 0$. The height of the second barrier at $x > L$ is finite and is equal to U_0. In the region $0 \leq x \leq L$ the potential energy is equal to zero: $U(x) = 0$ (see Fig. 3.3). What are the conditions on the value of $U_0 L^2$ for there to be at least one energy level or only one level in the well? If the energy level is equal to $E = U_0/2$, how does this condition change?

Problem 3.8. An electron is in a symmetric rectangular potential well with width $L = 10$ nm and with barriers of finite height: $U_0 = 10$ eV. The electron wavefunction is symmetric with respect to the coordinate origin placed at the center of the well (see Fig. 3.15), in contrast to the case shown in Fig. 3.5. For which values of quantum numbers, n, can we apply the approximation of barriers with infinite height? (Answer: $n \ll 52$.)

Problem 3.9. An electron with energy $E = 32$ eV, which is moving in the positive direction along the x-direction, encounters an obstacle – an infinitely long rectangular potential step. Find the reflection, R, and transmission, D, coefficients of the electron de Broglie waves for a given potential step in two cases: (a) the height of the potential step is lower than the electron energy ($U_1 = 30$ eV) and (b) the height of the potential step is higher than the electron energy ($U_2 = 34$ eV). (Answer: (a) $R \approx 0.36$ and $D \approx 0.64$; (b) $R \approx 1.0$ and $D \approx 0$.)

Problem 3.10. Estimate the transmission coefficient, D, of an electron with energy $E = 18$ eV for passage through a potential barrier with the following dependence of potential energy on coordinate:

$$U(x) = U_0 \left(1 - \frac{x^2}{L^2}\right) \tag{3.196}$$

in the region $|x| \leq L$ and

$$U(x) = 0 \tag{3.197}$$

in the region $|x| > L$. Here, $U_0 = 20$ eV and $L = 2$ nm. (Answer: $D \approx 0.24$.)

Chapter 4
Additional examples of quantized motion

In the previous chapter we have analyzed the peculiarities of quantized electron motion in layered structures with one-dimensional potential wells. From the mathematical point of view, the solution of the Schrödinger equation for one-dimensional potential profiles is much simpler. However, many quantum objects, such as atoms, molecules, and quantum dots, are three-dimensional objects. Thus, in order to analyze electron motion in such objects we need to find solutions of the Schrödinger equation for three-dimensional potential profiles. The electron motion in spaces with dimensionality higher than one, especially for rectangular potential profiles with infinite potential barriers, is not so difficult to analyze. At the same time we have to keep in mind that such potential profiles frequently represent some approximation of the more complex, real potential profiles. Depending on the type of structure and on the form of the potential profile, the electron motion may be limited in two directions (two-dimensional quantization) or in three directions (three-dimensional quantization). In this chapter we will show that the existence of discrete energy levels in the electron spectrum is an intrinsic feature of electron motion in potential wells of any form and dimensionality.

4.1 An electron in a rectangular potential well (quantum box)

In the previous chapter we studied the electron motion in one-dimensional potential wells. An electron's motion along one direction was confined by the potential profile and the momentum in this direction was quantized. In the two other directions the motion of the electron was free. For the specific potential profiles that we studied we were able to express the solution of the Schrödinger equation in the form of a product of three separate wavefunctions. In the direction of electron confinement the wavefunction is a *standing wave* and in the direction of free motion the wavefunctions have the form of *traveling waves*. Here, we will consider the main features of the electron energy spectrum in a rectangular potential box with impenetrable walls. The limitation of electron motion in each of the three directions is defined by the following conditions:

$$0 \le x \le L_x, \qquad 0 \le y \le L_y, \qquad 0 \le z \le L_z, \tag{4.1}$$

where L_x, L_y, and L_z are the dimensions of the box in the x-, y-, and z-directions, respectively. The above-mentioned conditions are realized if the electron potential energy satisfies the following conditions:

$$U(\mathbf{r}) = \begin{cases} 0, & 0 \le r_\alpha \le L_\alpha, \\ \infty, & r_\alpha < 0, r_\alpha > L_\alpha, \end{cases} \tag{4.2}$$

where $\alpha = x, y$, and z and $r_\alpha = \alpha$. In this case the Schrödinger equation inside the potential box, where $U(\mathbf{r}) = 0$, takes the form

$$-\frac{\hbar^2}{2m_e}\left(\frac{\partial^2}{\partial x^2} + \frac{\partial^2}{\partial y^2} + \frac{\partial^2}{\partial z^2}\right)\psi(x, y, z) = E\psi(x, y, z). \tag{4.3}$$

Since the electron cannot penetrate the region where the potential tends to infinity, for $r_\alpha < 0$ and $r_\alpha > L_\alpha$ the electron wavefunction, $\psi(x, y, z)$, is equal to zero in those regions. Therefore, at the boundaries of the box the following boundary conditions should hold:

$$\psi(0, y, z) = \psi(x, 0, z) = \psi(x, y, 0) = \psi(L_x, y, z) = \psi(x, L_y, z) = \psi(x, y, L_z) = 0. \tag{4.4}$$

For the potential profile of Eq. (4.2) and the boundary conditions of Eq. (4.4) the wavefunction can be written as a product of three wavefunctions, each of which depends on a single coordinate,

$$\psi(x, y, z) = \psi_x(x)\psi_y(y)\psi_z(z), \tag{4.5}$$

and the total electron energy, E, can be written as a sum of energies, E_α, in each of the three directions of confinement:

$$E = E_x + E_y + E_z. \tag{4.6}$$

Here, each of the wavefunctions $\psi_\alpha(r_\alpha)$ describes the electron motion along one specific coordinate r_α and this motion is independent from motion in the two other directions, $r_\beta \ne r_\alpha$. On substituting Eqs. (4.5) and Eqs. (4.6) into Eq. (4.3) and dividing by the wavefunction of Eqs. (4.5), we obtain

$$\frac{1}{\psi_x(x)}\left[\frac{\partial^2\psi_x(x)}{\partial x^2} + \frac{2m_e}{\hbar^2}E_x\psi_x(x)\right] + \frac{1}{\psi_y(y)}\left[\frac{\partial^2\psi_y(y)}{\partial y^2} + \frac{2m_e}{\hbar^2}E_y\psi_y(y)\right]$$
$$+ \frac{1}{\psi_z(z)}\left[\frac{\partial^2\psi_z(z)}{\partial z^2} + \frac{2m_e}{\hbar^2}E_z\psi_z(z)\right] = 0. \tag{4.7}$$

Each of the three terms on the left-hand side of Eq. (4.7) depends only on one independent variable, r_α. Since the sum of these three terms is equal to zero for any values of the variables x, y, and z, each of the terms must itself be equal to zero. Therefore, Eq. (4.7) may be presented as a system of three independent equations (compare with Eqs. (3.6)–(3.8)):

$$\frac{d^2\psi_x(x)}{dx^2} + k_x\psi_x(x) = 0, \tag{4.8}$$

$$\frac{d^2\psi_y(y)}{dy^2} + k_y\psi_y(y) = 0, \tag{4.9}$$

$$\frac{d^2\psi_z(z)}{dz^2} + k_z\psi_z(z) = 0, \tag{4.10}$$

where we introduced the parameters

$$k_\alpha = \frac{\sqrt{2m_e E_\alpha}}{\hbar}, \tag{4.11}$$

which are the projections of the wavevector **k** onto the coordinate axes, $\alpha = x, y$, and z. The solution of each of Eqs. (4.8)–(4.10), taking into account boundary conditions (4.4) as in the case of a one-dimensional potential well, has the form

$$\psi_x(x) = \sqrt{\frac{2}{L_x}}\sin(k_x x), \tag{4.12}$$

$$\psi_y(y) = \sqrt{\frac{2}{L_y}}\sin(k_y y), \tag{4.13}$$

$$\psi_z(z) = \sqrt{\frac{2}{L_z}}\sin(k_z z), \tag{4.14}$$

where

$$k_\alpha = \pm\frac{\pi n_\alpha}{L_\alpha} \tag{4.15}$$

and $n_\alpha = 1, 2, 3, \ldots$ It follows from Eqs. (4.11) and (4.15) that the eigenvalues of energy, E_α, are defined by the expression

$$E_\alpha = \frac{\pi^2\hbar^2 n_\alpha^2}{2m_e L_\alpha^2}. \tag{4.16}$$

As a result, the eigenfunctions of Eq. (4.3) have the form

$$\psi_{n_x n_y n_z}(x, y, z) = \sqrt{\frac{8}{L_x L_y L_z}}\sin\left(\frac{\pi n_x x}{L_x}\right)\sin\left(\frac{\pi n_y y}{L_y}\right)\sin\left(\frac{\pi n_z z}{L_z}\right). \tag{4.17}$$

The corresponding eigenvalues are

$$E = E_{n_x n_y n_z} = \frac{\pi^2\hbar^2}{2m_e}\left(\frac{n_x^2}{L_x^2} + \frac{n_y^2}{L_y^2} + \frac{n_z^2}{L_z^2}\right). \tag{4.18}$$

Here, the quantum numbers n_x, n_y, and n_z are positive integers. The wavefunctions (4.12)–(4.14) and (4.17) are orthonormal, which can be verified directly by substituting them into Eq. (2.144).

The most important property of the electron quantization in a three-dimensional potential box, as in the case of the one-dimensional confinement, is the quantization of the electron energy. The energy levels are defined by the set of three quantum numbers, n_x, n_y, and n_z (Eq. (4.18)). The minimum energy with the quantum numbers $n_x = n_y = n_z = 1$ corresponds to the ground state:

$$E_{\text{min}} = E_{111} = \frac{\pi^2 \hbar^2}{2m_e} \left(\frac{1}{L_x^2} + \frac{1}{L_y^2} + \frac{1}{L_z^2} \right). \tag{4.19}$$

This means that the state of the electron with minimum energy is not a state of rest. This property is of purely quantum origin and is the direct consequence of the uncertainty relations (2.118). Indeed, the state of rest must correspond to the electron energy and momentum being equal to zero. Since these values are defined exactly, the uncertainties in the momenta Δp_x, Δp_y, and Δp_z are equal to zero too. Thus, according to the uncertainty relations (2.118) the electron state at rest corresponds to complete uncertainty in coordinates, since in this case Δx, Δy, and Δz tend to infinity. But this contradicts the limited uncertainties of the electron coordinates, because we know for sure that the electron is in the box and the uncertainty in coordinates cannot be larger than the dimensions of the box, i.e., $\Delta x \approx L_x$, $\Delta y \approx L_y$, and $\Delta z \approx L_z$.

As a result, the electron motion in a limited space, in contrast to the case of the free electron in infinite space, cannot have zero momentum and, as a consequence, zero uncertainty of the momentum. From the uncertainty relations (2.118) it follows that for the electron confined in a box the uncertainty in the momenta is the following:

$$\Delta p_x \sim \frac{h}{L_x}, \qquad \Delta p_y \sim \frac{h}{L_y}, \qquad \Delta p_z \sim \frac{h}{L_z}. \tag{4.20}$$

Since projections p_α of momentum $\mathbf{p}$ should be of the order of the uncertainty Δp_α, we may conclude that Eqs. (4.15) are direct consequences of Eq. (4.20). It follows from Eqs. (4.15) and (4.18) that the smallest differences between discrete values of the momentum k_α and energy $E_{n_x n_y n_z}$ correspond to the α-axes with the largest value of the length L_α. For example, if $L_z > L_x, L_y$ then the set of quantum numbers $n_x = n_y = 1$ and $n_z = 2$ corresponds to the energy state next to E_{111}, namely E_{112}. In this case the energy E_{112} differs from the minimal energy E_{111} by ΔE:

$$\Delta E = E_{112} - E_{111} = \frac{3\pi^2 \hbar^2}{2m_e L_z^2}. \tag{4.21}$$

By changing the set of quantum numbers (n_x, n_y, n_z) we can calculate the energy of the confined electron for an arbitrary electron quantum state for the given dimensions of the box.

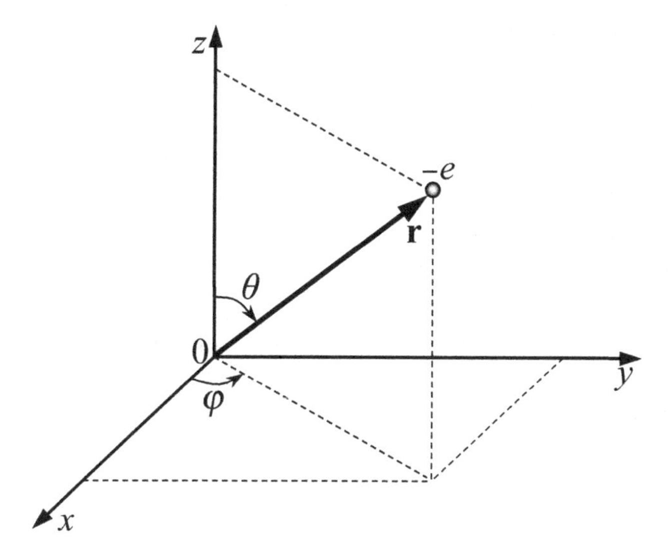

Figure 4.1 The spherical system of coordinates.

4.2 An electron in a spherically-symmetric potential well

4.2.1 The Schrödinger equation in spherical coordinates

Let us consider first the motion of an electron in a spherically-symmetric potential:

$$U = U(r), \tag{4.22}$$

where the potential $U(\mathbf{r})$ depends only on the distance $r = \sqrt{x^2 + y^2 + z^2}$ from the origin of the coordinate system ($r = 0$).

Taking into account the spherical symmetry of the potential, the Schrödinger equation may be solved using spherical coordinates, in which the position of the electron in space is given by the radial distance r and two angles – polar and azimuthal, θ and φ, respectively. The origin of coordinates, $r = 0$, coincides with the center of the potential profile $U(r)$; the azimuthal angle takes values in the interval $0 \leq \varphi \leq 2\pi$; and the polar angle takes values in the interval $0 \leq \theta \leq \pi$ (see Fig. 4.1). The wavefunction can be written as a function of the spherical coordinates, i.e., $\psi(\mathbf{r}) = \psi(r, \theta, \varphi)$.

The Laplace operator, ∇^2, in spherical coordinates includes radial and spherical parts, ∇_r^2 and $\nabla_{\varphi,\theta}^2$, respectively:

$$\nabla^2 = \nabla_r^2 + \frac{1}{r^2}\nabla_{\varphi,\theta}^2. \tag{4.23}$$

The radial part of the Laplace operator can be written as

$$\nabla_r^2 = \frac{1}{r^2}\frac{\partial}{\partial r}\left(r^2\frac{\partial}{\partial r}\right) = \frac{\partial^2}{\partial r^2} + \frac{2}{r}\frac{\partial}{\partial r} \tag{4.24}$$

and its spherical part can be written as

$$\nabla^2_{\varphi,\theta} = \frac{1}{\sin\theta}\frac{\partial}{\partial\theta}\left(\sin\theta\,\frac{\partial}{\partial\theta}\right) + \frac{1}{\sin^2\theta}\frac{\partial^2}{\partial\varphi^2}. \tag{4.25}$$

Thus, the Schrödinger equation for an electron in a spherically-symmetric potential well can be written as

$$\left[\nabla^2_r + \frac{1}{r^2}\nabla^2_{\varphi,\theta} + \frac{2m_e}{\hbar^2}(E - U(r))\right]\psi(r,\varphi,\theta) = 0. \tag{4.26}$$

Since the potential that acts on the electron, $U(r)$, does not depend on the angles φ and θ, the variables in Eq. (4.26) can be separated and the wavefunction can be expressed in the form of a product of two functions – a radial function, $X(r)$, which depends only on the coordinate r, and an angle-dependent function $Y(\varphi,\theta)$:

$$\psi(r,\varphi,\theta) = X(r)Y(\varphi,\theta). \tag{4.27}$$

4.2.2 Eigenvalues and eigenfunctions of the angular-momentum operators

The angular component of the wavefunction $Y(\varphi,\theta)$ of Eq. (4.27) in the case of the spherically-symmetric potential with arbitrary $U(r)$ is an eigenfunction of the operator for the square of the angular momentum, i.e., it is the solution of the equation

$$(\hat{L}^2 - L^2)Y(\varphi,\theta) = 0. \tag{4.28}$$

Here L^2 is the eigenvalue of the operator $\hat{L}^2$, which coincides up to a constant factor with the angular part of the Laplace operator (4.25):

$$\hat{L}^2 = \hat{L}_x\hat{L}_x + \hat{L}_y\hat{L}_y + \hat{L}_z\hat{L}_z = -\hbar^2\nabla^2_{\varphi,\theta}. \tag{4.29}$$

Operators of the projections of angular momentum $\hat{L}_\alpha$ $(\alpha = x, y, z)$ are defined by the following expressions for rectangular and spherical coordinates, respectively:

$$\hat{L}_x = -\mathrm{i}\hbar\left(y\frac{\partial}{\partial z} - z\frac{\partial}{\partial y}\right) = \mathrm{i}\hbar\left(\cos\varphi\cot\theta\frac{\partial}{\partial\varphi} + \sin\varphi\frac{\partial}{\partial\theta}\right), \tag{4.30}$$

$$\hat{L}_y = -\mathrm{i}\hbar\left(z\frac{\partial}{\partial x} - x\frac{\partial}{\partial z}\right) = \mathrm{i}\hbar\left(\sin\varphi\cot\theta\frac{\partial}{\partial\varphi} - \cos\varphi\frac{\partial}{\partial\theta}\right), \tag{4.31}$$

$$\hat{L}_z = -\mathrm{i}\hbar\left(x\frac{\partial}{\partial y} - y\frac{\partial}{\partial x}\right) = -\mathrm{i}\hbar\frac{\partial}{\partial\varphi}. \tag{4.32}$$

To write these expressions we used the equation for the angular momentum of a particle $\mathbf{L} = \mathbf{r}\times\mathbf{p}$ and found the projections of angular momentum by writing L as the determinant:

$$L_x = yp_z - zp_y, \qquad L_y = zp_x - xp_z, \qquad L_z = xp_y - yp_x. \tag{4.33}$$

Then, in Eq. (4.33) we substitute the variables by their corresponding operators. In the coordinate representation, that we use, the operator of coordinate is the coordinate itself, and the operators of the projections of momentum are defined by Eqs. (2.159):

$$\hat{p}_\alpha = -\mathrm{i}\hbar \frac{\partial}{\partial r_\alpha},$$

where $\alpha = x, y, z$ are Cartesian coordinates and $r_\alpha = \alpha$. In order to change Cartesian coordinates to spherical coordinates we used the relations

$$x = r \sin\theta \cos\varphi, \qquad y = r \sin\theta \sin\varphi, \qquad z = r \cos\theta. \tag{4.34}$$

The eigenfunctions in Eq. (4.28) are *spherical harmonics* $Y_{l,m}(\varphi, \theta)$, which are defined by two integer quantum numbers, l and m. The quantum number l, which is called the *orbital quantum number*, has the values $l = 0, 1, 2, \ldots$, and quantum number m, which is called the *orbital magnetic quantum number*, has the values

$$m = 0, \pm 1, \pm 2, \ldots, \pm l. \tag{4.35}$$

Using the normalizing conditions of Eq. (2.140) for spherical harmonics we have

$$\int_0^{2\pi} \int_0^{\pi} \left| Y_{l,m}(\varphi, \theta) \right|^2 \sin\theta \, \mathrm{d}\theta \, \mathrm{d}\varphi = 1. \tag{4.36}$$

We get the following expressions for the spherical harmonics for the first few values of quantum numbers l and m:

$$Y_{0,0} = \frac{1}{\sqrt{4\pi}}, \qquad Y_{1,0} = \sqrt{\frac{3}{4\pi}} \cos\theta, \qquad Y_{2,0} = \sqrt{\frac{5}{16\pi}} (3\cos^2\theta - 1),$$

$$Y_{1,\pm 1} = \mp \sqrt{\frac{3}{8\pi}} \sin\theta \, \mathrm{e}^{\pm \mathrm{i}\varphi}, \qquad Y_{2,\pm 1} = \mp \sqrt{\frac{15}{16\pi}} \sin(2\theta) \mathrm{e}^{\pm \mathrm{i}\varphi},$$

$$Y_{2,\pm 2} = \sqrt{\frac{15}{32\pi}} \sin^2\theta \, \mathrm{e}^{\pm 2\mathrm{i}\varphi}. \tag{4.37}$$

The spectrum of eigenvalues of the operator $\hat{L}^2$, which is defined by Eq. (4.28), is discrete:

$$L^2 = l(l+1)\hbar^2. \tag{4.38}$$

Each eigenvalue L^2 corresponds to $2l + 1$ eigenfunctions $Y_{l,m}(\varphi, \theta)$, where the orbital magnetic quantum number, m, is defined by Eq. (4.35). Thus, each state with a given value of the square of the orbital angular momentum, L^2, is $(2l + 1)$-fold degenerate.

4.2.3 The equation for the radial function

On substituting the wavefunction (4.27) into the Schrödinger equation (4.26), separating the radial part of this equation, and taking into account Eq. (4.38), we obtain

$$\left[\frac{d^2}{dr^2} + \frac{2}{r}\frac{d}{dr} + k^2 - \frac{2m_e U(r)}{\hbar^2} - \frac{l(l+1)}{r^2}\right] X(r) = 0, \tag{4.39}$$

where

$$k^2 = \frac{2m_e E}{\hbar^2}. \tag{4.40}$$

Let us consider the simplest case of $U(r)$:

$$U(r) = \begin{cases} 0, & r \leq a, \\ \infty, & r > a, \end{cases} \tag{4.41}$$

where r is the distance from the center of the sphere. It is clear that with such a potential the region outside the sphere is not available for the electron and its wavefunction at the boundary is equal to zero, i.e.,

$$X(r) = 0, \quad r \geq a. \tag{4.42}$$

The behavior of an electron inside of the region $r < a$ is described by the time-independent Schrödinger equation Eq. (3.1) with the potential $U(r) = 0$. Then, Eq. (4.39) with the help of the substitution

$$X(r) = \frac{R(r)}{r} \tag{4.43}$$

can be reduced to the following equation:

$$\left[\frac{d^2}{dr^2} + k^2 - \frac{l(l+1)}{r^2}\right] R(r) = 0. \tag{4.44}$$

The function $R(r)$ must satisfy the following two boundary conditions:

$$R(0) = 0 \quad \text{and} \quad R(a) = 0. \tag{4.45}$$

The first of these conditions is written by taking into account that the wavefunction $X(r)$ must be finite throughout the entire region of the electron's motion, in particular at $r = 0$. The second condition comes from the continuity of the wavefunction at the boundary of the potential well $r = a$ and the boundary condition (4.42).

Let us consider first the purely radial motion of the electron when its orbital angular momentum is equal to zero, i.e., $L = 0$. The orbital quantum number $l = 0$ corresponds to such a motion and to the following solution of Eq. (4.44):

$$R_0(r) = A_0 \sin(kr) + B_0 \cos(kr). \tag{4.46}$$

Note that for $l = 0$ Eq. (4.39) and its solution Eq. (4.46) coincide with the case of the one-dimensional quantum well considered earlier. Taking into account the boundary conditions (4.45), we find that $B_0 = 0$ and $k = n\pi/a$, where the radial quantum number n is a positive integer number: $n = 1, 2, 3, \ldots$ The energy of the corresponding quantum levels is equal to

$$E_{n0} = \frac{\pi^2\hbar^2}{2m_e a^2} n^2. \tag{4.47}$$

The normalization condition of the wavefunction $X_0(r)$ gives us the value of the constant $A_0 = \sqrt{2/a}$. As a result the radial function, $X_0(r)$, in the case of electron confinement being considered here has the form

$$X_0(r) = \sqrt{\frac{2}{a}} \frac{\sin(kr)}{r}. \tag{4.48}$$

In the case of electron motion with orbital angular momentum $L \neq 0$, the solution of Eq. (4.44) depends on the orbital quantum number l. This solution can be written in the form of special functions called *cylindrical Bessel functions* $J_{l+1/2}$:

$$R_l(r) = C_l \sqrt{r} J_{l+1/2}(kr). \tag{4.49}$$

In the general case when the index l is a non-integer number the Bessel functions cannot be reduced to well-known elementary mathematical functions, but in the case of integer values of l, the Bessel functions can be expressed in terms of trigonometric functions as follows:

$$J_{l+1/2}(x) = (-1)^l \sqrt{\frac{2x}{\pi}} x^l \left[\frac{\mathrm{d}}{x\,\mathrm{d}x}\right]^l \left(\frac{\sin x}{x}\right). \tag{4.50}$$

Note that $[\mathrm{d}/(x\,\mathrm{d}x)]$ in Eq. (4.50) is an operator, which must be applied to $(\sin x/x)$ in this equation l times. Taking into account Eqs. (4.49) and (4.50), the solutions of Eq. (4.44) for the first three values of the orbital quantum number $(l = 0, 1, 2)$ are the following functions of coordinate r:

$$\begin{aligned} R_0(r) &= C_0 \sqrt{\frac{2}{\pi k}} \sin(kr), \\ R_1(r) &= C_1 \sqrt{\frac{2}{\pi k}} \left(\frac{\sin(kr)}{kr} - \cos(kr)\right), \\ R_2(r) &= C_2 \sqrt{\frac{2}{\pi k}} \left[\frac{3}{kr}\left(\frac{\sin(kr)}{kr} - \cos(kr)\right) - \sin(kr)\right]. \end{aligned} \tag{4.51}$$

Here $C_0 = \sqrt{\pi k/a}$ and C_1 and C_2 can be found from the general normalization condition for the total wavefunction:

$$\iiint |\psi(r, \varphi, \theta)|^2 r^2 \,\mathrm{d}r \sin\theta \,\mathrm{d}\theta \,\mathrm{d}\varphi = 1, \tag{4.52}$$

where we integrate over the coordinate r from 0 to a, over the angle θ, from 0 to π, and over the angle φ, from 0 to 2π.

From Eq. (4.49), using the boundary condition (4.45) for $r = a$ we can determine the values of wavevector k that satisfy the condition $J_{l+1/2}(ka) = 0$:

$$k = \frac{1}{a} g_{nl}, \tag{4.53}$$

Table 4.1. *Values of g_{nl} for the first few states*

nl	10	11	12	20	13	21	22	30
g_{nl}	π	4.49	5.76	2π	6.992	7.73	9.09	3π

where the parameters g_{nl} are the roots of the corresponding Bessel functions. From Eqs. (4.53) and (4.40) we can find the expression for the energy of the allowed stationary states:

$$E_{nl} = \frac{\hbar^2}{2m_e a^2} g_{nl}^2. \tag{4.54}$$

For states with $l = 0$ the corresponding roots are $g_{n0} = n\pi$. Values of g_{nl} for the first few states are given in Table 4.1.

Thus, in a spherically-symmetric well described by the potential (4.41) the stationary states are defined by three quantum numbers, n, l, and m. The radial wavefunctions (4.43) correspond to these quantum numbers and the energy of the corresponding states is defined by Eq. (4.54).

Example 4.1. Consider an electron in a spherically-symmetric potential well with finite potential barriers (compare this with Eq. (4.41)):

$$U(r) = \begin{cases} 0, & 0 \le r \le a, \\ U_0, & r > a. \end{cases} \tag{4.55}$$

Solve Eq. (4.39) and find the normalized radial function $X(r)$. Also find the equation that defines the energy of stationary states with zero orbital angular momentum for $E < U_0$.

Reasoning. Zero orbital angular momentum corresponds to the orbital quantum number $l = 0$. Within the potential well where $U(r) = 0$, Eq. (4.39) coincides with Eq. (4.44) and its solutions for $l = 0$ have already been found (Eqs. (4.43) and (4.46)). As before, we have $B_0 = 0$. Now, we need to define the normalization constant A_0 that would be different from $A_0 = \sqrt{2/a}$ that was found for the quantum well with barriers of infinite height (see Eq. (4.48)).

Outside of the potential well where $U(r) = U_0$ the square of the wavenumber is a negative number. This is why we introduce a new parameter for $E < U_0$:

$$\kappa^2 = \frac{2m_e(U_0 - E)}{\hbar^2}, \tag{4.56}$$

and Eq. (4.39) can be written in the form

$$\left[\frac{d^2}{dr^2} + \frac{2}{r}\frac{d}{dr} - \kappa^2\right] X(r) = 0. \tag{4.57}$$

The solution of Eq. (4.57) in the region $r > a$ is an exponentially decreasing function. Thus, the expression for the radial wavefunction inside and outside the potential well is

$$X(r) = \begin{cases} A_0 \sin(kr)/r, & 0 < r < a, \\ A_1(1/r)\mathrm{e}^{-\kappa r}, & r \geq a. \end{cases} \tag{4.58}$$

From the condition of continuity of the wavefunction $X(r)$ and its derivative we obtain the following two equations:

$$A_0 \sin(ka) = A_1 \mathrm{e}^{-\kappa a}, \tag{4.59}$$

$$A_0[ka\cos(ka) - \sin(ka)] = -A_1(1 + \kappa a)\mathrm{e}^{-\kappa a}. \tag{4.60}$$

By solving this system of equations and taking into account the expressions for k and κ (Eqs. (4.40) and (4.56), respectively) we obtain the dispersion equation which defines the energy of an electron in the stationary states with $l = 0$:

$$\cot\left(\frac{\sqrt{2m_\mathrm{e}E}a}{\hbar}\right) = -\sqrt{\frac{U_0}{E} - 1}. \tag{4.61}$$

For the coefficients A_0 and A_1 we obtain the relationship

$$A_1 = A_0 \sin(ka)\mathrm{e}^{\kappa a} = \frac{A_0 k}{\sqrt{k^2 + \kappa^2}}\mathrm{e}^{\kappa a}. \tag{4.62}$$

The normalization condition of the radial function, $X(r)$ (see Eq. (4.58)), can be written as follows:

$$A_0^2\left[\int_0^a \sin^2(kr)\mathrm{d}r + \frac{k^2}{k^2 + \kappa^2}\mathrm{e}^{2\kappa a}\int_a^\infty \mathrm{e}^{-2\kappa r}\mathrm{d}r\right] = 1. \tag{4.63}$$

By carrying out the integration in Eq. (4.63) we obtain the following expression for the normalization constant A_0:

$$A_0 = \sqrt{\frac{\kappa}{1 + \kappa a}}. \tag{4.64}$$

Note that, for $U_0 \to \infty$, when $\kappa \to \infty$ the wavefunction coincides with Eq. (4.48).

4.3 Quantum harmonic oscillators

4.3.1 An electron in a parabolic potential well

Modern methods of fabrication of crystalline nanostructures using computerized technological regimes allow us to design potential wells with the desired profiles. Alongside the rectangular potential wells that we considered in the previous chapter, nowadays more complex dependences of $U(x, y, z)$ are realized. Of special interest are structures with parabolic potential wells, where, in contrast to Eqs. (4.16) and (4.47), the distance between levels is constant and is determined only by the parameter of the parabolic well. Moreover, in a structure with an arbitrary non-homogeneous potential profile $U(x, y, z)$ near the equilibrium position that corresponds to the minimum, the potential energy, $U(\mathbf{r})$, can be expanded in a Taylor series, taking into account the first three terms:

$$U(\mathbf{r}) = U_0 + \left.\frac{\mathrm{d}U}{\mathrm{d}\mathbf{r}}\right|_{\mathbf{r}=\mathbf{r}_0}(\mathbf{r}-\mathbf{r}_0) + \frac{1}{2}\left.\frac{\mathrm{d}^2U}{\mathrm{d}\mathbf{r}^2}\right|_{\mathbf{r}=\mathbf{r}_0}(\mathbf{r}-\mathbf{r}_0)^2. \tag{4.65}$$

Here, $U_0 = U(\mathbf{r}_0)$.

Let us assume that the origin of coordinates, (x_0, y_0, z_0), is at the bottom of a potential well, i.e., let us measure the energy of an oscillating particle from $U_0 = U(x_0, y_0, z_0) = 0$. The term with the first derivative for any potential profile at its minimum is equal to zero. Then, the particle's potential energy can be represented with the following function:

$$U(x, y, z) = \frac{1}{2}\left(\beta_x x^2 + \beta_y y^2 + \beta_z z^2\right), \tag{4.66}$$

where x, y, z are displacements from the position (x_0, y_0, z_0) since we put the origin of the new coordinate system onto $U(x_0, y_0, z_0)$. The coefficients β_α are defined as

$$\beta_\alpha = \frac{\partial^2 U}{\partial r_\alpha^2}. \tag{4.67}$$

Classical motion of a particle in such a potential takes place under the influence of the force

$$\mathbf{F} = -\nabla U = -\left(\mathbf{i}\beta_x x + \mathbf{j}\beta_y y + \mathbf{k}\beta_z z\right). \tag{4.68}$$

Note that $\mathbf{F}$ vanishes at the origin. In each of three independent directions a particle with mass m executes simple harmonic motion (see Eq. (A.73)) with angular frequency

$$\omega_\alpha = \sqrt{\frac{\beta_\alpha}{m}}, \tag{4.69}$$

where $\alpha = x, y, z$. The region within which this particle is allowed to move is limited by the interval of values of the corresponding coordinates $-A_\alpha \leq r_\alpha \leq A_\alpha$, where the amplitude of oscillations, A_α, is related to the energy of the particle's motion in this direction as

$$A_\alpha = \sqrt{\frac{2E_\alpha}{m\omega_\alpha^2}}. \tag{4.70}$$

At the turning points $r_\alpha = \pm A_\alpha$ the corresponding projection of the velocity v_α of the particle vanishes. Motion outside of the interval $(-A_\alpha, A_\alpha)$ implies that the kinetic energy of a particle is negative, which is meaningless in classical mechanics. A particle that moves in such a way is called a *classical harmonic oscillator*. Examples of such motion in real structures, alongside the electron oscillations, are the oscillations of atoms and ions in a crystalline lattice.

If the amplitude of the particle's oscillations is comparable to its de Broglie wavelength, then such a particle is called a *quantum oscillator*. An exact solution of the steady-state oscillatory motion of a quantum harmonic oscillator, which can be represented by a bound electron in a parabolic potential (Eq. (4.66)), must be based on the Schrödinger equation. In this case the Schrödinger equation has the form

$$\left[-\frac{\hbar^2}{2m_e}\left(\frac{\partial^2}{\partial x^2}+\frac{\partial^2}{\partial y^2}+\frac{\partial^2}{\partial z^2}\right)+\frac{1}{2}\beta_x x^2+\frac{1}{2}\beta_y y^2+\frac{1}{2}\beta_z z^2\right]\psi(x,y,z)=E\psi(x,y,z). \tag{4.71}$$

Since the potential energy is a sum of three terms, each of which depends on one coordinate only, by analogy to Eq. (4.3) the general solution of the Schrödinger equation (4.71) can be expressed in the form

$$\psi(x,y,z)=\psi_x(x)\psi_y(y)\psi_z(z), \tag{4.72}$$

where the wavefunctions $\psi_\alpha(r_\alpha)$ $(\alpha = x, y, z)$ describe the electron motion along the α-directions. The total energy can be written as a sum of individual energies: $E = E_x + E_y + E_z$. As a result Eq. (4.71) can be rewritten as a set of three independent differential equations (compare with Eqs. (4.7)–(4.10)):

$$\frac{\mathrm{d}^2\psi_\alpha(\alpha)}{\mathrm{d}r_\alpha^2}+\frac{2m_e}{\hbar^2}\left(E_\alpha-\frac{\beta_\alpha}{2}r_\alpha^2\right)\psi_\alpha(\alpha)=0, \quad \alpha=x,y,z. \tag{4.73}$$

Thus, the solution of the Schrödinger equation (4.71) that describes the motion of a three-dimensional harmonic oscillator is reduced to three equations of motion of one-dimensional harmonic oscillators.

4.3.2 A one-dimensional harmonic oscillator

Let us consider now the one-dimensional motion of an electron along the x-axis in a parabolic potential well $U(x) = \beta x^2/2$. This motion is described by Eq. (4.73), which for $\alpha = x$ takes the form

$$\left[\frac{\mathrm{d}^2}{\mathrm{d}x^2}+\frac{2m_e}{\hbar^2}\left(E-\frac{\beta}{2}x^2\right)\right]\psi(x)=0. \tag{4.74}$$

Here, the indices of ψ_x, E_x, and β_x are omitted for convenience. The wavefunction, $\psi(x)$, and its derivative, $\mathrm{d}\psi(x)/\mathrm{d}x$, must be finite and continuous for all values of x. For $x \to \pm\infty$ the electron potential energy tends to infinity. As a result, the wavefunction $\psi(x)$ at large distances from the equilibrium position tends to zero. Equation (4.74) can be rewritten by introducing dimensionless variable $\xi = x/x_0 = x\sqrt{m_e\omega/\hbar}$:

$$\frac{\mathrm{d}^2\psi(\xi)}{\mathrm{d}\xi^2}+\left(\frac{2E}{\hbar\omega}-\xi^2\right)\psi(\xi)=0. \tag{4.75}$$

Here, $x_0 = \sqrt{\hbar/m_e\omega}$ is the turning point for a classical oscillator in its ground state $E_0 = \hbar\omega/2$. Let us introduce into Eq. (4.75) the dimensionless parameter

$$\lambda = \frac{2E}{\hbar\omega}. \tag{4.76}$$

Then, Eq. (4.75) takes the form

$$\frac{d^2\psi(\xi)}{d\xi^2} + (\lambda - \xi^2)\psi(\xi) = 0. \tag{4.77}$$

We will seek the solution of this equation in the form of the following product of two functions:

$$\psi(\xi) = f(\xi)e^{-\xi^2/2}. \tag{4.78}$$

The unknown function $f(\xi)$ must behave in such a way that as $\xi \to \infty$ the wavefunction $\psi(\xi)$ is bounded. By substituting Eq. (4.78) into Eq. (4.77) we derive the following equation for the function $f(\xi)$:

$$\frac{d^2 f(\xi)}{d\xi^2} - 2\xi\frac{df(\xi)}{d\xi} + (\lambda - 1)f(\xi) = 0. \tag{4.79}$$

The solution of this differential equation can be found in the form of the following power series:

$$f(\xi) = \sum_{k=0} a_k\xi^k. \tag{4.80}$$

Let us find the first and second derivatives of the function $f(\xi)$:

$$\frac{df(\xi)}{d\xi} = \sum_{k=0} k a_k \xi^{k-1}, \tag{4.81}$$

$$\frac{d^2 f(\xi)}{d\xi^2} = \sum_{k=0} k(k-1)a_k\xi^{k-2}. \tag{4.82}$$

Let us substitute Eqs. (4.80)–(4.82) into Eq. (4.79). As a result we obtain

$$\sum_{k=0} k(k-1)a_k\xi^{k-2} - 2\xi\sum_{k=0} k a_k\xi^{k-1} + (\lambda - 1)\sum_{k=0} a_k\xi^k = 0. \tag{4.83}$$

In order for a power series of the form $\sum c_k\xi^k$ to be identically equal to zero, all the coefficients c_k of the series must be equal to zero. Let us find from Eq. (4.83) the coefficient c_k before ξ^k and equate it to zero:

$$c_k = (k+1)(k+2)a_{k+2} - (2k + 1 - \lambda)a_k = 0. \tag{4.84}$$

As a result we obtain the following recurrence formula, which relates coefficients a_{k+2} and a_k:

$$a_{k+2} = \frac{2k + 1 - \lambda}{(k+1)(k+2)}a_k. \tag{4.85}$$

The wavefunction (4.78) will be limited as $\xi \to \infty$ if the power series (4.80) has a finite number of terms. To satisfy this condition the power series (4.80) must end at some value of k. For example, for $a_k \neq 0$ the coefficient a_{k+2} must be equal to zero. Then, according to Eq. (4.85) the coefficients after a_{k+2} must be equal to zero too. Thus, the function $f(\xi)$ reduces to a kth-order polynomial. The condition $a_{k+2} = 0$, taking into account Eq. (4.85), gives us the expression for the nth energy level, E_n:

$$\lambda = 2n + 1, \tag{4.86}$$

or

$$\frac{2E_n}{\hbar\omega} = 2n + 1, \tag{4.87}$$

which we can rewrite as

$$E_n = \hbar\omega\left(n + \frac{1}{2}\right), \tag{4.88}$$

where the quantum number, n, has the values $n = 0, 1, 2, \ldots$ The quantum number n defines the energy of oscillatory motion of a quantum oscillator and it is called the *oscillatory quantum number*.

The nth-order polynomial, $f(\xi)$, is one of the so-called *Hermite polynomials*, which are usually denoted as $H_n(\xi)$. Then, the normalized wavefunction of a one-dimensional quantum oscillator in the nth energy state can be written as

$$\psi_n(\xi) = A_n f_n(\xi)e^{-\xi^2/2} = A_n H_n(\xi)e^{-\xi^2/2}. \tag{4.89}$$

The normalization constant A_n in this case depends on the quantum number n and is equal to

$$A_n = \frac{1}{\sqrt{2^n n!}}\left(\frac{m_e\omega}{\pi\hbar}\right)^{1/4}. \tag{4.90}$$

The Hermite polynomials, $H_n(\xi)$, in a general case can be defined as follows:

$$H_n(\xi) = (-1)^n e^{\xi^2}\frac{d^n}{d\xi^n}\left(e^{-\xi^2}\right). \tag{4.91}$$

For the first five values of the quantum number n the Hermite polynomials have the following values:

$$\begin{aligned} &H_0(\xi) = 1, \quad H_1(\xi) = 2\xi, \quad H_2(\xi) = 4\xi^2 - 2, \\ &H_3(\xi) = 8\xi^3 - 12\xi, \quad H_4(\xi) = 16\xi^4 - 48\xi^2 + 12. \end{aligned} \tag{4.92}$$

According to Eq. (4.88) the electron spectrum in a one-dimensional parabolic potential well consists of equidistant energy levels. The distance between adjacent energy levels is

$$E_{n+1} - E_n = \hbar\omega. \tag{4.93}$$

The energy of the electron ground state, which corresponds to the quantum number $n = 0$, is the lowest and equal to

$$E_0 = \frac{\hbar\omega}{2}. \tag{4.94}$$

Since $E_0 > 0$, then the lowest energy state is not the state at rest. This distinguishes the quantum oscillator from its classical counterpart and it is a direct consequence of Heisenberg's uncertainty principle. Indeed, the uncertainty of the electron coordinate in the ground energy state is of the order of the maximum value of the electron coordinate, i.e., $\Delta x \approx x_{\text{max}} = A_0$. At the turning point where $x = A_0$ the kinetic energy equals zero and the total energy coincides with the potential energy, which, in accordance with Eqs. (4.66) and (4.70), is

$$E = \frac{m_e \omega^2 A_0^2}{2}. \tag{4.95}$$

The potential energy becomes zero at $x = 0$ and the total energy is equal to the kinetic energy $E = p_{\text{max}}^2/(2m_e)$. Thus

$$p_{\text{max}} = m_e \omega A_0. \tag{4.96}$$

Thus, the uncertainty of the momentum is of the order of the maximum value of the momentum, i.e., $\Delta p \approx p_{\text{max}} = m_e \omega A_0$. If we write the uncertainty relation as $\Delta x \; \Delta p \geq \hbar$, then from the expression

$$\Delta x \; \Delta p \geq m_e \omega A_0^2 \geq \hbar, \tag{4.97}$$

we obtain the condition for the amplitude A_0:

$$A_0^2 \geq \frac{\hbar}{m_e \omega}. \tag{4.98}$$

Taking into account the inequality (4.98) and (4.95), we come to the expression

$$E \geq \frac{\hbar\omega}{2}, \tag{4.99}$$

which gives us the minimum energy of the quantum oscillator in the parabolic potential well:

$$E_{\text{min}} = E_0 = \frac{\hbar\omega}{2}. \tag{4.100}$$

The oscillations of a quantum oscillator for the lowest energy state are called *zeroth oscillations*. They have purely quantum origin and are not connected with the thermal energy of an electron. Such oscillations in real systems exist at temperatures even close to absolute zero.

The forms of the electron wavefunctions for the first three lowest energy levels are shown in Fig. 4.2. For the classical oscillator, which has a fixed amplitude of oscillation, the increase of energy of oscillations is connected according to Eq. (4.95) with the increase in amplitude, A_0. The interval within which the particle is allowed to move is $(-A_0, A_0)$. For the quantum oscillator the wavefunctions are not equal to zero outside of this interval. This fact demonstrates that there is a certain probability of finding the electron outside of the interval $(-A_0, A_0)$, where motion is forbidden classically. The distribution of the probability density of the particle's location in the ground state $(n = 0)$, which is characterized by $|\psi_0|^2$, is shown in Fig. 4.3.

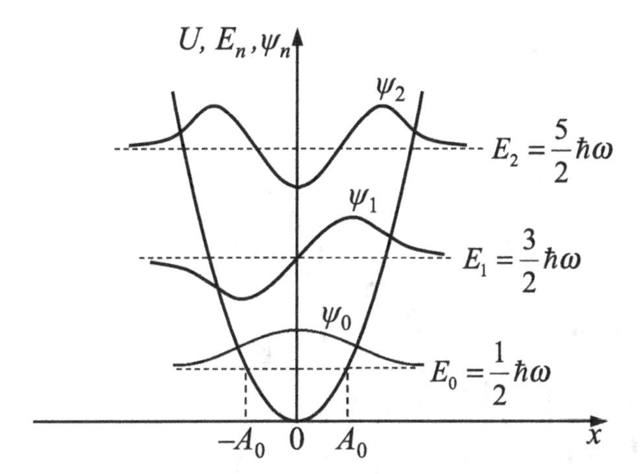

Figure 4.2 A quantum harmonic oscillator: the three lowest levels and corresponding wavefunctions are shown schematically.

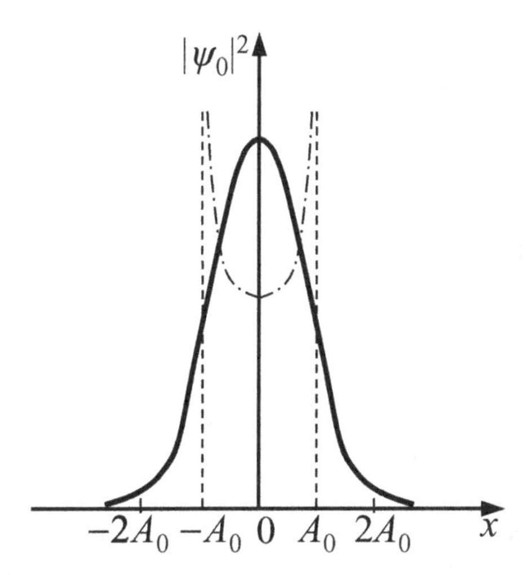

Figure 4.3 The probability density distribution, $|\psi_0|^2$, for the ground energy state, E_0. Classical turning points are located at $x = -A_0$ and $x = A_0$. The dash–dotted line shows the classical probability distribution of a harmonic oscillator with the same energy.

At high values of the quantum number, n, the behavior of a quantum oscillator increasingly resembles the behavior of the classical oscillator, for which the probability density changes smoothly from a minimum at $x = 0$ to infinity at the turning points, $-A_n$ and A_n (Fig. 4.4). Note that for a parabolic potential $A_n > A_0$ for $n > 0$. The behavior of the quantum oscillator becomes classical if its oscillation amplitude A_n becomes much greater than the amplitude of the zeroth oscillation, $A_0 = \sqrt{\hbar/(m_e\omega)}$:

$$A_n \gg \sqrt{\frac{\hbar}{m_e\omega}}. \tag{4.101}$$

Taking into account that the de Broglie wavelength for the maximum momentum on the nth level, $p_{\max} = m_e\omega A_n$, is equal to $\lambda_{\mathrm{Br}} = \hbar/p = \hbar/(m_e\omega A_n)$, we see

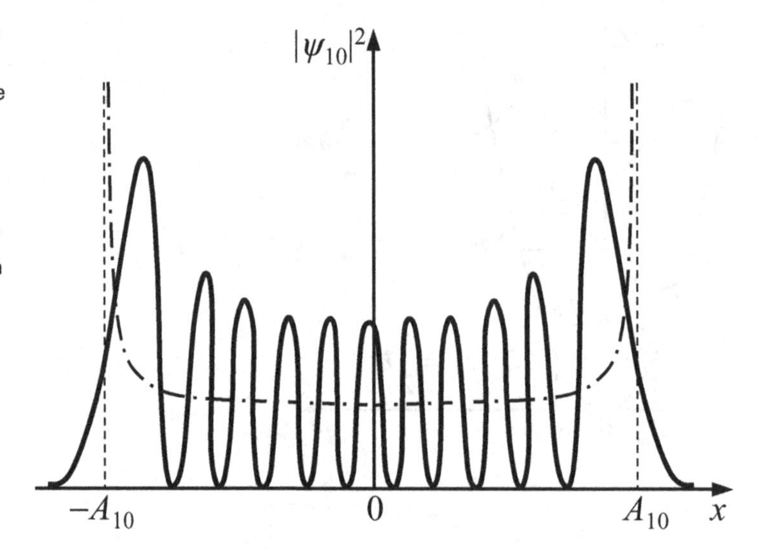

Figure 4.4 The probability density distribution, $|\psi_{10}|^2$, for the energy state with the principal quantum number $n = 10$. The dash–dotted line shows the classical probability distribution of a harmonic oscillator with the same energy (compare it with the one in Fig. A.8).

that Eq. (4.101) satisfies the condition that A_n is greater than λ_{Br}, i.e., an oscillator with large amplitude (4.101) is always classical.

4.3.3 Two-dimensional and three-dimensional harmonic oscillators

A two-dimensional harmonic oscillator

In the case of an isotropic two-dimensional quantum oscillator whose motion takes place in the xy-plane the harmonic coefficients are as follows: $\beta_x = \beta_y = \beta$ and $\beta_z = 0$. The Schrödinger equations (4.73) for the x- and y-directions are exactly the same as Eq. (4.74). The eigenvalue, E, is a sum of two energies, E_x and E_y. The wavefunctions which are the solutions of these two differential equations have the same form as in Eq. (4.89) and the energies E_x and E_y are given by Eq. (4.88). As a result, the energy eigenvalues are

$$E = \hbar\omega(N + 1), \tag{4.102}$$

where the angular frequency of the isotropic two-dimensional harmonic oscillator is $\omega = \sqrt{\beta/m_e}$, and the quantum number N is the sum of two quantum numbers, n_x and n_y: $N = n_x + n_y$. The quantum numbers n_x and n_y are positive integer numbers $0, 1, 2, \ldots$ The energy level $E = E_N$ with a given number $N = n_x + n_y$ corresponds to $N + 1$ degenerate wavefunctions. It follows that at a given value of N the corresponding quantum number takes the values $n_x = 0, 1, \ldots, N$ and $n_y = N - n_x$. Because the energy level E_N corresponds to $N + 1$ different wavefunctions $\psi(x, y)$, the order of degeneracy is

$$g_N = N + 1. \tag{4.103}$$

Note that we did not consider the solution of Eq. (4.74) for the z-coordinate since $\beta_z = 0$ corresponds to the case of a free particle, for which the solution was found earlier.

A three-dimensional harmonic oscillator

The harmonic coefficients of the three-dimensional spherical oscillator in the x-, y-, and z-directions are equal to each other:

$$\beta_x = \beta_y = \beta_z = \beta. \tag{4.104}$$

In this case the solutions of the Schrödinger equations (4.73) give

$$E = \hbar\omega\left(N + \frac{3}{2}\right), \tag{4.105}$$

where $N = n_x + n_y + n_z$, the frequency is $\omega = \sqrt{\beta/m_e}$, and each of the wavefunctions $\psi_\alpha(r_\alpha)$ in Eq. (4.72) has the form (4.89). The quantum numbers n_α are positive integer numbers $0, 1, 2, \ldots$ The order of degeneracy of the Nth energy level is equal to (see Example 4.4 later)

$$g_N = \frac{(N+1)(N+2)}{2}. \tag{4.106}$$

Thus, for a three-dimensional harmonic oscillator the ground state with $N = 0$ and $E_0 = 3\hbar\omega/2$ is non-degenerate (g_N for $N = 0$ is equal to unity). The energy state with $N = 1$ and $E_1 = 5\hbar\omega/2$ corresponds to three wavefunctions with the set of quantum numbers n_x, n_y, and n_z equal to (1, 0, 0), (0, 1, 0), and (0, 0, 1). Therefore, the order of degeneracy, g_1, is equal to three. The energy level $E_2 = 7\hbar\omega/2$ has the order of degeneracy g_2 equal to six (the quantum numbers are (1, 1, 0), (1, 0, 1), (0, 1, 1), (2, 0, 0), (0, 2, 0), and (0, 0, 2)), and so on.

The degeneracy of the energy levels is connected with the spherical symmetry of the oscillator. Indeed, for $\beta_x = \beta_y = \beta_z$ the three terms of the potential energy in Eq. (4.71) are equal to $U(r) = \beta r^2/2$. The Schrödinger equation for $U(r)$ in the spherical coordinate system has the following form:

$$\psi_{nlm}(r, \theta, \varphi) = X_{nl}(r)Y_{lm}(\theta, \varphi), \tag{4.107}$$

where the spherical harmonic functions $Y_{lm}(\theta, \varphi)$ are defined by Eq. (4.37). The quantum state of an electron in a spherically-symmetric potential, as we have already noted, is defined by a set of three quantum numbers (n, l, m): the principal, n, orbital, l, and orbital magnetic, m, quantum numbers. The Nth energy level (4.105) is defined by a certain combination of quantum numbers n and l: $N = 2n + l$, where n and l are positive integer numbers $0, 1, 2, \ldots$ Since at $N = 0$ the quantum numbers are $n = 0$ and $l = 0$, the ground energy state, E_0, of a spherical oscillator is non-degenerate. At $N = 1$ there can be three sets of quantum numbers (n, l, m): (0, 1, 1), (0, 1, 0), and (0, 1, −1). Therefore, the corresponding energy level is triply degenerate. The energy level $N = 2$

corresponds to six states with quantum numbers (1, 0, 0) and (0, 2, m), where $m = -2, -1, 0, 1, 2$.

Example 4.2. An electron is confined in a one-dimensional potential well, $U(x)$. The wavefunction in the stationary state is defined as

$$\psi(x) = C\mathrm{e}^{-\gamma x^2}, \tag{4.108}$$

where C and γ are constants and $\gamma > 0$. Find the energy of the electron, E, and the form of the function $U(x)$, if $U(0) = 0$.
Reasoning. Let us find the second derivative of the function $\psi(x)$:

$$\frac{\mathrm{d}^2\psi}{\mathrm{d}x^2} = 2C\gamma\mathrm{e}^{-\gamma x^2}(2\gamma x^2 - 1). \tag{4.109}$$

Let us write the one-dimensional Schrödinger equation taking into account the forms of the functions $\psi(x)$ and $\mathrm{d}^2\psi/\mathrm{d}x^2$:

$$\left[4\gamma^2x^2 - 2\gamma + \frac{2m_\mathrm{e}}{\hbar^2}(E - U)\right] C\mathrm{e}^{-\gamma x^2} = 0. \tag{4.110}$$

Let us write this equation for $x = 0$ and take into account $U(0) = 0$. As a result we get

$$E = \frac{\gamma\hbar^2}{m_\mathrm{e}}. \tag{4.111}$$

By substituting this expression for the energy, E, in Eq. (4.110) we find the expression for the potential energy $U(x)$:

$$U(x) = \frac{2\gamma^2\hbar^2}{m_\mathrm{e}}x^2. \tag{4.112}$$

On comparing the given wavefunction, $\psi(x)$, and the corresponding potential, $U(x)$, with Eqs. (4.89) and (4.66) we conclude that the electron is a quantum harmonic oscillator.

Example 4.3. For a one-dimensional quantum harmonic oscillator in its ground state calculate the probability of finding the particle outside of the classically allowed region, i.e., $|x| > x_0$, where x_0 is the amplitude of oscillations of the classical oscillator.
Reasoning. According to Eq. (4.88) the energy of a one-dimensional oscillator in the ground state is equal to $E_0 = \hbar\omega/2$. The wavefunction of this state, $\psi_0(x)$, according to Eq. (4.89) has the form

$$\psi_0(x) = \left(\frac{b}{\pi}\right)^{1/4} \mathrm{e}^{-bx^2/2}, \tag{4.113}$$

with $b = m\omega/\hbar$, where m is the mass of oscillating particle and ω is the frequency of the oscillations. The maximum displacement of the classical oscillator from its equilibrium position is defined by the equality of its kinetic energy to zero at the turning points (the points where the direction of the particle's motion changes to the opposite). As we indicated when discussing Eqs. (4.95) and (4.96), the kinetic energy at $x = 0$ is equal to the potential energy at $x = x_0$, i.e.,

$$\frac{\hbar\omega}{2} = \frac{m_e\omega^2 x_0^2}{2}, \tag{4.114}$$

$$x_0 = \sqrt{\frac{\hbar}{m_e\omega}}. \tag{4.115}$$

It follows from Eq. (4.115) that the parameter $b = 1/x_0^2$.

The probability of finding the oscillating particle inside of the potential well $|x| \leq x_0$ is as follows:

$$P_{\text{inside}} = \int_{-x_0}^{x_0} |\psi_0(x)|^2 \, dx = \left(\frac{b}{\pi}\right)^{1/2} \int_{-x_0}^{x_0} e^{-bx^2} \, dx. \tag{4.116}$$

Let us substitute the variable in this integral:

$$\sqrt{b}x = u, \qquad dx = \frac{du}{\sqrt{b}}. \tag{4.117}$$

As a result we arrive at the integral

$$P_{\text{inside}} = \frac{1}{\sqrt{\pi}} \int_{-1}^{1} e^{-u^2} \, du = \frac{2}{\sqrt{\pi}} \int_{0}^{1} e^{-u^2} \, du. \tag{4.118}$$

The integral in Eq. (4.118) is a probability integral, J, which in general form is defined as

$$J(\xi) = \frac{2}{\sqrt{\pi}} \int_{0}^{\xi} e^{-u^2} \, du. \tag{4.119}$$

Its values for various magnitudes of the upper limit of integration ξ are given in numerous handbooks. The particular value for $\xi = 1$ is $J(1) = 0.843$. Thus, $P_{\text{inside}} = 0.843$, and the probability of finding the particle in the ground state outside of the potential well is equal to

$$P_{\text{outside}} = 1 - P_{\text{inside}} = 0.157. \tag{4.120}$$

Example 4.4. Find the order of degeneracy of the Nth electron energy level in a spherically-symmetric parabolic potential well with the energy spectrum defined by Eq. (4.105).

Reasoning. For the given value of N the order of degeneracy is equal to the number of possible permutations of the three quantum numbers n_x, n_y, and n_z, whose sum is equal to N. Let us find the number of such permutations for the given number n_z. The number n_y will change within the range from 0 to $N - n_z + 1$ (the number n_x has the same values in the opposite order), i.e.,

$$g_N(n_z) = N - n_z + 1. \tag{4.121}$$

After summation of this expression over the possible values of n_z we find the order of degeneracy of the Nth energy level, i.e.,

$$\begin{aligned} g_N &= \sum_{n_z} g_N(n_z) = \sum_{n_z=0}^{N} (N - n_z + 1) \\ &= (N+1) + (N+1-1) + \cdots + (N+1-N) \\ &= (N+1) + N + (N-1) + (N-2) + \cdots + 2 + 1. \end{aligned} \tag{4.122}$$

Here we can use the formula for the sum, S_N, of an arithmetic sequence:

$$S_k = a_1 + (a_1 + d) + (a_1 + 2d) + \cdots + [a_1 + (k-1)d] = \frac{k(a_1 + a_k)}{2}. \tag{4.123}$$

For the arithmetic sequence (4.122) $d = -1$ and $k = N + 1$. Thus, the sum (4.122), g_N, is equal to

$$g_N = \frac{(N+1)(N+2)}{2}. \tag{4.124}$$

4.4 Phonons

4.4.1 Phonon gas

We show that in a one-dimensional crystalline lattice during the excitation of the wave motion the total mechanical energy can be represented as a sum of energies of independent harmonic oscillations (normal modes), which differ from each other by the values of the wavenumber q, i.e.,

$$E = \sum_q E_q. \tag{4.125}$$

At the same time the energy of an individual normal mode, expressed in terms of normal coordinates x_q and corresponding momenta p_q, is equal to

$$E_q = \frac{p_q^2}{2m} + \frac{m\omega_q^2 x_q^2}{2}. \tag{4.126}$$

The quantum-mechanical description of each normal mode can be considered as a set of quantum harmonic oscillators, whose energy is quantized and is equal to

$$E_q = \hbar\omega_q \left(n_q + \frac{1}{2} \right). \tag{4.127}$$

We see that the quantum oscillator can change its energy only by an amount equal to $\Delta E_q = \hbar\omega_q \, \Delta n_q$. In accord with the selection rules for the quantum transitions of a quantum oscillator, the quantum number n_q can change only by unity, i.e.,

$$\Delta n_q = \pm 1. \tag{4.128}$$

If $\Delta n_q = -1$, the crystalline lattice makes a transition to one of the allowed states, which has a lower energy level. The energy $\hbar\omega_q$ is transferred to the current carriers or to the surrounding medium. For $\Delta n_q = +1$, the crystalline lattice makes a transition to one of the allowed states, which has a higher energy level. Each individual quantum of energy, $\hbar\omega_q$, corresponds to an elementary wave excitation of the lattice, i.e., it corresponds to a quasiparticle with definite energy and momentum, which is called a *phonon*. The energy of such a quasiparticle is equal to

$$\varepsilon_q = \hbar\omega_q, \tag{4.129}$$

and this is related to its momentum by:

$$\mathbf{p}_q = \hbar\mathbf{q}. \tag{4.130}$$

The corresponding quantum transition of an oscillating lattice from one energy state to another is accompanied by the emission (*creation*) or absorption (*annihilation*) of a phonon. The transition from the set of normal modes to the set of wave excitations allows us to utilize the main idea of quantum mechanics – wave–particle duality. In this case we have a manifestation of the duality of quasiparticles (phonons) and wave excitations in a crystal. The velocity of such a quasiparticle is defined by the group velocity of the wave motion of the lattice, i.e.,

$$v_{\text{gr}} = \frac{\partial\omega_q}{\partial k_q} = \frac{\partial\varepsilon_q}{\partial p_q}. \tag{4.131}$$

During the interaction of other particles or quasiparticles with lattice oscillations their energy can change only by $\pm\hbar\omega_q$ and their momentum by $\hbar\mathbf{q}$. For this reason, for the interaction of phonons with electrons, i.e., for the description of the processes of creation and annihilation of phonons, it is convenient to use the laws of conservation of energy and momentum of the particles. Let us note that the phonon's momentum is usually called *quasimomentum*. This is because the quantities $\hbar\mathbf{q}$ and $\hbar(\mathbf{q} \pm 2\pi\mathbf{b}_g)$ are physically equivalent. Here $\mathbf{b}_g$ is the unit wavevector in the momentum space, which will be discussed in Chapter 7. Phonons can also interact with each other. This kind of interaction explains many of the thermal phenomena which take place in crystals. It is necessary to remember that in the processes of scattering of phonons by each other, during which an exchange of energy and momentum takes place, the number of phonons must be at least three.

Only those normal modes of a crystal which are excited higher than the crystal's lowest energy level

$$E_0 = \sum_q \frac{\hbar\omega_q}{2} \tag{4.132}$$

can interact with electrons or with each other. Namely, the excitations above the zero oscillations of Eq. (4.132) are called *phonons*. Zeroth oscillations of a normal mode establish a constant energy background, which during scattering of phonons by each other and by electrons does not participate in scattering processes. The total number of phonons in a crystal is defined by their sum over all normal modes of a crystal (i.e., over all quantum oscillators), i.e.,

$$N = \sum_q n_q. \tag{4.133}$$

During interactions the number of phonons changes and, in contrast to the case of a molecular gas, phonons can be created or annihilated. The energy state of a crystal is defined by the total number of phonons, which as a whole can be considered as a *phonon gas*. The energy of a phonon gas is defined by the energy excitations above the thermal energy background E_0, i.e.,

$$E_{\text{ph.gas}} = \sum_q \hbar\omega_q n_q. \tag{4.134}$$

A very important feature of a phonon gas is the dependence of the number of phonons on temperature. At thermodynamical equilibrium (at temperature T) the average number of phonons with wavenumber q in the unit volume of a crystal is given by the expression

$$\langle n_q(T)\rangle = \frac{1}{e^{\hbar\omega_q/(k_B T)} - 1}, \tag{4.135}$$

which coincides with the analogous expression (2.41) for the number of photons in the unit volume of a cavity filled by equilibrium thermal radiation. From Eq. (4.135) it follows that the number of phonons with energy $\varepsilon_q = \hbar\omega_q$ at high temperatures ($T \gg \hbar\omega_q/k_B$) is defined as

$$\langle n_q\rangle \approx \frac{k_B T}{\hbar\omega_q}, \tag{4.136}$$

whereas at low temperatures ($T \ll \hbar\omega_q/k_B$) it is defined as

$$\langle n_q\rangle \approx e^{-\hbar\omega_q/(k_B T)}. \tag{4.137}$$

Thus, at high temperatures, such that the thermal energy $k_B T$ is sufficiently high to excite a large number of modes with energy $\hbar\omega_q$, the number of phonons in a crystal is large. At low temperatures, for which the thermal energy $k_B T$ is not sufficiently high to excite lattice oscillations, i.e., $k_B T \ll \hbar\omega_q$, the number of phonons in a crystal is exponentially small, since $\langle n_q\rangle \approx e^{-\hbar\omega_q/(k_B T)} \ll 1$. Thus, at low temperatures the phonon gas can be considered an ideal gas, i.e., a gas of phonons that do not interact with each other. At high temperatures, on the other hand, because of the high concentration of phonons the interaction between phonons becomes more pronounced. In order to take into account the interaction between phonons it is necessary to take into consideration the anharmonicity in the potential energy, i.e., it is necessary to include the term proportional to the third power of the particle's displacement from its equilibrium position. The anharmonicity is responsible for the three-phonon interaction.

The temperature is considered high or low in comparison with the Debye temperature of the given material, T_D, which is defined as:

$$T_D = \frac{\hbar \omega_{max}}{k_B}. \quad (4.138)$$

Here the limiting frequency of oscillations is related to the parameters of the crystalline lattice by

$$\omega_{max} = \langle v_0 \rangle \left(\frac{6\pi^2 N}{V} \right)^{1/3}, \quad (4.139)$$

where $\langle v_0 \rangle$ is the mean velocity for longitudinal and transverse acoustic waves in the crystal. The Debye temperature, T_D, divides the entire temperature scale into two regions. For temperatures $T \ll T_D$ only long-wavelength waves, whose energy is small in comparison with $k_B T_D$, are excited. For $T > T_D$ all oscillations are excited, including the oscillations with the maximum frequency, whose energy is about $k_B T_D$.

4.4.2 Thermal properties

The concepts of thermal oscillations of a lattice as a phonon gas, which obeys quantum-mechanical laws, allow us to explain such very important and easily measured thermal phenomena as heat capacity and thermal conductivity, which we will briefly consider here.

Heat capacity

It is well established experimentally that at $T \to 0$ the heat capacity of all solids tends to zero. From the classical point of view this experimental fact cannot be explained. Moreover, the dependence of thermal conductivity on temperature cannot be explained either. The Dulong–Petit law according to which the molar heat capacity of simple solids does not depend on temperature and is equal to $3R$ is based on the classical concepts. In the framework of the phonon-gas model the heat capacity of a solid practically does not depend on temperature for $T > T_D$ and substantially decreases with decreasing temperature for $T < T_D$. The observed decrease of the heat capacity at low temperatures is due to a simple reason: the energy which corresponds to one oscillatory degree of freedom is too low to excite high-frequency phonons. Thus, the temperature for which the Dulong–Petit law is valid must be higher than the characteristic temperature T_D for a given material. The values of the Debye temperature for several crystalline materials are given in Table 4.2.

Figure 4.5 shows the dependence of the molar heat capacity, c, of simple crystalline materials on the dimensionless temperature T/T_D. This type of the dependence $c = f(T/T_D)$ is characteristic for most crystals.

Table 4.2. *The Debye temperature, T_D, for several materials*

Material	T_D (K)	Material	T_D (K)
Beryllium	1160	Zinc	308
Magnesium	406	Aluminum	418
Iron	467	Diamond	2000
Platinum	229	Silicon	658
Silver	225	Germanium	366
Gold	165	Lead	94

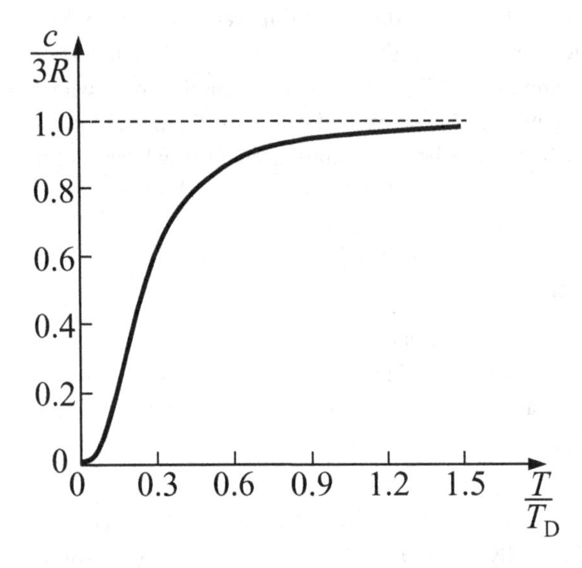

Figure 4.5 The dependence of the molar heat capacity, c, on the dimensionless temperature T/T_D.

By considering the problem of internal energy of crystals from the quantum-mechanical point of view, Debye proved that at temperatures close to absolute zero the internal energy is proportional to the fourth power of temperature:

$$E = \frac{3\pi^4 R}{5T_D^3} T^4, \tag{4.140}$$

where R is the universal gas constant. From this relation we can obtain the expression for the molar heat capacity of a simple (i.e., consisting of one type of atom) crystal at low temperatures:

$$c = \frac{dE}{dT} = \frac{12\pi^4 R}{5T_D^3} T^3. \tag{4.141}$$

Therefore, near absolute zero the heat capacity of a crystal is proportional to the third power of temperature. The region of such dependence covers the entire temperature interval from 0 to temperatures close to $0.1T_D$. At higher temperatures (up to $T = T_D$) there is an intermediate region. In this region the dependence changes from the law (4.141) to the Dulong–Petit law, which becomes valid at

$T > T_D$. Note that a small increase of heat capacity at $T > T_D$ and a deviation from the Dulong–Petit law are due to the high concentration of phonons and the fact that a phonon gas is not ideal. This means that at high temperatures it is necessary to take into account phonon–phonon interactions, which are absent in an ideal phonon gas.

Thermal conductivity

The phenomenon of thermal conductivity in crystals is related to phonon–phonon interactions, which lead to a scattering of phonons on phonons. If the two opposite ends of a crystalline specimen are kept at different temperatures, a continuous flux of heat occurs in this specimen. Quantitatively the heat flux, $\mathrm{d}Q/\mathrm{d}t$, through the cross-section of the rod S during unit time at a temperature gradient of $\mathrm{d}T/\mathrm{d}x$ is defined by the expression

$$\frac{\mathrm{d}Q}{\mathrm{d}t} = -\lambda \frac{\mathrm{d}T}{\mathrm{d}x} S, \tag{4.142}$$

where λ is the coefficient of thermal conductivity. In defining the coefficient of thermal conductivity it is necessary to take into account the quantum nature of the excitation of heat waves in a crystalline lattice. Let us imagine the excited state of a lattice as a gas of phonons, which can transfer energy in the direction opposite to the temperature gradient. Let us choose as a characteristic length in the x-direction the distance equal to the phonon mean-free-path length, l. The temperature difference in the given interval $(0, l)$ is defined as

$$\Delta T = \frac{\mathrm{d}T}{\mathrm{d}x} l. \tag{4.143}$$

For the resultant flux of energy created by phonons we obtain

$$\frac{\mathrm{d}Q}{\mathrm{d}t} = -\frac{1}{3} c \langle v_0 \rangle l \frac{\mathrm{d}T}{\mathrm{d}x} S, \tag{4.144}$$

where c is the heat capacity per unit volume of the phonon gas. Thus, the coefficient of thermal conductivity, λ, of a crystalline lattice is defined by the expression

$$\lambda = \frac{1}{3} c \langle v_0 \rangle l. \tag{4.145}$$

Deeper analysis shows that at sufficiently high temperatures the mean-free-path length of the phonon is inversely proportional to temperature. Therefore, the coefficient of thermal conductivity of the most non-conducting crystals at $T > T_D$ is inversely proportional to temperature since at $T > T_D$ the heat capacity of the phonon gas does not depend on temperature.

In pure crystals at temperatures close to absolute zero the dependence of the coefficient of thermal conductivity on the dimensions of the specimen becomes more pronounced. At low temperatures the concentration of phonons is negligible. Therefore, the probability of phonons scattering on phonons is small and their mean-free-path length, l, is limited by the dimensions of the specimen, L. Thus, in the interval of temperatures being considered here the coefficient of thermal conductivity is defined as

$$\lambda = \frac{1}{3} c \langle v_0 \rangle L. \tag{4.146}$$

4.4.3 The analogy between phonons and photons

Note that phonons and photons, which we discussed in Chapter 2, share a number of physical properties. If the electron can be considered as a harmonic oscillator, then during quantum transitions between adjacent energy levels a photon (quantum of electromagnetic field) is emitted or absorbed. Photon energy and momentum are defined by the expressions (2.35) and (2.39), which coincide with the corresponding expressions for phonons (4.129) and (4.130).

Photons and phonons belong to the same class of particles, called *bosons* – the particles which possess integer spin (we will talk about the particle property called *spin* in Section 6.3). Therefore, the average number of phonons in any given energy state, E, at thermodynamical equilibrium is given by the same expression:

$$\langle n(T) \rangle = \frac{1}{e^{E/(k_B T)} - 1}. \tag{4.147}$$

The main difference between a phonon and a photon is that the phonon is a quasiparticle and its introduction as a notion is convenient for the mathematical description of physical processes. In contrast to photons, which are real particles and can exist in media as well as in vacuum, phonons can exist only in crystals.

Example 4.5. Estimate the number of phonons in a linear chain of atoms which corresponds to a single normal oscillator with wavenumber q, if the average energy of the oscillator is defined as

$$\langle \varepsilon_q \rangle = \gamma k_B T, \tag{4.148}$$

where the parameter $\gamma > 1$. Assume that the temperature, T, is high.
Reasoning. According to Eq. (4.135) the average number of phonons which corresponds to a single normal oscillator is defined by the average quantum number,

$$\langle n_q \rangle = \frac{1}{e^{\hbar \omega_q/(k_B T)} - 1}, \tag{4.149}$$

and by the average energy of a single oscillator,

$$\langle \varepsilon_q \rangle = \hbar \omega_q \left(\langle n_q \rangle + \frac{1}{2} \right). \tag{4.150}$$

Suppose that the magnitude $\hbar\omega_q/(k_B T) \ll 1$, then

$$e^{\hbar\omega_q/(k_B T)} \approx 1 + \frac{\hbar\omega_q}{k_B T}. \tag{4.151}$$

Taking this into account, we get

$$\langle n_q \rangle = \frac{k_B T}{\hbar\omega_q}, \tag{4.152}$$

and

$$\langle \varepsilon_q \rangle = \frac{\hbar\omega_q}{2} + k_B T. \tag{4.153}$$

From Eq. (4.153) we find the energy of a single phonon:

$$\hbar\omega_q = 2\left(\langle \varepsilon_q \rangle - k_B T\right). \tag{4.154}$$

On substituting Eq. (4.154) into Eq. (4.152) for the average number of phonons we get the expression

$$\langle n_q \rangle = \frac{k_B T}{2\left(\langle \varepsilon_q \rangle - k_B T\right)}, \tag{4.155}$$

which in our case of relation (4.148) reduces to

$$\langle n_q \rangle = \frac{1}{2(\gamma - 1)}. \tag{4.156}$$

Note that for a high temperature ($\hbar\omega_q/(k_B T) \ll 1$) comparison of Eqs. (4.153) and (4.148) gives us $(\gamma - 1) \ll 1$. From Eqs. (4.152) and (4.156) we find that $\langle n_q \rangle \gg 1$.

4.5 Summary

1. Quantization of electron motion in two-dimensional or three-dimensional potential wells is defined by two or three quantum numbers. An energy level is degenerate if it corresponds to more than one set of quantum numbers (n_x, n_y) or (n_x, n_y, n_z).
2. In a square or cubic potential well, where $n_x = n_y$ or $n_x = n_y = n_z$, the energy levels are non-degenerate. All other levels are degenerate with the corresponding order of degeneracy.
3. In a spherically-symmetric potential well the bound-electron stationary states are defined by the set of three quantum numbers (n, l, m). The lack of dependence of the force field on angular variables allows presentation of the electron wavefunction in the form of a product of two functions – a radial function, which depends on the distance of the electron from the center of the well, and an angular function, which depends on the azimuthal and polar angles.
4. In a spherically-symmetric potential well for the electron motion the following quantities are conserved: the total energy, E, the square of the angular momentum $L^2 = \hbar^2 l(l+1)$, where l is the orbital quantum number, and the projection of angular momentum $L_z = \hbar m$, where m is the magnetic quantum number.

5. The peculiarities of the electron energy spectrum in a parabolic potential well, i.e., a harmonic oscillator, are that (a) the energy levels, separated by $\Delta E = \hbar\omega$, are equidistant; (b) the lowest energy level $E_0 = \hbar\omega/2$, which corresponds to the quantum number $n=0$, exists; and (c) there are so-called *zeroth oscillations*.
6. The energy of elastic waves in a crystalline lattice can be quantized and this energy can be considered as the energy of the normal modes, each of which consists of n_q phonons with energy $\hbar\omega_q$. Phonons are not real particles. They are members of the family of quasiparticles that do not possess mass. The total number of phonons is three times greater than the number of atoms in a crystal.
7. Phonons do not have momentum. Their motion in a crystal is described by a quasi-momentum $\mathbf{p} = \hbar\mathbf{q}$ (see Eq. (4.130)). Phonons possess integer spin and belong to the family of particles called bosons, and their statistics is defined by Eq. (4.135).

4.6 Problems

Problem 4.1. A free electron of mass m_e, moves in the direction defined by the polar angle θ and the azimuthal angle φ (see Fig. 4.6). Find the electron wavefunction $\Psi(\mathbf{r}, t)$ if the electron energy is equal to ε.

Problem 4.2. The dispersion equation Eq. (4.61) defines the allowed energy states of an electron in a spherically-symmetric potential well and in a general case it has to be solved numerically. Find the energy of stationary states in the limit of infinitely high potential barriers.

Problem 4.3. Find the energy levels and the wavefunctions of a positively charged harmonic oscillator in a homogeneous electric field **E**. The charge of the oscillating particle is e and its mass is m.

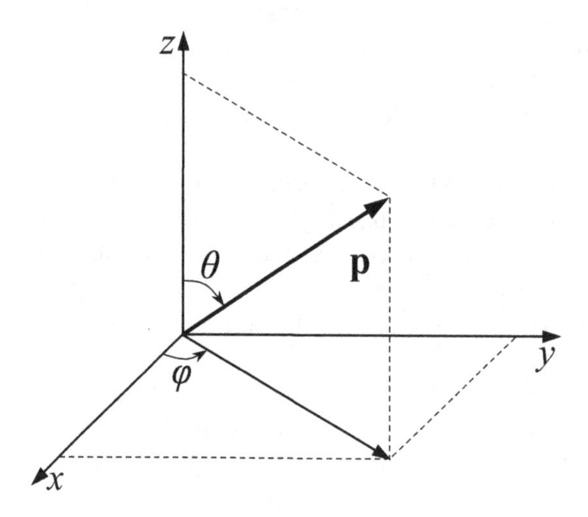

Figure 4.6 Momentum **p** in a spherical system of coordinates.

Problem 4.4. An electron is in its ground state in a cubic potential well with impenetrable barriers. Find the length, L, of the edge of the confining cube and the energy of the electron if the electron's highest probability density is equal to P_{m}.

Problem 4.5. An electron is in its ground state in a cubic potential well with impenetrable barriers and with edge length $L = 5$ nm. Estimate the pressure which the electron puts on the walls of the potential well. (Answer: $P = 0.35 \times 10^6$ Pa.)

Problem 4.6. At instant $t = 0$ the electron behavior is described by the following wavefunction:

$$\Psi(r, 0) = A\mathrm{e}^{-r^2/\alpha^2 + ikr}. \tag{4.157}$$

Find the normalization constant, A, the most probable value r_{pr}, and the radial part of the probability current, j.

Problem 4.7. An electron is in an infinitely deep spherically-symmetric potential well, i.e., $U(r) = 0$ at $r \leq r_0$ and $U(r) = \infty$ at $r > r_0$, where r_0 is the radius of the well. Find the form of the radial electron function, the electron energy in s-states ($l = 0$), the magnitudes $\langle r \rangle$ and $r_{\mathrm{sq}} = \sqrt{\langle r^2 \rangle}$, the most probable value r_{pr}, and the probability of the electron being in the region $r \leq r_{\mathrm{pr}}$.

Problem 4.8. For an electron in an infinitely deep spherically-symmetric potential well, i.e., $U(r) = 0$ at $r \leq r_0$ and $U(r) = \infty$ at $r > r_0$, where r_0 is the radius of the well, find the dispersion equation for s-states in the energy region $E < U_0$ and interval of values of $U_0 r_0^2$ when the well contains only one electron energy level.

Problem 4.9. The wavefunction of the ground state of a one-dimensional harmonic oscillator (particle with mass m in a one-dimensional parabolic potential $U(x) = m\omega^2 x^2/2$) has the following form:

$$\psi(x) = A\mathrm{e}^{-bx^2}. \tag{4.158}$$

Solve the Schrödinger equation and find the constant b and the energy of the oscillator in the ground state.

Problem 4.10. The wavefunction of the one-dimensional harmonic oscillator in a particular state has the following form:

$$\psi(x) = Ax\mathrm{e}^{-bx^2}. \tag{4.159}$$

Find the constant b and the energy of an oscillator in this state.

Problem 4.11. A particle with mass m moves in a three-dimensional parabolic potential $U(r) = m\omega^2 r^2/2$. Find the dependence of the oscillator's energy on the orbital quantum number.

Chapter 5
Approximate methods of finding quantum states

The solution of most problems associated with electron quantum states in physical systems and structures (atoms, molecules, quantum nanostructure objects, and crystals) is hard to find because of the mathematical difficulties of getting exact solutions of the Schrödinger equation. Therefore, approximate methods of solving such problems are of special interest. We will consider some of these methods, such as the *adiabatic approximation* now and later the *effective-mass method*, using real physical systems as examples. In this chapter we will consider several widely used approximation methods for finding the wavefunctions and energies of quantum states as well as the probabilities of transitions between quantum states. First of all, we will consider stationary and non-stationary perturbation theories. What is common to these two theories is that it is assumed that the perturbation is weak and that it changes negligibly the state of the unperturbed system. Stationary perturbation theory is used for the approximate description of a system's behavior if the Hamiltonian of the quantum system being considered does not directly depend on time. In the opposite case, non-stationary theory is used. Then, we will briefly consider the quasiclassical approximation, which is used for the problems of quantum mechanics which are close to analogous problems of classical mechanics.

5.1 Stationary perturbation theory for a system with non-degenerate states

This theory is used for the approximate calculation of the energy levels and the wavefunctions of stationary states of systems that are subjected to the influence of small perturbations. Let us assume that the Hamiltonian of the quantum system may be expressed in the following form:

$$\hat{H} = \hat{H}_0 + \hat{V}, \tag{5.1}$$

where $\hat{H}_0$ is the Hamiltonian of the unperturbed system and $\hat{V}$ is the perturbation operator. For the unperturbed Hamiltonian the solution of the Schrödinger equation,

$$\hat{H}_0 \psi_n^{(0)} = E_n^{(0)} \psi_n^{(0)}, \tag{5.2}$$

is assumed to be known, i.e., the complete set of eigenfunctions $\psi_n^{(0)}$ and eigenvalues $E_n^{(0)}$ of the unperturbed system is given. Let us assume, first, that the unperturbed Hamiltonian, $\hat{H}_0$, has a discrete spectrum and all states are non-degenerate. This means that each eigenvalue $E_n^{(0)}$ corresponds to only one eigenfunction $\psi_n^{(0)}$. The complete set of these functions is orthonormalized, i.e., unperturbed wavefunctions satisfy the condition (2.144):

$$\int \psi_n^{(0)*} \psi_l^{(0)} \, \mathrm{d}V = \delta_{nl}. \tag{5.3}$$

It is necessary to find the eigenvalues E_n and the eigenfunctions ψ_n of the operator $\hat{H}$, i.e., the solutions of the following Schrödinger equation:

$$\left(\hat{H}_0 + \hat{V}\right) \psi_n = E_n \psi_n. \tag{5.4}$$

This equation differs from Eq. (5.2) by the term $\hat{V}\psi_n$, which is assumed to be small in comparison with $\hat{H}_0\psi_n$, i.e., the state of the system must not change considerably as a result of perturbation.

Let us look for a solution of Eq. (5.4) in the form of a series expansion using the complete set of eigenfunctions of the unperturbed Hamiltonian, $\hat{H}_0$:

$$\psi_n = \sum_l C_l \psi_l^{(0)}. \tag{5.5}$$

Let us substitute this solution into Eq. (5.4) and use the fact that $\psi_l^{(0)}$ are the eigenfunctions of the unperturbed Hamiltonian, $\hat{H}_0$. As a result we get

$$\sum_l C_l \left(E_l^{(0)} + \hat{V}\right) \psi_l^{(0)} = \sum_l C_l E_n \psi_l^{(0)}. \tag{5.6}$$

Let us multiply both parts of this equation by $\psi_k^{(0)*}$ and integrate it over the entire volume available for the quantum system:

$$\sum_l C_l \left(E_n - E_l^{(0)}\right) \int \psi_k^{(0)*} \psi_l^{(0)} \, \mathrm{d}V = \sum_l C_l \int \psi_k^{(0)*} \hat{V} \psi_l^{(0)} \, \mathrm{d}V. \tag{5.7}$$

Taking into account the orthogonality condition (5.3), we arrive at the following expression:

$$\sum_l C_l \left(E_n - E_l^{(0)}\right) \delta_{kl} = \sum_l C_l V_{kl}, \tag{5.8}$$

where V_{kl} is the matrix element of the perturbation operator $\hat{V}$, in the basis of the unperturbed wavefunctions, $\psi_n^{(0)}$:

$$V_{kl} = \int \psi_k^{(0)*} \hat{V} \psi_l^{(0)} \, \mathrm{d}V. \tag{5.9}$$

The important feature of the Kronecker delta, δ_{kl}, is that it allows us to reduce the number of indices in the summation. Taking into account the property of the Kronecker delta, Eq. (5.8) reduces to

$$\left(E_n - E_k^{(0)}\right) C_k = \sum_l V_{kl} C_l. \tag{5.10}$$

Equation (5.10) constitutes a system of linear algebraic equations for the unknown coefficients C_l. This system is identical to the initial equation (5.4) since the set of coefficients C_l completely defines the wavefunction (5.5). Finding the exact solution of the system of Eqs. (5.10) is not much simpler than finding the solution of Eq. (5.4). However, the system of Eqs. (5.10) can be solved approximately using a series expansion over some small parameter. Let us present the perturbation operator, $\hat{V}$, in the form

$$\hat{V} = \epsilon \hat{W}. \tag{5.11}$$

Here, ϵ is a small dimensionless parameter and $\hat{W}$ is some perturbation operator. Using the fact that ϵ is small, we can present E_n and C_l in the form of a series over the small parameter ϵ:

$$E_n = E_n^{(0)} + \epsilon E_n^{(1)} + \epsilon^2 E_n^{(2)} + \cdots, \tag{5.12}$$

$$C_l = C_l^{(0)} + \epsilon C_l^{(1)} + \epsilon^2 C_l^{(2)} + \cdots. \tag{5.13}$$

Here the magnitudes $\epsilon E_n^{(1)}$ and $\epsilon C_l^{(1)}$ are of the same order as the perturbation $\hat{V}$, and the magnitudes $\epsilon^2 E_n^{(2)}$ and $\epsilon^2 C_l^{(2)}$ are of the second order of smallness, i.e., the order of smallness is $\hat{V}^2$, etc.

On substituting Eqs. (5.12) and (5.13) into (5.10) we get

$$\begin{aligned}&\left(E_n^{(0)} - E_k^{(0)}\right) C_k^{(0)} + \epsilon \left(E_n^{(0)} - E_k^{(0)}\right) C_k^{(1)} + \epsilon E_n^{(1)} C_k^{(0)} + \epsilon^2 E_n^{(2)} C_k^{(0)} + \cdots \\ &= \epsilon \sum_l W_{kl} C_l^{(0)} + \epsilon^2 \sum_l W_{kl} C_l^{(1)} + \cdots.\end{aligned} \tag{5.14}$$

Let us find the corrections to the energy $E_n^{(0)}$ and to the wavefunction $\psi_n^{(0)}$ for the nth quantum state of the system. Since in the zeroth approximation $\psi_n = \psi_n^{(0)}$, there is only one term in Eq. (5.14) that does not depend on ϵ, and it follows that $C_n^{(0)} = 1$ and all other coefficients are equal to zero, $C_k^{(0)} = 0$. Let us now consider separately equations for $k = n$ and $k \neq n$. For $k = n$ there is one term on the left-hand side and one term on the right-hand side of Eq. (5.14) of the first order of smallness, i.e., proportional to ϵ. As a result, in the first order of approximation the correction to the energy $E_n^{(0)}$ is defined as

$$\epsilon E_n^{(1)} = \epsilon W_{nn} = V_{nn} = \int \psi_n^{(0)*} \hat{V} \psi_n^{(0)} \, \mathrm{d}V = \langle V_n \rangle, \tag{5.15}$$

which is the average value of perturbation calculated using the corresponding unperturbed state. Let us note that V_{nn} are usually called the *diagonal matrix elements of the perturbation operator,* $\hat{V}$. For $k \neq n$ from Eq. (5.14) it follows that

$$\epsilon C_k^{(1)} = \frac{V_{nk}}{E_n^{(0)} - E_k^{(0)}}, \tag{5.16}$$

where the non-diagonal matrix elements of the perturbation operator, V_{nk}, are defined as

$$V_{nk} = \int \psi_n^{(0)*} \hat{V} \psi_k^{(0)} \, \mathrm{d}V, \tag{5.17}$$

and $C_n^{(1)} = 0$. Then, the corrections to the wavefunction in the first order of approximation are

$$\psi_n^{(1)} = \sum_{n \neq k} \frac{V_{nk}}{E_n^{(0)} - E_k^{(0)}} \psi_k^{(0)}. \tag{5.18}$$

To find the corrections to the energy for the second order of approximation, we equate the terms proportional to ϵ^2 on both sides of Eq. (5.14). As a result the correction $E_n^{(2)}$ is defined as

$$\epsilon^2 E_n^{(2)} = \sum_{n \neq k} \frac{|V_{nk}|^2}{E_n^{(0)} - E_k^{(0)}}. \tag{5.19}$$

As a rule, for the perturbed wavefunction it is sufficient to determine the first two terms of the expansion and for the energy, the first three terms of the expansion of these magnitudes. As a result the wavefunction and the energy of the nth perturbed state can be written as

$$\psi_n = \psi_n^{(0)} + \sum_{n \neq k} \frac{V_{nk}}{E_n^{(0)} - E_k^{(0)}} \psi_k^{(0)}, \tag{5.20}$$

$$E_n = E_n^{(0)} + \langle V_n \rangle + \sum_{n \neq k} \frac{|V_{nk}|^2}{E_n^{(0)} - E_k^{(0)}}. \tag{5.21}$$

These expressions are justified only when the next term of the expansion is considerably smaller than the previous one. This condition is satisfied if the matrix elements of the perturbation operator are considerably smaller than the distance between the unperturbed levels, i.e.,

$$|V_{nk}| \ll \left| E_n^{(0)} - E_k^{(0)} \right|, \quad n \neq k. \tag{5.22}$$

Example 5.1. An electron is confined in a one-dimensional potential well with infinite-height barriers of width L and is affected by a perturbation, which has the following coordinate dependence:

$$\hat{V}(x) = V_0 \cos^2\left(\frac{\pi x}{L}\right). \tag{5.23}$$

Find the change of electron energy levels in the potential well using the second-order approximation of the perturbation theory. Find the condition of applicability of the perturbation theory in the case considered here.

Reasoning. Let us find the matrix elements of the perturbation operator $\hat{V}(x)$ using the wavefunctions defined by Eq. (3.51),

$$\psi_n^{(0)}(x) = \sqrt{\frac{2}{L}} \sin\left(\frac{n\pi x}{L}\right),$$

of the unperturbed problem:

$$V_{nk} = \int_0^L \psi_n^{(0)*} \hat{V}(x) \psi_k^{(0)} \, dx = \frac{2V_0}{L} \int_0^L \cos^2\left(\frac{\pi x}{L}\right) \sin\left(\frac{n\pi x}{L}\right) \sin\left(\frac{k\pi x}{L}\right) dx. \tag{5.24}$$

As a result of integration we obtain for the matrix elements, V_{nk}, the following expression:

$$V_{nk} = \frac{V_0}{4} \begin{cases} 1, & n = k = 1, \\ 2, & n = k \neq 1, \\ 1, & n = k \pm 2, \\ 0, & n \text{ and } k \text{ are other than the above.} \end{cases} \tag{5.25}$$

The first-order correction with regard to energy is equal to $\epsilon E_n^{(1)} = V_{nn}$, and as follows from Eq. (5.25), in the case when $n = k = 1$ we get

$$\epsilon E_1^{(1)} = V_{11} = \frac{V_0}{4}. \tag{5.26}$$

In the case when $n = k \neq 1$ we get

$$\epsilon E_n^{(1)} = V_{nn} = \frac{V_0}{2}. \tag{5.27}$$

Taking into account Eq. (5.25), the second-order correction, defined according by Eq. (5.19), is equal to

$$\epsilon^2 E_n^{(2)} = \sum_{n \neq k} \frac{|V_{nk}|^2}{E_n^{(0)} - E_k^{(0)}} = -\frac{m_e L^2 V_0^2}{64\pi^2 \hbar^2} \begin{cases} 1, & n = 1, \\ 2/3, & n = 2, \\ -4/[(n+1)(n-1)], & n \geq 3, \end{cases} \tag{5.28}$$

where $E_n^{(0)}$ is defined by Eq. (3.44):

$$E_n^{(0)} = \frac{\pi^2 \hbar^2}{2m_e L^2} n^2.$$

In deriving Eq. (5.28) we took into account that for $n \neq k$ only V_{nk} with $n = k \pm 2$ are not equal to zero. The condition of applicability of the perturbation theory (5.22) in the current problem taking into account the unperturbed eigenvalues of energy (Eq. (3.44)) reduces to the following inequality:

$$V_0 \ll \frac{2\pi^2 \hbar^2}{m_e L^2} (n - 1). \tag{5.29}$$

5.2 Stationary perturbation theory for a system with degenerate states

Very often we deal with degenerate states of an unperturbed quantum system. Let us assume that the eigenvalues of the unperturbed Hamiltonian, $\hat{H}_0$, are degenerate and that the order of degeneracy of the nth energy level, $E_n^{(0)}$, is equal to s. In this case s orthogonal wavefunctions $\psi_{n1}^{(0)}, \ldots, \psi_{ns}^{(0)}$ correspond to a single eigenvalue $E_n^{(0)}$. The perturbation splits the degenerate energy level, $E_n^{(0)}$, into s closely spaced sublevels. Each of the sublevels is described by its own wavefunction, which is a linear combination of the unperturbed wavefunctions $\psi_{nk}^{(0)}$. Let us assume that the perturbation $\hat{V}$ is small. Therefore, we are expecting closely spaced energy levels in the first order of approximation of perturbation theory.

Let us consider the case of a doubly degenerate energy level $E_n^{(0)}$ with corresponding eigenfunctions, $\psi_{n1}^{(0)}$ and $\psi_{n2}^{(0)}$. The wavefunction of the perturbed state, ψ_n, can be written as a superposition of unperturbed wavefunctions:

$$\psi_n = C_1 \psi_{n1}^{(0)} + C_2 \psi_{n2}^{(0)}. \tag{5.30}$$

After the substitution of ψ_n into Eq. (5.4), multiplication of Eq. (5.4) by the complex-conjugate function ψ_n^*, and its integration (as we did in the previous section), we get the following system of algebraic equations with the unknown coefficients C_1 and C_2:

$$\begin{aligned} &\left(E_n^{(0)} + V_{11} - E_n\right) C_1 + V_{12} C_2 = 0, \\ &V_{21} C_1 + \left(E_n^{(0)} + V_{22} - E_n\right) C_2 = 0, \end{aligned} \tag{5.31}$$

where the matrix elements of the perturbation operator are defined by Eq. (5.17):

$$V_{lk}(x) = \int \psi_{ln}^{(0)*} \hat{V} \psi_{nk}^{(0)}. \tag{5.32}$$

Non-zero solutions for C_1 and C_2 exist only in the case when the determinant of the system of Eqs. (5.31) equals zero. By equating the determinant to zero we obtain the following quadratic equation with respect to energy E_n:

$$E_n^2 - \left(2E_n^{(0)} + V_{11} + V_{22}\right) E_n + \left(E_n^{(0)} + V_{11}\right)\left(E_n^{(0)} + V_{22}\right) - V_{12} V_{21} = 0. \tag{5.33}$$

The solution of this equation can be written as

$$E_n = E_n^{(0)} + E_n^{(1\pm)}, \tag{5.34}$$

where the corrections of the first order $E_n^{(1\pm)}$ to the unperturbed energy state $E_n^{(0)}$ are defined by the expression

$$E_n^{(1\pm)} = \frac{V_{11} + V_{22}}{2} \pm \sqrt{\frac{(V_{11} - V_{22})^2}{4} + |V_{12}|^2}, \tag{5.35}$$

and $V_{12}V_{21} = |V_{12}|^2$. The sign "±" in Eq. (5.35) indicates that in the presence of perturbation the degeneracy of the unperturbed quantum system is lifted and the energy levels split into two sublevels.

On substituting the expressions for the energy into the system of Eqs. (5.31) we get for the coefficients of expansion the following relation:

$$C_2 = C_1 \frac{E_n - E_n^{(0)} - V_{11}}{V_{12}} \equiv C_1 \frac{V_{21}}{E_n - E_n^{(0)} - V_{22}}. \tag{5.36}$$

In a system of homogeneous linear algebraic equations such as Eq. (5.31) one of the coefficients is always arbitrary. In our case we take the coefficient C_1 as arbitrary. We will find C_1 later, from the normalization condition of the wavefunction ψ_n. Taking into account the normalization and orthogonality conditions of the unperturbed wavefunctions, $\psi_{n1}^{(0)}$ and $\psi_{n2}^{(0)}$, the normalization condition of the wavefunction ψ_n takes the form

$$|C_1|^2 + |C_2|^2 = 1. \tag{5.37}$$

On substituting Eq. (5.36) into Eq. (5.37) we obtain the following expressions for the expansion coefficients C_1 and C_2:

$$C_1^{\pm} = \left(\frac{V_{12}}{2|V_{12}|} \left[1 \pm \frac{V_{11} - V_{22}}{\sqrt{(V_{11} - V_{22})^2 + 4|V_{12}|^2}} \right] \right)^{1/2}, \tag{5.38}$$

$$C_2^{\mp} = \left(\frac{V_{12}}{2|V_{12}|} \left[1 \mp \frac{V_{11} - V_{22}}{\sqrt{(V_{11} - V_{22})^2 + 4|V_{12}|^2}} \right] \right)^{1/2}. \tag{5.39}$$

Thus, in the case of doubly degenerate states the corresponding energy level splits into two sublevels, which correspond to the following wavefunctions:

$$\psi_n^{\pm} = C_1^{\pm} \psi_{n1}^{(0)} + C_2^{\mp} \psi_{n2}^{(0)}. \tag{5.40}$$

5.3 Non-stationary perturbation theory

Very often it is necessary to solve problems in which the perturbation which acts on the quantum-mechanical system depends on time. Let us assume that the initial state of a quantum system is a stationary state. It is necessary to find the state of the system at a later time t if at the initial instant, $t = 0$, the small perturbation, which depends on time, begins to act. In this case the first term, $\hat{H}_0$, in the Hamiltonian of the quantum system (5.1) does not depend on time, while the perturbation, $\hat{V}(t)$, depends on time. We have to find the change of the system's state with time. Generally this can be described by the non-stationary Schrödinger equation

$$\left[\hat{H}^{(0)} + \hat{V}(t) \right] \Psi(x, t) = i\hbar \frac{\partial \Psi(x, t)}{\partial t}. \tag{5.41}$$

The eigenfunctions $\Psi^{(0)}(x,t)$ and eigenvalues $E_n^{(0)}$ of the unperturbed Hamiltonian are assumed to be known, i.e., the solutions of the following equation are known:

$$\hat{H}^{(0)}\Psi^{(0)}(x,t) = \mathrm{i}\hbar\frac{\partial}{\partial t}\Psi^{(0)}(x,t). \tag{5.42}$$

For the nth quantum state they have the following form:

$$\Psi_n^{(0)}(x,t) = \mathrm{e}^{-\mathrm{i}E_n^{(0)}t/\hbar}\psi_n^{(0)}(x). \tag{5.43}$$

As follows from Eq. (5.41), the states of the perturbed system depend on time and the energy of the system is not a constant. Since the system does not have stationary states, the problem is not in finding these states but in the calculation of the system's wavefunction, which depends on time, and in finding the probability of the system's transition under the influence of the perturbation from the initial state to the final state. Therefore, the method of approximate solution of the Schrödinger equation, which is used for the case of a stationary perturbation, is not applicable here and must be modified.

Let us express the wavefunction $\Psi(x,t)$ in the form of an expansion over the complete set of unperturbed orthonormalized wavefunctions, $\Psi_k^{(0)}(x,t)$:

$$\Psi(x,t) = \sum_k b_k(t)\Psi_k^{(0)}(x,t), \tag{5.44}$$

where the expansion coefficients $b_k(t)$ are functions only of time, and the magnitude $|b_k(t)|^2$ defines the probability of the system being in the state with energy $E_k^{(0)}$ at the instant of time t. By substituting the expansion (5.44) into Eq. (5.41) we obtain the following system of differential equations:

$$\mathrm{i}\hbar\sum_k\left(\frac{\mathrm{d}b_k}{\mathrm{d}t}\Psi_k^{(0)}(x,t) + b_k\frac{\partial\Psi_k^{(0)}(x,t)}{\partial t}\right) = \sum_k b_k\left(\hat{H}^{(0)} + \hat{V}(t)\right)\Psi_k^{(0)}(x,t). \tag{5.45}$$

Let us multiply both sides of Eq. (5.45) by $\Psi_m^{(0)*}(x,t)$ and integrate over the entire space. Taking into account the form of the unperturbed wavefunctions (5.43) and their orthonormality, we get

$$\mathrm{i}\hbar\frac{\mathrm{d}b_m}{\mathrm{d}t} = \sum_k V_{mk}\mathrm{e}^{\mathrm{i}\omega_{mk}t}b_k, \tag{5.46}$$

where we introduced the frequency $\omega_{mk} = (E_m^{(0)} - E_k^{(0)})/\hbar$ and the matrix element of the perturbation operator,

$$V_{mk}(t) = \int \psi_m^{(0)*}(x)\hat{V}(t)\psi_k^{(0)}(x)\mathrm{d}x. \tag{5.47}$$

The system of equations (5.46) is exact and it is equivalent to the initial equation (5.41). In order to solve Eq. (5.46) approximately, let us assume that at $t \le 0$ the system was in the nth quantum state with the wavefunction $\Psi_n^{(0)}$. Then, for the initial instant of time all the coefficients with the index $k \ne n$

in the expansion (5.44) equal zero and the coefficient $b_n(0) = 1$. Beginning from the instant $t = 0$, the system is affected by the small perturbation and as a result the wavefunction of the initial state changes a little with time. Therefore, let us look for the coefficients $b_k(t)$ for the instant $t > 0$ in the form of expansion:

$$b_k(t) = b_k^{(0)}(t) + b_k^{(1)}(t) + b_k^{(2)}(t) + \cdots, \tag{5.48}$$

where $b_k^{(0)}(t) = b_k^{(0)}(0) = \delta_{kn}$. The correction $b_k^{(1)}(t)$ has the same order of smallness as the perturbation; $b_k^{(2)}(t)$ is quadratic with respect to the perturbation, and so on. On substituting the expansion (5.48) into Eq. (5.46), we obtain the equation for the correction of the first order of smallness:

$$\mathrm{i}\hbar \frac{\mathrm{d}b_m^{(1)}}{\mathrm{d}t} = \sum_k V_{mk} \mathrm{e}^{\mathrm{i}\omega_{mk}t} b_k^{(0)} = V_{mn} \mathrm{e}^{\mathrm{i}\omega_{mn}t}, \tag{5.49}$$

where we omitted all terms of second and higher orders of smallness with respect to the perturbation. By integrating the differential equation (5.49) we get

$$b_m^{(1)}(t) = \frac{1}{\mathrm{i}\hbar} \int_0^t V_{mn}(t) \mathrm{e}^{\mathrm{i}\omega_{mn}t}\, \mathrm{d}t. \tag{5.50}$$

The probability of finding the system at instant t in the state with energy $E_m^{(0)}$ in the first order of perturbation theory is defined as

$$P_{mn} = |b_m(t)|^2 = |b_m^{(1)}(t)| = \frac{1}{\hbar^2} \left| \int_0^t V_{mn}(t) \mathrm{e}^{\mathrm{i}\omega_{mn}t}\, \mathrm{d}t \right|^2. \tag{5.51}$$

Thus, knowing the dependence of the perturbation on time, we can find the probability of transition of the system from the initial state, $\Psi_n^{(0)}$, to the final state, $\Psi_m^{(0)}$. Equation (5.51) is very often written as

$$P_{mn} = \frac{4\pi^2}{\hbar^2} |V_{mn}(\omega_{mn})|^2, \tag{5.52}$$

where we introduced the Fourier transform of the matrix element of the perturbation operator (5.47), which corresponds to the frequency ω_{mn}:

$$V_{mn}(\omega_{mn}) = \frac{1}{2\pi} \int_{-\infty}^{\infty} V_{mn}(t) \mathrm{e}^{\mathrm{i}\omega_{mn}t}\, \mathrm{d}t. \tag{5.53}$$

One of the most important applied problems is the case of periodic perturbation when

$$\hat{V}(t) = \hat{V}_0 \mathrm{e}^{\mathrm{i}\omega t}. \tag{5.54}$$

In this case the probability of the corresponding transition during a sufficiently long time interval ($t \gg 2\pi/\omega$) is given by the expression

$$P_{mn} = \frac{2\pi}{\hbar} |(V_0)_{mn}|^2\, t\delta(E_m^{(0)} - E_n^{(0)} - \hbar\omega), \tag{5.55}$$

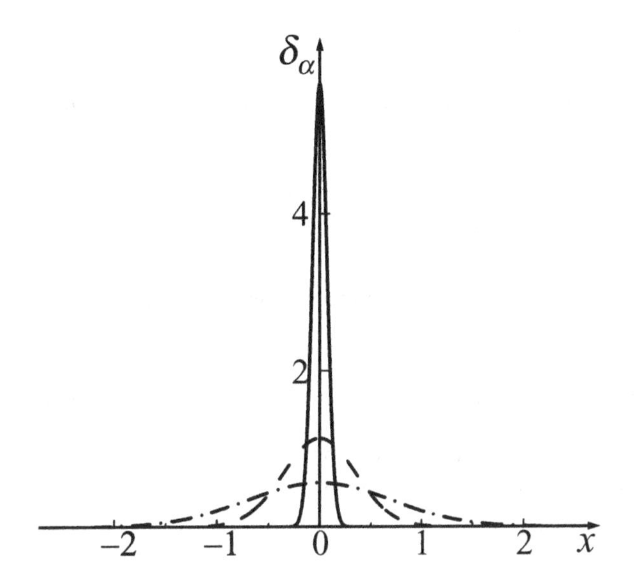

Figure 5.1 Dirac's δ-function can be thought of as a function $\delta_\alpha(x) = [1/(\alpha\sqrt{\pi})]e^{-x^2/\alpha^2}$ when α tends to zero. Then its amplitude tends to ∞, while its width tends to zero. The solid line corresponds to $\alpha = 0.1$, the dashed line to $\alpha = 0.5$, and the dash–dotted line to $\alpha = 1$.

where Dirac's δ-function, $\delta(E_m^{(0)} - E_n^{(0)} - \hbar\omega)$, shows the resonant character of the corresponding quantum transition, which is possible only at $\hbar\omega = E_m^{(0)} - E_n^{(0)}$. Therefore, the probability of the corresponding transition is not equal to zero if the magnitude $\hbar\omega$ is close to the energy difference $E_m^{(0)} - E_n^{(0)}$. Let us first introduce Dirac's δ-function and then continue our analysis.

Dirac's delta-function can be defined by the following two relationships:

$$\delta(x) = \begin{cases} 0, & x \neq 0, \\ \infty, & x = 0, \end{cases} \tag{5.56}$$

$$\int_{-\infty}^{\infty} f(x)\delta(x)\mathrm{d}x = f(0), \tag{5.57}$$

where $f(x)$ is an arbitrary function, which is continuous at the point $x = 0$. For convenience Dirac's δ-function can be presented in the form of the limit of the δ_α-function:

$$\delta(x) = \lim_{\alpha\to 0} \delta_\alpha(x), \tag{5.58}$$

where

$$\delta_\alpha(x) = \frac{1}{\alpha\sqrt{\pi}}e^{-x^2/\alpha^2}. \tag{5.59}$$

The function shown in Fig. 5.1 can be transformed into Dirac's δ-function by equating the parameter α to zero. Dirac's δ-function belongs to the mathematical class of so-called *generalized functions*, which can be considered as an extreme case for several continuous functions that depend on some accessory parameter.

Very often the δ-function is introduced as

$$\lim_{K\to\infty}\left(\int_{-K}^{K} \mathrm{e}^{\mathrm{i}kx}\,\mathrm{d}k\right) = 2\pi\,\delta(x) \tag{5.60}$$

or in shorter notation as

$$\frac{1}{2\pi}\int_{-\infty}^{\infty} \mathrm{e}^{\mathrm{i}kx}\,\mathrm{d}k = \delta(x). \tag{5.61}$$

Now let us continue our analysis of quantum transitions. For the rate of the transition probability the following expression is used:

$$\frac{\mathrm{d}P_{mn}}{\mathrm{d}t} = \frac{2\pi}{\hbar}\,|(V_0)_{mn}|^2\,\delta\left(E_m^{(0)} - E_n^{(0)} - \hbar\omega\right). \tag{5.62}$$

If the perturbation of an atom is caused by electromagnetic radiation, whose wavelength is much greater than the dimensions of the atom, then the perturbation operator, $\hat{V}$, can be expressed as

$$\hat{V} = -\mathbf{d}\cdot\mathbf{E}(t). \tag{5.63}$$

Here, $\mathbf{E}$ is the electric field and $\mathbf{d} = -e\mathbf{r}$ is the dipole moment of the electron in an atom, which is under the influence of the alternating electric field of the incident electromagnetic wave. Under the influence of this wave the atom can make a transition from the nth to the mth state with the following probability:

$$P_{mn} = \frac{4\pi^2}{\hbar^2}\,|d_{mn}|^2\,|\mathbf{E}(\omega_{mn})|^2, \tag{5.64}$$

where

$$\mathbf{E}(\omega_{mn}) = \frac{1}{2\pi}\int_{-\infty}^{\infty}\mathbf{E}(t)\mathrm{e}^{\mathrm{i}\omega_{mn}t}\,\mathrm{d}t. \tag{5.65}$$

The system can make such a transition if $\mathbf{E}(\omega_{mn}) \neq 0$, i.e., if there is a spectral line in the emission spectrum that corresponds to this transition, and if the corresponding matrix elements of the operator of dipole moment, $\hat{\mathbf{d}}$,

$$d_{mn}(t) = -\int \psi_m^{(0)*}(\mathbf{r})e\mathbf{r}(t)\psi_n^{(0)}(\mathbf{r})\mathrm{d}V, \tag{5.66}$$

are not equal to zero. These conditions define the so-called *selection rules* for the quantum systems to have allowed states.

Let us consider now the perturbation, which after switching on stays small and continues to act for an infinitely long time. In this case the expression for the probability of transition (5.51) is not valid since for the upper limit ($t \to \infty$) the integral diverges. We can show that in this case the probability of transition is defined by the expression

$$P_{mn} = \frac{1}{\hbar^2\omega_{mn}^2}\left|\int_0^t \frac{\mathrm{d}V_{mn}}{\mathrm{d}t}\,\mathrm{e}^{\mathrm{i}\omega_{mn}t}\,\mathrm{d}t\right|^2. \tag{5.67}$$

If the perturbation $\hat{V}(t)$ is small and weakly changes during the time interval $T_{mn} = 2\pi/\omega_{mn}$, then the magnitude of the integral in Eq. (5.51) and, correspondingly, the probability of transition are small. In the limit of infinitesimally slow changes of the perturbation, the probability of any transition between the energy levels tends to zero. In another limiting case, that of very fast (sudden) switching on of the perturbation, the derivatives $\mathrm{d}V_{mn}/\mathrm{d}t$ tend to infinity at the instant of switching on of the perturbation. Taking this into account, we can take the slowly changing factor $\mathrm{e}^{\mathrm{i}\omega_{mn}t}$ out of the integral in Eq. (5.67). After this the integral can be easily evaluated and as a result we obtain the following expression:

$$P_{mn} = \frac{1}{\hbar^2 \omega_{mn}^2} \left| V_{mn} \right|^2 . \tag{5.68}$$

The transition probability of a system at a sudden switching on of the perturbation can also be found in cases when the perturbation is not small. Let the system be in one of the quantum states $\psi_n^{(0)}$ of the unperturbed Hamiltonian $\hat{H}_0$. If the change of the Hamiltonian happens during a time that is small in comparison with the time of transition, $2\pi/\omega_{mn}$, then the wavefunction of the system does not have enough time to change and will stay the same as it was before the switching on of the perturbation. Thus, it will not be an eigenfunction of the new Hamiltonian, $\hat{H} = \hat{H}_0 + \hat{V}$, and, therefore, the initial state $\psi_n^{(0)}$ will not be stationary. The probability of the transition of a system to a new stationary state is defined according to Eq. (5.44) by the corresponding expansion coefficient of the function ψ_n, which is decomposed over the set of unperturbed eigenfunctions $\psi_n^{(0)}$:

$$P_{mn} = \left| \int \psi_m^* \psi_n^{(0)} \, \mathrm{d}x \right|^2 . \tag{5.69}$$

Example 5.2. Under the influence of a suddenly switched-on homogeneous electric field **E** the charged oscillator makes a transition from the ground state to the excited state. Find the probability of such a transition if the mass of the oscillator is equal to m and its charge is equal to e.

Reasoning. The potential energy of the charged oscillator in a homogeneous electric field can be written as follows:

$$U(x) = \frac{m\omega^2}{2}x^2 - e|\mathbf{E}|x = \frac{m\omega^2}{2}(x - x_0)^2 - \frac{e^2|\mathbf{E}|^2}{2m\omega^2}, \tag{5.70}$$

where the coordinate x_0 is equal to $x_0 = e|\mathbf{E}|/(m\omega^2)$. Equation (5.70) has the form of the potential energy of a linear oscillator with shifted equilibrium position x_0. If the wavefunction of the unperturbed oscillator has the form $\psi_n^{(0)}(x)$, then for the perturbed oscillator this function must have the form $\psi_n(x - x_0)$. The wavefunctions of the unperturbed harmonic oscillator (4.89) have the form

$$\psi_n^{(0)}(\xi) = A_n H_n(\xi) \mathrm{e}^{-\xi^2/2},$$

where $\xi = x\sqrt{m\omega/\hbar}$. The probability of transition from the ground state ($n = 0$) to the excited state ($n = k$) is defined according to Eq. (5.51) by the corresponding coefficient b_k of the expansion of the perturbed wavefunction in the series (5.44) over the unperturbed wavefunctions, i.e.,

$$P_{k0} = |b_k|^2 = \left| \int_{-\infty}^{\infty} \psi_0^{(0)*}(x)\psi_k^{(0)}(x - x_0)\mathrm{d}x \right|^2. \tag{5.71}$$

Using the explicit form of the wavefunctions, let us determine the expansion coefficient b_k:

$$b_k = \frac{(-1)^k}{\sqrt{2^k \pi k!}} \mathrm{e}^{-\xi_0^2/2} \int_{-\infty}^{\infty} \mathrm{e}^{-\xi\xi_0} \frac{\mathrm{d}^k}{\mathrm{d}\xi^k} \mathrm{e}^{-\xi^2+2\xi\xi_0}\, \mathrm{d}\xi, \tag{5.72}$$

where

$$\xi_0 = x_0\sqrt{\frac{m\omega}{\hbar}} = \frac{e|\mathbf{E}|}{\sqrt{m\hbar\omega^3}}.$$

By integrating Eq. (5.72) by parts k times we reduce the integral in Eq. (5.72) to the following form:

$$\int_{-\infty}^{\infty} \mathrm{e}^{-\xi\xi_0} \frac{\mathrm{d}^k}{\mathrm{d}\xi^k} \mathrm{e}^{-\xi^2+2\xi\xi_0}\, \mathrm{d}\xi = \xi_0^k \int_{-\infty}^{\infty} \mathrm{e}^{-\xi^2+\xi\xi_0}\, \mathrm{d}\xi = \xi_0^k \sqrt{\pi}\, \mathrm{e}^{\xi_0^2/4}. \tag{5.73}$$

On substituting this expression into Eq. (5.72) and then into Eq. (5.71) we obtain the following expression for the transition probability:

$$P_{k0} = \frac{\xi_0^{2k}}{2^k k!} \mathrm{e}^{-\xi_0^2/2}. \tag{5.74}$$

5.4 The quasiclassical approximation

If the de Broglie wavelength of a particle is small in comparison with the characteristic dimensions of the region within which the particle's motion takes place, then the behavior of a particle is approximately classical. The formal passage to the limit from the quantum description of the particle's behavior to the classical one can be brought about by making the reduced Planck constant, $\hbar$, tend to zero in the expression for the particle's wavefunction, which can be presented as

$$\Psi(\mathbf{r}, t) = A\mathrm{e}^{\mathrm{i}S(\mathbf{r},t)/\hbar}. \tag{5.75}$$

Here, $S(\mathbf{r}, t)$ is the action of the particle. As we noted in Section 2.3.3, the reduced Planck constant $\hbar$ is the *quantum of action*. For a free particle, the action is defined as

$$S(\mathbf{r}, t) = \mathbf{p} \cdot \mathbf{r} - Et. \tag{5.76}$$

The wavefunction of a free particle (2.145) can be written in the form (5.75). The expression for the wavefunction stays the same in the case of a particle's motion in a potential field $U(\mathbf{r})$. For the stationary state of the particle in the field, $U(\mathbf{r})$, the action can be written as

$$S(\mathbf{r}, t) = \int_L \mathbf{p} \cdot \mathrm{d}\mathbf{r} - Et = S_0(\mathbf{r}) - Et, \tag{5.77}$$

where the integration is carried out along the trajectory of the particle, L. The action, S_0, depends only on the particle's coordinates and thus S_0 is called the *reduced action*.

In the quasiclassical case the reduced Planck constant, $\hbar$, can be considered a small parameter and the physical magnitudes A and S can be considered independent of $\hbar$. On substituting Eq. (5.75) into the Schrödinger equation (2.182) and leaving only the terms proportional to $\hbar^0$ and $\hbar^1$ we arrive at the following two equations for the phase and amplitude of the wavefunction, $\Psi(\mathbf{r}, t)$:

$$\frac{\partial S}{\partial t} + \frac{1}{2m}(\nabla S)^2 + U = 0, \tag{5.78}$$

$$\frac{\partial A}{\partial t} + \frac{A}{2m} \nabla^2 S + \nabla S \nabla A = 0. \tag{5.79}$$

The first of these equations represents a well-known equation of classical mechanics – the *Hamilton–Jacobi equation for the particle's action*. The second equation, after its multiplication by $2A$, can be rewritten as

$$\frac{\partial A^2}{\partial t} + \operatorname{div}\left(A^2 \frac{\nabla S}{m}\right) = 0. \tag{5.80}$$

Here, $A^2 = |\Psi|^2$ is the probability density of the particle's location, and $\nabla S/m = \mathbf{p}/m$ is the classical velocity of the particle. Thus, Eq. (5.80) is the continuity equation, which shows that the probability density moves in each point of space according to the laws of classical mechanics with the velocity $\mathbf{p}/m$.

For stationary states, taking into account Eq. (5.77), the reduced action, S_0, and amplitude, A, satisfy the following equations:

$$\frac{1}{2m}(\nabla S_0)^2 + U = E, \tag{5.81}$$

$$\operatorname{div}\left(A^2 \nabla S_0\right)^2 = 0. \tag{5.82}$$

Let us write the wavefunction for stationary states in the case of a particle's one-dimensional motion in the field $U(x)$. Taking into account that $(\nabla S_0)^2 = (\mathrm{d}S_0/\mathrm{d}x)^2$, Eq. (5.81) reduces to

$$\frac{\mathrm{d}S_0}{\mathrm{d}x} = \pm\sqrt{2m[E - U(x)]}. \tag{5.83}$$

The solution of this equation can be written as

$$S_0 = \pm \int \sqrt{2m[E - U(x)]}\mathrm{d}x = \pm \int p(x)\mathrm{d}x, \tag{5.84}$$

where $p(x) = \sqrt{2m[E - U(x)]}$ is the momentum of the particle. Then, Eq. (5.82) and its solution can be written as

$$\frac{\mathrm{d}}{\mathrm{d}x}(A^2 p) = 0, \tag{5.85}$$

$$A^2 p = \text{constant}, \tag{5.86}$$

where for the amplitude of the wavefunction we get

$$A = \text{constant} \times \frac{1}{\sqrt{p}}. \tag{5.87}$$

Thus, in the quasiclassical approximation the wavefunction can be considered as the superposition of the wavefunctions of two states, which describe the motion of the particle along the x-axis in two possible directions:

$$\psi(x) = \frac{A_1}{\sqrt{p}} \mathrm{e}^{\mathrm{i}\int\sqrt{2m[E-U(x)]}\mathrm{d}x/\hbar} + \frac{A_2}{\sqrt{p}} \mathrm{e}^{-\mathrm{i}\int\sqrt{2m[E-U(x)]}\mathrm{d}x/\hbar}. \tag{5.88}$$

As we have already mentioned, the quasiclassical approximation is valid only for those regions of the particle's motion where its de Broglie wavelength, λ_{Br}, slightly changes at distances Δx of the order of λ_{Br} itself. This condition is not satisfied near the turning points where the direction of motion is reversed and the velocity of the particle tends to zero. Thus, at the turning points $p \to 0$ and therefore the de Broglie wavelength cannot be considered small.

Example 5.3. Estimate the probability of electron transmission through the potential barrier (3.182):

$$U(x) = \begin{cases} 0, & x < -a, \\ U_0(1 + x/a), & -a \le x \le 0, \\ U_0(1 - x/b), & 0 < x \le b, \\ 0, & x > b, \end{cases}$$

with a linear dependence of the potential in the regions of its increase and decrease. The electron is incident on the barrier from the negative part of the x-axis and has the energy $E < U_0$.

Reasoning. This probability can be estimated on the basis of the quasiclassical approximation if we assume that the electron potential energy, $U(x)$, and the momentum, $p(x)$, change slowly. The interval of electron motion under the barrier is limited by the coordinates x_1 and x_2, where

$$x_1 = a\left(\frac{E}{U_0} - 1\right), \tag{5.89}$$

$$x_2 = b\left(1 - \frac{E}{U_0}\right) \tag{5.90}$$

(see Example 3.9 and Fig. 3.11). In the classically unavailable region the electron wavefunction has the form of a damping exponent:

$$\sqrt{|p(x)|}\psi(x) = \mathrm{e}^{-\int_{x_1}^{x} |p(x)|\mathrm{d}x/\hbar}, \tag{5.91}$$

where $|p(x)| = \sqrt{2m_\mathrm{e}[U(x) - E]}\mathrm{d}x$. Thus, the amplitude of the electron wavefunction at the second boundary ($x_2 = b$) decreases by a factor of $\exp[\int_{x_1}^{x_2} |p(x)|\mathrm{d}x/\hbar]$. The probability of electron transmission through the barrier is equal to the square of its magnitude, i.e.,

$$D = \mathrm{e}^{-2\int_{x_1}^{x_2} \sqrt{2m_\mathrm{e}[U(x)-E]}\mathrm{d}x/\hbar}. \tag{5.92}$$

This formula can be used until the exponent's argument has become large in comparison with unity (compare it with Eq. (3.183)). On integrating Eq. (5.92) we arrive at the following expression:

$$D = \mathrm{e}^{-4\sqrt{2m_\mathrm{e}}(a+b)(U_0-E)^{3/2}/(3\hbar U_0)}, \tag{5.93}$$

which coincides with the exact solution (see Eq. (3.189)) that was obtained earlier to within a factor of order one.

5.5 Summary

1. The approximate methods are especially important for the solution of many problems concerning electron quantum states. Perturbation theory uses weak perturbation, which negligibly changes the state of the unperturbed system. If the Hamiltonian does not depend directly on time, stationary perturbation theory is used; for the opposite case non-stationary perturbation theory is used.
2. In the first order of approximation of stationary perturbation theory the correction to the unperturbed energy, $E_n^{(0)}$, is defined by the diagonal matrix element of perturbation, which is an average perturbation, i.e., $E_n^{(1)} = W_{nn} = \langle W_n \rangle$.
3. If all unperturbed quantum states are non-degenerate, then for the wavefunction and energy the following expressions are used:

$$\psi_n = \psi_n^{(0)} + \sum_{i \neq n} \frac{W_{in}}{E_n^{(0)} - E_i^{(0)}} \psi_i^{(0)},$$

$$E_n = E_n^{(0)} + \langle W_n \rangle + \sum_{i \neq n} \frac{|W_{in}|^2}{E_n^{(0)} - E_i^{(0)}}.$$

These relationships are valid if the matrix elements of the perturbation operator are significantly smaller than the distance between the unperturbed levels, i.e.,

$$|W_n| \ll \left| E_n^{(0)} - E_i^{(0)} \right|.$$

4. If the unperturbed energy level is s-fold degenerate, then it corresponds to s self-orthogonal wavefunctions $\psi_{n1}^{(0)}, \ldots, \psi_{ns}^{(0)}$. The perturbation splits a degenerate energy level into s close sublevels. Each of them corresponds to its own wavefunction, which is a linear combination of unperturbed wavefunctions $\psi_{nk}^{(0)}$.
5. The probability of transition of the system from the initial state to the final state is proportional to the square of the Fourier transform of the matrix element of the perturbation operator, which corresponds to the transition frequency, $\omega_{mn} = (E_m^{(0)} - E_n^{(0)})/\hbar$, i.e.,

$$P_{mn} = \frac{4\pi^2}{\hbar^2} |W_{mn}(\omega_{mn})|^2.$$

6. If the de Broglie wavelength of the particle is small compared with the characteristic size of the region within which the particle motion takes place, then the particle behavior can be described in the framework of the quasiclassical approximation. In this case the wavefunction can be presented as a superposition of two states, which describe the particle motion along and against the x-direction:

$$\psi = \psi^+ + \psi^-,$$

$$\psi^\pm = \frac{A^\pm}{\sqrt{p}} \mathrm{e}^{\mp \int \sqrt{2m(E-U(x))}\mathrm{d}x/\hbar}.$$

7. The quasiclassical probability of electron transmission through the barrier is defined by the approximate expression

$$D = \mathrm{e}^{-2\int_{x_1}^{x_2} \sqrt{2m_\mathrm{e}(U(x)-E)}\mathrm{d}x/\hbar}.$$

At the turning points, x_1 and x_2, the momentum of a particle tends to zero ($p \to 0$) and the de Broglie wavelength is not small. Therefore, these points must be excluded from the integration.

5.6 Problems

Problem 5.1. The unperturbed system has two close energy levels, $E_1^{(0)}$ and $E_2^{(0)}$, which are separated by an energy comparable to the energy of perturbation. Find the first-order corrections to the energy of these states.

Problem 5.2. A harmonic oscillator with charge e and mass m is placed in a homogeneous electric field $\mathbf{E}$ directed along the axis of oscillations. Considering the electric field as a perturbation, find the shift in the energy levels of the oscillator.

Problem 5.3. An electron in a one-dimensional rectangular potential well with the barriers of infinite height,

$$U(x) = \begin{cases} 0, & 0 \le x \le L, \\ \infty, & x < 0,\ x > L, \end{cases}$$

is subjected to a perturbation of the following form:

$$W(x) = \begin{cases} W_0, & b \le x \le L - b, \\ 0, & x < b,\ x > L - b. \end{cases}$$

Find the energy of the electron in stationary states, taking into account corrections arising from the first-order perturbation theory.

Problem 5.4. Find the energy of the ground state of a one-dimensional harmonic oscillator that is subjected to an anharmonic potential perturbation $W(x) = \gamma x^3$.

Problem 5.5. A charged one-dimensional oscillator is placed in a perturbation homogeneous electric field directed along the axis of oscillations. Find the energy of the oscillator, taking into account the first two orders of correction of perturbation theory. The charge of the particle is positive and equal to e and its mass is equal to m.

Problem 5.6. The electron is in a spherically-symmetric potential well (4.41):

$$U(r) = \begin{cases} 0, & r \le a, \\ \infty, & r > a. \end{cases}$$

The unperturbed magnitudes of the electron energy in the well are E_{nl}. Find, in the first order of perturbation theory, the electron energy and wavefunction in the magnetic field directed along the z-axis. The perturbation operator $\hat{V}$ is defined as

$$\hat{V} = \frac{\mathrm{i}\hbar e B}{2m_\mathrm{e}} \frac{\partial}{\partial \theta},$$

where θ is the polar angle in the plane perpendicular to the z-axis. Hint: use the spherical coordinate system.

Problem 5.7. An electron, which is in a one-dimensional potential $U(x) = \beta x^2/2$ (a linear harmonic oscillator), is placed in a homogeneous electric field. The electric field is directed along the axis of oscillations and changes with time as

$$\mathbf{E}(t) = \mathbf{E}_0 \mathrm{e}^{-t^2/\tau^2}.$$

Before the electric field is turned on ($t \to -\infty$) the electron was in one of its unperturbed states. Find, in the first order of perturbation theory, the probability of electron excitation to the higher states at $t \to \infty$.

Problem 5.8. The motion of a plane harmonic oscillator takes place in the xy-plane in the following potential:

$$U(x, y) = \frac{\beta(x^2 + y^2)}{2}.$$

Find, in the first order of perturbation theory, the splitting of the first excited energy level of the oscillator under the perturbation potential $W = \gamma xy$.

Problem 5.9. Find, in the quasiclassical approximation, the probability of electron transmission, P, through the potential barrier

$$U(x) = \begin{cases} 0, & x < 0, \\ U_0 \mathrm{e}^{-x/b}, & x \geq 0, \end{cases}$$

for an electron whose energy is $E < U_0$.

Chapter 6
Quantum states in atoms and molecules

The main characteristics of atoms and molecules are their *structure* and their *energy spectrum*. Under the term structure of an arbitrary particle we usually understand the size and the distribution of its mass and its charge in space. In quantum physics such a distribution is defined by the square of the modulus of the wavefunction of a particle. The particle itself may consist of a system of other particles that are bound by a certain type of coupling. For example, an atom consists of a nucleus and a system of interacting electrons, a molecule consists of a system of interacting atoms, and so on. If we consider an atom, the nucleus of the atom is assumed to be at rest (the so-called *adiabatic approximation*). This assumption can be made because the nucleus has a much larger mass than that of an electron (a proton's mass is 1836 times larger than the mass of an electron). Then, the square of the modulus of the wavefunction, $|\psi(\mathbf{r}_1, \mathbf{r}_2, \ldots, \mathbf{r}_n)|^2$, for the system of electrons defines the probability density of finding the jth electron at the point $\mathbf{r}_j$. Graphically it is very convenient to depict $|\psi|^2$ in the form of an *electron cloud*, which can be considered as an averaged distribution of matter added to the mass of the nucleus located at the center of an atom. Since electrons have not only mass, m_e, but also the electric charge, $-e$, the electron cloud should be considered as the averaged density of the negative electric charge, $\rho_e = -e|\psi(\mathbf{r}_1, \mathbf{r}_2, \ldots, \mathbf{r}_n)|^2$, which compensates for the nucleus' positive charge. The current chapter is devoted to the study of the wavefunction and the geometry of the electron cloud, as well as the energy spectra of the simplest atoms and molecules.

6.1 The hydrogen atom

6.1.1 An electron in a Coulomb potential

As has already been shown, the de Broglie wavelength of an electron in an atom is comparable to the atomic size. Therefore, we cannot neglect the wave properties of the electron and we have to describe its behavior in an atom on the basis of the Schrödinger equation.

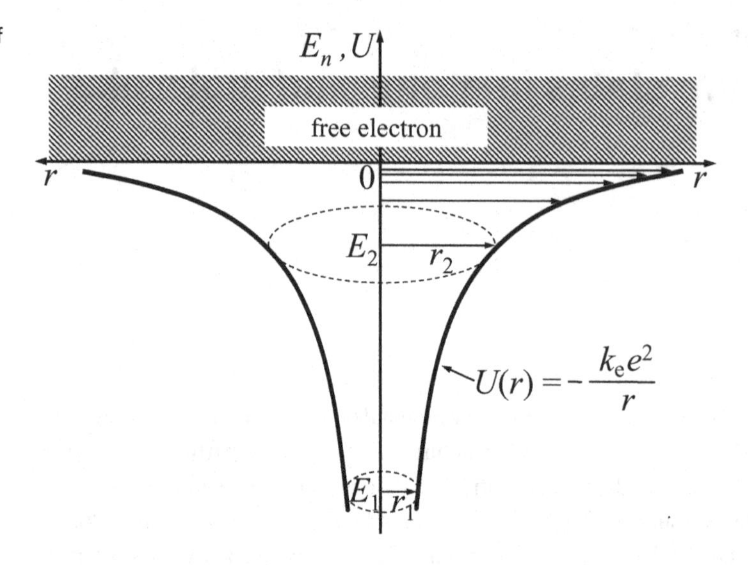

Figure 6.1 Quantization of an electron in the potential well $U(r)$.

The hydrogen atom is the simplest atomic system, where the confined electron with electric charge, $-e$, moves in the spherically-symmetric electrostatic field of a positively charged nucleus (proton) (the charge of the proton is $+e$). The potential energy of an electron in such a system is defined by its Coulomb interaction with the nucleus and is equal to

$$U(r) = -\frac{k_e e^2}{r}, \tag{6.1}$$

where r is the distance between the electron and the nucleus. The nucleus, in the first order of approximation, can be considered as a point (the proton size $a_p \approx 10^{-15}$ m while the distance between the electron and the proton is of the order of 10^{-10} m $= 1$ Å). The motion of an electron in the potential (6.1) can be considered to occur in a spherically-symmetric well of unlimited size (see Fig. 6.1). However, the size of the atom itself, i.e., the average diameter of an electron orbit in a stationary state, is finite and is of the order of 10^{-10} m.

For the description of possible electron quantum states in an atom we will first consider the hydrogen atom (H) and assume that its nucleus is motionless. Because of the large difference between proton and electron masses, m_p and m_e ($m_p/m_e = 1836$), their center of mass coincides with the proton's center of mass. The stationary Schrödinger equation (4.26) in the case of the potential defined by Eq. (6.1) can be written as

$$\nabla^2\psi(\mathbf{r}) + \frac{2m_e}{\hbar^2}\left(E + \frac{k_e e^2}{r}\right)\psi(\mathbf{r}) = 0. \tag{6.2}$$

Since the potential within which the electron is confined is spherically-symmetric, it is convenient to seek the solution of Eq. (6.2) in spherical coordinates, whose origin ($r = 0$) coincides with the center of mass of the nucleus. In such a coordinate system the electron wavefunction will depend on the distance of the electron from the atom's center, r, as well as on the azimuthal and polar angles, φ and θ, respectively.

We will write the wavefunction for Eq. (6.2) in the form of a product of two functions, as we have done already in the case of electron motion in a spherically-symmetric potential well (see Eq. (4.27)):

$$\psi(r, \varphi, \theta) = X(r)Y(\varphi, \theta). \tag{6.3}$$

According to Eq. (4.28), for a spherically-symmetric potential the function $Y_{lm}(\varphi, \theta)$ is the eigenfunction of the operator $\hat{L}^2$ with the eigenvalues of this operator equal to $l(l+1)\hbar^2$, where l is the orbital quantum number. This quantum number can have values $l = 0, 1, 2, \ldots$ and the orbital magnetic quantum number is equal to $m = 0, \pm 1, \ldots, \pm l$. The forms of several functions $Y_{lm}(\varphi, \theta)$ for $l = 0, 1, 2$ and $m = 0, \pm 1, \pm 2$ are given by the relations (4.37).

Let us substitute the wavefunction (6.3) into Eq. (6.2) and separate variables. As a result we find the following equation for the radial function, $X(r)$:

$$\frac{1}{r^2}\frac{\mathrm{d}}{\mathrm{d}r}\left(r^2\frac{\mathrm{d}X}{\mathrm{d}r}\right) + \frac{2m_\mathrm{e}}{\hbar^2}\left[E - \frac{l(l+1)\hbar^2}{2m_\mathrm{e}r^2} + \frac{k_\mathrm{e}e^2}{r}\right]X = 0. \tag{6.4}$$

The solution of this equation depends on two quantum numbers, l and n, and can be written as

$$X_{nl}(\rho) = \rho^l \mathrm{e}^{-\rho/n}\sum_{j=0}^{n-l-1} a_j\rho^j, \tag{6.5}$$

where we introduced the variable $\rho = r/r_1$, with

$$r_1 = \frac{\hbar^2}{k_\mathrm{e}m_\mathrm{e}e^2} = 0.53 \times 10^{-10}\,\mathrm{m} \tag{6.6}$$

the radius of the first Bohr orbit in the hydrogen atom for the ground state. The distinctive feature of Eq. (6.5) is that it has a finite number of terms $a_j\rho^j$. Only in this case will the exponential factor in Eq. (6.5) provide the vanishing of the square of the modulus of the wavefunction at infinity, i.e., at $\rho \to \infty$.

The principal quantum number, n, which defines the energy state of an electron in an atom, can take on the values $n = 1, 2, 3, \ldots$ The total energy of the electron is quantized (see Fig. 6.1) and is equal to

$$E_n = -\frac{k_\mathrm{e}^2 m_\mathrm{e}e^4}{2\hbar^2 n^2} = \frac{E_1}{n^2}, \tag{6.7}$$

where E_1 is defined as

$$E_1 = -\frac{k_\mathrm{e}^2 m_\mathrm{e}e^4}{2\hbar^2} = -13.6\,\mathrm{eV}. \tag{6.8}$$

Here, E_1 is the energy of the electron ground state in the hydrogen atom (i.e., the electron energy for the first Bohr orbit). The magnitude $|E_1| = 13.6$ eV is the *ionization energy* of the hydrogen atom. This energy is required to transfer the electron from the ground energy state, E_1 (shown schematically in Fig. 6.1), to the continuous energy spectrum (denoted by "free electron" in Fig. 6.1).

The coefficients a_j of the power series (6.5) for $j > 0$ can be found from the following relation (we will not show here how to derive this relation because of the complexity of the derivation):

$$a_{j+1} = 2a_j \frac{\sqrt{\epsilon}(j+l+1) - 1}{(j+l+2)(j+l+1) - l(l+1)}. \tag{6.9}$$

Here, $\epsilon = E/E_1$ is a positive dimensionless parameter for the confined states of an electron, i.e., for the region of the discrete spectrum with $E < 0$. Relation (6.9) is called *recursive* since it allows us to find the coefficient a_1 knowing the coefficient a_0, to find the coefficient a_2 knowing the coefficient a_1, and so on. The coefficient a_0 is found from the normalization of the total wavefunction of the hydrogen atom's ground state, i.e.,

$$\int_0^\infty \int_0^{2\pi} \int_0^\pi |\psi_{100}(r, \varphi, \theta)|^2 r^2 \sin\theta \, dr \, d\varphi \, d\theta = 1. \tag{6.10}$$

It follows from Eq. (6.5) that the last term of the power series has the index $j = n - l - 1$. Then, the coefficient a_{j+1} with $j = n - l - 1$ must be equal to zero. According to Eq. (6.9), a_{j+1} is equal to zero if the numerator in Eq. (6.9) is equal to zero:

$$\sqrt{\epsilon}(j+l+1) - 1 = 0, \quad \text{or } n\sqrt{\epsilon} - 1 = 0. \tag{6.11}$$

Thus,

$$n\sqrt{\epsilon} = 1, \quad \text{i.e., } n\sqrt{E/E_1} = 1. \tag{6.12}$$

From the last expression we get Eq. (6.7). The solution of Eq. (6.4) gives us the eigenfunctions, X, and the corresponding eigenvalues, E. Equation (6.5) shows that eigenfunctions X_{nl} depend on l for any particular n. In the general case each specific eigenvalue, E_{nl}, must correspond to a certain eigefunction, X_{nl}. However, the solution of Eq. (6.4) gives a different result and the eigenvalue E depends solely on n and does not depend on l at all (see Eq. (6.7)). Since the total energy, E_n, does not depend on the orbital quantum number, l, the total energy, E_n, corresponds to several radial wavefunctions X_{nl}, i.e., the energy level E_n is *degenerate*. We have considered so far an individual hydrogen atom and determined the energy levels for a single electron. If this hydrogen atom were to interact with other atoms or were placed in an external electric or magnetic field then this degeneracy would be lifted and the total energy E_{nl} would depend on the quantum number l as well.

For a given principal quantum number, n, the orbital quantum number, l, can take on the following values:

$$l = 0, 1, 2, \ldots, n - 1.$$

Table 6.1. *Notation of different quantum states*

Quantum number l	0	1	2	3	4	5
Symbol of state	s	p	d	f	g	h

The orbital magnetic quantum number, m, for each value of l can take on $2l+1$ values, and the total energy E_n depends only on the principal quantum number, n. Thus, all the electron states with $n > 1$ are degenerate, i.e., more than one eigenfunction corresponds to the nth energy state. The order of degeneracy, g, is defined by the number of different electron states for a given energy E_n:

$$g = \sum_{l=0}^{n-1} (2l+1) = 1 + 3 + 5 + \cdots + (2n-1) = n^2. \tag{6.13}$$

Thus, for the state described by the wavefunction ψ_{100} the order of degeneracy is equal to unity and therefore the energy E_1 is non-degenerate. For the energy E_2, g is equal to four and corresponds to four different electron states in the atom described by the wavefunctions

$$\psi_{200}, \quad \psi_{210}, \quad \psi_{21-1}, \quad \psi_{21+1}.$$

As will be shown later, the order of degeneracy of the nth energy level is equal to $g = 2n^2$ because the electron has intrinsic angular momentum (*spin*), which can take on two values (we will talk about spin in Section 6.3).

6.1.2 Symbols defining states and probability density

Different states of electrons in atoms, which correspond to different orbital quantum numbers, l, are denoted by lower-case Latin letters. Table 6.1 shows the correspondence of numbers and letters. When considering energy states we usually talk about s-states (or s-electrons) for $l = 0$, p-states (or p-electrons) for $l = 1$, and so on. In this notation of a quantum state the value of the principal quantum number, n, is given before the symbol of the state with a given orbital quantum number, l. Let us write the first four groups of states for the hydrogen atom: (1) 1s; (2) 2s, 2p; (3) 3s, 3p, 3d; and (4) 4s, 4p, 4d, 4f (see Table 6.1). Thus, an electron in state 4f has $n = 4$ and $l = 3$. The wavefunctions $\psi_{nlm}(r, \varphi, \theta)$ for the first two groups are given in Table 6.2.

The probability distribution of an electron's location in an atom in the corresponding stationary state is defined by the magnitude $|\psi_{nlm}|^2$. Its spatial distribution tells us about the form of the electron cloud.

Taking into account that an electron possesses not only mass but also charge, the parameter

$$\rho_e(\mathbf{r}) = -e|\psi_{nlm}|^2 \tag{6.14}$$

can be interpreted as the electron charge density. The spatial motion of an electron is equivalent to the existence of a current. The density of this current can be found

Table 6.2. *The wavefunctions $\psi_{nlm}(r, \varphi, \theta)$ for the first two groups of states (here, Z is the atomic number)*

Wavefunction	Expression	State
ψ_{100}	$\frac{1}{\sqrt{\pi}}\left(\frac{Z}{r_1}\right)^{3/2} e^{-\rho}$	1s
ψ_{200}	$\frac{1}{4\sqrt{2\pi}}\left(\frac{Z}{r_1}\right)^{3/2} (2-\rho)e^{-\rho/2}$	2s
ψ_{210}	$\frac{1}{4\sqrt{2\pi}}\left(\frac{Z}{r_1}\right)^{3/2} \rho e^{-\rho/2}\cos\theta$	2p
ψ_{21-1}	$\frac{1}{8\sqrt{\pi}}\left(\frac{Z}{r_1}\right)^{3/2} \rho e^{-\rho/2}\sin\theta\, e^{-i\varphi}$	2p
ψ_{21+1}	$\frac{1}{8\sqrt{\pi}}\left(\frac{Z}{r_1}\right)^{3/2} \rho e^{-\rho/2}\sin\theta\, e^{i\varphi}$	2p

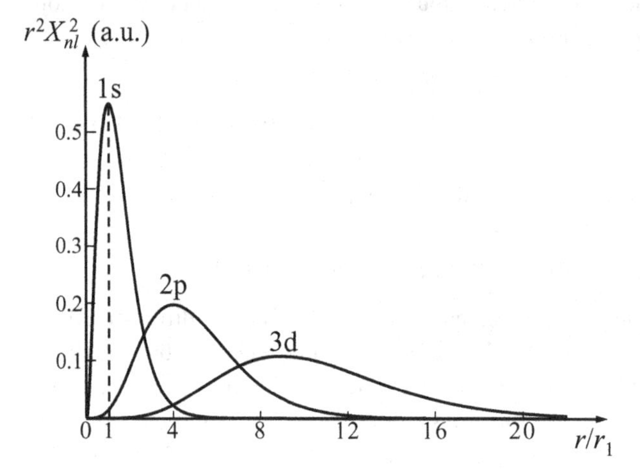

Figure 6.2 Distributions of the radial functions $r^2 X^2_{nl}$ for the 1s, 2p, and 3d quantum states of the hydrogen atom.

if we multiply the probability current density (2.190) in a given state by the charge of the electron:

$$\mathbf{j}_e = -e\frac{i\hbar}{2m_e}(\psi_{nlm}\,\nabla\psi^*_{nlm} - \psi^*_{nlm}\,\nabla\psi_{nlm}). \tag{6.15}$$

In the 1s-state the orbital quantum number, l, and angular momentum, L, are equal to zero. Therefore, the electron cloud in this state is spherically symmetric. From the classical point of view this corresponds to electron motion only along the radial direction – along $\mathbf{r}$. The electron would have to cross the region occupied by the nucleus, which is impossible in classical physics. To make this result of quantum theory more understandable, let us consider the probability of finding the electron within the volume dV:

$$dP = |\psi_{nlm}|^2\,dV.$$

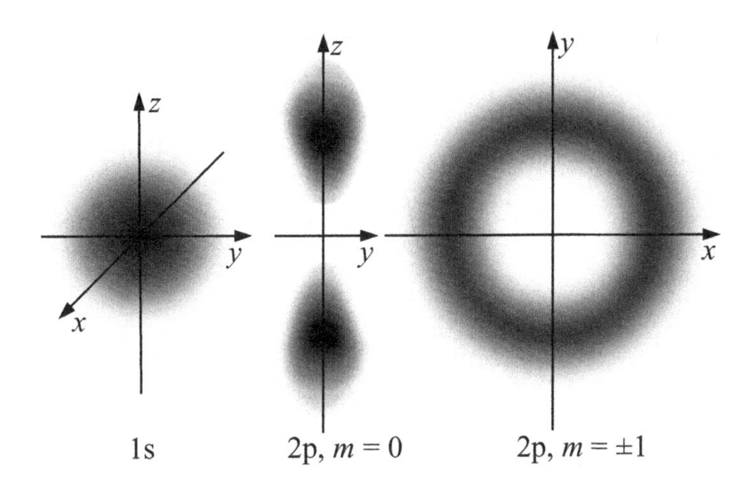

Figure 6.3 A pictorial representation of the electron clouds, $|\psi_{nlm}|^2$, for the 1s and 2p states in the hydrogen atom.

Let us choose a spherical shell of radius r, width $\mathrm{d}r$, and volume $\mathrm{d}V = 4\pi r^2\,\mathrm{d}r$. Then, for the 1s-shell the magnitude

$$\mathrm{d}P = 4\pi r^2 |\psi_{100}|^2\,\mathrm{d}r \tag{6.16}$$

defines the probability of finding an electron in this shell. The probability density, $\mathrm{d}P/\mathrm{d}r$, is the probability of finding an electron in the spherical shell around the radius r with width equal to unity. The radial dependences of $r^2 X_{nl}^2$ of the three states 1s, 2p, and 3d for a hydrogen atom are shown in Fig. 6.2. We see that the maximum probability density for the quantum state 1s corresponds to $r = r_1$, which coincides with the radius of the first Bohr orbit of the hydrogen atom. As r goes to zero the probability density, $\mathrm{d}P/\mathrm{d}r$, also tends to zero, which means that the electron is absent from the region occupied by the nucleus. The electron clouds, $|\psi_{nlm}|^2$, corresponding to the 1s and 2p states in the hydrogen atom are shown in Fig. 6.3. Note that $|\psi_{21\pm1}|^2$ has a doughnut shape and Fig. 6.3 shows it from the top.

Example 6.1. For the ground state of the hydrogen atom find the average, mean-square, and most probable distance between the electron and the nucleus.
Reasoning. The wavefunction of the ground state ($n = 1, l = 0, m = 0$) for the hydrogen atom ($Z = 1$) is given by the expression

$$\psi_{100}(r) = \frac{1}{\sqrt{\pi r_1^3}} \mathrm{e}^{-r/r_1} \tag{6.17}$$

(see Table 6.2). The average distance of an electron from the nucleus according to the quantum-mechanical definition of the average value of the physical magnitude (see Eq. (2.173)) can be obtained as follows:

$$\langle r \rangle = \int_0^\infty r|\psi_{100}|^2 4\pi r^2\,\mathrm{d}r = \frac{4}{r_1^3} \int_0^\infty r^3 \mathrm{e}^{-2r/r_1}\,\mathrm{d}r. \tag{6.18}$$

In order to calculate the integral, let us change variables from r to x:

$$x = \frac{2r}{r_1}. \tag{6.19}$$

As a result the integral takes the form

$$\langle r \rangle = \frac{4}{r_1^3}\left(\frac{r_1}{2}\right)^4 \int_0^\infty x^3 e^{-x}\, dx. \tag{6.20}$$

Taking into account that

$$\int_0^\infty x^n e^{-x}\, dx = n!, \tag{6.21}$$

we get the final result

$$\langle r \rangle = \frac{3r_1}{2}. \tag{6.22}$$

In order to find the mean-square distance between the electron and the nucleus, let us find first the following magnitude:

$$\left\langle r^2 \right\rangle = \int_0^\infty r^2 |\psi_{100}|^2 4\pi r^2\, dr = \frac{4}{r_1^3}\int_0^\infty r^4 e^{-2r/r_1}\, dr = 3r_1^2. \tag{6.23}$$

Thus, the mean-square distance, r_{ms}, is equal to

$$r_{ms} = \sqrt{\left\langle r^2 \right\rangle} = \sqrt{3}r_1. \tag{6.24}$$

Note that the integral in Eq. (6.23) was evaluated by using the same procedure as for Eq. (6.18). To find the most probable distance, r_{mp}, of an electron from the nucleus, it is necessary to equate the derivative of the probability density defined by Eq. (6.16) to zero:

$$\frac{d}{dr}\left(\frac{dP}{dr}\right) = \frac{d}{dr}\left(4\pi r^2 \frac{1}{\pi r_1^3} e^{-2r/r_1}\right) = \frac{8}{r_1^3} r e^{-2r/r_1}\left(1 - \frac{r}{r_1}\right) = 0. \tag{6.25}$$

From Eq. (6.25) it follows that $r_{mp} = r_1$. This distance corresponds to the maximum of the probability density, dP/dr.

6.1.3 Orbital magnetic moment and orbital magnetic quantum number

Since an electron is a charged particle, the current, I_e, must be connected with the electron motion around the atomic nucleus. The motion of an electron with velocity v along a circular orbit of radius r with period T is equivalent to a circular current, whose magnitude is given by the expression

$$I_e = -\frac{e}{T} = -\frac{ev}{2\pi r}, \tag{6.26}$$

where v is the orbital velocity of the electron. The minus sign in Eq. (6.26) indicates that the electron velocity and current have opposite directions because

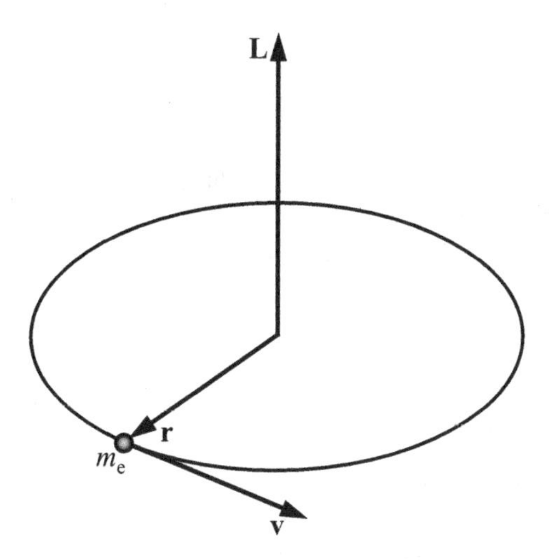

Figure 6.4 The electron angular momentum, **L**.

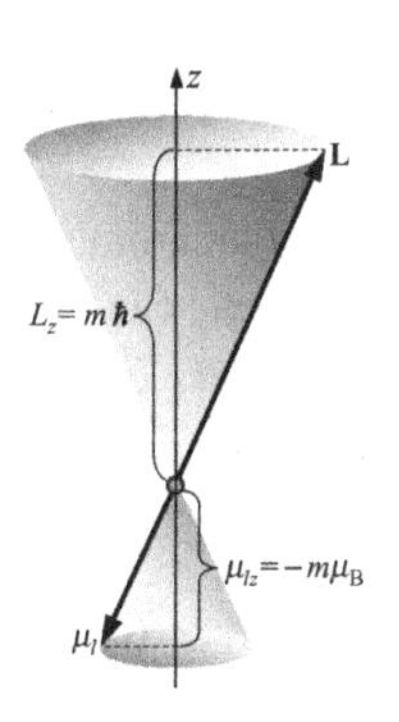

Figure 6.5 The electron orbital magnetic moment, μ_l, and angular momentum, **L**.

an electron possesses a negative electric charge. Such a circular current with the surface area $S = \pi r^2$ corresponds to an orbital magnetic moment with a magnitude

$$\mu_l = I_e S = -\frac{evr}{2}. \tag{6.27}$$

The orbital magnetic moment, $\boldsymbol{\mu}_l$, is a vector, whose direction is perpendicular to the plane of the electron orbit. The direction of the electron angular momentum, $\mathbf{L} = \mathbf{r} \times \mathbf{p}$ (see Fig. 6.4) is defined by the right-hand rule and the magnitude of the electron angular momentum is equal to

$$L = m_e vr. \tag{6.28}$$

The angular momentum, **L**, is also perpendicular to the plane of the electron orbit. Since the charge of the electron is negative, the direction of the vector of orbital magnetic moment, $\boldsymbol{\mu}_l$, is opposite to the direction of the vector **L** (see Fig. 6.5).

Let us introduce the so-called *gyromagnetic ratio*, γ, which defines the relation between the orbital magnetic moment and the electron angular momentum in an atom:

$$\gamma = \left|\frac{\mu_l}{L}\right| = \frac{e}{2m_e}. \tag{6.29}$$

Quantum-mechanical calculations of the gyromagnetic ratio, which will be considered in Example 6.2, lead to expression (6.29).

Thus, in any quantum-mechanical state the electron in the hydrogen atom has not only angular momentum, **L**, with magnitude

$$L = \hbar\sqrt{l(l+1)}, \tag{6.30}$$

but also orbital magnetic moment $\boldsymbol{\mu}_l$ with magnitude

$$\mu_l = \gamma L = \mu_B \sqrt{l(l+1)}. \tag{6.31}$$

Here, we introduced the constant μ_B, known as the *Bohr magneton*:

$$\mu_B = \frac{e\hbar}{2m_e} = 9.27 \times 10^{-24}\,\mathrm{J\,T^{-1}}, \tag{6.32}$$

which serves as a unit for measuring the orbital magnetic moment of an atom.

The physical meaning of the orbital magnetic quantum number, m, can be understood by taking into account that the wavefunction $\psi_{nlm}(r, \varphi, \theta)$ of an electron in the hydrogen atom is simultaneously the eigenfunction of the operator of the projection of angular momentum, $\hat{L}_z$:

$$\hat{L}_z \psi_{nlm} = m\hbar\psi_{nlm}. \tag{6.33}$$

It follows from Eq. (6.33) that the projection of angular momentum, L_z, onto the chosen direction (for example, the z-axis) can have only the following certain values:

$$L_z = m\hbar. \tag{6.34}$$

The axis direction is chosen taking into account the direction of an external magnetic or electric field. *The maximum value of $L_z = l\hbar$ is always smaller than L*, which is defined by Eq. (6.30) (see Fig. 6.6). The possible values of the projection of the atom's orbital magnetic moment, μ_l, along the z-axis, are also quantized and are determined by the orbital magnetic quantum number, m:

$$\mu_{lz} = -\gamma L_z = -m\mu_B. \tag{6.35}$$

Example 6.2. Find the orbital magnetic moment of the hydrogen atom μ_l, which occurs because of the orbital motion of an electron and find the gyromagnetic ratio, γ.

Reasoning. Let us use Eq. (2.190) for the probability current, **j**. Since the wavefunction given in spherical coordinates depends on variables r, θ, and φ, for the projections of the vector **j** onto the radial, polar, and azimuthal directions,

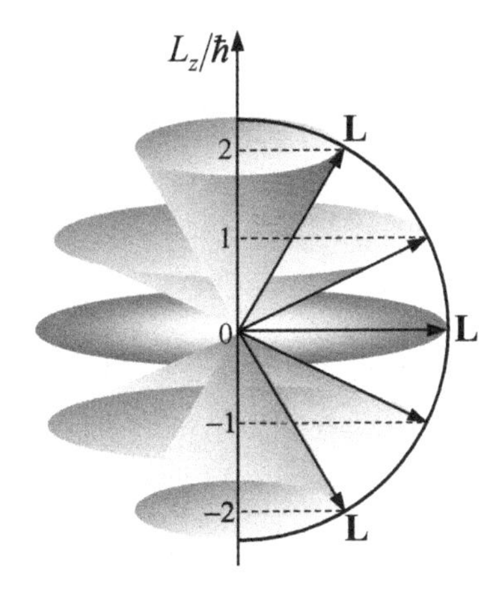

Figure 6.6 Projections of angular momentum, **L**, on the *z*-axis. The maximum value of L_z for $l = 2$ ($L_{z\max} = 2\hbar$) is smaller than the modulus of **L** ($|\mathbf{L}| = L = 2\hbar\sqrt{1.5}$).

respectively, we get

$$\begin{aligned} j_r &= \frac{i\hbar}{2m_e}\left(\psi_{nlm}\frac{d}{dr}\psi^*_{nlm} - \psi^*_{nlm}\frac{d}{dr}\psi_{nlm}\right) = 0, \\ j_\theta &= \frac{i\hbar}{2m_e r}\left(\psi_{nlm}\frac{d}{d\theta}\psi^*_{nlm} - \psi^*_{nlm}\frac{d}{d\theta}\psi_{nlm}\right) = 0, \\ j_\varphi &= \frac{i\hbar}{2m_e r\sin\theta}\left(\psi_{nlm}\frac{d}{d\varphi}\psi^*_{nlm} - \psi^*_{nlm}\frac{d}{d\varphi}\psi_{nlm}\right) = \frac{\hbar m}{m_e r\sin\theta}\left|\psi_{nlm}\right|^2. \end{aligned} \tag{6.36}$$

The projections j_r and j_θ are equal to zero because the parts of the wavefunction which depend on r and θ are real. In order to calculate the projection j_φ we used the explicit form of the angular part of the wavefunction (Eq. (4.37)). The probability current density, j, corresponds to the electric current density, j_e, as

$$\mathbf{j}_e = -e\mathbf{j}, \qquad j_e = -ej_\varphi. \tag{6.37}$$

Let us present the orbital magnetic moment of an atom as a sum of moments from the elementary circular currents with radius $r\sin\theta$ and cross-section ds. The magnitude of such an elementary current is

$$dI_e = j_e\,ds = -ej_\varphi\,ds. \tag{6.38}$$

The area of the circular orbit is $S = \pi r^2\sin^2\theta$. The orbital magnetic moment created by the elementary current is equal to

$$d\mu_l = S\,dI_e = -\pi r\sin\theta\,\frac{me\hbar}{m_e}\left|\psi_{nlm}\right|^2 ds. \tag{6.39}$$

The total orbital magnetic moment, μ_l, may be written as

$$\mu_l = -\frac{me\hbar}{2m_e}\int\left|\psi_{nlm}\right|^2 dV = -\frac{e\hbar}{2m_e}m, \tag{6.40}$$

where $dV = 2\pi r\sin\theta\,ds$ is the volume of a tube, and integration is carried out over the entire space. Taking into account that $L = \hbar m$, we find the gyromagnetic

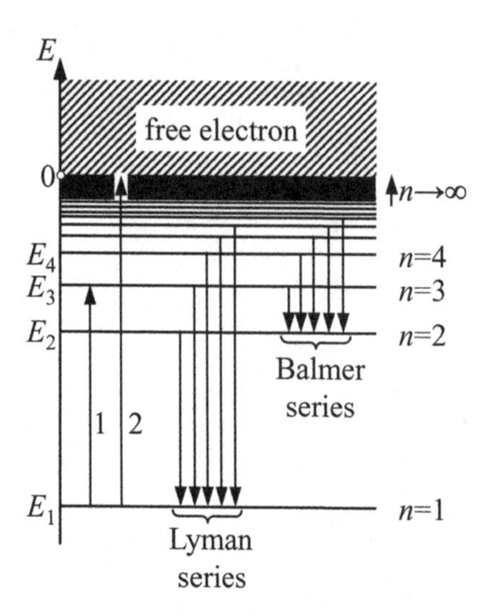

Figure 6.7 The electron energy spectrum in the hydrogen atom; E_1, E_2, E_3, and E_4 are the discrete energy levels of the electron in the potential well of the hydrogen atom.

ratio, γ:

$$\gamma = \left|\frac{\mu_l}{L}\right| = \frac{e}{2m_e}. \tag{6.41}$$

6.2 The emission spectrum of the hydrogen atom

Figure 6.7 shows the electron energy spectrum of the hydrogen atom, which was calculated using Eq. (6.7). If the electron has positive energy then it is free and its energy spectrum is continuous as shown in Fig. 6.7. In the region of negative values of the total energy the energy spectrum of the electron of the hydrogen atom is discrete. In the stationary state the electron can be in one of the energy states with $E_n < 0$. The undisturbed atom is in its ground state with energy E_1. If the atom is given additional energy, i.e., if the atom is excited, then it may make a transition to a state with higher energy $E_n > E_1$ (see Fig. 6.7, where transition 1 for the electron from level E_1 to level E_3 is shown). If the energy given to the atom is sufficiently high, then the electron may overcome the Coulomb attraction of the nucleus and become free. This process is called *ionization of the atom*. The minimal energy necessary for ionization, E_i, of a hydrogen atom (see transition 2 in Fig. 6.7) is $|E_i| = 13.6$ eV.

An atom may stay in the excited state only for a short time τ. This time is called the *lifetime of the atom in the given quantum state* and for different quantum states it can be of the order of $\tau = 10^{-6}$–10^{-10} s. During time τ the electron spontaneously makes transitions to levels with lower energy, including the ground state. During such a transition the atom emits an energy quantum in the form of electromagnetic radiation, i.e., *the atom emits a photon*. Experiments show that the emission from hydrogen atoms corresponds to certain regions of the electromagnetic spectrum and that it has the form of narrow spectral lines.

Knowing the structure of the energy levels, it is easy to find the structure of the optical emission spectrum of a hydrogen atom. During the transition of the electron from the mth energy level to the nth level ($m > n$, $E_m > E_n$), the electron loses energy $E_m - E_n$ by emitting a photon with angular frequency

$$\omega_{nm} = \frac{E_m - E_n}{\hbar}. \tag{6.42}$$

By substituting the expressions for energies E_m and E_n defined by Eq. (6.7) into Eq. (6.42) we obtain the equation for the emission frequency of the hydrogen atom for the different transitions from higher energy levels to lower:

$$\omega_{nm} = 2\pi c R_\infty \left(\frac{1}{n^2} - \frac{1}{m^2} \right), \quad m > n. \tag{6.43}$$

Here, we introduced the so-called *Rydberg constant* R_∞:

$$R_\infty = \frac{m_e}{4\pi \hbar c} \left(\frac{e^2}{4\pi \epsilon_0 \hbar} \right)^2 = 1.0974 \times 10^7 \, \mathrm{m}^{-1}, \tag{6.44}$$

whose value was first found experimentally; later it was obtained theoretically by Niels Bohr.

According to Eq. (6.43), there must be a series of spectral lines in the emission spectrum of a hydrogen atom corresponding to different parts of the optical spectrum. Such series were experimentally found a long time before they were theoretically predicted. The most well known of these series are the following.

The Lyman series

The Lyman series corresponds to transitions to the ground level $n = 1$ from higher energy levels with $m = 2, 3, 4, \ldots$ Its spectral lines are in the ultraviolet region. The minimal emission frequency in this series is $\omega_{12} = 1.55 \times 10^{16} \, \mathrm{s}^{-1}$, which corresponds to the maximal wavelength

$$\lambda_{12} = \frac{2\pi c}{\omega_{12}} = 121.4 \, \mathrm{nm}. \tag{6.45}$$

The maximal frequency, which corresponds to the short-wavelength boundary, i.e., when $m \to \infty$, is $\omega_{1\infty} = 2.07 \times 10^{16} \, \mathrm{s}^{-1}$, which corresponds to $\lambda_{1\infty} =$ 91 nm.

The Balmer series

The Balmer series has the following n and m numbers: $n = 2$ and $m = 3, 4, 5, \ldots$ The spectral lines of this series are in the visible range. The structure of these spectral lines is shown in Fig. 6.8. The symbols Hα, Hβ, Hγ, and Hδ label characteristic lines of this series. The spectral line Hr is the short-wavelength boundary of the series, which corresponds to $m \to \infty$ and maximal frequency $\omega_{2\infty} = 5.17 \times 10^{15} \, \mathrm{s}^{-1}$. The spectral line H$\alpha$ corresponds to the minimal frequency $\omega_{23} = 2.87 \times 10^{15} \, \mathrm{s}^{-1}$. The color of H$\alpha$ is red and Hδ is in the ultraviolet region.

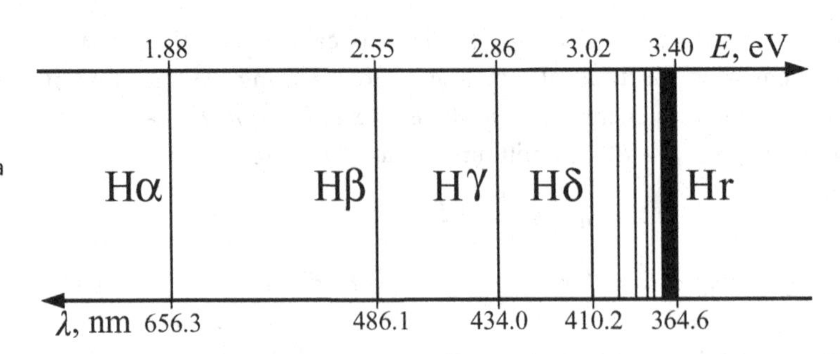

Figure 6.8 The Balmer series. The symbols Hα, Hβ, Hγ, and Hδ label characteristic spectral lines of this series. Hα is a red line, whereas Hδ is in the ultraviolet region.

The Paschen series

The Paschen series has the following n and m numbers: $n = 3$ and $m = 4, 5, 6, \ldots$ All spectral lines of this series are in the near-infrared region of the electromagnetic spectrum. The Brackett ($n = 4,\ m \geq 5$) and Pfund ($n = 5$, $m \geq 6$) series are in the same spectral region.

The absorption spectra have the same spectral rules. Absorption occurs when an electron in a lower energy state absorbs a photon and is promoted to a higher energy state. Thus, emission and absorption of energy quanta in the form of photons happen as a result of electron transitions from one energy level to another. However, not all the transitions in an atom are allowed. The only transitions allowed are those between quantum states whose orbital quantum numbers, l, differ by unity:

$$\Delta l = \pm 1, \tag{6.46}$$

i.e., transitions between s- and p-states, between p- and d-states, and so on are allowed. At the same time the orbital magnetic quantum number, m, must either stay the same or change by unity, i.e.,

$$\Delta m = 0, \pm 1. \tag{6.47}$$

Equations (6.46) and (6.47) are called the *selection rules*. They are derived from the calculations of transition probabilities between different stationary states with emission or absorption of a photon.

The existence of these rules for the quantum transitions of an electron is related to the conservation of the square of angular momentum, L^2, and its projection L_z. Indeed, as a result of the corresponding transition of an electron from one energy level to another a photon is emitted (or absorbed). The photon takes away (or transmits to the atom) not only an energy quantum, $\Delta E = \hbar\omega$, but also a quantum of angular momentum, $\Delta L = \hbar$. This fact leads to the condition defined by Eq. (6.46) for the orbital quantum number, l.

The selection rule for the orbital magnetic quantum number, $\Delta m = \pm 1$, shows that the emitted or absorbed photon may impart to the atom angular momentum with a certain projection on a chosen axis. This projection may be equal to $\pm\hbar$, which corresponds to the left-handed or right-handed circular polarization of the photon. The selection rule $\Delta m = 0$ shows that a photon having angular momentum $\hbar$ does not have a certain projection on the chosen axis and does not take away (or transmit) any projections of angular momentum. This happens in the case when the photon has linear polarization, which can be presented as a sum of two opposite circular polarizations – left-handed and right-handed.

The study of spectral lines using high-resolution spectrometers revealed that the spectral lines that correspond to the transitions between the energy levels with $l \geq 1$ are *doublets*. The energy difference between the lines constituting doublets is very small. Therefore, it is said that spectral lines have *fine structure*. Such a splitting of spectral lines may be connected only with the splitting of energy levels themselves. Indeed, the experimentally established value of the splitting of the energy level of the 2p-state of the hydrogen atom is equal to $\Delta E = 4.5 \times 10^{-5}$ eV and the corresponding splitting frequency of the main spectral line of the Lyman series is equal to $\Delta \nu = \Delta\omega/(2\pi) = 11$ GHz. However, such a splitting cannot be obtained from the solutions of the Schrödinger equation which we wrote for the electron in a spherically-symmetric potential. Understanding the atomic fine structure became possible only after the discovery that the electron has intrinsic angular momentum and an intrinsic magnetic moment not related to the electron orbital motion. In previous chapters we have mentioned this property called *spin*, and now we will consider it in more detail.

6.3 The spin of an electron

The following experimental facts that could not be explained by the quantum theory developed by Bohr and Schrödinger forced scientists to revise their concepts.

1. Analysis of the spectral lines of alkali metals shows that p-, d-, etc. terms are doublets, i.e., consist of two closely spaced spectral lines, whereas the s-term stays a singlet.
2. The Stern–Gerlach experiment showed that a beam of silver atoms splits into two beams in the presence of an inhomogeneous external magnetic field, a result that was not expected. Since the outermost electron in a silver atom is in its ground state its orbital quantum number, l, must be equal to zero and thus its orbital magnetic quantum number, m, must also be equal to zero, and the beam must not split.
3. Many atoms in the presence of an external magnetic field have *even* multiplets in the spectra, whereas it was expected that multiplets must be only *odd*: $2l + 1$ for all values of l is odd.

To explain these experimental facts George Uhlenbeck and Samuel Goudsmit suggested in 1925 that an electron, in addition to its orbital motion, also has intrinsic angular momentum, **S**, which they called *spin*. Later it was established that spin is not connected with the rotation of an electron around its own axis as was first suggested, but is a *quantum-mechanical and at the same time a relativistic internal property of the electron*. Together with the spin angular momentum, **S**, an electron has an *intrinsic spin magnetic moment*, $\boldsymbol{\mu}_s$. Similarly to angular momentum, the values of electron spin angular momentum, S, and spin magnetic moment, μ_s, are defined by the same expressions:

$$S = \hbar\sqrt{s(s+1)}, \tag{6.48}$$

$$\mu_s = 2\mu_B\sqrt{s(s+1)}, \tag{6.49}$$

where s is the spin quantum number, which is also called simply *spin*.

It has been established experimentally that for electrons $s = 1/2$, and therefore

$$S = \frac{\sqrt{3}}{2}\hbar, \tag{6.50}$$

$$\mu_s = \sqrt{3}\mu_B. \tag{6.51}$$

The spin gyromagnetic ratio, which relates the electron spin magnetic moment, $\boldsymbol{\mu}_s$, and spin angular momentum, **S**, as $\boldsymbol{\mu}_s = \gamma_S \mathbf{S}$, is equal to

$$\gamma_S = \frac{\mu_s}{S} = \frac{2\mu_B}{\hbar} = \frac{e}{m_e}. \tag{6.52}$$

Its magnitude is twice the orbital gyromagnetic ratio defined by Eq. (6.29).

As in the case of orbital motion, the magnitude of the spin angular momentum, S, may have the following values:

$$S^2 = \hbar^2 s(s+1). \tag{6.53}$$

Its projections onto an arbitrarily chosen axis (for example the z-axis) are equal to

$$S_z = \hbar m_s, \tag{6.54}$$

where the magnetic quantum number, m_s, is called the *spin magnetic quantum number* and may take $2s + 1$ values:

$$m_s = s, s-1, \ldots, -s+1, -s. \tag{6.55}$$

Since for an electron $s = 1/2$, two spin states are possible. These states correspond to two different projections with $m_s = \pm 1/2$ of spin angular momentum onto the chosen axis (see Fig. 6.9).

The existence of the electron spin magnetic moment, μ_s, and orbital magnetic moment, μ_l, leads to an additional interaction called *spin–orbit coupling*. Its energy is described by

$$U_{sl}(r) = \frac{\mu_0}{4\pi}\frac{\boldsymbol{\mu}_s \cdot \boldsymbol{\mu}_l}{r^3}, \tag{6.56}$$

where r is the radius of the corresponding orbit of the electron in the atom. The spin–orbit coupling is not taken into account in the Schrödinger equation (6.2). The existence of such an interaction leads to the splitting of energy levels that correspond to the states with orbital quantum number $l \geq 1$.

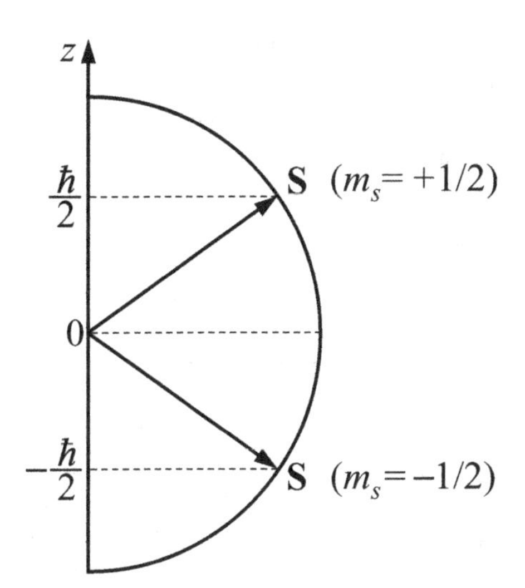

Figure 6.9 Projections of spin angular momentum, **S**, onto the chosen *z*-axis with two different values of m_s.

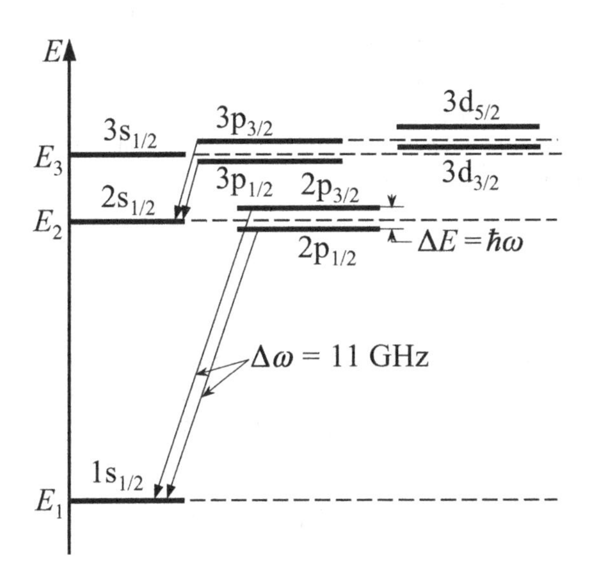

Figure 6.10 The splitting of energy levels due to spin–orbit coupling in the hydrogen atom. The quantum transitions that belong to the Lyman and Balmer series are shown.

Figure 6.10 shows the scheme of the splitting for the lower levels of the hydrogen atom. It also shows the quantum transitions which correspond to two doublet lines in the Lyman and Balmer series.

The quantum state of the electron in the hydrogen atom can be defined by the set of four quantum numbers whose values are given in Table 6.3.

Let us note again that many elementary particles have their own intrinsic angular momentum, i.e., spin. Particles whose spin is a half-integer are called *Fermi particles* or *fermions*. Electrons, protons, and neutrons are examples of fermions. For these particles $s = 1/2$. For other types of particles spin is an

Table 6.3. *The four quantum numbers, n, l, m, and m_s, which define the state of an electron in a hydrogen atom*

Quantum number	Symbol	Possible values
Principal	n	$1, 2, 3, \ldots$
Orbital	l	$0, 1, 2, \ldots, n-1$
Magnetic (orbital)	m	$-l, -l+1, \ldots, -1, 0, 1, \ldots, l-1, l$
Magnetic (spin)	m_s	$-1/2, +1/2$

integer and these particles are called *Bose particles* or *bosons*. Some of the mesons (π and K), photons, and the α-particle (the nucleus of a $_2He^4$ helium atom) are bosons. Ensembles of particles consisting of Fermi and Bose particles behave differently, and we will talk about this subject later.

Example 6.3. Estimate the splitting of energy level E_2 in the hydrogen atom which is due to spin–orbit coupling. Compare it with the splitting caused by the interaction of the spin magnetic moment of the electron, $\boldsymbol{\mu}_s$, with the spin magnetic moment of the nucleus (proton), $\boldsymbol{\mu}_p$.

Reasoning. The magnitude of splitting of the E_2 energy level has to be of the order of the energy of dipole–dipole interaction of the corresponding magnetic moments, which is defined by Eq. (6.56):

$$U_{sl}(r) = \frac{\mu_0}{4\pi}\frac{\boldsymbol{\mu}_s \cdot \boldsymbol{\mu}_l}{r^3}, \tag{6.57}$$

$$U_{sp}(r) = \frac{\mu_0}{4\pi}\frac{\boldsymbol{\mu}_s \cdot \boldsymbol{\mu}_p}{r^3}. \tag{6.58}$$

Here, r is the effective distance between the magnetic moments. This distance is of the order of the radius of the second Bohr orbit, $r_2 = 2.1 \times 10^{-10}$ m, for both types of interaction. The magnitudes of the corresponding magnetic moments are

$$\mu_s = \frac{e\hbar}{m_e}, \quad \mu_l = \frac{e\hbar}{2m_e}, \quad \mu_p = 2.8\frac{e\hbar}{2m_p}, \tag{6.59}$$

where $m_p = 1.67 \times 10^{-27}$ kg is the proton mass. Thus, the energy separations between the sublevels for the two cases are defined by the expressions

$$\Delta E_{sl} \approx \frac{\mu_0}{4\pi}\frac{e^2\hbar^2}{2m_e^2}\frac{1}{r_1^3} \approx 1.1 \times 10^{-5}\ \text{eV}, \tag{6.60}$$

$$\Delta E_{sp} \approx 1.4\frac{\mu_0}{4\pi}\frac{e^2\hbar^2}{m_e m_p}\frac{1}{r_1^3} \approx 1.25 \times 10^{-8}\ \text{eV}. \tag{6.61}$$

The estimated values agree to within an order of magnitude with the experimental data as well as with the theoretical results which were obtained on the basis of exact quantum-mechanical calculations. For example, the exact value of ΔE_{sl} is 4.5×10^{-5} eV. We see that ΔE_{sl} caused by the spin–orbital coupling is three orders of magnitude larger than ΔE_{sp}. Therefore, the splitting of the spectral lines due to spin–orbit coupling is called the *fine structure of spectral lines*, and the splitting caused by the interaction between the intrinsic magnetic moments of the electron and the nucleus is called the *hyperfine structure of spectral lines*.

6.4 Many-electron atoms

6.4.1 The Pauli exclusion principle and quantum states

Let us assume that in many-electron atoms the electron–electron interaction can be neglected. This means that each electron moves independently in the spherically-symmetric field of the nucleus:

$$U(r) = -\frac{Zk_e e^2}{r}, \tag{6.62}$$

where Z is the atomic number. In this approximation the energy of electron stationary states in a many-electron atom should have the same form as the energy (6.7) of the hydrogen atom:

$$E_n = E_1 \frac{Z^2}{n^2} = -13.6\frac{Z^2}{n^2} \text{ eV}. \tag{6.63}$$

If all electrons were in the state with $n = 1$, then the atom's ionization energy should be defined by the expression

$$E_i = 13.6Z^2 \text{ eV}.$$

Thus, for a mercury atom, Hg, with $Z = 80$ the ionization energy should be equal to $E_i = 8.7 \times 10^4$ eV, whereas the correct value of the ionization energy is $E_i = 10$ eV. Therefore, such a big difference between the calculated and experimental ionization energies shows that the approximation of non-interacting electrons is not valid. The drawbacks of such an approximation are determined by the fact that we do not take into account two important properties of the behavior of a system of electrons in a many-electron atom.

First of all, when the atomic number, Z, increases, and therefore the number of electrons increases, the filling of energy states by electrons occurs as follows. As a new electron is added it does not occupy the ground state but instead occupies the lowest of the available unoccupied energy states. The formation of many-electron atoms happens according to one of the main principles of quantum physics – the *Pauli exclusion principle*, according to which a *particular state, which is characterized by a given set of quantum numbers (in the case of an atom the quantum numbers are n, l, m, m_s), may be occupied only by one electron.* Because of this principle, the outer electrons, for example, in a mercury (Hg) atom can occupy only states with $n = 6$.

Second, it is necessary to take into account that the energy levels in this approximation are strongly degenerate, because their position depends only on the principal quantum number, n, and does not depend on the other quantum numbers. Electron repulsion lifts this degeneracy, and as a result the energy levels depend not only on the principal quantum number, n, but also on the orbital quantum number, l. Electrons with the same quantum number n may be grouped in shells, and electrons with a given n, which have the same quantum number l, may be grouped in subshells. Thus, the filling of energy levels by electrons in a many-electron atom is reduced to their distribution over the shells and subshells. Therefore, for $n = 1$ the values of the other quantum numbers are $l = 0$, $m = 0$, and $m_s = \pm 1/2$. For the $n = 1$ shell, which has only a single subshell with $l = 0$, we have that only two electrons with opposite spin projections may occupy it. At $n = 2$, two subshells can exist, with $l = 0$ and $l = 1$. The quantum numbers $m = 0$ and $m_s = \pm 1/2$ correspond to the first subshell, and two electrons may occupy it. The second subshell, with $l = 1$, corresponds to the quantum numbers $m = -1, 0, 1$ and each value of m corresponds to $m_s = \pm 1/2$. As a result this subshell may contain six electrons. Therefore, the total number of electrons in the shell with $n = 2$ is equal to 8. In the general case the capacity of the subshell with a given l is equal to $2(2l + 1)$ and for $l = 0, 1, 2, 3$, and 4 the capacity of subshells is equal to 2, 6, 10, 14, and 18 electrons, respectively. Correspondingly, the capacity of the shell is equal to

$$\sum_{l=0}^{n-1} 2(2l+1) = 2[1 + 3 + 5 + \cdots + (2n-1)] = 2n\frac{1 + (2n-1)}{2} = 2n^2. \quad (6.64)$$

For the principal quantum numbers $n = 1, 2, 3, 4$, and 5 the capacity of the shells is equal to 2, 8, 18, 32, and 50, respectively. Thus, in accordance with the Pauli exclusion principle, in many-electron atoms, after filling shells with lower n, the electrons are distributed over the shells and subshells with increasing n and l. The properties of atoms with closed shells drastically differ from those of atoms with half-filled shells. It is difficult to remove an electron from atoms with completely filled shells (for example, helium (He) or neon (Ne) with $Z = 2$ and 10, respectively). Atoms with one extra electron (for example, lithium (Li) and sodium (Na) with $Z = 3$ and 11, respectively) easily lose their outermost electron. Atoms lacking one electron for completion of their outer shell (for example, hydrogen (H) and fluorine (F) with $Z = 1$ and 9, respectively) easily acquire one more electron. In the next section we will consider the classification of the quantum states of many-electron atoms.

6.4.2 The total angular momentum of an atom

For many-electron atoms the angular momenta and spin angular momenta are defined by the expressions

$$M_L = \hbar\sqrt{L(L+1)}, \quad (6.65)$$

$$M_S = \hbar\sqrt{S(S+1)}, \quad (6.66)$$

where L and S are quantum numbers for the atomic angular momenta and spin angular momenta. The quantum number L is always either an integer or equal

Table 6.4. *Spectroscopic classification of quantum states*

L	0	1	2	3	4	5
Symbol	S	P	D	F	G	H

to zero. The quantum number S for an even number of electrons is an integer, whereas for an odd number of electrons it is a half-integer.

The total angular momentum of an atom, $\mathbf{M}_J$, is a result of quantum-mechanical summation of momenta, which depends substantially on the type of interaction between the momenta. For the most common *normal coupling* the angular momenta are added to give the resulting angular momentum, $\mathbf{M}_L$, and the spin angular momenta to give the resulting spin angular momentum, $\mathbf{M}_S$. The interaction of these momenta defines the total angular momentum of an atom, $\mathbf{M}_J$, whose magnitude is equal to

$$M_J = \hbar\sqrt{J(J+1)}. \tag{6.67}$$

Here, the quantum number J takes on one of the following values:

$$J = L + S, L + S - 1, \ldots, |L - S|. \tag{6.68}$$

This number takes on integer values if S is an integer, i.e., for an even number of electrons in an atom, and it takes on half-integer values if S is a half-integer, i.e., for an odd number of electrons.

The projection of the total angular momentum of an atom onto the chosen direction is defined by the formula of space quantization:

$$M_{J_z} = m_J\hbar, \tag{6.69}$$

where the quantum number m_J takes on $2J + 1$ values:

$$m_J = -J, -J + 1, \ldots, J - 1, J. \tag{6.70}$$

For the designation of different quantum states of many-electron atoms the spectroscopic classification of quantum states is used. This consists of the following: each value of the total orbital quantum number, L, corresponds to a certain capital letter of the Latin alphabet (see Table 6.4). As a superscript to the left of the Latin letter the multiplicity of states, which is equal to $2S + 1$ and defines the number of sublevels to which the level can be split, is written. The multiplicity is related to the spin quantum number S (do not confuse this with the symbol of state with $L = 0$). As a subscript to the right of the symbol the quantum number J, which defines the total angular momentum of the atom, is written. Thus, the ground state of the helium atom, He, is designated by the symbol $^1S_0$, for which $S = 0$, $L = 0$, and $J = 0$. In an atom with two electrons, two types of states with $S = 0$ (the electron spins are anti-parallel) and with $S = 1$ (electron spins are parallel) may be possible. In the first case $J = L = 0$ and the multiplicity is equal to $2S + 1 = 1$. In the second case the multiplicity is equal to $2S + 1 = 3$

Table 6.5. *Symbolic representation of quantum states of many-electron atoms*

S	0	0	0	1	1	1
L	0	1	2	0	1	2
J	0	1	2	1, 0	2, 1, 0	3, 2, 1
Symbol	$^1S_0$	$^1P_1$	$^1D_2$	$^3S_1, {}^3S_0$	$^3P_2, {}^3P_1, {}^3P_0$	$^3D_3, {}^3D_2, {}^3D_1$

Table 6.6. *The total number of states in shells*

Symbol of the shell	K	L	M	N	O
Quantum number n	1	2	3	4	5
Total number of states $2n^2$	2	8	18	32	50

and three values of J are possible: $L + 1$, L, and $|L - 1|$. The designations of the above-mentioned states are listed in Table 6.5.

According to the Pauli exclusion principle and Eq. (6.64), the quantum state with a given principal quantum number, n, corresponds to $2n^2$ states with different values of the quantum numbers l, m, and m_s. Those electrons in an atom with the same quantum number n form the so-called *shell*. In correspondence to the values of n, the shells are designated by capital Latin letters. The first five atomic shells are shown in Table 6.6. The shells are divided into subshells, which differ only by orbital quantum number, l. The number of different states in a subshell with different quantum numbers m and m_s is equal to $2(2l + 1)$. The subshells with $l = 0, 1, 2, 3, \ldots$ include the quantum states 2, 6, 10, 14, ..., respectively, and may be designated as

$$1\text{s},\ 2\text{s}\,2\text{p},\ 3\text{s}\,3\text{p}\,3\text{d},\ 4\text{s}\,4\text{p}\,4\text{d}\,4\text{f}, \ldots, \tag{6.71}$$

where the number designates the quantum number n, i.e., the affiliation with the corresponding shell. Completely filled shells and subshells have $L = 0$, $S = 0$, and $J = 0$. Each subsequent atom is built up from the previous one by changing the charge of the nucleus by adding a positive elementary charge, e, and by adding one electron, which according to the Pauli exclusion principle fills the next vacant state with the minimal energy. Thus, the next element after hydrogen (H), namely helium (He), fills the K shell by adding a second 1s-electron with spin opposite to the spin of the first electron. The third element, lithium (Li), has in addition to a filled K shell one more electron in the 2s subshell. The fourth element, beryllium (Be), has a filled K shell and 2s subshell.

The distribution of electrons over the states is called the *electron configuration*. Thus, the electron configuration of the ground state of silicon (Si), which contains

14 electrons, has the form

$$1s^2\ 2s^2\ 2p^6\ 3s^2\ 3p^2.$$

This means that the K and L shells and 3s subshell are completely filled, and there are two electrons in the 3p subshell.

6.4.3 Hund's rules

The fifth element of the Periodic Table of the elements, boron (B), has in its 2p subshell one electron with $L = 1$ and $S = 1/2$. These values correspond to values of the quantum number $J = 3/2$ and $1/2$, i.e., to the two quantum states of an atom $^2P_{3/2}$ and $^2P_{1/2}$. Which of these states is the ground state is decided by Hund's rules, which define the algorithm for the filling of subshells.

1. A given electron configuration has the smallest energy in the state with the largest possible S and L, which corresponds to the value of spin, S.
2. The quantum number J is equal to $J = |L - S|$ if the subshell is less than half-filled, and to $J = L + S$ if the subshell is more than half-filled.

Taking into account the second of Hund's rules, it is clear that for the boron atom (B) the ground state should be $^2P_{1/2}$ since with one electron in the 2p subshell the largest possible m_S and m_L are $m_S = 1/2$ and $m_L = 1$. Therefore, $S = 1/2$, $L = 1$, and $J = |L - S| = 1/2$. In Table 6.7 the result of the algorithm for filling shells and subshells with electrons is shown for the first 36 elements.

6.4.4 Optical spectra

Atoms of the alkali elements, i.e., metals of the first group (Li, Na, and K) of the Periodic Table of the elements (see Fig. 6.11), have one electron in their outer orbits, which is called the *valence electron*. These atoms have the simplest emission spectra. Their spectra, which are similar to the hydrogen-atom spectrum, consist of a great number of spectral lines. The systematization of these lines allowed one to group them in series, each of which is related to the transition of the excited atom to a certain energy level.

An atom of an alkali element has Z electrons, of which $Z - 1$ form a fairly strong *core* with the nucleus. The outer (valence) electron moves in the electric field of this core and is weakly connected to it. Therefore, atoms of the alkali elements can be considered as *hydrogen-like atoms*. However, in contrast to the hydrogen atom, where the electron moves in a spherically-symmetric potential, in the alkali elements the field of the core is not completely spherically symmetric. The inner electrons deform the electron core and break the spherical symmetry of its field. The violation of the potential's symmetry leads to bound states whose energy depends not only on the principal quantum number, n, as in the case of the hydrogen atom, but also on the orbital quantum number, l. The solution of

Table 6.7. *The filling of shells and subshells according to Hund's rules*

Number (Z)	Element	K (1s)	L (2s 2p)	M (3s 3p 3d)	N (4s 4p)	Ground state
1	H	1				$^2S_{1/2}$
2	He	2				$^1S_0$
3	Li	2	1			$^2S_{1/2}$
4	Be	2	2			$^1S_0$
5	B	2	2 1			$^2P_{1/2}$
6	C	2	2 2			$^3P_0$
7	N	2	2 3			$^4S_{3/2}$
8	O	2	2 4			$^3P_2$
9	F	2	2 5			$^2P_{3/2}$
10	Ne	2	2 6			$^1S_0$
11	Na	2	2 6	1		$^2S_{1/2}$
12	Mg	2	2 6	2		$^1S_0$
13	Al	2	2 6	2 1		$^2P_{1/2}$
14	Si	2	2 6	2 2		$^3P_0$
15	P	2	2 6	2 3		$^4P_{3/2}$
16	S	2	2 6	2 4		$^3P_2$
17	Cl	2	2 6	2 5		$^2P_{3/2}$
18	Ar	2	2 6	2 6		$^1S_0$
19	K	2	2 6	2 6	1	$^2S_{1/2}$
20	Ca	2	2 6	2 6	2	$^1S_0$
21	Sc	2	2 6	2 6 1	2	$^2D_{3/2}$
22	Ti	2	2 6	2 6 2	2	$^3F_2$
23	V	2	2 6	2 6 3	2	$^4F_{3/2}$
24	Cr	2	2 6	2 6 4	1	$^7S_3$
25	Mn	2	2 6	2 6 5	2	$^6S_{5/2}$
26	Fe	2	2 6	2 6 6	2	$^5D_4$
27	Co	2	2 6	2 6 7	2	$^4F_{9/2}$
28	Ni	2	2 6	2 6 8	2	$^3F_4$
29	Cu	2	2 6	2 6 10	1	$^2S_{1/2}$
30	Zn	2	2 6	2 6 10	2	$^1S_0$
31	Ga	2	2 6	2 6 10	2 1	$^2P_{1/2}$
32	Ge	2	2 6	2 6 10	2 2	$^3P_0$
33	As	2	2 6	2 6 10	2 3	$^4S_{3/2}$
34	Se	2	2 6	2 6 10	2 4	$^3P_2$
35	Br	2	2 6	2 6 10	2 5	$^2P_{3/2}$
36	Kr	2	2 6	2 6 10	2 6	$^1S_0$

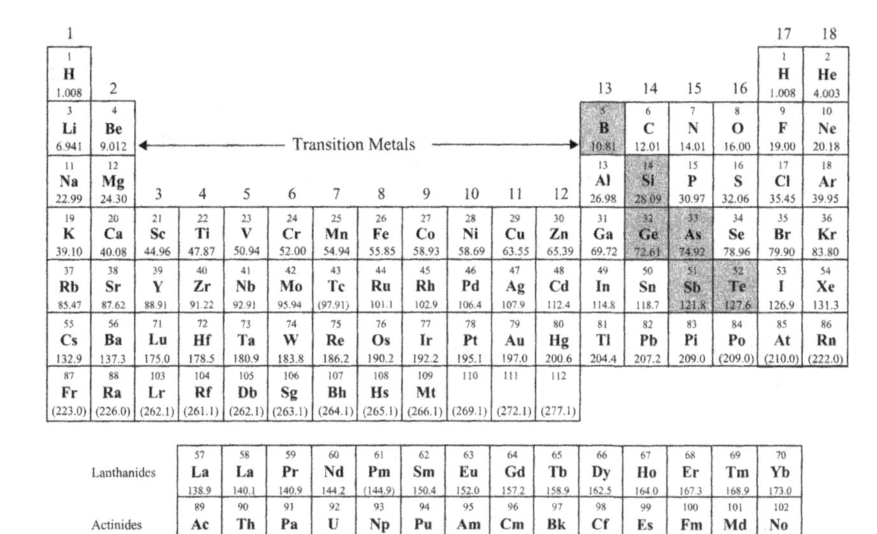

Figure 6.11 The Periodic Table of the elements.

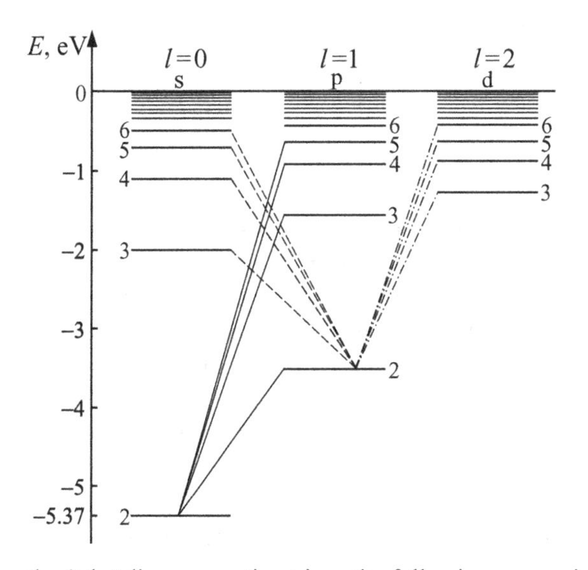

Figure 6.12 A schematic representation of energy levels and transitions for a lithium atom, Li: the principal series is shown by solid lines, the sharp series by dashed lines, and the diffuse series by dash–dotted lines.

the Schrödinger equation gives the following expression for this energy:

$$E_{nl} = -\frac{k_e^2 m_e e^4}{2\hbar^2 (n + b_l)^2} = \frac{E_1}{(n + b_l)^2}, \tag{6.72}$$

where b_l, which depends on l, is the quantum correction to the formula for the energy levels in a hydrogen atom (see Eq. (6.7)).

The dependence of the electron energy in the alkali elements on the orbital quantum number, l, is the principal difference of the energy spectrum of the alkali elements from the spectrum of the hydrogen atom. This dependence means that for the alkali elements the degeneracy of the orbital quantum number, l, is lifted.

Figure 6.12 shows schematically the energy levels and the corresponding transitions for lithium (Li). The ground state for lithium is the state 2s, for which $n = 2$

because the state with $n = 1$ is occupied by two electrons, which form the lithium core. Three series of spectral lines of the lithium atom are shown in Fig. 6.12: *principal*, *sharp*, and *diffuse*. The lines of the principal series correspond to the transitions from p-levels to the 2s-level. The frequencies of the spectral lines of the principal series, taking into account Eq. (6.72), can be found as follows:

$$\omega = 2\pi c R_\infty \left[\frac{1}{(2 + b_s)^2} - \frac{1}{(n + b_p)^2} \right], \quad n = 2, 3, 4, \ldots, \tag{6.73}$$

where R_∞ is the Rydberg constant. The lines of the sharp series correspond to transitions from the s-levels to the 2p-level, and the frequencies of the spectral lines of this series are given by the expression

$$\omega = 2\pi c R_\infty \left[\frac{1}{(2 + b_p)^2} - \frac{1}{(n + b_s)^2} \right], \quad n = 3, 4, 5, \ldots \tag{6.74}$$

The lines of the diffuse series correspond to the transitions from the d-levels to the 2p-level. The frequencies of the spectral lines of this series are given by the expression

$$\omega = 2\pi c R_\infty \left[\frac{1}{(2 + b_p)^2} - \frac{1}{(n + b_d)^2} \right], \quad n = 3, 4, 5, \ldots \tag{6.75}$$

We note that the quantum corrections b for each series are practically the same, but they may change slightly from series to series.

As was the case for the hydrogen atom, the spectral lines of the lithium atom have fine doublet structure, which is related to the splitting of its energy levels due to the existence of the electron spin. The splitting of energy levels is caused by the spin–orbit coupling, i.e., by interaction of magnetic moments μ_l and μ_s. Since the core of the lithium atom has a completely filled 1s-level, its angular momentum is equal to zero. Therefore, the angular momentum of the atom in general is equal to the angular momentum of its external electron, which is defined by the orbital quantum number l. The total momentum of this electron is equal to the sum of the orbital magnetic and spin magnetic moments and is defined by the quantum number $j = l \pm s$, where l and $s = 1/2$ are the orbital and spin quantum numbers. Thus, each energy level of p-states ($l = 1$) splits into two sublevels with $j = 1/2$ and $3/2$.

Each energy level of d-states ($l = 2$) splits into sublevels with $j = 3/2$ and $5/2$. The levels of s-states ($l = 0$) do not split since they correspond to the single value of $j = 1/2$.

Example 6.4. Find the maximum possible total angular momentum and corresponding spectral symbol for the atomic state with the electron configuration $1s^2\,2p\,3d$.

Reasoning. The total maximal angular momentum will be formed from the maximal orbital magnetic and spin magnetic moments. Since for s-electrons

$l = 0$, for p-electrons $l = 1$, and for d-electrons $l = 2$,

$$L_{max} = 1 + 2 = 3.$$

For the filled $1s^2$ shell spin $s = 0$, and therefore

$$S_{max} = \frac{1}{2} + \frac{1}{2} = 1.$$

As a result, for the maximum quantum number which defines the total angular momentum, we get

$$J_{max} = L_{max} + S_{max} = 4. \tag{6.76}$$

The total maximal angular momentum of the atom is equal to

$$M_{max} = \hbar\sqrt{J_{max}(J_{max} + 1)} = \sqrt{20}\hbar. \tag{6.77}$$

Since the multiplicity of the state is $2S_{max} + 1 = 3$, the symbol of this state is $^3F_4$.

Example 6.5. Give the spectral symbol of the atomic state whose degeneracy with respect to J is equal to seven, with multiplicity equal to five, and for which the value of the orbital quantum number is maximal.
Reasoning. From the multiplicity of the state,

$$2S + 1 = 5,$$

we find spin quantum number $S = 2$.

From the order of degeneracy,

$$2J + 1 = 7,$$

we find $J = 3$. The quantum number J can take all the integer values from $L + S$ to $|L - S|$. Therefore, the values of J and S are $J = 3$ and $S = 2$. The requirement for L to be maximal corresponds to

$$J = L - S.$$

Therefore, $L = J + S = 5$ and the spectral symbol of this state is $^5H_3$.

6.5 The wavefunction of a system of identical particles

For particles that obey the laws of classical physics, it is not important whether they are identical or not. If we assign numbers to these particles we can track their motion along each particle's trajectory. In quantum physics the situation radically changes since microscopic particles do not follow trajectories and thus it is impossible to track their motion. Because of the uncertainty principle, identical particles in the same region become indistinguishable and their numbering does not make sense. Therefore, the description of the behavior of a system of identical particles is subject to an additional requirement. Besides normalization and orthogonality, for the wavefunction of a system of particles we must take into account the fact that they are indistinguishable.

To illustrate this, let us consider a system of two identical particles. For a complete description we need the set of variables that define the position of each particle in space, r_j, and the projection of spin on the chosen axis, σ_j. The wavefunction of the system, $\psi(\mathbf{r}_1, \sigma_1; \mathbf{r}_2, \sigma_2)$, must depend on these variables. Since the particles are indistinguishable, any permutation of the particles would have no effect on the physical results. This means that permutation of the arguments $(\mathbf{r}_1, \sigma_1)$ and $(\mathbf{r}_2, \sigma_2)$ can change only the phase factor of the wavefunction, i.e.,

$$\psi(\mathbf{r}_1, \sigma_1; \mathbf{r}_2, \sigma_2) = \mathrm{e}^{\mathrm{i}\theta}\psi(\mathbf{r}_2, \sigma_2; \mathbf{r}_1, \sigma_1). \tag{6.78}$$

As a result of the second permutation of arguments on the right-hand side of this equation, the wavefunction takes the initial form but with a new additional phase factor, i.e.,

$$\psi(\mathbf{r}_1, \sigma_1; \mathbf{r}_2, \sigma_2) = \mathrm{e}^{2\mathrm{i}\theta}\psi(\mathbf{r}_1, \sigma_1; \mathbf{r}_2, \sigma_2). \tag{6.79}$$

Since the second permutation of particles (or arguments) returns the system to its initial physical state, the following equalities have to be satisfied:

$$\mathrm{e}^{2\mathrm{i}\theta} = 1 \quad \Rightarrow \quad \mathrm{e}^{\mathrm{i}\theta} = \pm 1. \tag{6.80}$$

Therefore, the permutation of the wavefunction's arguments requires the satisfaction of the following condition:

$$\psi(\mathbf{r}_1, \sigma_1; \mathbf{r}_2, \sigma_2) = \pm\psi(\mathbf{r}_2, \sigma_2; \mathbf{r}_1, \sigma_1). \tag{6.81}$$

It follows from Eq. (6.81) that the wavefunction of a system of two identical particles either does not change after permutation of its particles or changes its sign. In the first case the wavefunction is called *symmetric* and in the second, *antisymmetric*. Thus, because the particles are indistinguishable, the wavefunction of the system of particles must have some kind of symmetry with respect to permutations.

Relativistic quantum theory shows that the symmetry of the wavefunction is unambiguously defined by the spin of the particles. A system of particles with integer spins (bosons) must be described by a symmetric wavefunction, whereas a system of particles with half-integer spins (fermions) must be described by an antisymmetric wavefunction. Since electrons have half-integer spin, the wavefunction of a system of electrons must be antisymmetric.

Let us consider the algorithm for finding the wavefunction which satisfies the above-mentioned condition, using the example of two free non-interacting electrons. The wavefunction of a free electron without taking into account its spin has the form (3.18). Taking into account the spin variables, the wavefunction of the system of two electrons with momenta $\mathbf{p}_1$ and $\mathbf{p}_2$ must be represented (without taking into account its dependence on time) as

$$\psi(\mathbf{r}_1, \sigma_1; \mathbf{r}_2, \sigma_2) = A\mathrm{e}^{\frac{\mathrm{i}}{\hbar}(\mathbf{p}_1\mathbf{r}_1 + \mathbf{p}_2\mathbf{r}_2)}\varphi_1(\sigma_1)\varphi_2(\sigma_2), \tag{6.82}$$

where σ_1 and σ_2 are the projections of the spin of each of the electrons onto the chosen axis. Together with the solution (6.82) of the corresponding Schrödinger equation, three additional wavefunctions, which are also solutions of the Schrödinger equation, may be obtained from Eq. (6.82) by permutations of spatial variables $\mathbf{r}_1 \leftrightarrow \mathbf{r}_2$ and spin variables $\sigma_1 \leftrightarrow \sigma_2$. These four functions give us two linear combinations, which have the correct type of symmetry – two linear combinations are made antisymmetric. These wavefunctions can be written in the following form:

$$\psi_s(\mathbf{r}_1, \sigma_1; \mathbf{r}_2, \sigma_2) = \psi_s(\mathbf{r}_1, \mathbf{r}_2)\varphi_{as}(\sigma_1, \sigma_2), \tag{6.83}$$

$$\psi_{as}(\mathbf{r}_1, \sigma_1; \mathbf{r}_2, \sigma_2) = \psi_{as}(\mathbf{r}_1, \mathbf{r}_2)\varphi_s(\sigma_1, \sigma_2). \tag{6.84}$$

Here, the symmetric and antisymmetric parts of the coordinate dependence of the wavefunction of the system of two free electrons are given by the expressions

$$\psi_s(\mathbf{r}_1, \mathbf{r}_2) = A\left[e^{\frac{i}{\hbar}(\mathbf{p}_1\mathbf{r}_1+\mathbf{p}_2\mathbf{r}_2)} + e^{\frac{i}{\hbar}(\mathbf{p}_1\mathbf{r}_2+\mathbf{p}_2\mathbf{r}_1)}\right], \tag{6.85}$$

$$\psi_{as}(\mathbf{r}_1, \mathbf{r}_2) = A\left[e^{\frac{i}{\hbar}(\mathbf{p}_1\mathbf{r}_1+\mathbf{p}_2\mathbf{r}_2)} - e^{\frac{i}{\hbar}(\mathbf{p}_1\mathbf{r}_2+\mathbf{p}_2\mathbf{r}_1)}\right], \tag{6.86}$$

and the symmetric and antisymmetric parts of the spin dependence by

$$\varphi_s(\sigma_1, \sigma_2) = \varphi_1\left(\frac{1}{2}\right)\varphi_2\left(\frac{1}{2}\right), \tag{6.87}$$

$$\varphi_{as}(\sigma_1, \sigma_2) = \varphi_1\left(\frac{1}{2}\right)\varphi_2\left(-\frac{1}{2}\right). \tag{6.88}$$

Thus, in the symmetric spin state the spins of the two electrons are parallel to each other and the total projection of the spin of the system is equal to $S_z = 1$. In the antisymmetric spin state the spins of the two electrons are anti-parallel to each other and $S_z = 0$. The procedure for finding the correct wavefunctions is sometimes called *symmetrization*. The operation of symmetrization must be carried out in all cases of systems with identical particles.

If there are more than two identical particles, then the system has such states, whose wavefunctions, for particles with integer spins, are symmetric with respect to permutations of any pair of particles (i.e., the wavefunctions do not change their sign), and, for particles with half-integer spins, are antisymmetric with respect to the permutations of any pair of particles (i.e., the wavefunctions change their sign to the opposite). One of the formulations of the Pauli exclusion principle which relates to a system of electrons is that the *total wavefunction of the system must be antisymmetric, since electrons are particles with half-integer spin*. This formulation is closely connected with the fact that it is not possible to have more than one electron occupying the same quantum state.

For complex particles (nuclei, atoms, and molecules) the following rule approximately applies: if a particle consists of an odd number of fermions (electrons, protons, or neutrons), then its spin is half-integer and it behaves as a fermion. If the complex particle consists of an even number of fermions, then its spin is integer and it behaves as a boson. Looking at a particular example, individual helium atoms of types ${}_2\mathrm{He}^3$ and ${}_2\mathrm{He}^4$ are chemically indistinguishable. The atom ${}_2\mathrm{He}^3$ is a fermion since it contains two electrons and a nucleus, which consists of two protons and one neutron. The atom ${}_2\mathrm{He}^4$ is a boson since its nucleus contains one additional neutron. This difference substantially affects the properties of a system consisting of a large number of the considered isotopes of helium. Liquid helium of type ${}_2\mathrm{He}^4$ at temperature $T \approx 2$ K becomes a superfluid, but liquid helium of type ${}_2\mathrm{He}^3$ does not exhibit such a property. This is because the creation of a so-called *Bose–Einstein* condensate is possible only in a system of bosons. Its main property is the transition of the entire ensemble of particles to the lowest energy level when the phase-transition temperature is reached, which for ${}_2\mathrm{He}^4$ is approximately equal to 2 K.

Example 6.6. Show that, if at some time a quantum system that consists of identical particles is in the state described by the symmetric wavefunction Ψ_S, then it will be described by the symmetric wavefunction for all subsequent times.
Reasoning. Let us write the time derivative of the wavefunction as

$$\frac{\mathrm{d}\Psi_\mathrm{s}}{\mathrm{d}t} \to \frac{\Delta\Psi_\mathrm{s}}{\Delta t}, \tag{6.89}$$

where by $\Delta\Psi_\mathrm{s}$ we understand the change of the wavefunction during the time Δt. Let us substitute expression (6.89) into the time-dependent Schrödinger equation (2.182), which describes the initial state of the system:

$$\Delta\Psi_\mathrm{s} = \frac{1}{\mathrm{i}\hbar}\hat{H}\Psi_\mathrm{s}\,\Delta t. \tag{6.90}$$

Since the Hamiltonian $\hat{H}$ is symmetric with respect to the system of coordinates of the particles, the wavefunction $\hat{H}\Psi_\mathrm{s}$ is also a symmetric function of coordinates. Therefore, in the process of evolution of the wavefunction, which is defined by Eq. (6.90), its symmetry does not change. Let us note that the preservation of the symmetry of the wavefunction is a universal property. The same argument applies also to an antisymmetric wavefunction.

6.6 The hydrogen molecule

Let us consider one more important example where the Pauli exclusion principle considerably affects the formation of stationary quantum states. This is the system of two atoms of hydrogen, which form a *hydrogen molecule*. The wavefunctions of each of the electrons in their atoms substantially differ from zero only in a small area close to their nucleus. If the atoms are separated from each other by such a distance that they can be considered to be independent, the wavefunctions of electrons do not overlap and for this reason it is meaningless to make the total wavefunction of the system of two electrons antisymmetric. If the atoms are brought together to a distance such that the wavefunctions of individual electrons

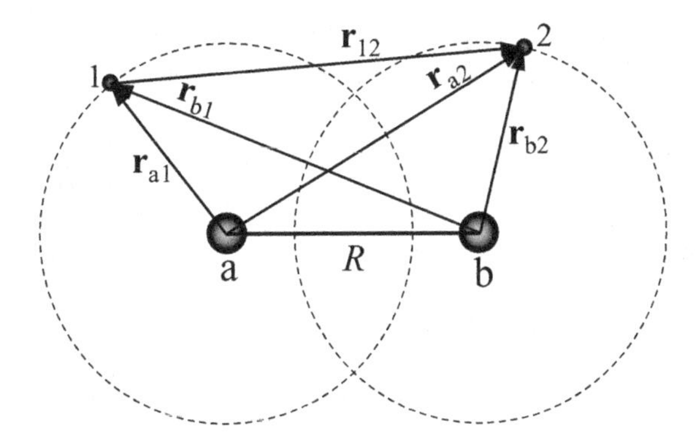

Figure 6.13 The scheme of the hydrogen molecule. The nuclei are designated as a and b, the electrons as 1 and 2.

overlap, then the quantum states of such a system of electrons must be described by a total antisymmetric wavefunction.

Let us place the nuclei a and b of two hydrogen atoms such that they are separated by a distance R (see Fig. 6.13). Let us use an adiabatic approximation, whereby the kinetic energy of the nuclei can be neglected and we can consider the nuclei as motionless. In the zeroth approximation we can consider the atoms as isolated and therefore the interaction between them can be neglected. Electron 1 near nucleus a and electron 2 near nucleus b are described by the atomic wavefunctions of the 1s-state given in Table 6.2. Let us denote these functions as

$$\psi_a(\mathbf{r}_1) \equiv \psi_a(1), \tag{6.91}$$

$$\psi_b(\mathbf{r}_2) \equiv \psi_b(2). \tag{6.92}$$

These functions satisfy the following equations:

$$\hat{H}_a^{(0)} \psi_a(1) = E_1 \psi_a(1), \tag{6.93}$$

$$\hat{H}_b^{(0)} \psi_b(2) = E_1 \psi_b(2), \tag{6.94}$$

where the Hamiltonians of the unperturbed atoms are as follows:

$$\hat{H}_a^{(0)} = -\frac{\hbar^2}{2m_e} \nabla_1^2 - \frac{k_e e^2}{r_{a1}}, \tag{6.95}$$

$$\hat{H}_b^{(0)} = -\frac{\hbar^2}{2m_e} \nabla_2^2 - \frac{k_e e^2}{r_{b2}}. \tag{6.96}$$

Here E_1, which is defined by Eq. (6.8), is the energy of the electron in the ground state of the hydrogen atom. The total energy of the electron in the hydrogen molecule in the zeroth approximation is equal to twice the energy of the electron in the hydrogen atom, i.e.,

$$E_e^{(0)} = 2E_1 = -\frac{k_e^2 m_e e^4}{\hbar^2}. \tag{6.97}$$

If we consider that the atoms are not isolated we have to take into account all the interactions in the hydrogen molecule: interaction of electron 1 with nucleus b, interaction of electron 2 with nucleus a, and interaction of electron 1 with electron 2. In this case the Schrödinger equation for the hydrogen molecule will be the following:

$$[\hat{H}^{(0)} + \hat{W}]\psi(1,2) = E_e\psi(1,2), \tag{6.98}$$

where the unperturbed Hamiltonian, $\hat{H}^{(0)}$, is defined as

$$\hat{H}^{(0)} = \hat{H}_a^{(0)} + \hat{H}_b^{(0)}. \tag{6.99}$$

The perturbation term, $\hat{W}$, is the potential energy of the interactions mentioned above:

$$\hat{W} = W(1,2) = \left(\frac{1}{r_{12}} - \frac{1}{r_{a2}} - \frac{1}{r_{b1}}\right) k_e e^2. \tag{6.100}$$

Since the nuclei are considered motionless, their interaction is not included in the total Hamiltonian, and therefore we will take into account this term by adding e^2/R to the total energy.

The wavefunctions of the unperturbed Hamiltonian of the hydrogen molecule, $\hat{H}^{(0)}$, can be constructed from two one-particle wavefunctions (6.3). In order for the total wavefunction of the system of two electrons to be antisymmetric, the radial part of this function must be symmetric in the case of the total spin equal to $S = 0$ and antisymmetric in the case of the total spin equal to $S = 1$, i.e.,

$$\psi_s^{(0)}(1,2) = \psi_a(1)\psi_b(2) + \psi_a(2)\psi_b(1), \tag{6.101}$$

$$\psi_{as}^{(0)}(1,2) = \psi_a(1)\psi_b(2) - \psi_a(2)\psi_b(1). \tag{6.102}$$

These functions have the following property: after permutation of coordinates (replacement $\mathbf{r}_1 \leftrightarrow \mathbf{r}_2$, i.e., $1 \leftrightarrow 2$) the symmetric function stays the same and the antisymmetric one changes its sign:

$$\psi_s^{(0)}(1,2) = \psi_s^{(0)}(2,1), \tag{6.103}$$

$$\psi_{as}^{(0)}(1,2) = -\psi_{as}^{(0)}(2,1). \tag{6.104}$$

Using the methods of perturbation theory for degenerate stationary states (see Section 5.2), we can show that under the influence of perturbation, $\hat{W}$, the unperturbed degenerate level E_e splits into two. The one of these levels with lower energy, E_e^s, corresponds to the symmetric wavefunction, $\psi_s^{(0)}(1,2)$, and the other one, with higher energy, E_e^{as}, corresponds to the antisymmetric wavefunction, $\psi_{as}^{(0)}(1,2)$. The expressions for the energies of the symmetric and antisymmetric states have the following forms:

$$E_e^{as} = 2E_1 + W_{as}(R), \tag{6.105}$$

$$E_e^s = 2E_1 + W_s(R), \tag{6.106}$$

where

$$W_s = \frac{Q + A}{1 + F^2} \qquad (6.107)$$

and

$$W_{as} = \frac{Q - A}{1 - F^2}. \qquad (6.108)$$

Here we introduced the following quantities:

$$Q = \int |\psi_a(1)|^2 W(1,2) |\psi_b(2)|^2 dV_1\, dV_2, \qquad (6.109)$$

$$A = \int \psi_a^*(1)\psi_b(2) W(1,2) \psi_a^*(2)\psi_b(1) dV_1\, dV_2, \qquad (6.110)$$

$$F = \int \psi_a^*(1)\psi_b(1) dV_1 = \int \psi_a^*(2)\psi_b(2) dV_2. \qquad (6.111)$$

The quantity Q is called the *Coulomb integral* since it includes the energy of the electrostatic interaction of the electron clouds with each other and with the atomic nuclei. The electron clouds have the following electric charge density:

$$\rho_a(1) = -e|\psi_a(1)|^2, \qquad (6.112)$$

$$\rho_b(2) = -e|\psi_b(2)|^2. \qquad (6.113)$$

The quantity A, which is called the *exchange integral*, is closely connected with the wave nature of the electron. In this integral there are two "exchange charge densities":

$$\rho_{ex}(1) = -e\psi_a^*(1)\psi_b(1), \qquad (6.114)$$

$$\rho_{ex}(2) = -e\psi_a^*(2)\psi_b(2), \qquad (6.115)$$

which can be interpreted as if each of the electrons belonged to nucleus a and nucleus b simultaneously.

The quantity F is called the *overlap integral* and it is a measure of the overlap of the wavefunctions of atoms (i.e., overlap of electron clouds) at a given distance between them, R. All the above-mentioned quantities are functions of this distance, R, and consequently the corrections to the energy, E_e (Eqs. (6.107) and (6.108)), also depend on R.

In analyzing the expressions (6.105)–(6.111) for the energy of the system of electrons E_e^s and E_e^{as} we emphasize that, when atoms approach each other, the energy level $E_e = 2E_1$ defined by Eq. (6.97) splits into two sublevels. The magnitude of the splitting depends on the distance between nuclei and is equal to

$$\Delta E(R) = E_e^{as}(R) - E_e^s(R) = W_{as}(R) - W_s(R) = -2\frac{A - QF^2}{1 - F^4}. \qquad (6.116)$$

Under the assumption of the adiabatic approximation, the energy of the electrons plays a role as one of the components of the potential energy of interaction of the

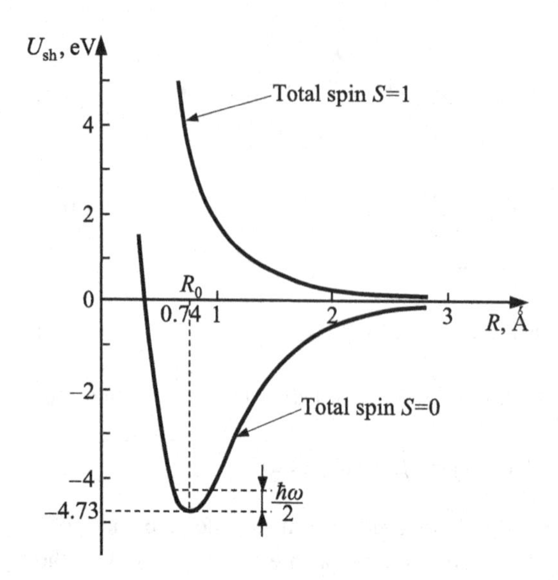

Figure 6.14 Two dependences of the potential energy U on R: (1) the dependence with the correction W_{as} (total spin equal to $S = 1$) and (2) the dependence with the correction W_s (total spin equal to $S = 0$). Here $U_{sh} = U(R) - 2E_1$ and $R_0 = 0.74$ Å.

atoms. The total potential energy of interaction of the hydrogen atoms, including the energy of the Coulomb repulsion of the nuclei, is defined by the following expression:

$$U(R) = 2E_1 + \frac{k_e e^2}{R} + \begin{cases} W_{as}(R), \\ W_s(R). \end{cases} \tag{6.117}$$

The total potential energy of interaction of the hydrogen atoms strongly depends on the total spin of the system of electrons. Figure 6.14 shows two dependences of the potential energy $U(R)$: (1) the dependence that corresponds to the correction W_{as} and to the total spin $S = 1$ and (2) the dependence that corresponds to the correction W_s and to the total spin $S = 0$. The monotonic decrease of energy with increasing R in the state with parallel spins means that atoms repel each other. Thus, it is impossible to form the hydrogen molecule with $S = 1$. The minimum potential energy for the hydrogen molecule, $U(R_0) = -4.73$ eV, which occurs at $R_0 = 0.74$ Å $\approx 1.4r_1$, occurs only for the state with opposite spins. Therefore, for this state ($S = 0$) the bonding between hydrogen atoms, which is called *covalent bonding*, occurs. This type of bonding plays a significant role in the formation of molecules as well as of numerous crystals.

Example 6.7. Estimate the oscillation frequency of the nuclei of the hydrogen molecule in the ground state and the dissociation energy of the hydrogen molecule (i.e., the energy required for the breakup of the hydrogen molecule into separate atoms).

Reasoning. The adiabatic approximation can be used for the solution of the problem of oscillations of nuclei in the hydrogen molecule. To find the wavefunction of the nuclei $\Phi(R)$ in a given quantum state we will take the energy of the electrons, $E_e(R)$, as the potential energy of the interaction of nuclei. Let us consider the motion of the nuclei in the system of their center of mass. Then, the wavefunction of the nuclei, $\Phi(a, b)$, will depend only on the relative distance, R, between them and define their relative motion under the influence of internal forces. To find this function it is necessary to solve the following Schrödinger equation:

$$\left[-\frac{\hbar^2}{2m_{ab}}\frac{d^2}{dR^2} + U(R)\right]\Phi(R) = E\Phi(R), \tag{6.118}$$

where the reduced mass of two identical nuclei, m_{ab}, is equal to the half-mass of the proton,

$$m_{ab} = \frac{m_a m_b}{m_a + m_b} = \frac{m_p}{2}, \tag{6.119}$$

and E is the total energy of the hydrogen molecule, taking into account the motion of the nuclei. The magnitude $U(R)$ is the total potential energy of the nuclei and it is defined by Eq. (6.117). It includes the energy of the Coulomb interaction of the nuclei and the energy of the electrons, $E_e(R)$, in the given quantum state. To find the frequency of the small-amplitude oscillations of the nuclei let us expand the function $U(R)$ near the equilibrium position of the nuclei R_0 in series over the displacement $R - R_0$ and let us keep only the first two terms of the expansion:

$$U(R) = U(R_0) + \frac{\beta}{2}(R - R_0)^2, \tag{6.120}$$

where

$$\beta = \left(\frac{d^2U}{dR^2}\right)_{R=R_0}. \tag{6.121}$$

Taking into account the relation of the elastic constant with frequency, we get

$$\omega = \sqrt{\frac{\beta}{m_{ab}}} = \sqrt{\frac{2}{m_p}\left(\frac{d^2U}{dR^2}\right)_{R=R_0}}. \tag{6.122}$$

The expression for the energy $U(R)$ can be rewritten in the form

$$U(R) = U(R_0) + \frac{m_{ab}\omega^2}{2}(R - R_0)^2. \tag{6.123}$$

On choosing R_0 as the origin of coordinates and $U(R_0)$ as the reference point for energy, i.e., taking $R_0 = 0$ and $U(R_0) = 0$, we get the equation

$$\left[-\frac{\hbar^2}{2m_{ab}}\frac{d^2}{dR^2} + \frac{m_{ab}\omega^2}{2}R^2\right]\Phi(R) = E\Phi(R), \tag{6.124}$$

which coincides with the one-dimensional Schrödinger equation for an electron in the parabolic potential (4.74). On solving this equation we get the quantized harmonic oscillator states for the system of the nuclei, whose energy according to Eq. (4.88) is given by the expression

$$E_{\text{osc}} = \hbar\omega\left(n + \frac{1}{2}\right), \tag{6.125}$$

where n is the quantum number of the oscillator. The ground state of the hydrogen molecule corresponds to the energy of the $n = 0$ oscillator state:

$$E_{\text{osc}} = \frac{\hbar\omega}{2} = 0.27 \text{ eV}. \tag{6.126}$$

This value of oscillatory motion energy was established from measurements of the optical spectra of the hydrogen molecule. From Eq. (6.126) it follows that the oscillations of the nuclei happen with frequency $\omega = 8.6 \times 10^{14}\ \text{s}^{-1}$. The oscillatory motion of the nuclei splits the electron levels into closely spaced oscillatory sublevels. Then, the energy levels of the hydrogen molecule are given by the expression

$$E(R) = E_{\text{e}}(R_0) + \frac{k_{\text{e}}e^2}{R_0} + \hbar\omega\left(n + \frac{1}{2}\right). \tag{6.127}$$

The energy of dissociation, E_{dis}, of the hydrogen molecule is defined by the minimal energy necessary for dividing the hydrogen molecule into two individual atoms:

$$E_{\text{dis}} = \left|U(R_0) + \frac{\hbar\omega}{2}\right| = |-4.73 + 0.27|\ \text{eV} = 4.46\ \text{eV}. \tag{6.128}$$

We note that a more precise solution of the problem for the hydrogen molecule requires the taking into account of two more factors – rotational motion of the molecule and the fact that the nuclei are identical. Under the influence of rotational motion the oscillatory energy levels split into a system of rotatory sublevels. The distance between these sublevels is of the order of 8×10^{-3} eV.

6.7 Summary

1. The electron state in the hydrogen atom is defined by the set of four quantum numbers: the principal quantum number, n, orbital quantum number, l, orbital magnetic quantum number m, and spin magnetic quantum number m_s. The electron energy spectrum of the hydrogen atom, which is defined only by the principal quantum number, n, is discrete and all quantized levels are in the region of negative energy values.
2. One of the main principles of quantum physics is the Pauli exclusion principle, according to which a *particular state, which is characterized by a given set of quantum numbers (in the case of an atom the quantum numbers are n, l, m, m_s), may be occupied by only one electron.*
3. Considering the system of identical particles, we have to take into account that in a quantum description the particles are distinguishable. Systems of particles with integer spin (bosons) must be described by symmetric wavefunctions, whereas systems

of particles with half-integer spin (fermions) must be described by antisymmetric wavefunctions. Since electrons have spin $s = 1/2$, the wavefunction of the electron system must be antisymmetric.

4. All electrons of a many-electron atom with the same principal quantum number, n, form a shell, which has $2n^2$ states. Shells consist of subshells, which differ by the orbital quantum number l. The number of different states in a subshell with different quantum numbers m and m_s is equal to $2(2l + 1)$. Completely filled shells and subshells have the resulting quantum numbers of the orbital, spin, and total moment equal to zero: $L = 0$, $S = 0$, and $J = 0$.
5. The order of filling by electrons of the atom's subshells is defined by Hund's rules: (1) the electron configuration has the minimal energy in the state with the largest spin S and with the largest (for such S) L; (2) for a less than half-filled subshell the quantum number J is equal to $J = |L - S|$, whereas for a more than half-filled subshell J is equal to $J = L + S$.
6. The covalent bonding between hydrogen atoms is caused by the exchange interaction, and this type of bonding occurs for the state with anti-parallel spins ($S = 0$). This type of bonding plays an important role in the formation of molecules and numerous crystals.

6.8 Problems

Problem 6.1. Using Bohr's model of the hydrogen atom and his postulates, find the radii of electron stationary orbits, r_n, electron velocities, v_n, orbital periods, T_n, the electron energy on the corresponding orbit, and the circular frequency of a photon emitted during electron transition from the nth to the mth orbit. Find the radius and velocity, r_1 and v_1, for the first electron orbit. Find the angular frequency, ω_{21}, of the electron transition from the second to the first orbit. (Answer: $r_1 \approx 0.53$ Å, $v_1 \approx 2.19 \times 10^6\ \mathrm{m\,s^{-1}}$, and $\omega_{21} \approx 1.53 \times 10^{16}\ \mathrm{s^{-1}}$.)

Problem 6.2. Find the minimum, $\lambda_{\min}$, and maximum, $\lambda_{\max}$, wavelength of the hydrogen spectral lines in the visible range of the spectrum. What is the minimum speed, $v_{\min}$, of electrons incident on a hydrogen atom to excite these spectral lines? (Answer: $\lambda_{\min} \approx 365$ nm, $\lambda_{\max} \approx 656$ nm, and $v_{\min} \approx 8.2 \times 10^6\ \mathrm{m\,s^{-1}}$.)

Problem 6.3. The electron in the hydrogen atom is in the ground state described by the following wavefunction:

$$\psi_{1s}(r) = \left(\pi r_1^3\right)^{-1/2} e^{-r/r_1}. \tag{6.129}$$

Find the average electrostatic potential at the distance r from the nucleus.

Problem 6.4. Find the spectrum of the vibrational levels of a diatomic molecule. For the diatomic CO molecule, estimate the distance between vibrational energy levels, $\Delta E_{n+1,n}$, and the emission wavelength, λ, during the corresponding transitions. The carbon atom mass $m_1 = 1.99 \times 10^{-26}$ kg, the oxygen atom mass $m_2 = 2.66 \times 10^{-26}$ kg, and the force constant $\beta = 190$ kg s^{-2}. (Answer: $\Delta E_{n+1,n} \approx 8.44 \times 10^{-2}$ eV and $\lambda \approx 24$ μm. Hint: use Eq. (6.122).)

Problem 6.5. Find for the first-order perturbation theory the energy, E_1, of the ground state of the helium, He, atom. Use as the perturbation the energy of interaction between electrons. (Answer: $E_1 \approx -18.7$ eV.)

Problem 6.6. In the alkaline atoms the nucleus with charge eZ and the first $Z-1$ electrons form a core with positive charge equal to the elementary charge, e. A valence electron rotates in the field of this core. Find the energy spectrum of the valence electron if its potential energy is defined by the following expression:

$$U(r) = -\frac{k_e e^2}{r}\left(1 + \frac{C}{r}\right), \tag{6.130}$$

where we can consider $C/r_1 \ll 1$ (r_1 is the Bohr radius).

Problem 6.7. Find the angular momentum for electrons in an atom in 3s, 4d, and 5g states and the maximum number of electrons in an atom with the same sets of quantum numbers n, l, and m.

Chapter 7
Quantization in nanostructures

In Chapters 3 and 4 we have discussed electron behavior in potential wells of various profiles and dimensionalities. We have established that localization of electrons in such potential wells, regardless of their form, leads to the discretization of the electron energy spectrum whereby the distance between energy levels substantially depends on the geometrical size of the potential wells. If this size is macroscopic then the distance between the energy levels is so small that we can consider the energy spectrum to be practically continuous (or *quasicontinuous*). Electrons in metallic samples of macroscopic sizes have this kind of energy spectrum. Another limiting case is that of small clusters consisting of just a few atoms, where the distance between energy levels is of the order of electron-volts. Gradual decrease of one or several geometrical dimensions of the potential well from macroscopic to about 1 μm practically does not change the form of the electron energy spectrum. Very often macroscopic materials (or macroscopic crystals) are referred to as *bulk materials* or *bulk crystals*. Changes happen only when the size of structures is of the order of or less than 100 nm. Such structures are called *nanostructures*. The change of the electron spectrum from quasicontinuous to discrete implies changes in most of the physical properties of nanostructures compared with those in bulk crystals. In this chapter we will consider the main peculiarities of the electron energy spectrum in nanostructures of various dimensionalities.

7.1 The number and density of quantum states

Electric, optical, thermal, and other properties of *macroscopic crystalline materials* significantly depend on the energy states of electrons. Let us define two quantities, which are important for describing the features of the electron energy spectrum in these materials. The first quantity, $N(E)$, is the *number of quantum states* corresponding to the energy interval $(0, E)$. The second quantity, $g(E)$, called the *density of states*, defines the number of quantum states, N, corresponding to the unit energy interval in the vicinity of E. In order to calculate these quantities we consider first a three-dimensional potential well, which represents bulk crystal. Such a structure is called *three-dimensional* because of the number of dimensions along which the electron can freely move. In such a case the electron momentum space is also three-dimensional. For the analysis of electron behavior in a crystal in an external field an approximate method, which is called the *effective-mass approximation*, is widely used. In this approximation the Schrödinger equation for an electron with free electron mass, m_e, in the periodic inner-crystal potential is reduced to the Schrödinger equation which describes the behavior of free electrons with mass equal to the so-called *effective mass*, m^*. We will show in Section 7.6.3 how to define the effective mass of an electron.

To find how electrons are distributed over the allowed quantum states in a crystal with linear dimensions L_x, L_y, and L_z, where they can freely move, we will assume the boundary conditions (4.4): $\psi(0, y, z) = \psi(x, 0, z) = \psi(x, y, 0) = \psi(L_x, y, z) = \psi(x, L_y, z) = \psi(x, y, L_z) = 0$. The components of momentum according to Eq. (4.15) can take the following values:

$$p_x = \frac{\pi\hbar}{L_x} n_x, \qquad p_y = \frac{\pi\hbar}{L_y} n_y, \qquad p_z = \frac{\pi\hbar}{L_z} n_z, \tag{7.1}$$

where $n_\alpha = 1, 2, 3, \ldots$ and $\alpha = x, y$, and z. We will consider first the case of large L_x, L_y, and L_z. Thus, the spacing between components of the momentum, $\Delta p_\alpha = \pi\hbar/L_\alpha$, is small and we can consider the momentum, p, to be *quasicontinuous*. Therefore, in order to calculate the number of states we can use integration over p instead of summation.

Let us choose an infinitesimally small volume of momentum space, $\mathrm{d}V_p$, around some momentum $\mathbf{p}$:

$$\mathrm{d}V_p = \mathrm{d}p_x \,\mathrm{d}p_y \,\mathrm{d}p_z. \tag{7.2}$$

It is easy to carry out calculations for a *macroscopic crystal*, i.e., for macroscopic values of L_x, L_y, and L_z. In this case the change of momentum projections, p_x, p_y, and p_z, is practically continuous and the electron motion can be considered as a classical motion. Taking into account Eq. (7.1), we obtain the following expression for the quantum numbers n_α, which define the number of quantum states:

$$n_\alpha = \frac{L_\alpha}{\pi\hbar} p_\alpha, \tag{7.3}$$

where $\alpha = x$, y, and z. On differentiating Eq. (7.3), we find the number of quantum states $\mathrm{d}n_\alpha$ in the interval $\mathrm{d}p_\alpha$:

$$\mathrm{d}n_\alpha = \frac{L_\alpha}{\pi\hbar} \mathrm{d}p_\alpha. \tag{7.4}$$

The number of quantum states is equal to the number of sets of n_x, n_y, and n_z in this three-dimensional interval of numbers $\mathrm{d}n = \mathrm{d}n_x \,\mathrm{d}n_y \,\mathrm{d}n_z$. We have to take into account the fact that each value of n_α corresponds to the two values of p_α with the same absolute value, but with opposite signs. Therefore, the total number of states in the three-dimensional interval of momentum space is equal to

$$dn = \frac{L_x L_y L_z}{(2\pi\hbar)^3}\, dp_x\, dp_y\, dp_z. \quad (7.5)$$

In this expression the physical quantity $V = L_x L_y L_z$ represents the volume in real space which is available for an electron, and the quantity $dV_p = dp_x\, dp_y\, dp_z$ represents the differential volume element in the momentum space available for an electron. In the case of a free electron the momentum space is *isotropic* since all directions are equivalent. Therefore, for the differential volume of the momentum space, we can use spherical coordinates and write

$$dV_p = dp_x\, dp_y\, dp_z = 4\pi p^2\, dp. \quad (7.6)$$

Taking into account all the above, the expression (7.5) can be rewritten as

$$dn(p) = \frac{V}{2\pi^2\hbar^3} p^2\, dp. \quad (7.7)$$

The total number of states in the momentum space of radius p is defined as follows:

$$N(p) = 2\int dn(p) = \frac{V}{\pi^2\hbar^3}\int_0^p p^2\, dp = \frac{V}{3\pi^2\hbar^3} p^3. \quad (7.8)$$

Note that we put a factor of 2 in front of the integral in Eq. (7.8). This is because the set of quantum numbers n_x, n_y, and n_z does not completely define the state of an electron. Because the electron has its own internal angular momentum (spin) (see Section 6.3), the quantum state of the electron has to be defined by an additional quantum number m_s:

$$m_s = \pm\frac{1}{2}. \quad (7.9)$$

According to the Pauli exclusion principle, two electrons with opposite spins can be in a state with the same momentum **p**. Therefore, the number of quantum states is twice as large as it otherwise would be.

The number of states that corresponds to the unit interval of momentum space is given by the *density of states*, $g(p)$:

$$g(p) = \frac{dN(p)}{dp} = \frac{p^2 V}{\pi^2\hbar^3}. \quad (7.10)$$

We see that with increasing momentum, p, the number of states, $N(p)$, increases proportionally to the momentum raised to the third power, and that the density of states, $g(p)$, increases proportionally to the square of the momentum.

In many cases it is more convenient to consider energy, E, rather than momentum, **p**, as a variable. In order to do this, we have to perform a transformation of the equations from the momentum space to the energy space. Taking into account that the energy of a free electron has a quadratic dependence on the momentum, the electron surface of equal energy in the momentum space, **p**, is a sphere of radius p:

$$E = \frac{p^2}{2m^*}, \tag{7.11}$$

$$\mathrm{d}E = \frac{p}{m^*}\,\mathrm{d}p, \tag{7.12}$$

and

$$\mathrm{d}p = \sqrt{\frac{m^*}{2E}}\,\mathrm{d}E. \tag{7.13}$$

By substituting Eqs. (7.11) and (7.13) into Eq. (7.7) we obtain the following expression for the number of states in the energy interval $\mathrm{d}E$:

$$\mathrm{d}n(E) = \frac{\sqrt{2}Vm^{*3/2}}{2\pi^2\hbar^3}\sqrt{E}\,\mathrm{d}E. \tag{7.14}$$

The total number of quantum states in the interval $(0, E)$ is defined by the integral

$$N(E) = 2\int \mathrm{d}n(E) = 2\frac{\sqrt{2}Vm^{*3/2}}{3\pi^2\hbar^3}E^{3/2}. \tag{7.15}$$

For the density of states, $g(E)$, i.e., the number of states that corresponds to the unit energy interval, we obtain

$$g(E) = \frac{\mathrm{d}N(E)}{\mathrm{d}E} = \frac{\sqrt{2}Vm^{*3/2}}{\pi^2\hbar^3}\sqrt{E}. \tag{7.16}$$

We see that the number of states, $N(E)$, for the electron in a three-dimensional space of macroscopic volume V increases with increasing energy proportionally to $E^{3/2}$, and that the density of states, $g(E)$, increases proportionally to $\sqrt{E}$. Both quantities depend linearly on the volume, V, and, therefore, they can be defined per unit volume. In this case, the volume V in Eqs. (7.15) and (7.16) is canceled out. Figure 7.1 shows the electron's surface of equal energy in the momentum space, $E(p)$, and Fig. 7.2 shows the dependences of the number of states, N, and the density of states, g, on energy, E, for the electron's three-dimensional motion in a macroscopic crystal.

Note that Eqs. (7.2)–(7.16) can be used only in the case of crystals with macroscopic volume V. All three dimensions, L_x, L_y, and L_z, are so large that the differences between different values of electron momentum and energy are small, and thus the magnitudes $\mathbf{p}$ and E can be considered as continuous (or quasicontinuous) variables.

If there are N electrons in the volume V, then at zero temperature they occupy the lowest energy states and on each level there are two electrons with opposite spins defined by Eq. (4.18).

The distribution of electrons among the energy levels at an arbitrary temperature, T, is described by *Fermi–Dirac quantum statistics* for particles with spin $m_s = \pm 1/2$. The probability of finding an electron in the state with energy E is

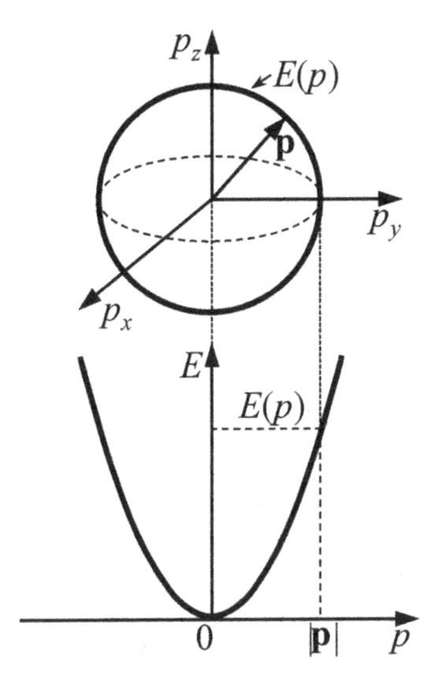

Figure 7.1 The surface of equal energy in the momentum space for the electron's three-dimensional motion. Here, $|\mathbf{p}| = \sqrt{p_x^2 + p_y^2 + p_z^2}$.

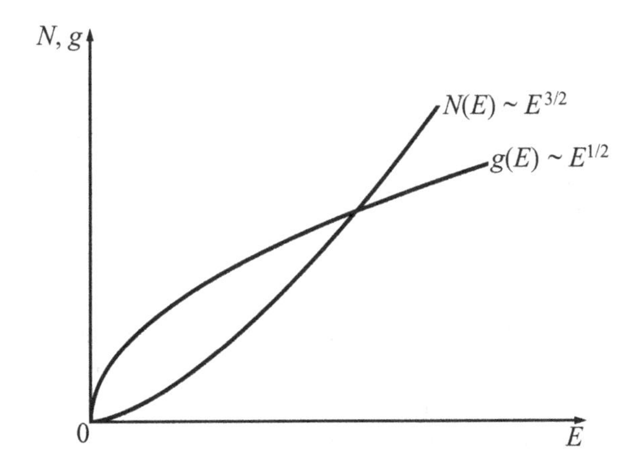

Figure 7.2 The dependences of the number, $N(E)$, and the density of states, $g(E)$, on energy, E, in the case of an electron's three-dimensional motion in a crystal.

given by the *Fermi–Dirac distribution function*, $f(E)$:

$$f(E) = \frac{1}{e^{(E-E_F)/(k_B T)} + 1}, \tag{7.17}$$

where E_F is the *Fermi energy* and k_B is Boltzmann's constant. From Eq. (7.17) it follows that at $T = 0$ K the probability of occupation of all the states with the energy $E \leq E_F$ is equal to unity, and the probability of occupation of the states with energy $E > E_F$ is equal to zero (see Fig. 7.3). Thus, the Fermi energy, E_F, defines the *energy up to which all the energy states at $T = 0$ K are occupied.* Energy states higher than E_F at $T = 0$ K are unoccupied.

Given the importance of the notion of the Fermi energy, let us determine E_F at $T = 0$ K. Let us assume that there are N electrons in the volume V. Since

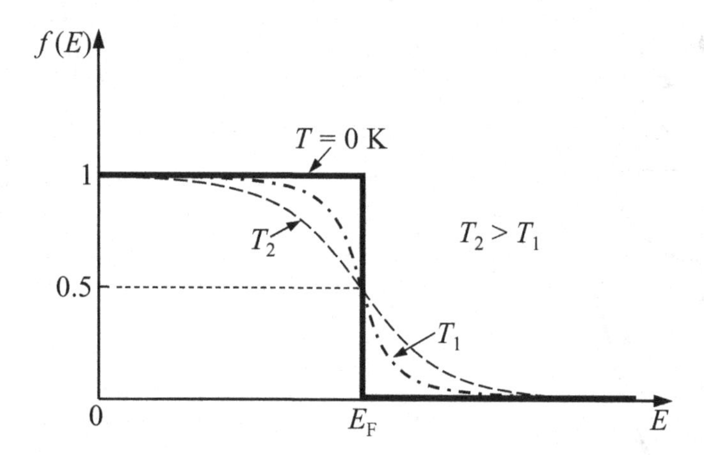

Figure 7.3 The Fermi–Dirac distribution function, $f(E)$, for various temperatures. The distribution functions at temperatures higher than 0 K are shown by dashed and dash–dotted lines. Here, $T_2 > T_1$.

two electrons with opposite spins occupy each available state, the number of electrons, N, that can be placed in all states from $E = 0$ to $E = E_F$ is equal to the number of states $N(E_F)$. According to Eq. (7.15) this number is related to the Fermi energy, E_F, by

$$N = N(E_F) = 2\frac{\sqrt{2}Vm^{*3/2}}{3\pi^2\hbar^3}E_F^{3/2}. \tag{7.18}$$

From the above equation we obtain the following expression for the Fermi energy:

$$E_F = \frac{\hbar^2}{2m^*}\left(\frac{3\pi^2 N}{V}\right)^{2/3}. \tag{7.19}$$

Let us estimate E_F for the conduction electrons in a metal, for example in copper. For copper the electron effective mass is close to the free electron mass: $m^* \approx m_e$. The electron density for copper is equal to $\rho = N/V \approx 10^{29}\ \text{m}^{-3}$. Thus, the Fermi energy is about several electron-volts and the distance between energy levels is infinitesimally small compared with E_F. Therefore, the energy spectrum of the electron gas in metals can be considered to be *quasicontinuous*.

Example 7.1. An electron is confined in a cubic box with impenetrable walls and with the edge lengths $L = 1\ \mu\text{m}$. Find the distance between the levels E_{113} and E_{122}, and compare this energy with the electron thermal energy, E_{th}, at room temperature:

$$E_{th} = \frac{3}{2}k_B T \approx 38.4\ \text{meV}. \tag{7.20}$$

Determine whether the quantum-mechanical approach is required or the classical approach with a quasicontinuous spectrum is applicable. Assume that the mass of the electron in the box is equal to the electron mass in vacuum, $m_e = 9.1 \times 10^{-31}$ kg.

Reasoning. According to Eq. (4.18), the energy level E_{122} has the same magnitude as the other two states with the sets of quantum numbers (2, 1, 2) and (2, 2, 1):

$$E_{122} = E_{212} = E_{221} = \frac{\pi^2\hbar^2}{2m_e L^2}(1^2 + 2^2 + 2^2). \tag{7.21}$$

The energy level E_{113} has the same magnitude as the other two states with the sets of quantum numbers (1, 3, 1) and (3, 1, 1):

$$E_{113} = E_{131} = E_{311} = \frac{\pi^2\hbar^2}{2m_e L^2}(1^2 + 1^2 + 3^2). \tag{7.22}$$

The distance between the energy levels E_{113} and E_{122} is

$$E_{113} - E_{122} = \frac{\pi^2\hbar^2}{m_e L^2} = 8.75 \times 10^{-7}\ \text{eV} = 8.75 \times 10^{-4}\ \text{meV}. \tag{7.23}$$

According to Eq. (7.23), in the cubic box of dimensions 1 μm × 1 μm × 1 μm, the distance between energy levels is so small in comparison with the thermal energy at room temperature of 38.4 meV (see Eq. (7.20)) that we can consider the electron energy spectrum to be quasicontinuous. It follows from the above expression that in order to increase the distance between the adjacent energy levels it is necessary to decrease the size of the region within which the electron is confined.

7.2 Dimensional quantization and low-dimensional structures

As we have shown previously, the quantum properties of an electron become apparent when the electron de Broglie wavelength becomes comparable to the size of the region within which the electron motion takes place. For estimates of this region we have to take into account that electron motion takes place not in vacuum but in a certain medium at a certain temperature. Therefore, we have to replace the mass of the electron in vacuum, m_e, by its effective mass in a medium, m^*. In Section 7.6.3 we will present a detailed discussion about the effective mass of an electron, but for now we note that the electron motion in a crystalline material is not completely free since it takes place under the influence of the inner-crystal potential of atoms. An electron effective mass is introduced to take into account the crystal potential and, in many practical cases, to describe the motion of an electron in a medium, under the influence of external fields, using the conventional equations in which the free-electron mass, m_e, is replaced by the effective mass, m^*, of an electron in the particular medium.

We will use the effective de Broglie wavelength, λ^*, of an electron which is connected with its effective mass, m^*, in a crystalline material, as a criterion for

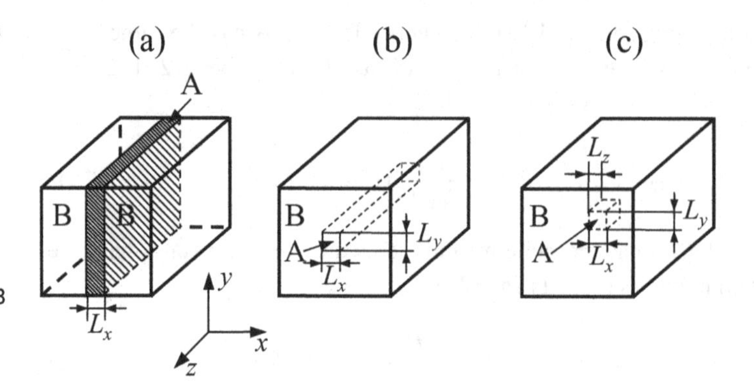

Figure 7.4 Different types of nanostructures: (a) a quantum well, (b) a quantum wire, and (c) a quantum dot. A and B denote the materials which constitute the nanostructures, with A being the material of the nanostructure itself and B being the material of the surrounding matrix. The nanostructure is referred to as *free-standing* if the surrounding matrix, B, is vacuum.

a space quantization:

$$\lambda^* = \frac{h}{p} = \frac{2\pi\hbar}{m^* v} = \frac{2\pi\hbar}{\sqrt{3m^* k_B T}}. \tag{7.24}$$

In the case when electron motion along *only one direction* is limited (for example, along the x-direction) (see Fig. 7.4(a)), i.e., when the condition

$$L_x \le \lambda^* \tag{7.25}$$

is satisfied, the electron energy states corresponding to the motion in this direction are quantized. At the same time the electron motion in the other two directions (y- and z-directions) stays free, with the continuous energy spectrum. Such a structure is called a *quantum well* (QW).

In the case when the electron motion is restricted along *two directions* (for example, along the x- and y-directions), i.e., when the conditions

$$L_x \le \lambda^* \quad \text{and} \quad L_y \le \lambda^* \tag{7.26}$$

are satisfied, the energy which corresponds to these two directions of motion is quantized. Such a structure is called a *quantum wire* (QWR) (see Fig. 7.4(b)).

In the case when the electron motion is restricted along *all three directions*, i.e., when the conditions

$$L_x \le \lambda^*, \quad L_y \le \lambda^*, \quad \text{and} \quad L_z \le \lambda^* \tag{7.27}$$

are satisfied, the energy spectrum of such a structure, which is called a *quantum dot* (QD) (see Fig. 7.4(c)), is totally quantized. Quantum dots can take the form of a cube, a short cylinder, a disk, or a sphere of nanometer size, and they have the potential to become in the future the main building blocks of nanoelectronics. For example, from a set of quantum dots we can make artificial crystals with desired parameters. In semiconductor crystals, which are used for building nanostructures, the ratio of the electron mass in vacuum, m_e, to its effective mass in a semiconductor, m^*, is greater than unity. For example, for GaAs, which is one of the most widely used semiconductors in nanoelectronics, this ratio is approximately equal to 10: $m_e/m^* \approx 10$. Therefore, at room temperature the de Broglie wavelength of an electron in GaAs is of the order of 10 nm according to Eq. (7.24). This defines in accordance with Eqs. (7.25)–(7.27) the size of structures at which the quantization of the electron spectrum becomes noticeable. Note that on lowering the temperature by a factor of 100, i.e., at temperatures of about 3 K, λ^* increases by one order of magnitude. Therefore, at 3 K the quantization is important in structures with sizes of the order of 100 nm.

The simplest nanostructure with the quantum limitation along one of the directions, the *quantum well*, is a thin semiconductor film with a thickness of about $L_x \sim \lambda^*$. The electrons are mostly localized in the film and cannot escape the film because the energy of an electron outside of the film is higher than the energy of an electron within the film, i.e., there is a potential barrier for the electrons outside of the film. The height of the potential barrier that forms a quantum well for electrons in the film is equal to the work function, A_{wf}, which for most semiconductors is about 3–5 eV. This energy is two orders of magnitude higher than the energy of thermal motion of free electrons, $E_{\mathrm{th}} = 3k_{\mathrm{B}}T/2$. At room temperature, E_{th} is about 4×10^{-2} eV and E_{th} decreases linearly with decreasing temperature, T.

We can estimate the ground-state energy of an electron in a quantum well using the uncertainty relationships. In a direction of restricted motion the electron momentum projection is equal to $p_x \sim \Delta p_x \sim h/L_x$. Therefore, for the energy of the ground state we get

$$E_1 \sim \frac{p_x^2}{2m^*} \sim \frac{h^2}{2m^* L_x^2}. \tag{7.28}$$

In a GaAs quantum well with $L_x = 20$ nm and with the effective mass of the electron equal to $m^* = 0.067m_e$ the energy of the ground state is equal to $E_1 \approx 0.1$ eV. The distances between the adjacent lower energy levels are of the same order of magnitude. The potential well formed in the x-direction does not affect the electron motion in the yz-plane of the semiconductor film. The electron energy spectrum corresponding to this motion is continuous and quadratic with respect to momentum, as it is in the case of free electron motion. The total energy of an electron in such a thin semiconductor film has a mixed discrete–continuous character:

$$E = E_n + \frac{p_y^2 + p_z^2}{2m^*}, \tag{7.29}$$

where E_n is the energy of the nth level of the quantized motion in the x-direction. Because of the continuity of the spectrum along the other two directions (the y- and z-directions) electrons that belong to the nth quantum state may have total energy, E, equal to any value within the interval $E_n \le E < \infty$. Such a set of quantum states for the given n is usually called the *subband of dimensional quantization*.

In order to observe the quantization of the energy spectrum experimentally, the distance between the adjacent energy levels has to be sufficiently large. First of all it must be larger than the thermal energy of electrons in a quantum well, i.e.,

$$E_{n+1} - E_n > k_B T. \tag{7.30}$$

This condition must be satisfied in order to exclude the possible thermal transitions of the electron between quantized energy levels which will hinder the observation of the *quantum-dimensional effects* (QDEs). Second, for observation of QDEs the mean free path of the electron, l_e, in the medium must be substantially larger than the size of the region of quantized motion. This is because the quantization takes place only if the electron wavefunction has the form of a standing wave in the region of electron motion. Such a regularity of the wavefunction is possible only in the case of weak electron scattering on vibrations of atoms of the films as well as on defects, which are always present in real structures and destroy the coherent character of the electron motion. To have such a regularity, an electron in the quantum state with momentum p_n must have a chance to make several flights from one wall of the quantum well (quantum wire or quantum dot) to another during its time of free flight, i.e., during the time between two collisions:

$$\tau_e = \frac{l_e}{v_n}, \tag{7.31}$$

where v_n is the characteristic velocity that corresponds to momentum p_n, given by

$$v_n = \frac{p_n}{m^*}.$$

This may be possible only in the case of $l_e \gg L$, where L is the characteristic size of a QW, QWR, or QD. Let us note that criterion (7.30) can be applied to semiconductors in which the electron concentration is low and a small number of levels E_n is filled. Metallic nanostructures are not suitable for the observation of QDEs because, due to the large number of electrons, the energy of conduction electrons in typical metals is several electron-volts, which is much larger than the energy distance between the levels of dimensional quantization. Therefore, nowadays semiconductor thin-film structures are used for the observation of QDEs. It is worth noting that the first experiments in which the quantization of the electron spectrum was observed were carried out with *semimetallic films of bismuth* (first of all the effective mass of electrons in bismuth, Bi, is very small and second, it is easy to grow thin films of bismuth by laser ablation).

Alongside structures containing quantum wells, structures containing a potential barrier with a width of about λ^* are also of great interest. An electron incident at the potential barrier can be either reflected or transmitted through the barrier depending on the ratio of the potential-barrier height and the electron kinetic energy. Structures with quantum wells as well as barriers have numerous applications in nanoelectronic devices.

The logical development of one-well and one-barrier nanostructures is the nanostructures which contain several quantum wells and barriers, as well as periodic nanostructures with quantum wells separated by narrow barriers. In the case when such a periodic structure has sufficiently narrow barriers, electrons can relatively easily tunnel from one quantum well to another and in this way can pass through the entire structure. Such a periodic structure is called a *superlattice*. Usually, the width of the barriers in a superlattice is several nanometers. In the case of a superlattice we observe more or less three-dimensional behavior of electrons since the motion of electrons is possible both along the layers and perpendicular to the layers. At the same time the energy spectrum of a superlattice is equivalent neither to the structure of energy levels of an individual quantum well nor to the energy spectrum of the bulk materials which constitute the quantum wells and potential barriers. As a result of the Pauli exclusion principle the energy levels of the individual quantum wells develop into one-dimensional energy minibands. The width and distance between the forbidden and allowed energy minibands substantially depend on the width of the potential wells as well as on the width and height of the potential barriers. The system of electron energy states formed in this way can be controlled also by external fields. The forbidden miniband is sometimes called a *minigap*.

In low-dimensional structures the operational range of the external electric fields which control the energy spectrum is substantially greater than for bulk materials. Also the temperature at which it is possible to observe most of the quantum effects is greater. These and many other unique properties of low-dimensional nanostructures provide the material basis for the development of modern nanoelectronics.

In the current chapter we will discuss the peculiarities of the discrete energy spectrum of the electron, placing emphasis on its motion along the direction of its quantization.

Example 7.2. Using the semiclassical Bohr theory of the hydrogen atom, find the radius of an electron orbit in the ground state of an ion with positive charge $q = Ze$ placed in a medium with dielectric constant equal to ϵ.

Reasoning. According to Bohr's postulates, only the orbits which satisfy the following condition for the angular momentum, $m_e v r_n$, can exist in the hydrogen atom:

$$m_e v r_n = n\hbar, \tag{7.32}$$

where $n = 1, 2, 3, \ldots$ is the orbital quantum number. The equation which describes the rotational motion of the electron around the nucleus is the equation of classical mechanics and is defined by Newton's second law:

$$m_e \frac{v^2}{r_n} = \frac{k_e Z e^2}{r_n^2}, \tag{7.33}$$

where m_e is the mass of a free electron, v^2/r_n the electron's centripetal acceleration, v the electron velocity, r_n the radius of the nth circular orbit, e the charge of the electron, $k_e = 1/(4\pi\epsilon_0)$, and $\epsilon_0 = 8.854 \times 10^{-12}$ F m^{-1}. Excluding the electron velocity, v, from Eq. (7.33) and taking into account Eq. (7.32), we get the following expression for the allowed orbits' radii, r_n:

$$r_n = \frac{\hbar^2}{k_e m_e Z e^2} n^2. \tag{7.34}$$

The radius of the first orbit (for $Z = 1$), which is called the *Bohr radius*, is equal to

$$r_1 = \frac{\hbar^2}{k_e m_e e^2} = 5.3 \times 10^{-2} \text{ nm}. \tag{7.35}$$

If an atom is placed in a medium with dielectric constant ϵ, then the expression for the Coulomb force on the right-hand side of Eq. (7.33) becomes

$$F_e = \frac{k_e Z e^2}{\epsilon r_n^2}. \tag{7.36}$$

The decrease of F_e by a factor of ϵ increases the radius of the first orbit by a factor of ϵ. Note that in a medium the dynamics of electron motion is defined not by the mass of a free electron, m_e, but by its effective mass, m^*. Thus, the radius of the first orbit of an electron in the ion placed in a medium, r_1^*, is defined as

$$r_1^* = \epsilon \frac{m_e}{Z m^*} r_1 = 5.3 \times 10^{-2} \epsilon \frac{m_e}{Z m^*} \text{ nm}. \tag{7.37}$$

For a typical semiconductor, whose dielectric constant, ϵ, is of the order of 10 and in which the effective mass of the electron is $m^* \approx 0.1 m_e$, the radius of the first Bohr orbit, r_1^*, for the hydrogen-like impurity with $Z = 1$ is, according to Eq. (7.37), equal to

$$r_1^* \approx 5.3 \text{ nm}, \tag{7.38}$$

which is a *hundred times* larger than the radius of the first Bohr orbit in a hydrogen atom placed in vacuum.

7.3 Quantum states of an electron in low-dimensional structures

7.3.1 Quantum dots

If the electron motion is restricted in all three directions, i.e., Eq. (7.27) is satisfied, then the electron energy spectrum becomes totally discrete as in an isolated atom. Such a system is called a *zero-dimensional system* or *quantum dot* (we can also call it a *quantum box* if the shape of the nanostructure is close to a parallelepiped). A quantum dot is analogous to an *artificial atom*, though it can consist of a large number of real atoms. As in the case of a real atom, the quantum dot may contain one or several electrons. It is necessary to keep in mind one substantial difference between quantum dots and atoms. As we have discussed in Section 6.4, an atom consists of positively charged protons, an equal number of negatively charged electrons, and neutrons. The positive charge of the nucleus, which consists of protons and neutrons, creates an attractive potential that confines electrons. Very often we refer to such electrons as *bound*, since they belong to individual atoms. A quantum dot consists of atoms and the wavefunctions from different atoms overlap as we discussed in Section 6.6. It is necessary to distinguish between electrons whose wavefunctions practically do not overlap (or overlap insignificantly; these electrons can be considered as bound to the individual atoms) and electrons with substantial overlapping of wavefunctions. The latter belong simultaneously to a large number of atoms and can be considered as *unbound* electrons. The confinement potential for the unbound electrons is created by the barrier between the quantum dot and the surrounding material. If a quantum dot contains only one unbound electron, then we can consider it as an artificial hydrogen atom, if two, as an artificial helium atom, and so on. The main characteristic of a quantum dot is its *energy spectrum*, which depends on many factors, such as the geometrical form of the dot, the surrounding material that creates potential barriers for the electron confinement, and the number of unbound electrons.

In Chapter 4 we have already discussed the character of the energy spectrum of an electron in three-dimensional rectangular and spherically-symmetric potential wells with barriers of infinite height. The electron energy spectrum in a quantum dot of cubic form (a *cubic box*) with the edge length equal to L according to Eq. (4.18) contains an infinite number of levels defined by the equation

$$E_{n_x n_y n_z} = \frac{\pi^2 \hbar^2}{2m^* L^2}\left(n_x^2 + n_y^2 + n_z^2\right). \tag{7.39}$$

The discrete energy levels are determined by the set of three quantum numbers n_x, n_y, and n_z. The ground state, $E^{(1)}$, corresponds to the lowest electron energy with $n_x = n_y = n_z = 1$:

$$E^{(1)} = E_{\min} = E_{111} = \frac{3\pi^2 \hbar^2}{2m^* L^2}. \tag{7.40}$$

The next energy level corresponds to three sets of quantum numbers, (112, 121, 211), with the same value for the energy, $E^{(2)}$:

$$E^{(2)} = E_{112} = E_{121} = E_{211} = \frac{3\pi^2 \hbar^2}{m^* L^2}. \tag{7.41}$$

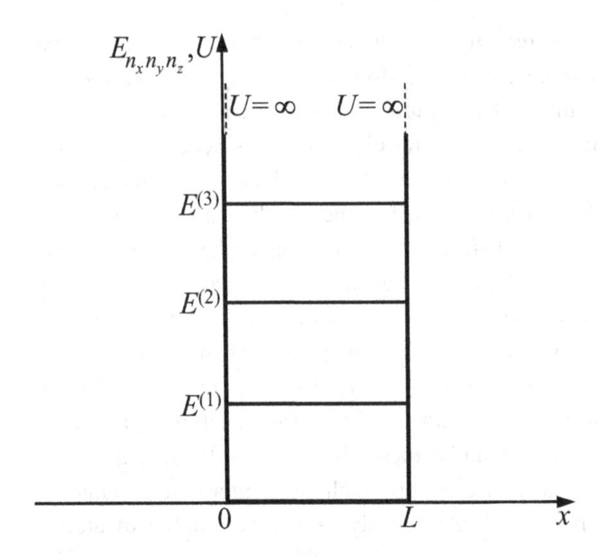

Figure 7.5 Energy levels in a quantum dot.

Such a level is called *degenerate* and has a *degeneracy* of three, corresponding to three sets of quantum numbers, whereas the ground state corresponds to a *non-degenerate* level with the lowest energy $E_{\min}$.

If the size of the edge of a quantum box is equal to $L = 5$ nm, then the ground state has energy equal to $E^{(1)} \approx 0.04$ eV, and the first excited state has energy equal to $E^{(2)} \approx 0.08$ eV. Let us note that in the case of semiconductor nanostructures in the above-mentioned expressions we have to use the electron effective mass, m^*, which in general is smaller than the mass of a free electron, m_{e}, and leads to a noticeable increase in $E^{(1)}$ and $E^{(2)}$. The distance between lower energy levels for real quantum dots may reach several hundreds of meV.

The energy levels in a quantum dot are shown schematically in Fig. 7.5, where for the sake of simplicity we use only one index to denote the energy level instead of three, i.e., $E^{(1)}$ is the lowest allowed energy in the quantum dot and the index n increases with increasing energy $E^{(n)}$. If the height of the potential barrier is infinite we have an infinite number of levels in the dot. In reality only a finite number of energy levels can exist, because the height of the potential barriers is finite. Note that $E^{(1)}$ has a non-zero finite value, i.e., in a quantum dot the electron cannot be at the bottom of the potential well, so it cannot have energy equal to zero. The smaller the dot size, the higher is the energy of the ground level, $E^{(1)}$ (see Eq. (7.40)). Only for $L \to \infty$ do we get $E^{(1)} = E_{\min} \to 0$. The distance between levels $E^{(n)}$ and $E^{(n+1)}$ is finite and it increases as L decreases (see Eq. (7.39)). For the example given above, the spacing between the energy levels $E^{(1)}$ and $E^{(2)}$ is equal to 0.04 eV. Thus, an electron can absorb finite portions of energy or "quanta" of energy in order for it to be transferred from lower to higher energy levels. The wavefunction of an electron in a cubic quantum box corresponds to a standing wavefunction (see Eq. (4.17) with $L_x = L_y = L_z = L$):

$$\psi_{n_x n_y n_z}(x, y, z) = \sqrt{\frac{8}{L^3}} \sin\left(\frac{\pi n_x x}{L}\right) \sin\left(\frac{\pi n_y y}{L}\right) \sin\left(\frac{\pi n_z z}{L}\right), \qquad (7.42)$$

i.e., in this case there is no classical motion of the electron in any direction. The electron momentum cannot change continuously and the electron kinetic energy cannot be a continuous function of momentum. Instead we have discrete values of momentum and discrete (quantized) values of energy.

7.3.2 Quantum wires

If along one of the directions of a nanostructure (for example, along the z-axis) the size $L_z \to \infty$, then the electron motion along this direction becomes free. Along the two other directions the electron motion stays quantized. As a result we get the type of nanostructure called a *one-dimensional system* or a *quantum wire*. We can find the energy spectrum and wavefunction of an electron moving in such a quantum wire from Eqs. (4.17) and (4.18) if we substitute the wavefunction and the energy of quantized motion along the z-axis with the term that describes the free motion of the electron. Suppose that in the x- and y-directions the electron motion is restricted due to the small size of the structure, i.e., inequalities (7.26) are satisfied and the rectangular potential barriers are of infinite height, i.e., $U_0 = \infty$. For an electron in such a quantum wire we can write

$$\psi_{n_x n_y}(x, y, z) = \sqrt{\frac{4}{L_x L_y}} \sin\left(\frac{\pi n_x x}{L_x}\right) \sin\left(\frac{\pi n_y y}{L_y}\right) e^{ik_z z}, \qquad (7.43)$$

$$E_{n_x n_y}(k_z) = \frac{\hbar^2}{2m^*}\left(\frac{\pi^2 n_x^2}{L_x^2} + \frac{\pi^2 n_y^2}{L_y^2}\right) + \frac{\hbar^2 k_z^2}{2m^*}, \qquad (7.44)$$

where $\hbar k_z = p_z$ is the corresponding component of the momentum, $\mathbf{p}$, along the direction of free motion of the electron.

Figure 7.6 shows the energy dependences $E_{11}(p_z)$ and $E_{12}(p_z)$ on the momentum $p_z = \hbar k_z$ as well as the quantized levels $E_{11}(p_z = 0)$ and $E_{12}(p_z = 0)$. *Note that, in order to keep the notations less cumbersome, we will henceforth write $E_n(\mathbf{p})$ for the energy that depends on momentum, $\mathbf{p}$, while for the energy of the nth level we will write E_n without an argument, e.g., $E_{n_x,n_y}(p_z)$ and E_{n_x,n_y}, correspondingly, in the case of a quantum wire* (see Fig. 7.6).

7.3.3 Quantum wells

If we remove the restriction on electron motion also in the y-direction, i.e., consider L_y and L_z infinitely large, then we will obtain the type of nanostructure called a *two-dimensional system* or a *quantum well*. The forms of the wavefunction and energy spectrum of an electron in a rectangular quantum well with

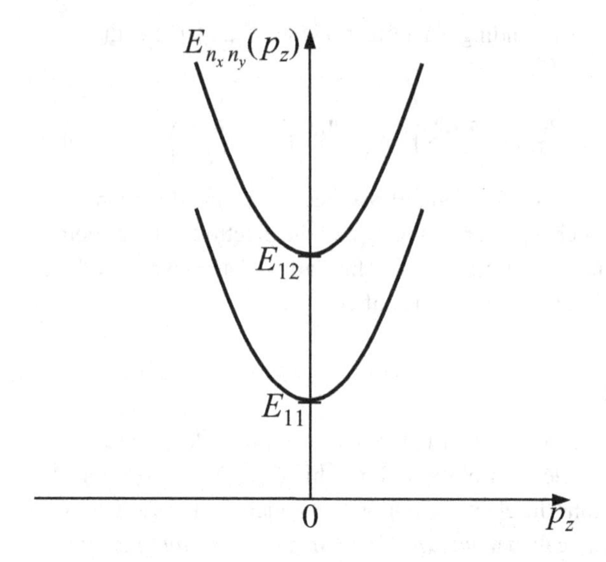

Figure 7.6 Quantum wire: the quantized energy levels, E_{11} and E_{12}, as well as the dependences of the electron energy $E_{n_x n_y}$ on the momentum p_z defined by Eq. (7.44).

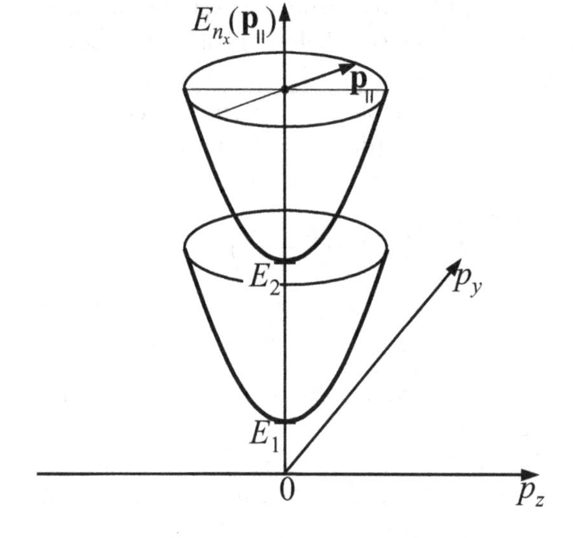

Figure 7.7 Quantum well: quantized energy levels, E_1 and E_2, and the dependences of the electron energy, E_{n_x}, on the momentum $|\mathbf{p}_{\|}| = \hbar\sqrt{k_y^2 + k_z^2}$ defined by Eq. (7.46).

barriers of infinite height are defined by the expressions

$$\psi_{n_x}(x, y, z) = \sqrt{\frac{2}{L_x}} \sin\left(\frac{\pi n_x x}{L_x}\right) e^{i(k_y y + k_z z)}, \tag{7.45}$$

$$E_{n_x}(k_y, k_z) = \frac{\hbar^2}{2m^*}\left(\frac{\pi^2 n_x^2}{L_x^2}\right) + \frac{\hbar^2(k_y^2 + k_z^2)}{2m^*}. \tag{7.46}$$

Figure 7.7 shows the dependences of the electron energy, $E_{n_x}(\mathbf{p}_{\|})$ on the lateral momentum $|\mathbf{p}_{\|}|$ as well as the quantized levels $E_1 = [\hbar^2/(2m^*)](\pi^2/L_x^2)$ and $E_2 = 4E_1$.

Equations (7.44) and (7.46) show that the energy spectra of the electron in a quantum wire and in a quantum well consist of branches called *subbands*. The total energy in these subbands is the sum of the energy of dimensional quantization along two or one of the axes and the kinetic energy along the other one or two directions.

Example 7.3. For an electron in the ground state in a quantum well with infinite barriers, find the probabilities of finding the electron in the central region with coordinates $L_x/4 \leq x \leq 3L_x/4$ and in the peripheral regions with coordinates $0 \leq x \leq L_x/4$ and $3L_x/4 \leq x \leq L_x$ if the width of the quantum well is equal to L_x. The central and peripheral regions are of the same width, which is equal to $L_x/2$.
Reasoning. According to Eq. (2.139) the probability of finding an electron in the ground state in the peripheral regions, P_1, can be found as

$$P_1 = \int_0^{L_x/4} |\psi_1(x, y, z)|^2 \, dx + \int_{3L_x/4}^{L_x} |\psi_1(x, y, z)|^2 \, dx, \tag{7.47}$$

where the wavefunction ψ_1 has the form (7.45) and

$$|\psi_1(x, y, z)|^2 = \frac{2}{L_x} \sin^2\left(\frac{\pi x}{L_x}\right).$$

Owing to the symmetry of the wavefunction $\psi_1(x, y, z)$ the integrals in Eq. (7.47) are equal and the expression for the probability, P_1, takes the form

$$P_1 = 2\left(\frac{2}{L_x}\right)\int_0^{L_x/4} \sin^2\left(\frac{\pi x}{L_x}\right) dx = 2\left(\frac{2}{L_x}\right)\left\{\frac{1}{2}\int_0^{L_x/4}\left[1 - \cos\left(\frac{2\pi x}{L_x}\right)\right] dx\right\}$$
$$= \frac{1}{2}\left(1 - \frac{2}{\pi}\right). \tag{7.48}$$

In the central region the probability of finding the electron will be defined by the same integral with the new limits of integration $L_x/4$ and $3L_x/4$. As a result we get

$$P_2 = \frac{1}{2}\left(1 + \frac{2}{\pi}\right). \tag{7.49}$$

Thus, in the ground state the probability, P_2, of finding the electron in the central part of the quantum well is larger than the probability, P_1, of finding it at the well edge.

Example 7.4. Find the energies of the first five levels and the order of their degeneracy for an electron confined in a cubic quantum dot with edge size $L = 3$ nm. Assume that the potential barriers in the quantum dot are of infinite height and that the mass of the electron is equal to the free electron mass, m_e.
Reasoning. According to Eq. (7.39) the ground state corresponds to a single set of quantum numbers (1, 1, 1). Thus, the degeneracy of the ground state, g_1,

is equal to 1. The first excited state corresponds to the three possible sets of quantum numbers (1, 1, 2), (1, 2, 1), and (2, 1, 1), i.e., the second energy level has the order of degeneracy, g_2, equal to 3. The next excited state, i.e., the third energy level, corresponds to the set of quantum numbers (1, 2, 2), (2, 1, 2), and (2, 2, 1), and thus its order of degeneracy, g_3, is equal to 3. The fourth energy level corresponds to the set of quantum numbers (1, 1, 3), (1, 3, 1), and (3, 1, 1), and the order of degeneracy is $g_4 = 3$. Finally, the fifth energy level corresponds to a single set of quantum numbers, (2, 2, 2), and therefore the order of degeneracy is $g_5 = 1$.

On substituting the corresponding combinations of quantum numbers into Eq. (7.39) we get for the energies of the first five levels the following numbers:

$$E^{(1)} = 0.123 \text{ eV}; \qquad E^{(2)} = 0.246 \text{ eV}; \qquad E^{(3)} = 0.369 \text{ eV};$$

$$E^{(4)} = 0.451 \text{ eV}; \qquad E^{(5)} = 0.492 \text{ eV}.$$

The energy levels, $E^{(n)}$, correspond to the following energy levels $E_{n_x n_y n_z}$:

$$E^{(1)} = E_{111}; \qquad E^{(2)} = E_{112} = E_{121} = E_{211}; \qquad E^{(3)} = E_{122} = E_{212} = E_{221};$$

$$E^{(4)} = E_{113} = E_{131} = E_{311}; \qquad E^{(5)} = E_{222}.$$

7.4 The number of states and density of states for nanostructures

Let us explore how the number of states, $N(E)$, and the density of states, $g(E)$, change when one, two, or three dimensions of the potential well become comparable to the de Broglie wavelength of the electron, i.e., when the electron motion in these directions becomes confined and the electron energy spectrum cannot be considered quasicontinuous.

7.4.1 Quantum wells

Let us consider the case when electron motion takes place in a two-dimensional system – a quantum well. In this case the electron motion is continuous in two directions (for example, in the y- and z-directions) with macroscopic sizes of the system, L_y and L_z, and is quantized only in the x-direction. The size of the quantum well in the x-direction, L_x, is much smaller than L_y and L_z: $L_x \ll L_y, L_z$. This means that there is a one-dimensional potential, $U(x)$, which limits electron motion in the x-direction:

$$U(x) = \begin{cases} 0, & 0 \leq x \leq L_x, \\ \infty, & x < 0, x > L_x. \end{cases} \tag{7.50}$$

The region of quantization is $0 \leq x \leq L_x$. Such electron motion can be considered *quasi-two-dimensional*. In this case the electron wavefunctions in the quantum well are defined by Eq. (7.45) and the energy spectrum by Eq. (7.46), which we will present in the form

$$E_{n_x}(k_y, k_z) = \frac{\hbar^2}{2m^*}\left(\frac{\pi^2 n_x^2}{L_x^2}\right) + \frac{\hbar^2\left(k_y^2 + k_z^2\right)}{2m^*},$$

$$E_{n_x}(p_y, p_z) = \frac{\hbar^2}{2m^*}\left(\frac{\pi^2 n_x^2}{L_x^2}\right) + \frac{p_y^2 + p_z^2}{2m^*}. \tag{7.51}$$

In writing this expression we took into account that the wavevector components are defined by the expressions $p_y = \hbar k_y$ and $p_z = \hbar k_z$. From Eq. (7.51) it follows that the electron energy spectrum in the case of quasi-two-dimensional motion consists of overlapping subbands with a parabolic dispersion relation as shown in Fig. 7.7. The lowest energy in each of the subbands corresponds to the subbands' bottoms with $p_y = p_z = 0$ and energy

$$E_{n_x} = E_{n_x}(0, 0) = \frac{\pi^2\hbar^2}{2m^*}\left(\frac{n_x}{L_x}\right)^2. \tag{7.52}$$

The lowest energy that an electron can have in a quantum well corresponds to $n_x = 1$ and is equal to

$$E_1 = E_1(0, 0) = \frac{\pi^2\hbar^2}{2m^*}\left(\frac{1}{L_x}\right)^2. \tag{7.53}$$

Electron states with $E < E_1$ are forbidden. Therefore, for $E < E_1$ the number of states is equal to zero, $N(E) = 0$, and the density of states is also equal to zero, $g(E) = 0$. In the energy region $E_1 < E < E_2$, where the lowest energy value in the second subband is equal to

$$E_2 = E_2(0, 0) = \frac{\pi^2\hbar^2}{2m^*}\left(\frac{2}{L_x}\right)^2, \tag{7.54}$$

there are states that belong only to the first subband, $E_1(p_y, p_z)$.

Let us calculate the number, N, and density of states, g, in each of the subbands. In order to do this, let us draw the surfaces of equal energy (or *isoenergetic surfaces*) for this two-dimensional system in **p**-space. Let us take into account that in the case of free three-dimensional electron motion the surface of equal energy has the form of a sphere of radius $p = \sqrt{2m^*E}$. The quantum limitation in one of the directions, for example in the x-direction, corresponds to a constant x-component of the momentum $p_x = n_x\pi\hbar/L_x$ which is caused by the quantized motion of the electron in the x-direction. Thus, the spherical surface is divided into a set of circular cross-sections, which are perpendicular to the axis of quantization (the x-axis) and correspond to the specific values of the quantum number n_x. For this kind of confinement the surface of equal energy is a circle of radius p with a given n_x (see Fig. 7.8). To estimate the number of quantum states, let us choose a circular segment with radius p and width $\mathrm{d}p$, which has an area

$$\mathrm{d}S_p = 2\pi p\,\mathrm{d}p. \tag{7.55}$$

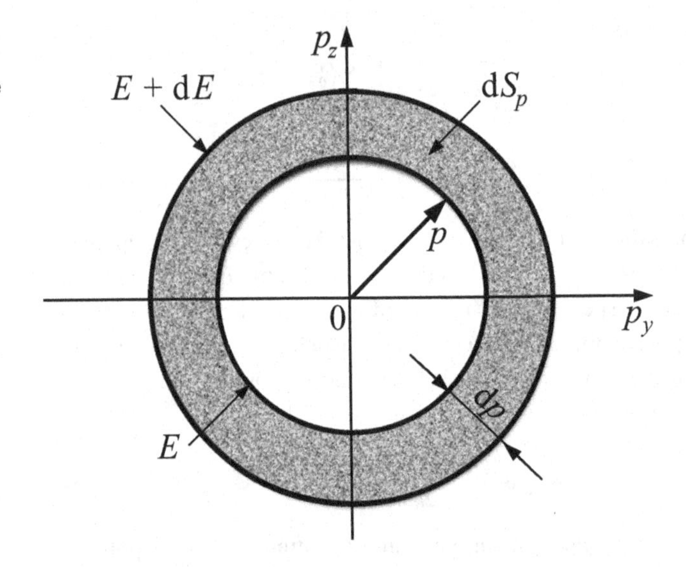

Figure 7.8 A surface of equal energy, E, in a two-dimensional **p**-space for a given n_x.

The number of quantum states corresponding to this area, $\mathrm{d}S_p$, and to the area available for an electron, $S_\mathrm{r} = L_y L_z$, is given by the expression

$$\mathrm{d}n(p) = \frac{S_\mathrm{r} \times 2\pi p \,\mathrm{d}p}{(2\pi\hbar)^2} = \frac{S_\mathrm{r} p \,\mathrm{d}p}{2\pi\hbar^2}. \tag{7.56}$$

Taking into account that the energy of a free electron inside the potential well depends on the momentum quadratically, $E = p^2/(2m^*)$, we can get from Eq. (7.56)

$$\mathrm{d}n(E) = \frac{S_\mathrm{r} m^* \,\mathrm{d}E}{2\pi\hbar^2}. \tag{7.57}$$

Now, we can find the total number of states in the energy interval $(0, E)$ by integration of Eq. (7.57) within the limits of this interval, 0 and E. After integration we arrive at the following expression:

$$N_0(E) = 2\int \mathrm{d}n(E) = 2\int_0^E \frac{S_\mathrm{r} m^*}{2\pi\hbar^2}\,\mathrm{d}E = \frac{m^* L_y L_z}{\pi\hbar^2} E. \tag{7.58}$$

In Eq. (7.58) we multiplied the integral by 2 because the spin of an electron can have two values, $+1/2$ and $-1/2$. Taking into account the definition of the density of states,

$$g_0(E) = \frac{\mathrm{d}N_0(E)}{\mathrm{d}E},$$

we find

$$g_0(E) = \frac{m^* L_y L_z}{\pi\hbar^2}. \tag{7.59}$$

From Eqs. (7.58) and (7.59) it follows that in the case of two-dimensional electron motion the number of quantum states increases linearly with the energy and that the density of states does not depend on energy at all.

Now let us take into account that such motion takes place also in the other energy subbands. The number of states, $N_0(E)$, and density of states, $g_0(E)$, in the first subband are defined by Eqs. (7.58) and (7.59). Taking into account that there are no states below energy state E_1, they can be written as

$$N(E) = N_0(E - E_1) = \frac{m^* L_y L_z}{\pi \hbar^2}(E - E_1), \tag{7.60}$$

$$g(E) = \frac{\mathrm{d}N(E)}{\mathrm{d}E} = \frac{m^* L_y L_z}{\pi \hbar^2}. \tag{7.61}$$

In the energy interval $E_2 < E < E_3$ the number and density of states are equal to the sum of states of the first two subbands, i.e.,

$$N(E) = N_0(E - E_1) + N_0(E - E_2) = \frac{m^* L_y L_z}{\pi \hbar^2}(2E - E_1 - E_2), \tag{7.62}$$

$$g(E) = \frac{2m^* L_y L_z}{\pi \hbar^2}. \tag{7.63}$$

For an energy in the interval $E_n < E < E_{n+1}$ each of the n subbands will have its input in the magnitudes of $N(E)$ and $g(E)$. Therefore, within this energy interval, $N(E)$ and $g(E)$ can be defined as follows:

$$N(E) = \frac{m^* L_y L_z}{\pi \hbar^2}(nE - E_1 - E_2 - \cdots - E_n), \tag{7.64}$$

$$g(E) = \frac{nm^* L_y L_z}{\pi \hbar^2}. \tag{7.65}$$

Thus, the number of states, $N(E)$, is a continuous function consisting of linear segments with slope increasing from region to region. The angle of slope, α_n, is defined as

$$\alpha_n = \arctan\left(\frac{nm^* L_y L_z}{\pi \hbar^2}\right).$$

The density of states, $g(E)$, is a step-function, which undergoes jumps,

$$\Delta g_n = \frac{m^* L_y L_z}{\pi \hbar^2},$$

each time the energy of the electron becomes equal to the bottom of the next subband, i.e., when $E = E_n = n^2 E_1$. Figure 7.9 shows the dependences $N(E)$ and $g(E)$ for the quantum well.

Very often these quantities are defined for the unit area of the two-dimensional system. In this case the obtained expressions for the number and density of states must be divided by the magnitude $S_r = L_y \times L_z$:

$$N(E) = \frac{m^*}{\pi \hbar^2}(nE - E_1 - E_2 - \cdots - E_n), \tag{7.66}$$

$$g(E) = \frac{nm^*}{\pi \hbar^2}. \tag{7.67}$$

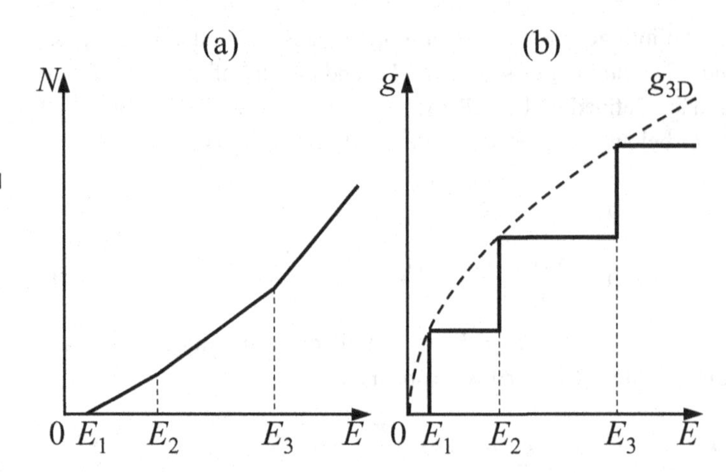

Figure 7.9 The dependences of (a) the number of states, $N(E)$, and (b) the density of states, $g(E)$, in a rectangular quantum well with barriers of infinite height. Here g_{3D} is the electron density of states in bulk (three-dimensional) material.

7.4.2 Quantum wires

Let us now consider electron motion in a *one-dimensional system – a quantum wire*, where electron motion is continuous along only one of the directions (for example, along the z-direction). The size of the region where the continuous motion takes place, L_z, is macroscopic; in the other two directions (along the x- and y-directions) the motion is quantized. Therefore, the length of the quantum wire, L_z, satisfies the following inequality: $L_x, L_y \ll L_z$. This means that the electron motion takes place in the inner-crystal periodic potential with an additional two-dimensional potential, $U(x, y)$, which limits electron motion in a crystal in the x- and y-directions:

$$U(x, y) = \begin{cases} 0, & 0 \leq x \leq L_x, 0 \leq y \leq L_y, \\ \infty, & x < 0, y < 0, \\ \infty, & x > L_x, y > L_y. \end{cases} \tag{7.68}$$

Such electron motion can be considered *quasi-one-dimensional*. The wavefunctions are defined by Eq. (7.43) and the electron energy spectrum by Eq. (7.44), which we will present in the following form:

$$E_{n_x n_y}(k_z) = \frac{\hbar^2 \pi^2}{2m^*}\left(\frac{n_x^2}{L_x^2} + \frac{n_y^2}{L_y^2}\right) + \frac{\hbar^2 k_z^2}{2m^*} = \frac{\hbar^2 \pi^2}{2m^*}\left(\frac{n_x^2}{L_x^2} + \frac{n_y^2}{L_y^2}\right) + \frac{p_z^2}{2m^*}. \tag{7.69}$$

The electron spectrum in a quantum wire consists of subbands with a one-dimensional parabolic dispersion relation $E_{n_x n_y}(p_z)$ (see Fig. 7.6). The electron motion along the z-direction is "free" and its dynamic properties are defined by the effective mass, m^*. The minimal energy in each of the subbands corresponds to their bottom with $p_z = 0$:

$$E_{n_x n_y} = E_{n_x n_y}(0) = \frac{\hbar^2 \pi^2}{2m^*}\left(\frac{n_x^2}{L_x^2} + \frac{n_y^2}{L_y^2}\right). \tag{7.70}$$

The lowest energy which an electron can have in a quantum wire is equal to

$$E_{11} = E^{(1)} = E_{11}(0) = \frac{\pi^2\hbar^2}{2m^*}\left(\frac{1}{L_x^2} + \frac{1}{L_y^2}\right). \tag{7.71}$$

Electron states are forbidden for energies lower than $E^{(1)}$. Therefore, for all energies $E < E^{(1)}$ the number and density of states in a quantum wire are equal to zero, i.e., $N(E) = 0$ and $g(E) = 0$. In the energy interval $E^{(1)} < E < E^{(2)}$, where the minimum energy in the second subband is equal to

$$E^{(2)} = E_{12}(0) = E_{21}(0) = \frac{\pi^2\hbar^2}{2m^*}\left(\frac{1}{L_x^2} + \frac{2^2}{L_y^2}\right) = \frac{\pi^2\hbar^2}{2m^*}\left(\frac{2^2}{L_x^2} + \frac{1}{L_y^2}\right), \tag{7.72}$$

there exist states that belong only to the first subband $E_{11}(p_z)$.

Let us find the number and density of states in each of the subbands. For quasi-one-dimensional "free" motion in the z-direction in the considered potential well with infinite barriers, the electron energy spectrum can be considered practically continuous (see Example 7.5). Taking into account Eq. (7.69), the expression for the z-component of the electron momentum in the corresponding subband can be written as follows:

$$p_z = \sqrt{2m^*[E - E_{n_x n_y}(0)]}. \tag{7.73}$$

On differentiating Eq. (7.73) we get

$$\mathrm{d}p_z = \sqrt{\frac{m^*}{2}}\frac{\mathrm{d}E}{\sqrt{E - E_{n_x n_y}(0)}}. \tag{7.74}$$

Since the electron momentum in the potential well with infinite barriers is defined as

$$p_z = \frac{\pi\hbar n_z}{L_z},$$

from Eq. (7.73) we find the number of energy levels:

$$n_z = \frac{L_z}{\pi\hbar}p_z = \frac{L_z}{\pi\hbar}\sqrt{2m^*[E - E_{n_x n_y}(0)]}. \tag{7.75}$$

We have already mentioned that the total number of quantum states is twice as large as the number of energy levels, i.e., $N(E) = 2n_z$. This is because of Pauli's principle for particles with half-integer spin, according to which on each energy level there can be two electrons with opposite spins. Thus,

$$N(E) = 2n_z = \frac{2L_z}{\pi\hbar}\sqrt{2m^*\left[E - E_{n_x n_y}(0)\right]}. \tag{7.76}$$

Let us find now the density of states, $g(E)$, i.e., the number of states in the energy interval $\mathrm{d}E$:

$$g(E) = \frac{\mathrm{d}N(E)}{\mathrm{d}E} = \frac{L_z}{\pi\hbar}\sqrt{\frac{2m^*}{E - E_{n_x n_y}(0)}}. \tag{7.77}$$

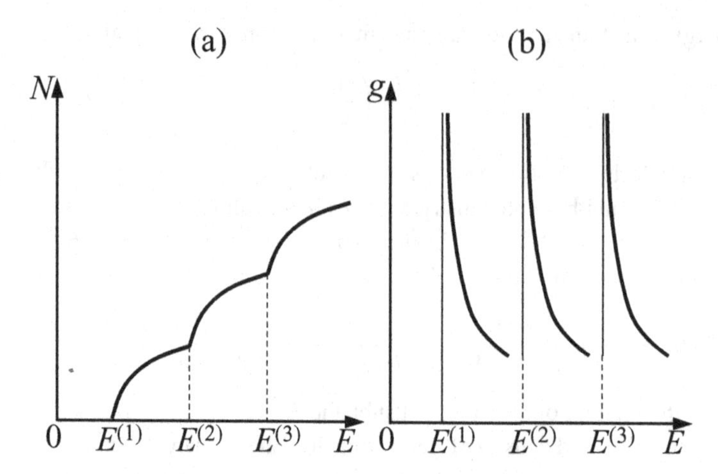

Figure 7.10 The dependences of (a) the number of states, $N(E)$, and (b) the density of states, $g(E)$, in the rectangular quantum wire with barriers of infinite height.

Thus, in the energy interval $E^{(1)} < E < E^{(2)}$ the number and density of states are defined only by the states of the first subband and are given by the expressions

$$N(E) = \frac{2L_z}{\pi\hbar}\sqrt{2m^*(E - E_{11})}, \tag{7.78}$$

$$g(E) = \frac{L_z}{\pi\hbar}\sqrt{\frac{2m^*}{E - E_{11}}}. \tag{7.79}$$

In the general case, for an energy E we get

$$N(E) = \frac{2L_z}{\pi\hbar}\sum_{n_x,n_y}\sqrt{2m^*(E - E_{n_xn_y})} \times \Theta(E - E_{n_xn_y}), \tag{7.80}$$

$$g(E) = \frac{L_z}{\pi\hbar}\sum_{n_x,n_y}\sqrt{\frac{2m^*}{E - E_{n_xn_y}}} \times \Theta(E - E_{n_xn_y}). \tag{7.81}$$

Here we introduced the *Heaviside function* Θ (or the *step-function*):

$$\Theta(t) = \begin{cases} 0, & t < 0, \\ 1, & t \geq 0. \end{cases} \tag{7.82}$$

The function $N(E)$ is continuous, but at the points $E = E_{n_xn_y}$ its derivative, i.e., the function $g(E)$, becomes discontinuous. The number of states within the limited energy interval is always limited. Figure 7.10 shows the dependences $N(E)$ and $g(E)$ for electron motion in a quantum wire.

7.4.3 Quantum dots

Let us consider now the electron motion in a *zero-dimensional system* or a *quantum dot*, i.e., in a potential well whose dimensions in all three directions are comparable to the de Broglie wavelength. In this case the electron energy spectrum is purely discrete. For a quantum dot with $L_x \neq L_y \neq L_z$ the number

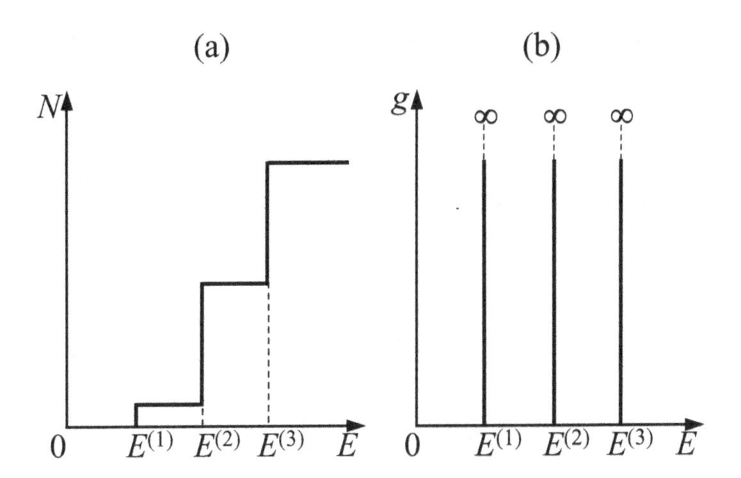

Figure 7.11 The dependences of (a) the number of states, $N(E)$, and (b) the density of states, $g(E)$, in a rectangular box with potential barriers of infinite height.

of states doubles (taking into account the spin degeneracy) every time the energy becomes equal to the energy level $E_{n_x n_y n_z}$. If the quantum dot has the form of a cubic box ($L_x = L_y = L_z$) with potential barriers of infinite height, the energy levels with different n_α are degenerate ($\alpha = x, y$, and z). In this case the jump in the number of states, $N(E)$, at $E = E_{n_x n_y n_z}$ is equal to twice the degeneracy, g_{n_x,n_y,n_z}, of the corresponding level.

The general expressions for the number, $N(E)$, and density of states, $g(E)$, are

$$N(E) = 2 \sum_{n_x,n_y,n_z} g_{n_x,n_y,n_z} \times \Theta(E - E_{n_x n_y n_z}), \tag{7.83}$$

$$g(E) = 2 \sum_{n_x,n_y,n_z} g_{n_x,n_y,n_z} \times \delta(E - E_{n_x n_y n_z}), \tag{7.84}$$

where $\delta(E - E_{n_x n_y n_z})$ is Dirac's δ-function, which was introduced by Eq. (5.56), $\Theta(E - E_{n_x n_y n_z})$ the Heaviside function, and g_{n_x,n_y,n_z} the degeneracy of the (n_x, n_y, n_z) state in a quantum dot. Figure 7.11 shows both dependences, $N(E)$ and $g(E)$. The function $N(E)$ for a quantum dot has the form of a step-function and the density of states $g(E)$ is a set of infinitely narrow and infinitely high peaks.

Example 7.5. For a quantum well with infinite barriers and width $L_x = 10$ μm, find the number of levels that will be filled up to the energy of $E = 1$ eV as well as the distance between the levels in the vicinity of the electron energy equal to $E = 1$ eV. Compare the distance between energy levels with the thermal energy at room temperature, which is of the order of $k_B T$. Assume that the electron effective mass is equal to the electron mass in vacuum: $m^* = m_e$.

Reasoning. According to Eq. (7.52) the electron energy spectrum in a quantum well is defined as

$$E_{n_x} = \frac{\pi^2\hbar^2}{2m^*}\left(\frac{n_x}{L_x}\right)^2. \tag{7.85}$$

From Eq. (7.85) we get that in the vicinity of energy $E_{n_x} = 1$ eV the level number, n_x, is equal to

$$n_x = \sqrt{2m^*E_{n_x}}\,\frac{L_x}{\pi\hbar} \approx 10^4. \tag{7.86}$$

The distance between adjacent energy levels with quantum numbers n_x and $n_x + 1$ is defined by the expression

$$\Delta E_{n_x} = E_{n_x+1} - E_{n_x} = \frac{\pi^2\hbar^2}{2m^*L_x^2}(2n_x+1) = E_{n_x}\frac{2n_x+1}{n_x^2}. \tag{7.87}$$

For $E_{n_x} = 1$ eV and $n_x = 10^4$ we obtain

$$\Delta E_{n_x} = 1\ \text{eV} \times \frac{2\times 10^4 + 1}{10^8} \approx 2\times 10^{-4}\ \text{eV}. \tag{7.88}$$

The distance between the adjacent energy levels, ΔE_{n_x}, is many orders of magnitude less than the thermal energy, $k_B T$, at room temperature:

$$\frac{\Delta E_{n_x}}{k_B T} = \frac{\pi^2\hbar^2(2n_x+1)}{2m^*L_x^2 k_B T} \approx 2\times 10^{-7} \times (2n_x+1). \tag{7.89}$$

Therefore, the change in electron energy in the quantum well considered here is practically continuous and the classical description can be used.

7.5 Double-quantum-dot structures (artificial molecules)

In the previous section we considered the behavior of an electron in an isolated quantum dot. The main peculiarity of the energy spectrum of an electron confined in the potential of a quantum dot is its discreteness, which is caused by the quantization of the electron motion in all three directions.

Modern technological capabilities of epitaxial growth allow us to fabricate types of nanostructures even with more complicated potential profiles. For example, it is possible to fabricate nanostructures that contain two or more coupled low-dimensional nanoobjects (a structure with size less than or about 10 nm is called a *nanoobject* or a *nanostructure*). It becomes possible to control the energy spectrum of electrons in such structures not only by changing the form of an individual nanoobject, but also by changing the distance and barrier height between neighboring nanoobjects. As we have mentioned already, if individual nanoobjects are separated by low and narrow potential barriers, electrons can easily tunnel from one nanoobject to another. This significantly affects the character of the electron energy spectrum in such a structure. In structures consisting of several quantum dots, where electrons are able to tunnel between neighboring

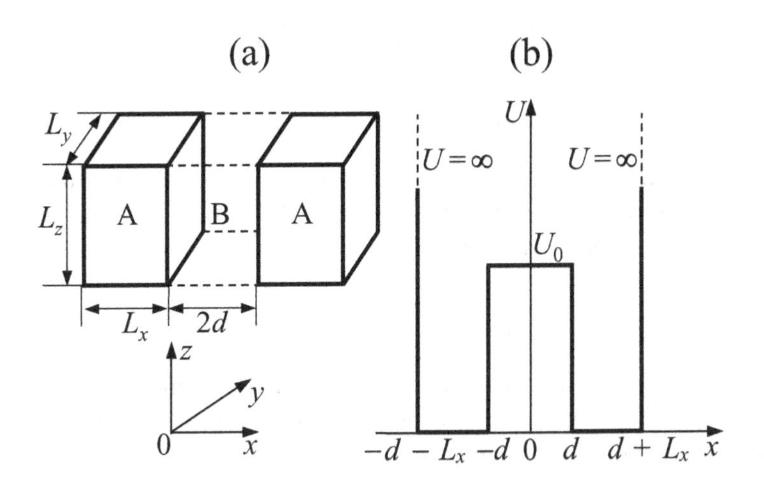

Figure 7.12 A schematic picture of a double-quantum-dot structure (a) and its potential profile $U(x)$ (b). A and B denote different semiconductor materials, which constitute quantum dots and barriers between them.

quantum dots, the main physical properties are similar to those of a molecule. The energy spectrum of such an *artificial molecule* will be determined not only by the parameters of the individual quantum dots but also by the parameters of the potential barriers between them. Double-quantum-dot structures are nowadays the building blocks of many electronic and optoelectronic devices. Let us consider the electron energy spectra formed by joining two nanoobjects such as quantum dots (or quantum wires or quantum wells) into a single nanostructure.

7.5.1 Double-quantum-dot structures: qualitative analysis

Let us assume that a quantum-dot structure consists of two quantum dots separated by a rectangular potential barrier, U_0 (see Fig. 7.12). The region that is available for electrons in the y- and z-directions is given by the inequalities

$$0 \le y \le L_y \quad \text{and} \quad 0 \le z \le L_z. \tag{7.90}$$

The rectangular distribution of the potential takes place along the x-direction. The general dependence of the potential $U(x, y, z)$ can be described by the following expression:

$$U(x, y, z) = \begin{cases} \infty, & y < 0,\ z < 0, \\ U(x), & 0 \le y \le L_y, 0 \le z \le L_z, \\ \infty, & y > L_y, z > L_z, \end{cases} \tag{7.91}$$

where

$$U(x) = \begin{cases} U_0, & |x| \le d, \\ 0, & d \le |x| \le d + L_x, \\ \infty, & |x| > d + L_x. \end{cases} \tag{7.92}$$

In the wavefunction of such a potential we can separate variables and write the total wavefunction as

$$\psi(x, y, z) = \psi_x(x)\psi_y(y)\psi_z(z), \tag{7.93}$$

where the wavefunctions ψ_y and ψ_z describe the states of the electron in the rectangular potential wells with infinitely high potential barriers. According to Eq. (3.51), these wavefunctions have the following form:

$$\psi_y(y) = \sqrt{\frac{2}{L_y}} \sin\left(\frac{n_y \pi y}{L_y}\right), \qquad \psi_z(z) = \sqrt{\frac{2}{L_z}} \sin\left(\frac{n_z \pi z}{L_z}\right). \tag{7.94}$$

In the x-direction the quantum dots are separated by a potential barrier of height U_0 and width $2d$. The x-component of the wavefunction, $\psi_x(x)$, must satisfy the one-dimensional Schrödinger equation with the potential $U(x)$.

The total electron energy in the double-quantum-dot structure must be a sum of the energies of quantum confinement along the y- and z-directions (defined by Eq. (3.44)) and the energy of electron motion along the x-axis, E_x, in the potential $U(x)$:

$$E = E_x + \frac{\pi^2 \hbar^2}{2m^*}\left(\frac{n_y^2}{L_y^2} + \frac{n_z^2}{L_z^2}\right). \tag{7.95}$$

Thus, we have reduced the problem of finding the quantum states of an electron in a structure consisting of two three-dimensional quantum dots to the one-dimensional Schrödinger equation (3.31) with the potential $U(x)$ defined by Eq. (7.92).

Before finding the exact solution of the Schrödinger equation, let us analyze qualitatively the dependences of the wavefunction ψ_x and energy E_x on the distance between quantum dots, $2d$.

If the quantum dots are far from each other, then the wavefunction $\psi_x(x)$ at the point $x = 0$ between them is practically equal to zero and the solution of the Schrödinger equation for energies below U_0 for the double-quantum-dot structure must practically coincide with the solution of the one-dimensional Schrödinger equation (3.51) for an individual quantum dot. The only difference between these two solutions will be that the magnitude $|\psi_x|^2$ for the new solution is reduced by a factor of 2 compared with the wavefunction of the individual quantum dot because of the normalization condition that takes into account the possibility of the electron being in each of the quantum dots. From those two wavefunctions for the individual quantum dots it is possible to construct a wavefunction that we will call a *symmetric solution*, ψ_s. The wavefunction, ψ_s, for the lowest energy state of a double quantum dot is shown in Fig. 7.13 by a solid line. The individual wavefunctions enter into the combined function, ψ_s, with the same signs. However, for the given potential profile there may exist another solution of the Schrödinger equation. It differs from the symmetric solution, ψ_s, by the signs of the wavefunctions of the individual quantum dots which constitute the resulting

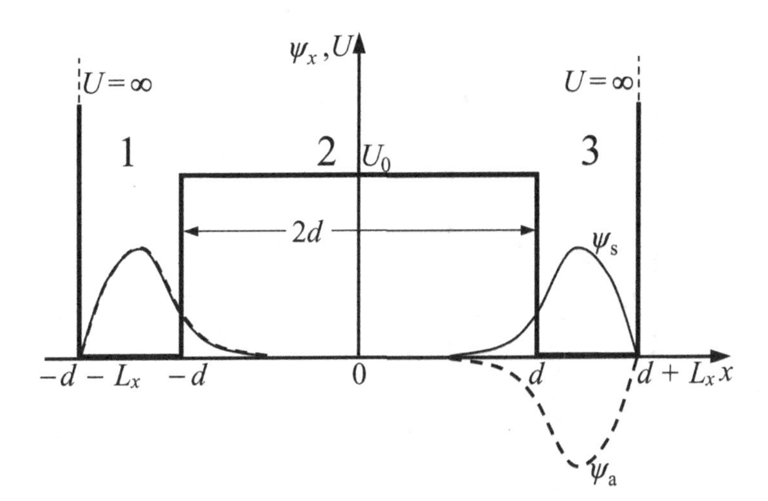

Figure 7.13 The wavefunctions of a double-quantum-dot structure, ψ_s and ψ_a, when the quantum dots are far from each other.

wavefunction. The corresponding wavefunction, ψ_a, is called *antisymmetric* and is shown in Fig. 7.13 by the dashed line. The eigenvalues $E_x = E_s$ and $E_x = E_a$ of the two solutions are practically the same. This follows from the equality of the two solutions for the average kinetic energy, which is proportional to $|d\psi_{s,a}/dx|^2$, and average potential energy, which is proportional to $U(x)|\psi_{s,a}|^2$.

Bringing the quantum dots closer to each other changes the form of the wavefunctions. The state with the symmetric wavefunction, ψ_s, corresponds to a smaller energy E_s than the energy E_a, which corresponds to the state with the antisymmetric wavefunction, ψ_a. This is due to the fact that the average value of the kinetic energy for the symmetric state is smaller than for the antisymmetric state because

$$\overline{\left|\frac{d\psi_s}{dx}\right|^2} < \overline{\left|\frac{d\psi_a}{dx}\right|^2}, \tag{7.96}$$

where the overbar means averaging over the width of the entire structure. The average values of the potential energy for the two states are practically the same.

In the limiting case when the quantum dots approach each other, the barrier between the quantum dots disappears ($2d = 0$). Thus, a potential well with infinitely high barriers and width twice as large, $2L_x$, is established (see Fig. 7.14). The symmetric wavefunction, ψ_s, is transformed into the wavefunction of the ground state, ψ_1, for the quantum dot with width $2L_x$, and the antisymmetric wavefunction, ψ_a, is transformed into the wavefunction of the first excited state with $n_x = 2$, i.e., into ψ_2. The energy levels in the quantum-dot structure with the width of the potential well equal to $2L_x$ are, according to Eq. (3.44),

$$E_x = \frac{\pi^2\hbar^2}{2m^*(2L_x)^2}n_x^2 = E_x^{(0)}\frac{n_x^2}{4}, \tag{7.97}$$

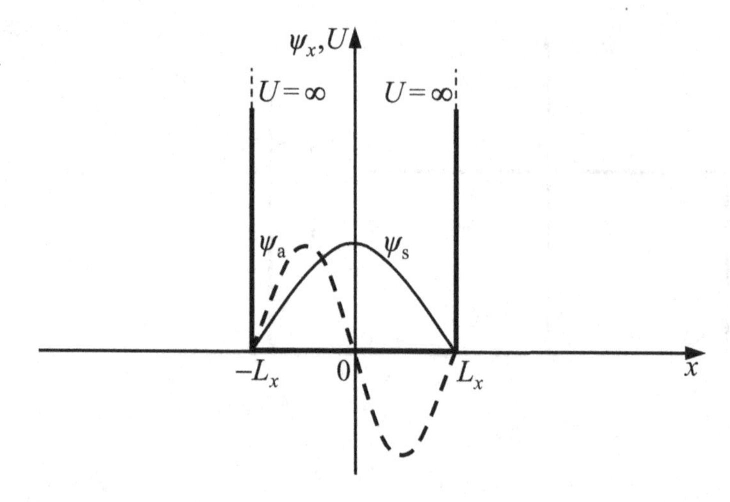

Figure 7.14 The wavefunctions ψ_s and ψ_a, when the width of the barrier between the two quantum dots is equal to zero.

where $E_x^{(0)} = \pi^2\hbar^2/(2m^*L_x^2)$ is the energy of the ground state in the individual quantum dots. The quantum number $n_x = 1$ corresponds to the state ψ_s and $n_x = 2$ to the state ψ_a. The energy E_x of the ground state ($n_x = 1$) in the double quantum dot formed in this way is four times smaller than the energy $E_x^{(0)}$, which is the initial ground state of the individual quantum dots separated by a large distance. The energy of the quantum state with ψ_a is exactly equal to the energy of the initial state $E_x^{(0)}$ of the individual quantum dots. Thus, by bringing the quantum dots closer to each other, we change the energy system of the electrons. Let us assume that in each of the separated quantum dots in the ground state there are only two electrons with opposite spins, and after joining of the quantum dots these electrons fill the levels with $n_x = 1$ and $n_x = 2$. In the initial state, when the quantum dots were separated, the total energy of the electrons was equal to $4E_x^{(0)}$. After the joining of the quantum dots, two electrons will be in the lowest state ψ_s and two electrons in the state ψ_a. Then, the energy of the electrons becomes equal to

$$\left(2E_x^{(0)} \times \frac{1}{4}\right) + \left(2E_x^{(0)} \times \frac{4}{4}\right) = \frac{5}{2} \times E_x^{(0)}.$$

As a result, after the joining of the quantum dots the energy of the system of electrons decreases by

$$\Delta E_x = 4E_x^{(0)} - \left(\frac{5}{2} \times E_x^{(0)}\right) = \frac{3}{2} \times E_x^{(0)}.$$

The qualitative dependences of the energies $E_x^{(1)}$ and $E_x^{(2)}$ on the distance, d, between the quantum dots are shown in Fig. 7.15. We see that, as in the case of the coupled quantum dots, i.e., when the magnitude of d is small enough, the level $E_x^{(0)}$ corresponding to the individual quantum dot splits into two levels, $E_x^{(1)}$ and $E_x^{(2)}$. The energy of splitting, ΔE_x, increases with decreasing d. If the electrons

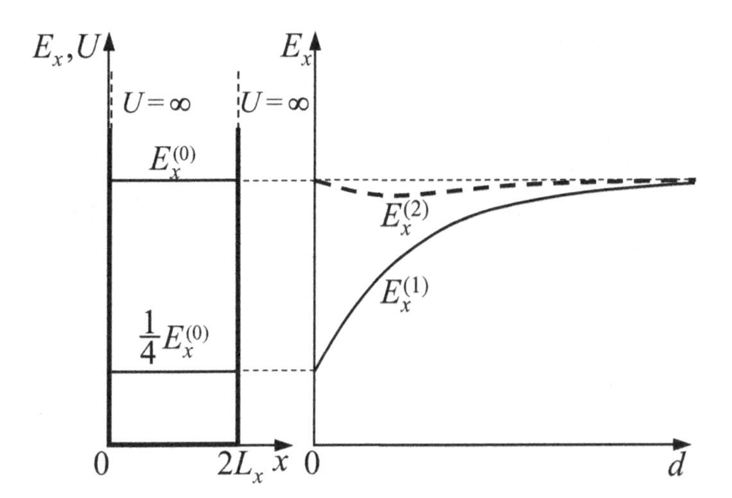

Figure 7.15 The dependences of the energies of the symmetric, $E_x^{(1)}$, and antisymmetric, $E_x^{(2)}$, states of the double-quantum-dot structure on the width of the barrier, *d*, separating individual quantum dots.

in the structure with two coupled quantum dots are in the state of lowest energy, then the wavefunction must be symmetric. The electrons in quantum dots in this state are considered to be *in phase*. If the electrons are in the state with higher energy, then the wavefunction must be antisymmetric. The electrons in quantum dots in such a state are considered to be *out of phase*.

7.5.2 Double-quantum-dot structures: exact solution

According to Fig. 7.13, for each of the three regions of the symmetric structure we can write the Schrödinger equation and its solution. Here, to simplify the calculations, we assume that the *electron effective mass, m^*, in the potential wells and that in the barriers are the same.* We will use this assumption for all subsequent calculations. For the regions 1 and 3, where $d \le |x| \le d + L_x$, we have

$$-\frac{\hbar^2}{2m^*}\frac{\mathrm{d}^2\psi_j(x)}{\mathrm{d}x^2} = E_x\psi_j(x), \tag{7.98}$$

$$\psi_j(x) = A_j\mathrm{e}^{ikx} + B_j\mathrm{e}^{-ikx}, \tag{7.99}$$

where $j = 1, 3$ and

$$k = \sqrt{\frac{2m^*E_x}{\hbar^2}}. \tag{7.100}$$

Let us assume that for the region $|x| \le d$ with $j = 2$ the electron energy, E_x, is smaller than the height of the potential barrier, U_0. Then, the Schrödinger equation and its solution can be written in the form

$$-\frac{\hbar^2}{2m^*}\frac{\mathrm{d}^2\psi_2(x)}{\mathrm{d}x^2} + U_0\psi_2(x) = E_x\psi_2(x), \tag{7.101}$$

$$\psi_2(x) = A_2\mathrm{e}^{\kappa x} + B_2\mathrm{e}^{-\kappa x}, \tag{7.102}$$

where we introduced the parameter κ:

$$\kappa = \sqrt{\frac{2m^*(U_0 - E_x)}{\hbar^2}}. \tag{7.103}$$

The boundary conditions for this structure are

$$\psi_1(-d - L_x) = \psi_3(d + L_x) = 0, \tag{7.104}$$

$$\psi_1(-d) = \psi_2(-d), \tag{7.105}$$

$$\psi_3(d) = \psi_2(d), \tag{7.106}$$

$$\left(\frac{\mathrm{d}\psi_1}{\mathrm{d}x} - \frac{\mathrm{d}\psi_2}{\mathrm{d}x}\right)\bigg|_{x=-d} = 0, \tag{7.107}$$

$$\left(\frac{\mathrm{d}\psi_3}{\mathrm{d}x} - \frac{\mathrm{d}\psi_2}{\mathrm{d}x}\right)\bigg|_{x=d} = 0. \tag{7.108}$$

Equation (7.104) reflects the inability of the electron to escape the considered region and Eqs. (7.105)–(7.108) reflect the continuity of the wavefunction and its derivative at the boundaries of the potential barrier U_0. On substituting the expressions for the wavefunction (7.99) and (7.102) into Eqs. (7.104)–(7.108) we obtain two dispersion relations:

$$\tan(kL_x) + \frac{k}{\kappa}\coth(\kappa d) = 0, \tag{7.109}$$

$$\tan(kL_x) + \frac{k}{\kappa}\tanh(\kappa d) = 0, \tag{7.110}$$

where we introduced the hyperbolic functions tanh and coth:

$$\tanh\alpha = \frac{1}{\coth\alpha} = \frac{\mathrm{e}^{\alpha} - \mathrm{e}^{-\alpha}}{\mathrm{e}^{\alpha} + \mathrm{e}^{-\alpha}}. \tag{7.111}$$

The dispersion relation (7.109) corresponds to the symmetric states and the dispersion relation (7.110) to the antisymmetric states. For sufficiently large distances between the quantum dots, i.e., at $\kappa d \gg 1$, dispersion relations are reduced to one, namely to

$$\tan(kL_x) \approx -\frac{k}{\kappa} = -\sqrt{\frac{E_x}{U_0 - E_x}}. \tag{7.112}$$

For the first two deepest levels at $E_x \ll U_0$ we obtain from Eq. (7.112)

$$kL_x \approx \pi \tag{7.113}$$

and, on substituting Eq. (7.100) into Eq. (7.113), we get

$$E_x^{(1,2)} \approx \frac{\pi^2\hbar^2}{2m^* L_x^2} = E_x^{(0)}. \tag{7.114}$$

Therefore, when the potential wells are positioned far away from each other, i.e., when $d \to \infty$, the levels that correspond to the symmetric wavefunction, ψ_s, and the antisymmetric wavefunction, ψ_a, merge, as shown in Fig. 7.15.

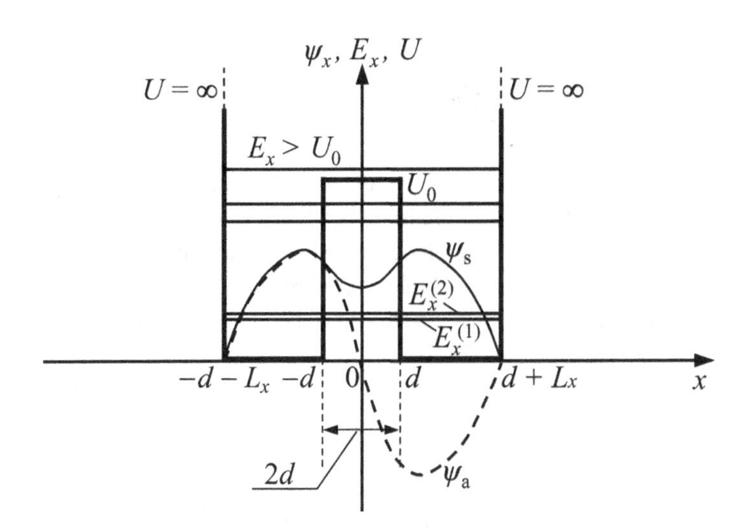

Figure 7.16 The wavefunctions of a double-quantum-dot structure, ψ_s and ψ_a, when the quantum dots are close to each other.

In the case of a narrow barrier (see Fig. 7.16), i.e., at $\kappa d \ll 1$, the dispersion equations (7.109) and (7.110) become

$$\tan(kL_x) = -\frac{k}{\kappa^2 d}, \tag{7.115}$$

$$\tan(kL_x) = -kd. \tag{7.116}$$

Equation (7.115) corresponds to the symmetric state and Eq. (7.116) to the antisymmetric state. Let us rewrite Eq. (7.116) as

$$kL_x = \arctan(-kd). \tag{7.117}$$

In the case of small d (i.e., at $kd \ll 1$) we can obtain from Eq. (7.117) the dependence of the energy of the antisymmetric state on the distance between the quantum dots, d. We can expand the right-hand side of Eq. (7.117) in a Taylor series taking into account only its first two terms. Using this approximation, $\arctan t$ can be written as

$$\arctan t = \pi + t. \tag{7.118}$$

Taking into account Eq. (7.118), we can rewrite Eq. (7.117) as

$$kL_x = \pi - k\text{d}. \tag{7.119}$$

By substituting into Eq. (7.119) the expression for k from Eq. (7.100), we find

$$E_x^{(2)} \approx \frac{\pi^2\hbar^2}{2m^* L_x^2}\frac{1}{(1 + d/L_x)^2}. \tag{7.120}$$

We see that at $d = 0$ the energy of the antisymmetric state becomes equal to the energy of the first excited level of an individual quantum dot with width $2L_x$ and the electron wavefunction becomes an antisymmetric function, ψ_a, of the individual quantum dot of width $2L_x$ (see Fig. 7.14):

$$E_x^{(2)} \approx \frac{\pi^2 \hbar^2}{2m^* L_x^2} = E_x^{(0)}. \tag{7.121}$$

As the distance between quantum dots, d, increases the level $E_x^{(2)}$ slightly decreases, but with further increase of d the level $E_x^{(2)}$ tends to $E_x^{(0)}$ according to Eq. (7.114). The dependence of $E_x^{(1)}$, which corresponds to the symmetric state, on d is shown in Fig. 7.15, but it can be obtained only by numerical solution of Eq. (7.109). In the limiting case when $d = 0$, from Eq. (7.115) it follows that

$$kL = \frac{\pi}{2} \tag{7.122}$$

and

$$E_x^{(1)} = \frac{E_x^{(0)}}{4}. \tag{7.123}$$

Let us consider now the case when the electron energy of motion along the x-axis in the double-quantum-dot structure is greater than the height of the potential barrier, U_0, which separates the potential wells ($E_x > U_0$). The solutions of the Schrödinger equation for regions 1 and 3 have the same form (7.99), and the solution in region 2 can be written as

$$\psi_2(x) = A_2 e^{ik_2 x} + B_2 e^{-ik_2 x}, \tag{7.124}$$

where

$$k_2 = \sqrt{\frac{2m^*(E_x - U_0)}{\hbar^2}}. \tag{7.125}$$

Compare Eqs. (7.124) and (7.125) with the corresponding Eqs. (7.102) and (7.103). After substitution of the wavefunctions $\psi_j(x)$ into the boundary conditions (7.104)–(7.108) we arrive at the following dispersion relations, which define the energy spectrum of an electron in the quantum-dot structure:

$$\tan(kL_x) - \frac{k}{k_2}\cot(k_2 d) = 0, \tag{7.126}$$

$$\tan(kL_x) + \frac{k}{k_2}\tan(k_2 d) = 0, \tag{7.127}$$

where Eq. (7.126) corresponds to the allowed symmetric states and Eq. (7.127) to the allowed antisymmetric states. Compare Eqs. (7.126) and (7.127) with the corresponding Eqs. (7.109) and (7.110).

Finding the general analytical solutions of Eqs. (7.126) and (7.127) is impossible because these are *transcendental equations*. Therefore, let us consider one of the possible particular solutions of Eqs. (7.126) and (7.127). Let us assume that in the region of the quantum dots, 1 and 3, as well as in the region of the barrier, 2, the electron waves are de Broglie standing waves. For the formation of these waves the following conditions must be satisfied:

$$k_2 d = m\pi \quad \text{and} \quad kL_x = \left(n + \frac{1}{2}\right)\pi \tag{7.128}$$

for the symmetric states and

$$k_2 d = m\pi \quad \text{and} \quad kL_x = n\pi \tag{7.129}$$

for the antisymmetric states. Therefore, if condition (7.128) is satisfied the energy of the stationary symmetric state must be equal to

$$E_x^{\mathrm{s}} = \frac{\pi^2\hbar^2}{2m^* L_x^2}\left(n + \frac{1}{2}\right)^2 = U_0 + \frac{\pi^2\hbar^2}{2m^* d^2} m^2. \tag{7.130}$$

Here, we took into account that the equation for kinetic energy is defined as

$$E = \frac{\hbar^2 k^2}{2m^*}.$$

We see that $m = 1$ corresponds to the lowest above-barrier energy state. This state may exist if the height of the potential barrier is equal to

$$U_0 = \frac{\pi^2\hbar^2}{2m^*}\left[\frac{(n + 1/2)^2}{L_x^2} - \frac{1}{d^2}\right]. \tag{7.131}$$

For the energy of antisymmetric states, from condition (7.129) we get

$$E_x^{\mathrm{a}} = \frac{\pi^2\hbar^2}{2m^* L_x^2} n^2 = U_0 + \frac{\pi^2\hbar^2}{2m^* d^2} m^2. \tag{7.132}$$

The lowest energy value corresponds to $m = 1$. This state can be realized if the barrier height is equal to

$$U_0 = \frac{\pi^2\hbar^2}{2m^*}\left(\frac{n^2}{L_x^2} - \frac{1}{d^2}\right). \tag{7.133}$$

Thus, different barrier heights correspond to symmetric and antisymmetric above-barrier states (see Eqs. (7.131) and (7.133)). If the conditions of Eqs. (7.131) and (7.133) are satisfied only one of these states can be realized. From the discussion above it follows that by controlling the height and width of the barrier we can shift the allowed energy levels and change the distance between them.

From Eqs. (7.130) and (7.132) it follows that for a particular set of nanostructure parameters the realization of the above-barrier regime with symmetric or antisymmetric states, defined by quantum numbers n and m, is possible only if one of the following conditions is satisfied:

$$d > \frac{m}{n + 1/2} L_x, \quad \text{or} \quad d > \frac{m}{n} L_x. \tag{7.134}$$

Example 7.6. Apart from under conditions (7.128) and (7.129), the stationary above-barrier states in a double-quantum-dot structure are possible also at

$k_2 d = (m + 1/2)\pi$. Find the lowest energy for the above-barrier symmetric and antisymmetric states.

Reasoning. For symmetric states, according to Eq. (7.129), the condition $kL_x = n\pi$ must be satisfied. Then, we have

$$E_x^{\mathrm{s}} = \frac{\pi^2\hbar^2}{2m^* L_x^2} n^2 = U_0 + \frac{\pi^2\hbar^2}{2m^* d^2}\left(m + \frac{1}{2}\right)^2. \tag{7.135}$$

For antisymmetric states the condition $kL_x = (n + 1/2)\pi$ must be satisfied. Then, we have

$$E_x^{\mathrm{a}} = \frac{\pi^2\hbar^2}{2m^* L_x^2}\left(n + \frac{1}{2}\right)^2 = U_0 + \frac{\pi^2\hbar^2}{2m^* d^2}\left(m + \frac{1}{2}\right)^2. \tag{7.136}$$

From Eqs. (7.135) and (7.136) it follows that in the case considered here the realization of the above-barrier regime is possible only if one of the following conditions is satisfied:

$$d > \frac{m + 1/2}{n} L_x \quad \text{or} \quad d > \frac{m + 1/2}{n + 1/2} L_x \tag{7.137}$$

for symmetric and antisymmetric states, respectively. The lowest possible energy of the symmetric states corresponds to $m = 0$ and $n = 1$, and that of antisymmetric states corresponds to $m = 0$. The above-barrier symmetric state has energy greater than that of the antisymmetric state.

Example 7.7. Write the expressions for the wavefunction and the electron energy for double-quantum-wire and double-quantum-well structures.

Reasoning. From the nanostructures of two coupled quantum dots considered above we can remove the restriction along one of the directions of quantization. For example, along the y-direction we can remove the restriction for electron motion by increasing infinitely the length L_y. Then, the electron motion in this direction will be free. As a result we will obtain a nanostructure consisting of two coupled quantum wires. For an electron in such a structure the wavefunction and energy can be written as

$$\psi(x, y, z) = \sqrt{\frac{2}{L_z}} \sin\left(\frac{\pi n_z z}{L_z}\right) \mathrm{e}^{\mathrm{i}k_y y} \psi_x(x), \tag{7.138}$$

$$E = E_x + \frac{\hbar^2 k_y^2}{2m^*} + \frac{\hbar^2\pi^2 n_z^2}{2m^* L_z^2}, \tag{7.139}$$

where $\psi_x(x)$ and E_x are defined by the solutions of the one-dimensional Schrödinger equation for the potential $U(x)$ which we have already obtained.

If we remove one more restriction for electron motion along the z-direction by increasing infinitely the length L_z, then we will obtain a nanostructure consisting of two coupled quantum wells. The electron wavefunction and the energy for such a nanostructure have the forms

$$\psi(x, y, z) = \psi_x(x)\mathrm{e}^{\mathrm{i}(k_y y + k_z z)}, \tag{7.140}$$

$$E = E_x + \frac{\hbar^2}{2m^*}\left(k_y^2 + k_z^2\right). \tag{7.141}$$

The electron motion remains quantized only along the x-axis and electron motion in the yz-plane is free. The forms of the one-dimensional wavefunction $\psi_x(x)$ and energy E_x are completely defined by the form of the potential $U(x)$. For the dependence $U(x)$ defined by Eq. (7.92) the wavefunction $\psi_x(x)$ and energy E_x have already been found earlier in this section.

7.6 An electron in a periodic one-dimensional potential

The most important structures which are suitable for practical applications whilst still being simple enough for theoretical analysis are periodic structures consisting of coupled nanoobjects such as quantum dots, quantum wires, or quantum wells. There is a conventional name for periodic structures of repeating quantum wells. They are called *superlattices*. The origin of this name is that each of the quantum wells which are the elements of the periodic structure has its own lattice and in the superlattice the ordering of quantum wells is superimposed on the lattice of quantum wells. Therefore it is not simply a lattice but a *superlattice*. We can generalize this notion and understand under the term "superlattice" those periodic structures which include not only quantum wells but, instead of quantum wells, other nanoobjects such as nanowires or quantum dots. From the point of view of finding solutions of the Schrödinger equation, one-dimensional superlattices, which can be formed by placing nanoobjects regularly along one of the directions, are the simplest. In this section we will consider the general properties of electron behavior in a one-dimensional superlattice, which do not depend on the particular form of a periodic potential.

7.6.1 Bloch's theorem

Let us consider the general form of the wavefunction of an electron moving in a periodic potential (of period D) of a one-dimensional superlattice. The condition of periodicity of the potential, which has the form

$$U(x + nD) = U(x), \tag{7.142}$$

leads to the condition of *translational invariance* for the electron wavefunction in the superlattice:

$$|\psi(x + nD)|^2 = |\psi(x)|^2. \tag{7.143}$$

Here n is an arbitrary integer number. The simplest example of a step-like one-dimensional periodic potential is shown in Fig. 7.17. The last expression shows that the probability density of finding an electron at the point with the coordinate x is a periodic function with the *period D*.

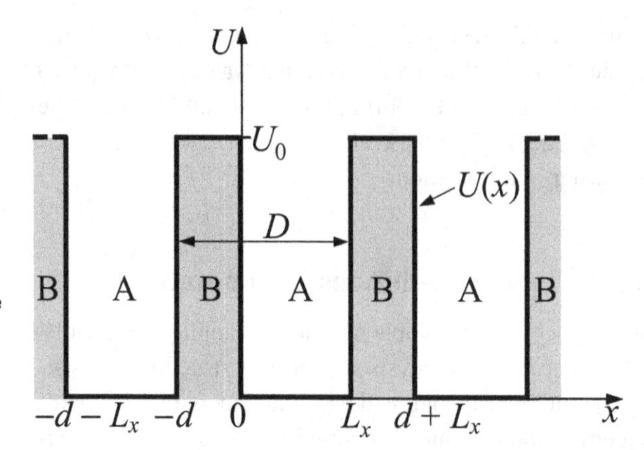

Figure 7.17 The simplest example of a one-dimensional step-like periodic potential. A denotes the material of the potential wells and B the material of the barriers which separate potential wells. D is the superlattice period, U_0 the height of the barriers, d the width of the barriers, and L_x the width of the quantum wells.

The wavefunctions $\psi(x)$ and $\psi(x + nD)$ satisfy the same Schrödinger equation with the potential (7.142). Therefore, these functions must be linearly dependent:

$$\psi(x + nD) = C(nD)\psi(x). \tag{7.144}$$

From condition (7.143) it follows that

$$|C(nD)|^2 = 1, \tag{7.145}$$

and from Eq. (7.145) we have

$$C(nD) = \mathrm{e}^{\mathrm{i}K_x nD}, \tag{7.146}$$

where K_x is a real parameter, which has the dimension of the wavenumber. Thus, the shift of the wavefunction's argument, x, by nD changes the electron wavefunction as follows:

$$\psi(x + nD) = \mathrm{e}^{\mathrm{i}K_x nD}\psi(x). \tag{7.147}$$

Let us make the following transformations in Eq. (7.147):

$$\begin{aligned}\psi(x) &= \mathrm{e}^{-\mathrm{i}K_x nD}\psi(x + nD) = \mathrm{e}^{-\mathrm{i}K_x(x+nD)}\psi(x + nD)\mathrm{e}^{\mathrm{i}K_x x} \\ &= \varphi(x + nD)\mathrm{e}^{\mathrm{i}K_x x} = \varphi(x)\mathrm{e}^{\mathrm{i}K_x x}.\end{aligned} \tag{7.148}$$

Here we introduced the periodic function

$$\varphi(x + nD) = \mathrm{e}^{-\mathrm{i}K_x(x+nD)}\psi(x + nD), \tag{7.149}$$

which satisfies the following condition:

$$\varphi(x + nD) = \varphi(x). \tag{7.150}$$

Note that, according to Eq. (7.150), $\varphi(x)$ is a periodic function, whereas only the square of the modulus of the function $\psi(x)$, i.e., $|\psi(x)|^2$, is periodic (see Eq. (7.143)). The wavefunction $\psi(x)$ itself is a product of a periodic function $\varphi(x)$ and a free-electron wavefunction, $e^{iK_x x}$.

Thus, the *electron wavefunction, $\psi(x)$, in the periodic potential of a superlattice always can be presented as a wavefunction of a free electron, $e^{iK_x x}$, modulated by a periodic function, $\varphi(x)$*, i.e.,

$$\psi(x) = \varphi(x)e^{iK_x x}, \tag{7.151}$$

where the parameter K_x is an effective wavenumber (and in the general case it is an effective wavevector) of an electron in a superlattice. This statement constitutes *Bloch's theorem*. The wavefunction (7.151) is called *Bloch's wavefunction*. The wavefunction (7.151) must satisfy the following cyclic boundary conditions:

$$\psi(x + \mathcal{L}_x) = \psi(x), \tag{7.152}$$

where $\mathcal{L}_x$ is the superlattice length in the x-direction. The superlattice length, $\mathcal{L}_x$, contains a sufficiently large number of superlattice periods, D. From this condition, taking into account Eq. (7.151), it follows that the wavenumber, K_x, must have discrete values:

$$K_x = \frac{2\pi}{\mathcal{L}_x} l, \quad l = 0, \pm 1, \pm 2, \ldots \tag{7.153}$$

The momentum of the particle $\mathbf{p}$ is related to the effective wavevector $\mathbf{K}$ as $\mathbf{p} = \hbar\mathbf{K}$, and in the case of one-dimensional motion we have

$$p_x = \hbar K_x = \frac{2\pi\hbar}{\mathcal{L}_x} l. \tag{7.154}$$

7.6.2 Quasimomentum

For free electron motion the magnitudes k and p are well defined, in contrast to the case for electron motion in a superlattice, for which k and p are not unambiguously defined, which is a consequence of the periodicity of $U(x)$. Let us change the wavenumber K by $K' = K + 2\pi l/D$, where l is an integer number. Here we have omitted the subscript x from the wavevector K_x for convenience. Let us translate the wavefunction, $\psi(x)$ by nD according to Eq. (7.147):

$$\psi(x + nD) = e^{iKnD}\psi(x) = e^{i[K' - 2\pi l/D]nD}\psi(x) = e^{iK'nD}\psi(x)e^{-i2\pi ln} = e^{iK'nD}\psi(x), \tag{7.155}$$

where we took into account that the product ln is an integer number and

$$e^{-i2\pi ln} = 1. \tag{7.156}$$

From a comparison of Eqs. (7.155) and (7.147) it follows that the electron quantum states with wavenumbers K and $K' = K + 2\pi l/D$ in the superlattice correspond to the same physical state. Because of this equivalence for the electron in the superlattice the magnitude $\mathbf{K}$ is called the *quasiwavevector* rather than just the wavevector. We can also talk about the physical equivalence of the states with magnitudes of momentum in the superlattice p and $p + 2\pi l/D$. Therefore, **p** is called the *quasimomentum*.

In Chapter 2 it was shown that for a free particle the principle of conservation of momentum **p** holds. This is due to the translational homogeneity of free space (of vacuum). In the superlattice the homogeneity of space is broken because the potential energy of the electron becomes a periodic function. The existence of a periodic potential in the superlattice leads to the conservation *not of the momentum* of the electron but of its *quasimomentum*. Thus, for the electron's quasimomentum Newton's second law is valid:

$$\frac{d\mathbf{p}}{dt} = \mathbf{F}_{\text{ext}}, \tag{7.157}$$

where the force $\mathbf{F}_{\text{ext}}$ is related to the external fields and does not include the periodic force $\mathbf{F}_{\text{SL}}$ related to the superlattice potential, $U(\mathbf{r})$, by

$$\mathbf{F}_{\text{SL}} = -\nabla U(\mathbf{r}). \tag{7.158}$$

The interval of quasimomenta which contains all physically non-equivalent states is symmetric relative to the origin of the coordinates, i.e.,

$$-\frac{\pi\hbar}{D} \le p \le \frac{\pi\hbar}{D}. \tag{7.159}$$

This interval is called the *first Brillouin zone*. The second zone is located symmetrically with respect to the first zone with respect to the origin of coordinates. It includes two intervals:

$$\left(-\frac{2\pi\hbar}{D}, -\frac{\pi\hbar}{D}\right) \quad \text{and} \quad \left(\frac{\pi\hbar}{D}, \frac{2\pi\hbar}{D}\right). \tag{7.160}$$

Then, the third Brillouin zone is located similarly and so on. The width of the interval which corresponds to any Brillouin zone is always equal to $2\pi\hbar/D$.

Let us now discuss what values the quasiwavevector $\mathbf{K} = \mathbf{p}/\hbar$, which defines the boundaries of the Brillouin zone, may have. Let us assume that an electron with wavevector **K** is moving along the x-direction and that the period of the superlattice in this direction is equal to D. Let us assume that the *interaction of the electron with an individual nanoobject of the superlattice is weak*. This means that during propagation of the electron wave in the periodic potential of the superlattice the amplitude of the reflected wave from an individual nanoobject is much smaller than the amplitude of the incident electron wave. Since the superlattice contains many nanoobjects, the phase relationships between the reflected waves play a significant role. If the electron de Broglie wavelength exceeds the distance between nanoobjects, i.e., exceeds the period D, then the phase difference between the waves reflected from neighboring objects is small.

For

$$2D = n\lambda_{\mathrm{Br}}, \tag{7.161}$$

where n is an integer number, the phase difference of reflected waves is a multiple of 2π. The waves reflected from different nanoobjects are in phase with each other and their amplitudes are added. Because of this, even in the case of small amplitudes of the waves reflected from the individual nanoobjects, the electron is totally reflected from the superlattice as a whole. Therefore, if condition (7.161) is satisfied the electron wave cannot propagate in the superlattice.

The condition of total reflection (7.161) coincides with Bragg's law, which was obtained for the diffraction of an electromagnetic wave on a crystalline lattice. If we take the diffraction angle equal to $\theta = \pi/2$ and write the condition for total reflection for the wavevector of an electron in a one-dimensional superlattice as

$$k = \frac{2\pi}{\lambda_{\mathrm{Br}}}, \tag{7.162}$$

then we find the equation which defines the boundaries of the Brillouin zone, i.e.,

$$K_n = \pm\frac{n\pi}{D}. \tag{7.163}$$

Therefore, the boundaries of each Brillouin zone correspond to magnitudes of the wavevector at which the electron wave in a superlattice cannot propagate. In this case we can conclude that the electron wave, as a result of diffraction by a superlattice, undergoes 180° Bragg reflection. As K approaches K_n the superlattice slows the traveling electron wave down more and more, and, when K and K_n coincide, the *traveling wave* becomes a *standing wave*. We can show that the group velocity of the electron wave,

$$v_{\mathrm{gr}} = \frac{\mathrm{d}\omega}{\mathrm{d}K} = \frac{1}{\hbar}\frac{\mathrm{d}E}{\mathrm{d}K} = \frac{\mathrm{d}E}{\mathrm{d}p}, \tag{7.164}$$

becomes equal to zero in the case of $K = K_n$. This behavior of an electron in the superlattice is principally different from the behavior of a free electron in vacuum.

7.6.3 Effective mass

According to Eq. (7.164) the electron velocity in the superlattice is defined by the dispersion relation $E(p)$. Let us find the acceleration of an electron in a one-dimensional superlattice:

$$a = \frac{\mathrm{d}v}{\mathrm{d}t} = \frac{\mathrm{d}}{\mathrm{d}t}\left(\frac{\mathrm{d}E}{\mathrm{d}p}\right) = \frac{\mathrm{d}}{\mathrm{d}p}\left(\frac{\mathrm{d}E}{\mathrm{d}t}\right) = \frac{\mathrm{d}}{\mathrm{d}p}\left(\frac{\mathrm{d}E}{\mathrm{d}p}\frac{\mathrm{d}p}{\mathrm{d}t}\right) = \frac{\mathrm{d}^2E}{\mathrm{d}p^2}\frac{\mathrm{d}p}{\mathrm{d}t}$$
$$= \frac{\mathrm{d}^2E}{\mathrm{d}p^2}F_{\mathrm{ext}} = \frac{F_{\mathrm{ext}}}{m^*}, \tag{7.165}$$

where we took into account that

$$\frac{\mathrm{d}}{\mathrm{d}p}\left(\frac{\mathrm{d}p}{\mathrm{d}t}\right) = \frac{\mathrm{d}}{\mathrm{d}t}\left(\frac{\mathrm{d}p}{\mathrm{d}p}\right) = 0. \tag{7.166}$$

The electron acceleration linearly depends on the external force, F_{ext}, and on the magnitude

$$\frac{1}{m^*} = \frac{\mathrm{d}^2 E(p)}{\mathrm{d}p^2} = \frac{1}{\hbar^2}\frac{\mathrm{d}^2 E(K)}{\mathrm{d}K^2}, \tag{7.167}$$

which defines the *electron effective mass* m^* in a superlattice. The equation

$$m^* a = F_{\mathrm{ext}} \tag{7.168}$$

formally coincides with Newton's second law, which describes the motion of a particle with mass m^* in free space under the influence of an external force, F_{ext}. The physical meaning of the electron effective mass for a superlattice is the following: the internal forces (the forces of interaction of the electron with the superlattice) become apparent in the formation of the dispersion relation $E(p)$ and they define the behavior of an electron under the influence of external forces through the effective mass. With the help of Eq. (7.168) the unknown internal forces which act on an electron by virtue of its presence in the superlattice are included in the definition of the effective mass. For an electron in vacuum, i.e., for a free electron, $m^* = m_{\mathrm{e}}$ because

$$\frac{1}{m^*} = \frac{\mathrm{d}^2 E(p)}{\mathrm{d}p^2} = \frac{\mathrm{d}^2}{\mathrm{d}p^2}\left(\frac{p^2}{2m_{\mathrm{e}}}\right) = \frac{1}{m_{\mathrm{e}}}. \tag{7.169}$$

The use of the effective mass as a physical parameter that defines the behavior of an electron in a superlattice is justified in those regions where m^* weakly depends (or does not depend at all) on the quasimomentum p. These conditions are usually satisfied in the vicinity of points p_0 where the electron energy $E(p)$ reaches a maximum or a minimum. Let us expand $E(p)$ in a Taylor series near an extremum point $p = p_0$:

$$E(p) = E(p_0) + \left(\frac{\mathrm{d}E}{\mathrm{d}p}\right)_{p=p_0}(p - p_0) + \frac{1}{2!}\left(\frac{\mathrm{d}^2 E}{\mathrm{d}p^2}\right)_{p=p_0}(p - p_0)^2 + \frac{1}{3!}\left(\frac{\mathrm{d}^3 E}{\mathrm{d}p^3}\right)_{p=p_0}(p - p_0)^3 + \cdots, \tag{7.170}$$

where the derivatives are calculated at $p = p_0$. Limiting ourselves to a small region around the point p_0 and taking into account that at the extremum the derivative $(\mathrm{d}E/\mathrm{d}p)_{p=p_0} = 0$, the dispersion relation can be written as

$$E(p) = E(p_0) + \frac{1}{2m^*}(p - p_0)^2. \tag{7.171}$$

In Eq. (7.171) all terms of order greater than two were disregarded. We introduced the effective mass of an electron, m^*, which does not depend on the

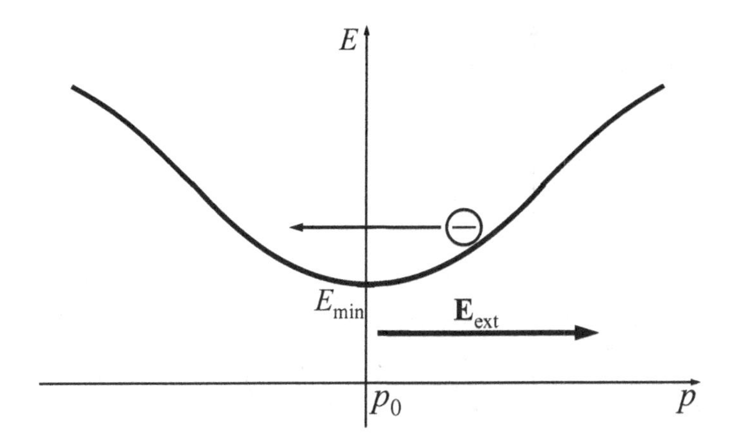

Figure 7.18 The electron motion near the energy minimum is defined by a parabolic dispersion relation. An electron near the energy minimum moves in a direction opposite to the applied external field $\mathbf{E}_{ext}$.

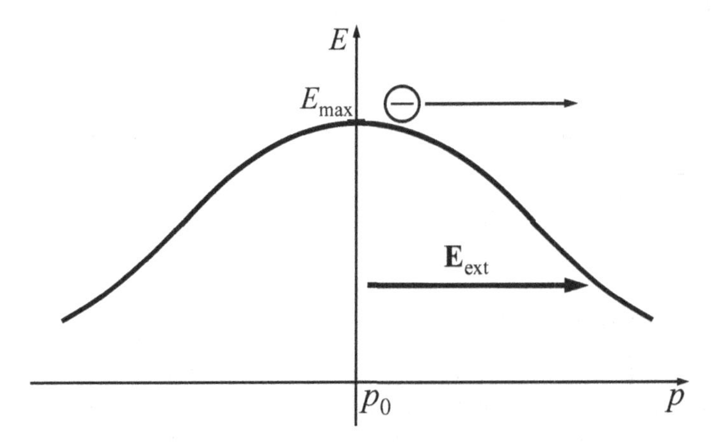

Figure 7.19 The electron motion near the energy maximum is defined by a parabolic dispersion relation. An electron near the energy maximum moves in the same direction as the applied external field $\mathbf{E}_{ext}$.

quasimomentum, p:

$$m^* = \left(\frac{d^2 E}{dp^2} \right)^{-1}_{p=p_0}. \tag{7.172}$$

The magnitude m^* is positive in the region around the *energy minimum* (see Fig. 7.18) and is negative near the *energy maximum* (see Fig. 7.19). In the first case (near the energy minimum) the electron behavior in a superlattice is similar to that of a free electron. Thus, in the external electric field, $\mathbf{E}_{ext}$, the velocity of the electron increases in the direction opposite to the direction of the electric field since the electron is *negatively charged*. In the second case (near the energy maximum) the electron accelerates in the direction of the external electric field, $\mathbf{E}_{ext}$, i.e., it behaves as a particle with positive charge and positive mass (see Fig. 7.19).

Moving far away from the extremum point $p = p_0$, i.e., with a large increase of the difference $(p - p_0)$, we have to take into account in the series (7.170)

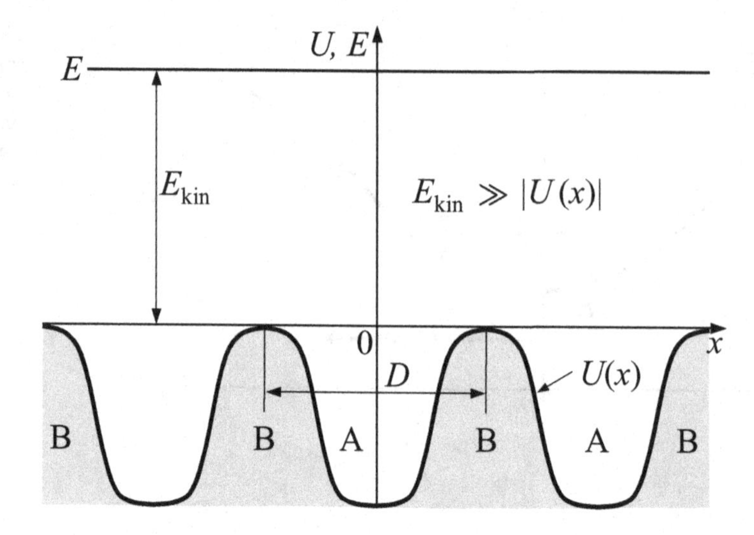

Figure 7.20 A one-dimensional superlattice with an arbitrary periodic potential.

not only the quadratic term but also higher-order terms. The effective mass, m^*, becomes dependent on the electron quasimomentum, p, if we were to write the dispersion relation for large magnitudes of $(p - p_0)$ in the form of Eq. (7.171).

7.6.4 The energy spectrum

Let us consider the main properties of the electron energy spectrum in a one-dimensional superlattice with an arbitrary periodic potential $U(x)$. For simplicity we will assume that the electron potential energy, $U(x)$, is small in comparison with its kinetic energy, E_{kin} (see Fig. 7.20). This allows us to solve the problem of electron motion in the periodic potential of the superlattice by using the methods of perturbation theory.

Let us express the electron energy and the wavefunction in the superlattice in the form of the following expansions:

$$E(K) = E^{(0)}(K) + E^{(1)}(K) + E^{(2)}(K) + \cdots, \tag{7.173}$$

$$\psi_K(x) = \psi_K^{(0)}(x) + \psi_K^{(1)}(x) + \psi_K^{(2)}(x) + \cdots \tag{7.174}$$

According to stationary perturbation theory (see Section 5.1), in order to find the corresponding corrections to $E(K)$ and $\psi_K(x)$ for the periodic potential $U(x)$ it is necessary to find the following matrix elements (for the unperturbed states):

$$U_{K'K} = \int_0^{\mathcal{L}_x} \psi_{K'}^{(0)*}(x) U(x) \psi_K^{(0)}(x) \mathrm{d}x, \tag{7.175}$$

where $\mathcal{L}_x$ is the length of a superlattice. The first-order correction to the energy coincides with the diagonal matrix element U_{KK}, i.e.,

$$U_{KK} = E^{(1)}(K) = \int_0^{\mathcal{L}_x} \psi_K^{(0)*}(x)U(x)\psi_K^{(0)}(x)\mathrm{d}x. \tag{7.176}$$

The wavefunction in the zeroth approximation describes the motion of a free electron in vacuum in the interval $(0, \mathcal{L}_x)$:

$$\psi_K^{(0)}(x) = \frac{1}{\sqrt{\mathcal{L}_x}}\mathrm{e}^{\mathrm{i}Kx}. \tag{7.177}$$

Substituting Eq. (7.177) into Eq. (7.176) we obtain that the first-order correction is equal to the following average value of the electron potential energy:

$$E^{(1)} = \frac{1}{\mathcal{L}_x}\int_0^{\mathcal{L}_x} U(x)\mathrm{d}x = \langle U(x)\rangle\,. \tag{7.178}$$

From this expression it follows that the correction $E^{(1)}$ does not depend on the wavevector (quasimomentum). This correction is negative since $U(x)$ itself is negative (see Fig. 7.20) and it shifts the parabolic dispersion of the free electron by $\langle U(x)\rangle$, decreasing the electron energy in the superlattice in comparison with that of a free electron in vacuum.

In the second-order perturbation theory, according to Eq. (5.19), the correction $E^{(2)}$ to the electron energy in a superlattice can be written in the form

$$E^{(2)}(K) = \sum_{K_n'} \frac{|U_{K_n'K}|^2}{E^{(0)}(K) - E^{(0)}(K_n')} = \sum_n \frac{|C_n|^2}{E^{(0)}(K) - E^{(0)}(K - 2\pi n/D)}, \tag{7.179}$$

where we introduced the coefficients, C_n,

$$C_n = \frac{1}{D}\int_0^D U(x)\mathrm{e}^{-\mathrm{i}2\pi nx/D}\,\mathrm{d}x, \tag{7.180}$$

of expansion of the periodic potential $U(x)$ in the Fourier series:

$$U(x) = \sum_{n=-\infty}^{\infty} C_n \mathrm{e}^{\mathrm{i}2\pi nx/D}, \tag{7.181}$$

where $n = 0, \pm 1, \pm 2, \ldots$ Taking into account the zeroth and the first-order approximations, the electron wavefunction can be written as

$$\psi_K(x) = \psi_K^{(0)}(x) + \sum_{K_n'} \frac{C_n \psi_{K_n'}^{(0)}(x)}{E^{(0)}(K) - E^{(0)}(K_n')}, \tag{7.182}$$

where $K_n' = K - 2\pi n/D$. The calculation of the corrections using perturbation theory becomes incorrect if the difference $(E^{(0)}(K) - E^{(0)}(K_n'))$ becomes small or comparable to $|C_n|$. We can show that this energy difference tends to zero at the boundaries of the Brillouin zone (see Example 7.8). In this case the corrections to the energy and the wavefunction become extremely large, i.e., the electron motion near the Brillouin-zone boundaries undergoes a strong perturbation. Therefore, at the Brillouin-zone boundaries at exact equality, $E^{(0)}(K) = E^{(0)}(K_n')$, the energy level $E^{(0)}(K)$ becomes degenerate since it corresponds to two different wavefunctions $\psi^{(0)}(K)$ and $\psi^{(0)}(K_n')$. Thus, near Brillouin-zone boundaries, even in the zeroth approximation we have to take degeneracy into account. As a result we obtain the following expressions using the perturbation theory for the degenerate states (see Section 5.2):

$$\psi_K(x) = \psi_K^{(0)}(x) + \frac{U_{21}}{E^{(0)}(K_n') + \langle U \rangle - E}\,\psi_{K_n'}^{(0)}(x), \tag{7.183}$$

$$E(K) = \langle U \rangle + \frac{E^{(0)}(K) + E^{(0)}(K_n')}{2} \pm \frac{1}{2}\sqrt{[E^{(0)}(K) - E^{(0)}(K_n')]^2 + 4|C_n|^2}. \tag{7.184}$$

From Eq. (7.184) it follows that the electron energy in the superlattice becomes discontinuous at the Brillouin-zone boundaries . When approaching, for example, the Brillouin-zone boundary $K = \pi n/D$ from the left and from the right, the electron energy takes the values

$$E^-\left(\frac{\pi n}{D}\right) = \langle U \rangle + E^{(0)}\left(\frac{\pi n}{D}\right) - |C_n|, \tag{7.185}$$

$$E^+\left(\frac{\pi n}{D}\right) = \langle U \rangle + E^{(0)}\left(\frac{\pi n}{D}\right) + |C_n|. \tag{7.186}$$

Thus, the energy discontinuity, E_{ng}, at the corresponding Brillouin-zone boundary is equal to

$$E_{ng} = E^+ - E^- = 2|C_n|. \tag{7.187}$$

The discontinuities in the dispersion relation $E(K)$ indicate the appearance of forbidden energy minibands, which alternate with the allowed energy minibands, in the electron energy spectrum of a superlattice.

As has already been noted, all Brillouin zones contain physically equivalent quantum states, which correspond to the same energy level. The analytical dependence of E on K corresponds to the dispersion relation, which can be graphically presented in the form of a parabola with discontinuities $E_{ng} = 2|C_n|$ at the Brillouin-zone boundaries (see Fig. 7.21). Taking into account the physical equivalence of the states in different Brillouin zones (see Eq. (7.147)), the dispersion relation can be graphically plotted in the first Brillouin zone (the *reduced Brillouin zone*, see Fig. 7.22). Then, the electron energy becomes a *multivalued periodic function* of the quasimomentum with period $2\pi\hbar/D$. The one-to-one dependence of energy on momentum occurs only in the limits of an individual subband of the allowed energies.

The analysis given above corresponds to the case of weak coupling of an electron with individual nanoobjects of the superlattice, when the periodic potential, $U(x)$, is small in comparison with the total energy of the electron, E. The periodic potential introduces weak perturbations in the free electron motion.

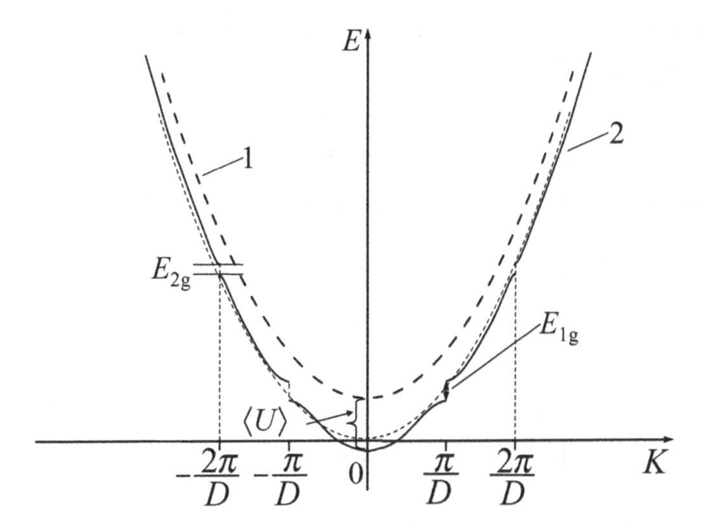

Figure 7.21 The parabolic dispersion relation of a free electron (denoted as 1) is shifted down by $\langle U \rangle$ in a superlattice. The curve with discontinuities (denoted as 2) is the electron dispersion relation in a superlattice.

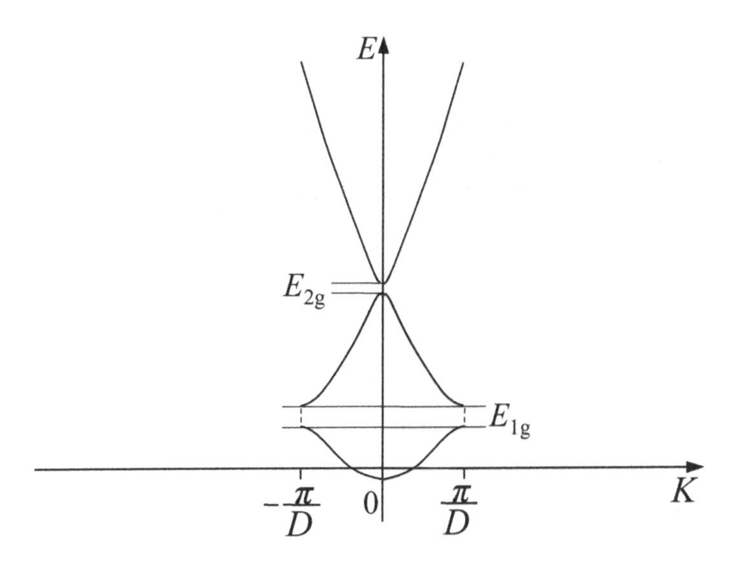

Figure 7.22 The electron dispersion relation in the reduced Brillouin zone.

The coefficients C_n in the expansion of the potential $U(x)$ in the Fourier series (Eq. (7.181)) are small in comparison with the energy of an electron at the corresponding Brillouin-zone boundary, i.e., $|C_n| \ll E^{(0)}(n\pi/D)$. Therefore the width of the forbidden minibands, E_{ng} (Eq. (7.187)), is small in comparison with the width of subbands with allowed energies. As we will show in the next section, a similar property is characteristic for electrons in quantum-dot superlattices.

Example 7.8. Show that at the boundaries of the Brillouin zone the energy difference, $E^{(0)}(K) - E^{(0)}(K'_n)$, and the group velocity of the electron wave tend to zero (here $K'_n = K - 2\pi n/D$).

Reasoning. The electron energy in the unperturbed states with wavenumbers K and K'_n can be written as

$$E^{(0)}(K) = \frac{\hbar^2 K^2}{2m^*} \quad \text{and} \quad E^{(0)}(K'_n) = \frac{\hbar^2}{2m^*}\left(K - \frac{2\pi n}{D}\right)^2. \tag{7.188}$$

Let us equate the difference of these two quantities to zero:

$$\frac{\hbar^2 K^2}{2m^*} - \frac{\hbar^2}{2m^*}\left(K - \frac{2\pi n}{D}\right)^2 = 0. \tag{7.189}$$

From Eq. (7.189) we obtain

$$\frac{4\pi n}{D}\left(K - \frac{\pi n}{D}\right) = 0. \tag{7.190}$$

From Eq. (7.190) it follows that Eq. (7.189) is satisfied for wavenumbers

$$K_n = \frac{\pi n}{D} \tag{7.191}$$

or for quasimomenta

$$p_n = \frac{\hbar\pi n}{D}.$$

The wavenumbers from Eq. (7.191) correspond to the boundaries of the first Brillouin zone of a one-dimensional superlattice.

Let us rewrite Eqs. (7.185) and (7.186) for the wavenumbers in the immediate vicinity of the Brillouin-zone boundaries:

$$\begin{aligned} E^{\pm}(K) &= \langle U \rangle + \frac{E^{(0)}(K) + E^{(0)}(K'_n)}{2} \pm |C_n| \\ &= \langle U \rangle \pm |C_n| + \frac{\hbar^2 K^2}{2m^*} - \frac{\pi n \hbar^2}{m^* D}\left(K - \frac{\pi n}{D}\right). \end{aligned} \tag{7.192}$$

The group velocity of the electron wave in this case is defined by the following expression:

$$v_{\mathrm{gr}}^{\pm} = \frac{1}{\hbar}\frac{\mathrm{d}E^{\pm}(K)}{\mathrm{d}K} = \frac{\hbar}{m^*}\left(K - \frac{\pi n}{D}\right). \tag{7.193}$$

According to Eq. (7.191), at the Brillouin-zone boundaries the magnitudes of wavenumbers are equal to

$$K = K_n = \frac{\pi n}{D}.$$

Thus, from Eq. (7.192) we see that at the Brillouin-zone boundaries the group velocity of an electron wave tends to zero and the graphical dependence $E(K)$ must have a horizontal tangent at Brillouin-zone boundaries (see Fig. 7.22). For a two-dimensional superlattice the Brillouin-zone boundaries are straight lines and for three-dimensional superlattices they are planes in the space of wavevectors where the surface of equal energy has discontinuities and is perpendicular to them.

7.7 A one-dimensional superlattice of quantum dots

In Section 7.5 we have shown that in the case of a nanostructure with two coupled quantum dots the energy level that corresponds to an individual quantum dot splits into two sublevels. With increasing number of nanoobjects the number of sublevels also increases. The magnitude of the splitting depends on the barrier between the nanoobjects, i.e., it depends on the width and height of the potential barrier. For a large number of nanoobjects *the energy level of an individual nanoobject* transforms into *an energy miniband.* The type of these minibands is defined mostly by the type of nanoobjects, by the type of the structure itself, and by the periodicity of the nanoobjects' distribution in the structure. In the previous section we considered the general properties of the electron energy spectrum in a one-dimensional superlattice with an arbitrary profile of a periodic potential, $U(x)$. As a special case in Section 7.6.4 we have assumed that the potential, $U(x)$, was small in comparison with the total energy, E. Here we will discuss the properties of electron motion in superlattices with a specific rectangular potential profile.

Let us consider a periodic structure consisting of a set of quantum dots regularly distributed along the x-direction with period D. Let us assume that along the y- and z-directions the electron is strictly confined to the region defined by the intervals $(0, L_y)$ and $(0, L_z)$, i.e., the height of the potential barriers in these directions is infinite. In this case the potential energy $U(x, y, z)$ is defined by the expression Eq. (7.91), where the potential profile in the direction of periodicity has the form shown in Fig. 7.17. This periodic potential profile in the limits of the superlattice period, $D = d + L_x$, is defined as follows:

$$U(x) = \begin{cases} U_0, & -d \leq x \leq 0, \\ 0, & 0 \leq x \leq L_x. \end{cases} \tag{7.194}$$

Here L_x and d are the widths of quantum wells and barriers in the x-direction, respectively. Such a potential distribution for an electron in a one-dimensional periodic structure is called the *Kronig–Penney model.* The potential profile (7.194) is written for one superlattice period only. This potential profile repeats itself over the entire superlattice along the x-direction. In the Schrödinger equation for the potential $U(x, y, z)$ under consideration we can separate variables and present the total wavefunction, $\psi(x, y, z)$, as the following product:

$$\psi(x, y, z) = \psi_x(x)\psi_y(y)\psi_z(z).$$

According to Eq. (7.94), the wavefunctions $\psi_y(y)$ and $\psi_z(z)$ are equal to

$$\psi_y(y) = \sqrt{\frac{2}{L_y}} \sin\left(\frac{n_y \pi y}{L_y}\right)$$

and

$$\psi_z(z) = \sqrt{\frac{2}{L_z}} \sin\left(\frac{n_z \pi z}{L_z}\right).$$

The energy according to Eq. (7.95) is defined by the expression

$$E = E_x + \frac{\pi^2 \hbar^2}{2m^*}\left(\frac{n_y^2}{L_y^2} + \frac{n_z^2}{L_z^2}\right).$$

We will consider two different cases. First, we will deal with the case when the electron energy, E, is higher than the potential-barrier height, U_0. Then, we will consider the opposite case for which $E < U_0$.

Let us first consider the case with electron above-barrier motion, i.e., when $E \geq U_0$. The wavefunction $\psi_x(x)$ in the interval $-d \leq x \leq L_x$ can be presented as follows:

$$\psi_x(x) = \begin{cases} A_1 \mathrm{e}^{\mathrm{i}k_1 x} + B_1 \mathrm{e}^{-\mathrm{i}k_1 x}, & -d \leq x \leq 0, \\ A \mathrm{e}^{\mathrm{i}kx} + B \mathrm{e}^{-\mathrm{i}kx}, & 0 \leq x \leq L_x. \end{cases} \tag{7.195}$$

Here the wavenumbers in the barrier region, k_1, and in the well region, k, are defined as

$$k_1 = \frac{\sqrt{2m_\mathrm{e}(E - U_0)}}{\hbar}, \tag{7.196}$$

$$k = \frac{\sqrt{2m_\mathrm{e}E}}{\hbar}. \tag{7.197}$$

Both wavenumbers are written for the above-barrier motion of the electron, when its energy in the superlattice is greater than U_0, i.e., when $E \geq U_0$. Only a certain type of wavefunction will satisfy the condition of the superlattice potential periodicity

$$U(x + nD) = U(x). \tag{7.198}$$

Here n is an integer number. According to Bloch's theorem formulated for the electron wavefunctions in a periodic potential, this function must have the form

$$\psi_x(x) = \varphi(x)\mathrm{e}^{\mathrm{i}K_x x}. \tag{7.199}$$

Here $\varphi(x)$ is a periodic function with a superlattice period D and K_x is the effective (Bloch) wavenumber in the x-direction, which depends on the parameters of the nanostructure and the electron energy, E. In our further calculations we will omit the subscript x from K_x and L_x. From Eq. (7.199) the following important property follows:

$$\psi_x(x + D) = \psi_x(x)\mathrm{e}^{\mathrm{i}KD}, \tag{7.200}$$

i.e., the wavefunction, ψ_x, when shifted by one period, D, is multiplied by the phase factor $\mathrm{e}^{\mathrm{i}KD}$. Using Eq. (7.200) and the conditions of continuity of the wavefunction and its derivative at the boundaries $x = 0$ and $x = L$, we arrive at the following system of four linear equations for the unknown coefficients A, B,

A_1, and B_1 from Eq. (7.195):

$$\begin{aligned} A_1 + B_1 &= A + B, \\ k_1(A_1 - B_1) &= k(A - B), \\ e^{iKD}(A_1 e^{-ik_1 d} + B_1 e^{ik_1 d}) &= A e^{ikL} + B e^{-ikL}, \\ k_1 e^{iKD}(A_1 e^{-ik_1 d} - B_1 e^{ik_1 d}) &= k(A e^{ikL} - B e^{-ikL}). \end{aligned} \tag{7.201}$$

This system of equations has a non-zero solution only if the determinant of this system is equal to zero. This condition leads to the following dispersion equation, which relates the parameter K to the known parameters k_1, k, L, and d:

$$y^2 - 2Fy + 1 = 0, \tag{7.202}$$

where we introduced the variable

$$y = e^{iKD} \tag{7.203}$$

and

$$F = \cos(kL)\cos(k_1 d) - \frac{k^2 + k_1^2}{2kk_1}\sin(kL)\sin(k_1 d). \tag{7.204}$$

For two roots of the quadratic equation (7.202) the following equalities (Viète's formulae) apply:

$$y_1 y_2 = 1 \tag{7.205}$$

and

$$y_1 + y_2 = 2F. \tag{7.206}$$

From Eq. (7.205) it follows that

$$y_1 = e^{iKD} \tag{7.207}$$

and

$$y_2 = e^{-iKD}. \tag{7.208}$$

Using Eqs. (7.202) and (7.206)–(7.208), the dispersion equation for the superlattice may be written as follows:

$$\cos(KD) = \cos(kL)\cos(k_1 d) - \frac{k^2 + k_1^2}{2kk_1}\sin(kL)\sin(k_1 d). \tag{7.209}$$

Equation (7.209) describes the *above-barrier regime of electron motion in the superlattice* since it was derived for $E \geq U_0$ and for real wavenumbers k_1. We considered the general properties of such motion in Section 7.6.4 for a small-amplitude periodic potential of arbitrary profile, i.e., in the case of weak coupling of the electron with the individual quantum dots.

The opposite situation in which the electron has the energy $E < U_0$ and may be localized in the region of an individual quantum dot is much more interesting. As quantum dots are moved closer to each other they form a superlattice. Owing to tunneling through the potential barrier between neighboring quantum

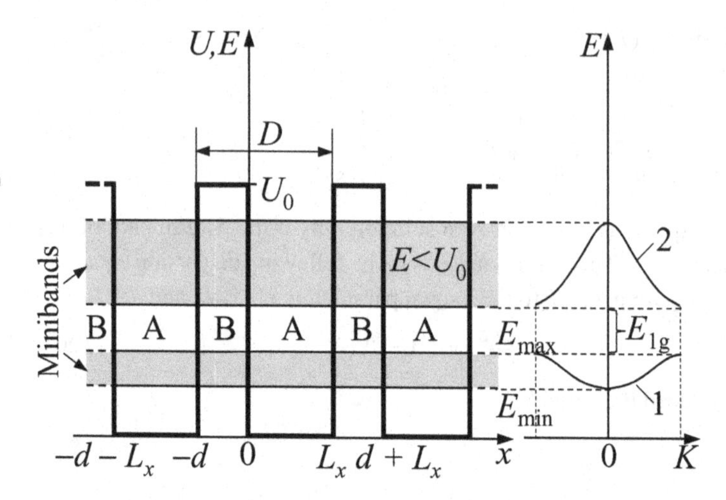

Figure 7.23 The potential profile of the superlattice in the direction of the periodicity (the x-direction). U_0 is the height of the barrier which separates neighboring quantum dots, d is the width of the potential barrier, and L_x is the width of the potential well. A denotes the material of the quantum dot and B denotes the material of the barriers that separate quantum dots. The first energy miniband is labeled 1 and the second is labeled 2.

dots, complete delocalization of the electron is possible, and as a result the electron belongs to the entire superlattice. If the electron energy $E < U_0$, then the wavenumber k_1 becomes imaginary (see Eq. (7.196)). Therefore, in the expression for the wavefunction (7.195) and in the dispersion equation (7.209) it is convenient to substitute k_1 by $\mathrm{i}\kappa$, where the real parameter κ is equal to

$$\kappa = \frac{\sqrt{2m^*(U_0 - E)}}{\hbar}. \tag{7.210}$$

As a result the dispersion equation in this case takes the form

$$\cos(KD) = \cos(kL)\cosh(\kappa d) - \frac{k^2 - \kappa^2}{2k\kappa}\sin(kL)\sinh(\kappa d). \tag{7.211}$$

Actually, $\cos(k_1 d)$ and $\sin(k_1 d)$ in Eq. (7.209) are replaced by $\cosh(\kappa d)$ and $\sinh(\kappa d)$ in Eq. (7.211).

The analysis of Eq. (7.211) shows that the energy spectrum of the superlattice at $E < U_0$ is divided into a set of minibands with allowed and forbidden energy values (see Fig. 7.23). The minibands with allowed energies correspond to values of energy for which the magnitude of the right-hand side of Eq. (7.211) lies within the interval $(-1, 1)$ and the effective wavenumber K is real. The energy regions where $|\cos(KD)| > 1$ do not correspond to the real wavenumbers K. Therefore, they are the minibands with forbidden energies. Electrons with imaginary wavenumbers, K, and corresponding energies, E, according to Eq. (7.199), cannot propagate for long distances in a superlattice.

If the superlattice has wide potential barriers, the lowest minibands with allowed energies are very narrow and lie near the energy levels of an individual quantum dot. In the limiting case when $d \to \infty$

$$\cosh(\kappa d) \approx \sinh(\kappa d) \approx \frac{1}{2}\mathrm{e}^{\kappa d}. \tag{7.212}$$

Then, Eq. (7.211) is transformed into

$$\cos(KD) = \frac{1}{2}e^{\kappa d}\left(\cos(kL) - \frac{k^2-\kappa^2}{2k\kappa}\sin(kL)\right). \tag{7.213}$$

Since the left-hand side of Eq. (7.213) is in the interval $(-1, 1)$ and the right-hand side contains the large factor $\exp(\kappa d)$, the expression within the large brackets should be small; in the first approximation it must be equal to zero:

$$\cot(kL) = \frac{k^2-\kappa^2}{2k\kappa}. \tag{7.214}$$

Equation (7.214) defines the energy spectrum of an electron in an individual symmetric potential well (see Section 3.3.2). To analyze the dispersion relation in the intermediate region of values of d, for which the coupling between quantum dots can be considered weak, the dispersion relation (7.211) can be rewritten in more suitable form. Let us denote the right-hand side of Eq. (7.211) as

$$F_1(E) = \cos(kL)\cosh(\kappa d) - \frac{k^2-\kappa^2}{2k\kappa}\sin(kL)\sinh(\kappa d), \tag{7.215}$$

and the discrete energy levels in an individual well, defined by Eq. (7.214), as E_n. Let us expand the function $F_1(E)$ in a Taylor series in the region around the energy E_n and let us limit ourselves to two terms of the expansion:

$$F_1(E) = F_1(E_n) + \left(\frac{dF_1}{dE}\right)_{E=E_n}(E - E_n), \tag{7.216}$$

where the derivative, dF_1/dE, is calculated at $E = E_n$. Let us substitute $F(E)$ in Eq. (7.216) by $F_1(E)$ from Eq. (7.215) and use Eqs. (7.207) and (7.208) to find the energy E:

$$E = E_n - F_1(E_n)\left(\frac{dF_1}{dE}\right)^{-1}_{E=E_n} + \left(\frac{dF_1}{dE}\right)^{-1}_{E=E_n}\cos(KD). \tag{7.217}$$

On introducing into this equation the notations

$$W_n = F_1(E_n)\left(\frac{dF_1}{dE}\right)^{-1}_{E=E_n} \tag{7.218}$$

and

$$A_n = \frac{1}{2}\left(\frac{dF_1}{dE}\right)^{-1}_{E=E_n}, \tag{7.219}$$

we obtain an expression for the energy in the miniband which corresponds to the nth level of an individual quantum dot:

$$E(K) = E_n - W_n + 2A_n\cos(KD). \tag{7.220}$$

Thus, by bringing the quantum dots in the superlattice closer to each other, we make the energy levels corresponding to an individual quantum dot shift by

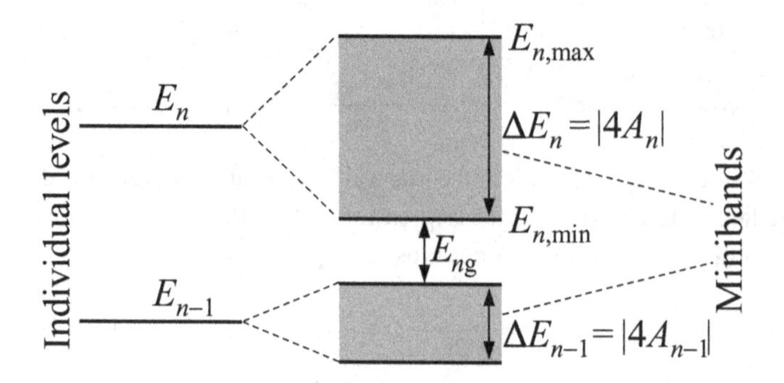

Figure 7.24 The energy levels of individual quantum dots, E_n and E_{n-1}, spread into energy minibands $\Delta E_n = E_{\max} - E_{\min}$ and ΔE_{n-1}. Here, $E_{n,\max} = E_n - W_n + 2|A_n|$, $E_{n,\min} = E_n - W_n - 2|A_n|$, and E_{ng} is the energy minigap.

a magnitude W_n. In the system of N non-interacting (isolated) quantum dots the energy level E_n is N-fold degenerate. The interaction of quantum dots, which takes place on bringing them closer to each other, lifts this degeneracy and creates a miniband of finite width. The width of the nth miniband is

$$\Delta E_n = |4A_n|. \tag{7.221}$$

The analysis of $F_1(E)$ shows that its derivative, $\mathrm{d}F_1/\mathrm{d}E$, in the first miniband is always negative (i.e., $A_1 < 0$). Thus, the lowest-energy state in the first miniband is at $K = 0$:

$$E_{1,\min} = E_1(0) = E_1 - W_1 + 2A_1. \tag{7.222}$$

This state corresponds to the lowest energy of the miniband since $\cos(KD) = 1$. The highest-energy states in the first miniband are at the boundaries of the first Brillouin zone: $K = \pm\pi/D$. Therefore,

$$E_{1,\max} = E_1(\pm\pi/D) = E_1 - W_1 - 2A_1. \tag{7.223}$$

These states correspond to the top of the miniband (see Fig. 7.24) since $\cos(KD) = -1$ at $K = \pm\pi/D$.

With decreasing barrier width, d, the minibands with allowed energy states become wider. In the limit $d \to 0$ the miniband character of the electron energy spectrum in the superlattice is transformed into the parabolic dispersion relation of a free electron (see Fig. 7.21).

From the dispersion equations (7.209) and (7.211) it follows that the substitution of the effective wavenumber K by $K + 2\pi l/D$, where l is an integer number, does not lead to any new solutions of these equations. Therefore, the analysis of the dispersion relation of electrons in a superlattice must be done for physically different states, i.e., for those values of wavenumber K which correspond to the first Brillouin zone:

$$-\frac{\pi}{D} \leq K \leq \frac{\pi}{D}. \tag{7.224}$$

Let us analyze Eq. (7.211) for small values of parameters L and d satisfying the following conditions:

$$kL \ll 1 \qquad \text{and} \qquad \kappa d \ll 1. \tag{7.225}$$

Let us assume that the bottom of the first miniband of allowed energies is located at the center of the Brillouin zone ($K = 0$). Then, for small wavenumbers K, i.e., for $KD \ll 1$, the left-hand side of Eq. (7.211) can be presented in the form

$$\cos(KD) \approx 1 - \frac{(KD)^2}{2}. \tag{7.226}$$

Here we expanded the function $\cos(KD)$ in a Taylor series and took into account terms including the quadratic term of the expansion. On expanding the right-hand side of Eq. (7.211) in a Taylor series (we took into account conditions (7.225)), we come to the following dispersion relation:

$$E(K) = E_0 + \frac{\hbar^2 K^2}{2m^*}, \tag{7.227}$$

where

$$E_0 = \frac{d}{L+d} U_0. \tag{7.228}$$

The magnitude E_0 represents the average value of the potential $U(x)$ throughout the entire superlattice, i.e.,

$$E_0 = \langle U(x) \rangle = \frac{d}{D} U_0. \tag{7.229}$$

From Eq. (7.227) it follows that, near the bottom of a miniband with allowed energies

$$E_{\min} = E_0, \tag{7.230}$$

the electron behaves as a free particle with effective mass m^*. The energy of the bottom of the first miniband with allowed energies, E_0, at $U_0 d \to 0$ becomes equal to zero.

The maximum value of energy, i.e., the top of the allowed energy miniband, is at the Brillouin-zone boundary at $K = \pi/D$. By expanding in a Taylor series the function $\cos(KD)$ near this point we obtain

$$\cos(KD) \equiv -1 + \frac{1}{2}(KD - \pi)^2. \tag{7.231}$$

Finally, we arrive at the following form of the dispersion relation in this region:

$$E(K) = E_{\max} - \frac{\hbar^2}{2m^* D^2}(KD - \pi)^2, \tag{7.232}$$

where the energy at the top of the miniband is equal to

$$E_{\max} = E_0 + \frac{2\hbar^2 \pi^2}{m^* D^2}. \tag{7.233}$$

Therefore, the width of the first miniband with allowed energies ΔE_1 is defined as

$$\Delta E_1 = E_{\max} - E_{\min} = 4|A_1| = \frac{2\hbar^2\pi^2}{m^* D^2}. \tag{7.234}$$

Figure 7.23 shows the dependence of electron energy on the effective wavenumber, K, for the first miniband. Taking into account the higher terms in the Taylor-series expansion of Eq. (7.211), we can get a more accurate expression for the width of the first energy miniband, ΔE_1,

$$\Delta E_1 = \frac{2\hbar^2\pi^2}{m^* D^2}\left(1 + \frac{m^* D}{3\hbar^2} U_0 d\right)^{-1}, \tag{7.235}$$

and the following expression for the electron effective mass, m_{SL}, near the bottom of the miniband:

$$m_{\mathrm{SL}} = m^*\left(1 + \frac{m^* D}{3\hbar^2} U_0 d\right). \tag{7.236}$$

From Eqs. (7.235) and (7.236), it follows that with increasing height and width of the barrier the width of the miniband with allowed energies decreases while the electron effective mass in the superlattice, m_{SL}, increases.

Thus, we showed that the lowest energy level of an electron in an individual quantum dot splits into an energy miniband of the superlattice, which consists of a system of a large number of coupled quantum dots. An analogous splitting happens with the higher energy levels of an individual quantum dot. Therefore, the energy spectrum of an electron in a superlattice consists of a set of minibands composed of the levels of space quantization of individual quantum dots (see Fig. 7.24).

Let us assume that the length of the structure becomes unlimited in the y-direction ($L_y \to \infty$). If in the other two directions (along the x- and z-axes) the electron is spatially quantized and in addition to this there is a periodicity of the structure (and correspondingly a periodicity of the potential $U(x, y, z)$) along the x-direction, then we obtain a superlattice formed from quantum wires (Fig. 7.25). If the length of the structure is unlimited along the y- and z-axes , i.e., $L_y \to \infty$ and $L_z \to \infty$, and we have space quantization and periodicity along the x-axis, then a superlattice is formed from quantum wells (see Fig. 7.26). For a superlattice formed from quantum wires, the wavefunctions and energy spectra can be described by Eqs. (7.138) and (7.139), where the functions $\psi_x(x)$ and $E_x(K)$ are defined by the type of the periodic potential $U(x)$. For superlattices formed from quantum wells the wavefunctions and energy spectra are described by Eqs. (7.140) and (7.141).

Example 7.9. Assuming that the average value of a one-dimensional periodic potential, $\langle U \rangle$, is small in comparison with the kinetic energy of an electron, find the dependence $E(K)$ near the Brillouin-zone boundaries for the Kronig–Penney model.

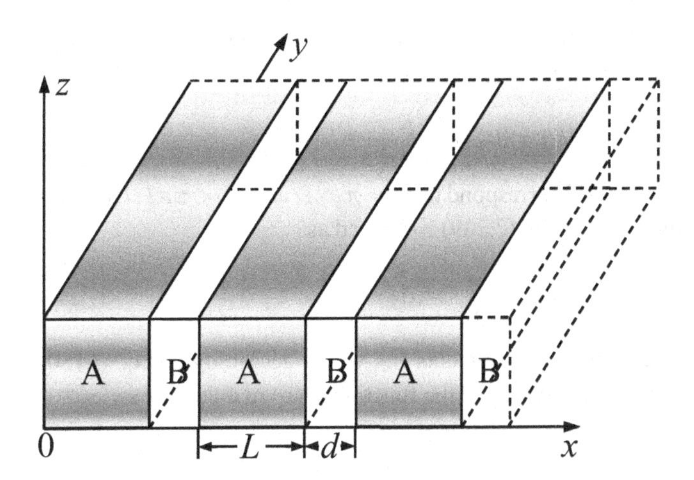

Figure 7.25 A superlattice composed from quantum wires. A denotes the material of the quantum wires and B the material of the barriers which separate the quantum wires.

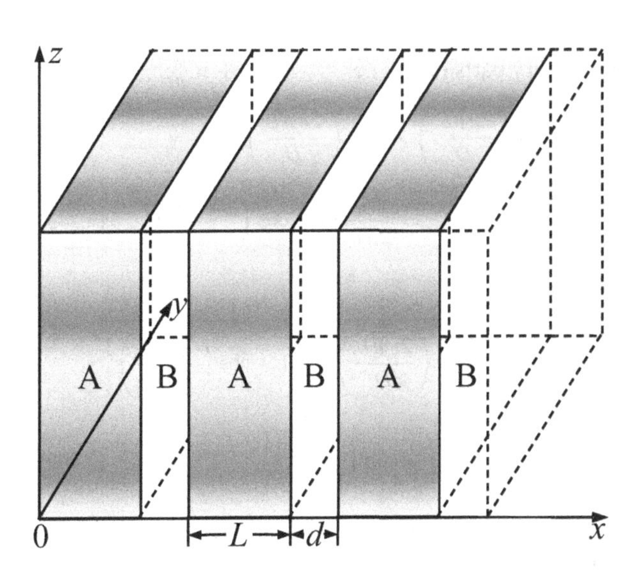

Figure 7.26 A superlattice composed from quantum wells. A denotes the material of the quantum wells and B the material of the barriers, which separate the quantum wells.

Reasoning. According to Eq. (7.184), in the framework of perturbation theory the expression for the energy near the boundaries $K = \pi n/D$ has the form

$$E(K) = \langle U \rangle + \frac{\hbar^2 K^2}{2m^*} - \frac{\pi n \hbar^2}{m^* D}\left(K - \frac{\pi n}{D}\right) \pm \frac{1}{2}\sqrt{\left[\frac{2\pi n \hbar^2}{m^* D}\left(K - \frac{\pi n}{D}\right)\right]^2 + 4|C_n|^2}. \tag{7.237}$$

Here we took into account that

$$E^{(0)}(K) = \langle U \rangle + \frac{\hbar^2 K^2}{2m^*}. \tag{7.238}$$

This expression can be simplified if we take into account the smallness of $(K - \pi n/D)$ near the boundaries of the Brillouin zone, $K = \pm\pi n/D$. Then,

$$E(K) = \langle U \rangle + \frac{\hbar^2 K^2}{2m^*} \pm |C_n|, \tag{7.239}$$

where the signs "+" and "−" correspond to $K > \pi n/D$ and $K < \pi n/D$, respectively. The magnitude $\langle U \rangle$ in Eq. (7.239) is defined as

$$\langle U \rangle = \frac{1}{D}\int_0^D U(x)\mathrm{d}x = \frac{1}{D}\int_L^D U_0\,\mathrm{d}x = \frac{D-L}{D}U_0 = \frac{d}{D}U_0 \tag{7.240}$$

(see Fig. 7.23). Let us find the coefficients C_n, which according to Eq. (7.180) are the coefficients of the expansion of the function $U(x)$ in a Fourier series:

$$C_n = \frac{1}{D}\int_0^D U(x)\mathrm{e}^{-\mathrm{i}2\pi nx/D}\,\mathrm{d}x = \frac{1}{D}\int_L^D U_0\mathrm{e}^{-\mathrm{i}2\pi nx/D}\,\mathrm{d}x = \mathrm{i}\frac{U_0}{2\pi n}\mathrm{e}^{-\mathrm{i}2\pi nx/D}\Big|_L^D$$

$$= \mathrm{i}\frac{U_0}{2\pi n}\left[1 - \mathrm{e}^{-\mathrm{i}2\pi nL/D}\right]. \tag{7.241}$$

Let us find the real and imaginary parts of C_n, i.e., $C_n = C_n' + \mathrm{i}C_n''$:

$$\mathrm{e}^{-\mathrm{i}2\pi nL/D} = \cos\left(\frac{2\pi nL}{D}\right) - \mathrm{i}\sin\left(\frac{2\pi nL}{D}\right), \tag{7.242}$$

$$C_n' = -\frac{U_0}{2\pi n}\sin\left(\frac{2\pi nL}{D}\right), \qquad C_n'' = \frac{U_0}{2\pi n}\left[1 - \cos\left(\frac{2\pi nL}{D}\right)\right]. \tag{7.243}$$

Let us find now the modulus of C_n:

$$|C_n| = \sqrt{(C_n')^2 + (C_n'')^2} = \frac{U_0}{2\pi n}\sqrt{\sin^2\left(\frac{2\pi nL}{D}\right) + \left[1 - \cos\left(\frac{2\pi nL}{D}\right)\right]^2}$$

$$= \frac{U_0}{2\pi n}\sqrt{2 - 2\cos\left(\frac{2\pi nL}{D}\right)} = \frac{U_0}{\pi n}\left|\sin\left(\frac{\pi nL}{D}\right)\right|, \tag{7.244}$$

where $n = 1, 2, 3, \ldots$ Thus, the discontinuities at the boundaries of the Brillouin zone, i.e., the minigaps, are equal to

$$E_{ng} = 2|C_n| = \frac{2U_0}{\pi n}\left|\sin\left(\frac{\pi nL}{D}\right)\right|. \tag{7.245}$$

7.8 A three-dimensional superlattice of quantum dots

The picture of miniband spectra of superlattices we obtained in the previous sections generally stays the same in the case of superlattices of higher dimensionality. For an ideal three-dimensional superlattice consisting of identical quantum dots the potential energy is spatially periodic:

$$U(\mathbf{r} + \mathbf{a}_n) = U(\mathbf{r}). \tag{7.246}$$

Here, the superlattice translation vector, $\mathbf{a}_n$, is equal to

$$\mathbf{a}_n = n_x\mathbf{D}_x + n_y\mathbf{D}_y + n_z\mathbf{D}_z. \tag{7.247}$$

The vectors $\mathbf{D}_\alpha = \mathbf{e}_\alpha D_\alpha$ are the basis vectors whose length is equal to the period of the superlattice in the respective direction ($\alpha = x, y,$ and z). Let us apply the methods of perturbation theory to find the energy spectrum of an electron in such a superlattice.

Let us assume that the solution of the Schrödinger equation for the electron in an individual quantum well with potential $U_0(\mathbf{r})$ is known. The Schrödinger equation for an electron in the superlattice can be written in the following form:

$$\hat{H}\psi(\mathbf{r}) = E\psi(\mathbf{r}), \tag{7.248}$$

where the Hamiltonian, $\hat{H}$, is equal to

$$\hat{H} = -\frac{\hbar^2}{2m_e}\nabla^2 + \sum_n U_0(\mathbf{r} - \mathbf{a}_n) + W(\mathbf{r}). \tag{7.249}$$

Here, $W(\mathbf{r})$ is the perturbation which takes into account the coupling between the quantum dots and $\sum_n U_0(\mathbf{r} - \mathbf{a}_n)$ is the potential energy of the electron in the superlattice formed by the non-interacting quantum dots. We can represent the electron wavefunction in the superlattice which corresponds to the jth energy state in the form of a linear combination of electron wavefunctions $\psi_{qj}(\mathbf{r} - \mathbf{a}_n)$ for the individual quantum dots:

$$\psi_{0j}(\mathbf{r}) = \sum_n \exp(\mathrm{i}\mathbf{K}\cdot\mathbf{a}_n)\psi_{qj}(\mathbf{r} - \mathbf{a}_n). \tag{7.250}$$

On multiplying Eq. (7.248) by the complex-conjugate wavefunction $\psi_{0j}^*(\mathbf{r})$ and integrating this equation over the entire volume of the superlattice we find the average energy of an electron in the superlattice:

$$E(\mathbf{K}) = \frac{\int \psi_{0j}^*(\mathbf{r})\hat{H}\psi_{0j}\,\mathrm{d}V}{\int \psi_{0j}^*\psi_{0j}\,\mathrm{d}V}. \tag{7.251}$$

On substituting the wavefunction (7.250) into Eq. (7.251) and taking into account that the electron wavefunctions of the individual quantum dots, $\psi_{qj}(\mathbf{r} - \mathbf{a}_n)$, are normalized, we obtain the following expression for the electron energy:

$$E(\mathbf{K}) = E_j^{(0)} + E_j^{(1)} + \sum_l A(\mathbf{a}_l)\mathrm{e}^{\mathrm{i}\mathbf{K}\cdot\mathbf{a}_l}. \tag{7.252}$$

Here, $E_j^{(0)}$ is the energy of an electron in an individual quantum dot in the jth quantum state found for the zeroth approximation. The correction to the energy in the first approximation does not depend on the wavevector $\mathbf{K}$ and is negative:

$$E_j^{(1)} = \int \psi_{qj}^*(\mathbf{r})[U(\mathbf{r} - \mathbf{a}_n) + W(\mathbf{r})]\psi_{qj}\,\mathrm{d}V < 0. \tag{7.253}$$

The third term in Eq. (7.252) originates from the exchange interaction between two quantum dots at a distance $\mathbf{a}_l$:

$$A(\mathbf{a}_l) = \int \psi_{qj}^*(\mathbf{r}) \left[\sum_{\mathbf{r} \neq \mathbf{a}_l} U(\mathbf{r} - \mathbf{a}_n) + W(\mathbf{r}) \right] \psi_{qj}(\mathbf{r} - \mathbf{a}_l) \mathrm{d}V. \qquad (7.254)$$

The exchange interaction occurs between any pair of quantum dots. It is related to the overlap of their wavefunctions $\psi_{qj}(\mathbf{r})$ and $\psi_{qj}(\mathbf{r} - \mathbf{a}_l)$. At the same time the electron becomes non-localized. Taking this into account, the exchange interaction leads to the splitting of the energy levels of the individual quantum dot into a miniband whose width is proportional to the exchange integral $A(\mathbf{a}_l)$.

The exponential factor in Eq. (7.252) takes into account the real structure of the superlattice. Thus, for a simple cubic superlattice with period D and with six nearest-neighboring quantum dots, whose wavevectors $\mathbf{a}_l$ have the components $(\pm D, 0, 0)$, $(0, \pm D, 0)$, $(0, 0, \pm D)$, using Eq. (7.252) we obtain the following expression for the electron energy:

$$\begin{aligned} E(\mathbf{K}) &= E_j^{(0)} + E_j^{(1)} + A_j(D) \sum_\alpha \left(\mathrm{e}^{\mathrm{i}K_\alpha D} + \mathrm{e}^{-\mathrm{i}K_\alpha D} \right) \\ &= E_j^{(0)} + E_j^{(1)} + 2A_j(D) \left[\cos(K_x D) + \cos(K_y D) + \cos(K_z D) \right], \end{aligned} \qquad (7.255)$$

where $\alpha = x, y$, and z. Inside of the miniband the electron energy is a periodic function of the quasiwavevector with period $2\pi/D$. The sign of the exchange integral A_j can be positive as well as negative. For the lowest miniband usually $A_j < 0$. Therefore, the electron energy has its minimum value at the center of the first Brillouin zone $K_\alpha = 0$, and the maximum value at the boundaries $K_\alpha = \pm\pi D$, where $\alpha = x, y$, and z. The minimum and maximum values of the electron energy are defined as

$$E_{\mathrm{min}} = E(0) = E_j^{(0)} + E_j^{(1)} - 6|A_j(D)|, \qquad (7.256)$$

$$E_{\mathrm{max}} = E\left(\pm\frac{\pi}{D}\right) = E_j^{(0)} + E_j^{(1)} + 6|A_j(D)|. \qquad (7.257)$$

The width of the minibands is proportional to the exchange integral $A_j(D)$, i.e.,

$$\Delta E_j = E_{\mathrm{max}} - E_{\mathrm{min}} = 12|A_j(D)|. \qquad (7.258)$$

The magnitude of the exchange integral depends strongly on the distance between quantum dots, i.e., on the period of the superlattice, D. The overlap of the wavefunctions of electrons belonging to different quantum dots increases when the distance between the quantum dots decreases, which leads to an increase of the width of the minibands. Owing to the high degree of overlap of the wavefunctions the exchange interaction and the width of minibands are large for the higher-lying minibands, i.e., with increasing electron energy the width of the minibands increases (see Fig. 7.22).

The effective mass of an electron in the cubic three-dimensional superlattice is a scalar, i.e., it does not depend on the direction of the **K**-vector:

$$\frac{1}{m_j^*} = \pm\frac{2}{\hbar^2}A_j D^2. \tag{7.259}$$

In the general case of a superlattice with different periods D_α along the directions of the main axes the inverse effective mass becomes a tensor:

$$\left(\frac{1}{m^*}\right)_{\alpha\beta} = \frac{1}{\hbar^2}\frac{\partial^2 E}{\partial k_\alpha\, \partial k_\beta}. \tag{7.260}$$

Example 7.10. Show that the electron motion in the three-dimensional superlattice in an external electric field, $\mathbf{E}_{\text{ext}}$, can be described by Newton's second law if the inverse mass of the electron is considered as a tensor with the components defined by Eq. (7.260).

Reasoning. The expression for the group electron velocity in the case of three-dimensional motion can be written as follows:

$$\mathbf{v}_{\text{gr}} = \mathbf{e}_x\frac{\partial\omega}{\partial k_x} + \mathbf{e}_y\frac{\partial\omega}{\partial k_y} + \mathbf{e}_z\frac{\partial\omega}{\partial k_z} = \nabla_k\omega, \tag{7.261}$$

where

$$\nabla_k = \mathbf{e}_x\frac{\partial}{\partial k_x} + \mathbf{e}_y\frac{\partial}{\partial k_y} + \mathbf{e}_z\frac{\partial}{\partial k_z}. \tag{7.262}$$

Taking into account the relation of the electron energy to the frequency of the electron wave, $E(\mathbf{k}) = \hbar\omega$, the group velocity of the wave packet which represents the electron can be written through its energy as

$$\mathbf{v}_{\text{gr}} = \frac{1}{\hbar}\nabla_k E(\mathbf{k}). \tag{7.263}$$

If the superlattice is in an external electric field, $\mathbf{E}_{\text{ext}} = \mathbf{e}_x E_x + \mathbf{e}_y E_y + \mathbf{e}_z E_z$, the change in the electron energy, E, in unit time is defined as

$$\frac{\mathrm{d}E}{\mathrm{d}t} = \mathbf{F}\cdot\mathbf{v} = e\mathbf{E}_{\text{ext}}\cdot\mathbf{v} = e\mathbf{E}_{\text{ext}}\cdot\left(\frac{1}{\hbar}\nabla_k E(\mathbf{k})\right), \tag{7.264}$$

where we considered the group velocity of the electron, $\mathbf{v}_{\text{gr}}$, as the electron wave-packet velocity, $\mathbf{v}$. The electric force applied to the electron is equal to $\mathbf{F} = e\mathbf{E}_{\text{ext}}$. Using Eqs. (7.263) and (7.264), let us find the acceleration, $\mathbf{a}$, which the electron gains in the external field, $\mathbf{E}_{\text{ext}}$:

$$\begin{aligned}\mathbf{a} &= \frac{\mathrm{d}\mathbf{v}}{\mathrm{d}t} = \frac{1}{\hbar}\frac{\mathrm{d}}{\mathrm{d}t}(\nabla_k E(\mathbf{k})) = \frac{1}{\hbar}\nabla_k\left(\frac{\mathrm{d}E(\mathbf{k})}{\mathrm{d}t}\right) = \frac{e}{\hbar^2}\nabla_k(\mathbf{E}_{\text{ext}}\cdot\nabla_k E(\mathbf{k}))\\ &= \frac{e}{\hbar^2}\nabla_k\left(E_x\frac{\partial E(\mathbf{k})}{\partial k_x} + E_y\frac{\partial E(\mathbf{k})}{\partial k_y} + E_z\frac{\partial E(\mathbf{k})}{\partial k_z}\right).\end{aligned} \tag{7.265}$$

Let us write the x-projection of the acceleration, $\mathbf{a}$:

$$a_x = \frac{dv_x}{dt} = \frac{e}{\hbar^2}\left(E_x \frac{\partial^2 E(\mathbf{k})}{\partial k_x \partial k_x} + E_y \frac{\partial^2 E(\mathbf{k})}{\partial k_x \partial k_y} + E_z \frac{\partial^2 E(\mathbf{k})}{\partial k_x \partial k_z}\right). \tag{7.266}$$

Analogous expressions can be written for the other two projections of the acceleration, $\mathbf{a}$. These expressions, according to Newton's second law in the general form, can be presented as

$$a_\alpha = \sum_\beta \left(\frac{1}{m^*}\right)_{\alpha\beta} F_\beta, \tag{7.267}$$

where the summation is done over all three projections onto the coordinate axes $\beta = x, y$, and z, and F_β are the projections of the force $\mathbf{F}$. On comparing the expressions obtained, we find that the magnitudes

$$\left(\frac{1}{m^*}\right)_{\alpha\beta} = \frac{1}{\hbar^2}\frac{\partial^2 E}{\partial k_\alpha \, \partial k_\beta} \tag{7.268}$$

are the components of the inverse effective mass tensor.

7.9 Summary

1. In a potential well of macroscopic size, the electron energy spectrum is quasicontinuous. With increasing energy the number of electron states in such a well increases proportionally to $E^{3/2}$ and the density of states increases proportionally to $\sqrt{E}$. These dependences hold only in the case of a quadratic dependence of energy on momentum.
2. A structure of size less than or about 100 nm is called a *nanoobject* or a *nanostructure*. In a nanostructure the electron energy spectrum radically changes, which leads to changes of the physical properties of an object. For a nanostructure that is formed from a few atoms and placed in vacuum the potential barriers are high and the spacing between levels is about several electron-volts.
3. In quantum wells (two-dimensional structures) the electron motion in one direction is limited and the energy that corresponds to this motion is quantized. The motion in the two other directions stays free and is characterized by a continuous energy spectrum. Therefore, the electron spectrum in a quantum well is a set of two-dimensional subbands, which have the form of paraboloids.
4. In quantum wires (one-dimensional structures) electron motion is limited in two directions, with the corresponding energy quantization. The electron motion along a wire (the third direction) is free and the energy spectrum is continuous. The electron spectrum in a quantum wire is a set of one-dimensional subbands, which have the form of parabolas.
5. In quantum dots (zero-dimensional structures) the electron motion is limited in all three directions. The energy spectrum of such motion is completely quantized.

6. The density of states of a quantum well in any of its subbands is constant and does not depend on energy. Each subband has the same input into the total density of states, which is defined by the electron effective mass and the width of the quantum well.
7. Superlattices are structures with periodic positioning of quantum-dimensional objects along one, two, or three directions. The electron wavefunction in a superlattice must have the form of the wavefunction of a free electron modulated by a periodic superlattice function.
8. The periodicity of the superlattice potential leads to the conservation not of electron momentum, but of quasimomentum. For quasimomentum Newton's second law in the form $d\mathbf{p}/dt = \mathbf{F}_{ext}$ is valid. Here $\mathbf{F}_{ext}$ is connected with the external fields and does not include the periodic force of a superlattice, $\mathbf{F}_{SL}$, which is related to the superlattice potential by $\mathbf{F}_{SL} = -\vec{\nabla} U$, with U being the potential of the superlattice.
9. The total external force is related to the electron acceleration by $\mathbf{F}_{ext} = m^*\mathbf{a}$, where the effective mass of the electron, m^*, reflects the character of the electron energy dispersion in a superlattice. Internal forces (potentials) define the energy dispersion, $E(p)$, and, through the effective mass, m^*, define the electron motion under the influence of external forces.
10. The interval of quasimomenta $(-\pi\hbar/D \leq p \leq \pi\hbar/D)$ that contains all physically equivalent electron states in a one-dimensional superlattice as in the case of oscillations of a chain of atoms is called the first Brillouin zone. The second Brillouin zone consists of two quasimomentum intervals, $(-2\pi\hbar/D, -\pi\hbar/D)$ and $(\pi\hbar/D, 2\pi\hbar/D)$, and so on. The total length of each Brillouin-zone interval is equal to $2\pi\hbar/D$.
11. The boundaries of Brillouin zones correspond to wavevectors at which the electron wave cannot propagate in the superlattice. At the boundaries of Brillouin zones the electron energy in a superlattice has a discontinuity, which indicates the occurrence in the electron energy spectrum of forbidden minibands, which alternate with allowed minibands.
12. Each individual level of the identical quantum dots splits into the superlattice miniband if the quantum dots are assembled as a superlattice.

7.10 Problems

Problem 7.1. Find the expressions for the wavefunction and electron energy in a symmetric rectangular quantum well with the following potential profile:

$$U(r) = \begin{cases} U_0, & |x| > L_x, \\ 0, & |x| \leq L_x. \end{cases} \tag{7.269}$$

The electron motion along the y- and z-directions is free.

Problem 7.2. Using the solution of Problem 7.1, consider the limiting cases of very deep and very shallow quantum wells. Show that in a shallow quantum well there is always at least one stationary state. Derive an approximate expression for the energy, E_1, of this state and estimate it for $U_0 = 2.5 \times 10^{-3}$ eV and $L = 2$ nm. (Answer: $E_1 \approx 2.375$ meV.)

Problem 7.3. In an asymmetric rectangular quantum well the motion of an electron in the yz-plane is free and motion in the x-direction is limited by the following potential:

$$U(x) = \begin{cases} U_1, & x < 0, \\ 0, & 0 \le x \le L_x, \\ U_2, & x > L_x, \end{cases} \tag{7.270}$$

where $U_2 < U_1$. Find the condition for which discrete quantum states with energy $E \le U_2$ exist. Consider the limiting cases $U_1 \to \infty$ and $U_1 = U_2$.

Problem 7.4. Estimate the energy of the electron ground state in a quantum well, where electron motion in the yz-plane is free but motion along the x-direction is limited by the potential $U(x)$:

$$U(x) = \begin{cases} \infty, & x < 0, \\ bx, & x \ge 0. \end{cases} \tag{7.271}$$

Problem 7.5. Find the energy and the wavefunctions of the stationary states of an electron in a quantum wire. The electron motion in the z-direction is free and motion in the plane perpendicular to the z-direction is limited by the potential $U(\rho)$ ($\rho^2 = x^2 + y^2$):

$$U(\rho) = \begin{cases} 0, & \rho \le a, \\ \infty, & \rho > a. \end{cases} \tag{7.272}$$

Problem 7.6. In a quantum wire the electron motion along the z-direction is free and that in the plane perpendicular to the z-direction is limited by the potential $U(x, y)$:

$$U(x, y) = \frac{\beta(x^2 + y^2)}{2}. \tag{7.273}$$

Find the energy, the wavefunctions, and the magnitudes of various projections of the electron angular momentum onto the z-direction and occupation probabilities for the states with quantum numbers $n_x = 1$ and $n_y = 1$.

Problem 7.7. The electron motion in a spherical quantum dot happens in the following potential:

$$U(r) = U_0 e^{-r/b}, \quad U_0 < 0. \tag{7.274}$$

Find the average electron energy of a ground state in the given quantum well.

Problem 7.8. Electron motion takes place in an infinitely deep double-quantum-dot structure and is limited by the potential $U(x, y, z)$:

$$U(x, y, z) = U(x)U(y)U(z), \tag{7.275}$$

where

$$U(x) = \begin{cases} \infty, & x < 0,\ x > L_x, \\ 0, & 0 < x < b,\ L_x - b < x < L_x, \\ U_0, & b \le x \le L_x - b, \end{cases} \tag{7.276}$$

$$U(y) = \begin{cases} \infty, & y < 0,\ y > L_y, \\ 0, & 0 \le y \le L_y, \end{cases} \tag{7.277}$$

$$U(z) = \begin{cases} \infty, & z < 0,\ z > L_z, \\ 0, & 0 \le z \le L_z. \end{cases} \tag{7.278}$$

Find using first-order perturbation theory the expression for the electron energy states in a double-quantum-dot structure.

Problem 7.9. Prove that in isotropic three-dimensional systems the density of states per unit volume of momentum space (the number of states per unit interval of momentum modulus) is proportional to the square of momentum.

Problem 7.10. The electron motion in a quantum wire is free along the z-direction and that in the xy-plane is limited in the region $0 < x < L_x$ and $0 < y < L_y$ by barriers of infinite height. Find the number and density of states in the lowest-energy subband of the quantum wire.

Chapter 8
Nanostructures and their applications

Nanotechnology is based on the ability to manipulate individual atoms and molecules in order to assemble them into bigger structures. Such artificial nanoscale structures, usually fabricated using self-assembly phenomena, possess new physical, chemical, and biological properties. The fabrication of various types of nanostructures and study of their properties require new technological means and new principles.

Nanotechnology has initiated a new so-called *bottom-up* technology. The bottom-up technology is based on the *self-assembly* phenomenon, i.e., the process of formation of complex ordered structures from simpler ones. The main idea of this technology is in the development of the controlled self-assembly of the atoms, molecules, and molecular chains into nanoscale objects. The bottom-up technology allows the fabrication of nanoobjects, such as quantum dots, quantum wires, and superlattices.

The bottom-up approach is opposite in principle to the traditional approach, which may be called the *top-down* approach, which is based on the sequential decrease of the object's size by means of mechanical or chemical processing for the fabrication of objects of nanoscale size (nanoobjects). Thus, for example, some of the nanoparticles can be obtained by grinding material consisting of particles of micrometer or larger size in a special grinder. The traditional technologies include laser methods for the processing of semiconductor surfaces and making masks of various configurations and sizes for photolithography.

In this chapter we will briefly consider the main fabrication and characterization techniques for nanostructures and will give some examples of applications of nanostructures in modern nanoelectronics.

8.1 Methods of fabrication of nanostructures

In this section we consider the main techniques of fabrication of nanostructures.

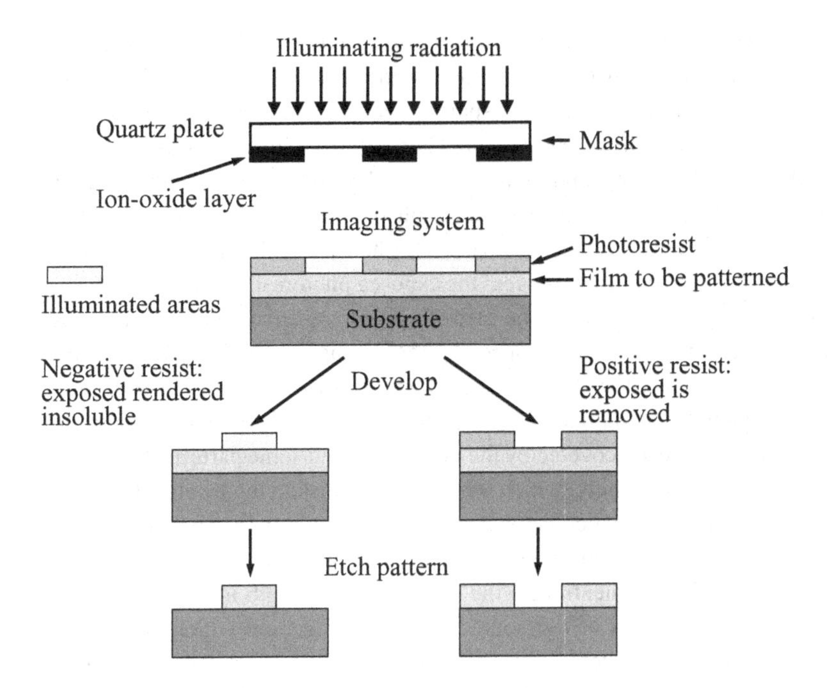

Figure 8.1 Standard steps of a photolithographic process.

8.1.1 Photolithography

Lithographic methods are the main representatives of the top-down techniques in the fabrication of devices and integrated circuits (ICs). Under the term "lithography" one usually understands a set of physicochemical processes of layer-by-layer formation of the patterns of ICs and of elements of nanostructures. The decrease of microchip sizes can be achieved by decreasing the pattern size which is formed with the help of lithography on the surface of solid-state materials.

From optics it is known that the resolution of an optical system increases with decreasing wavelength of the light. Therefore, to fabricate smaller elements by lithographic methods it is necessary to switch to shorter wavelengths. Depending on the wavelength and the methods of generating radiation, one can distinguish among four types of photolithography that are in use: (1) optical lithography, (2) X-ray lithography, (3) ion-projection lithography, and (4) electron-beam lithography. Optical lithography, which is often also called *photolithography*, is a set of methods of formation of the given relief on the sample surface with the help of a focussed light beam.

In the simplest scenario the photolithography process consists of several steps (see Fig. 8.1). In the first step the surface of the solid-state material (*substrate*) is covered with a layer of photosensitive material. The structure of the photosensitive layer changes under the influence of optical radiation. This photosensitive layer is called the *photoresist*. A *photomask* is placed on the surface of the photoresist or somewhere between the photoresist and source of light. The photomask consists of transparent and non-transparent regions. The next step of the photolithographic process is called *exposure*. The surface of the solid-state material covered by the photoresist and by the photomask is subjected to radiation from the optical or X-ray source. As a result of photochemical reactions underneath the transparent regions of the photomask, the photoresist changes its properties. In the case of a *positive-resist* process the exposed photoresist is removed during the so-called *etching process*. In the case of a *negative-resist* process the exposed region becomes resistant to the special chemical agents, called *etchants*, and the unexposed parts of the photoresist are removed by the etching. In both cases we will end up with a certain image on the surface of the substrate, such that some parts of the substrate are covered by the photoresist and some parts are open. The next step is chemical etching, which is based on the dissolving by etchants of the part of the sample area which is not protected by the photoresistive layer. Thus, patterns of complex configurations can be developed on the substrate.

As we have already mentioned, the radiation wavelength is one of the most important characteristics of the source of optical radiation which is used for the illumination of the photoresist. Because of the diffraction phenomenon the wavelength used cannot be larger than the feature size that we would like to obtain by photolithography. If we use a source of radiation with wavelength 1 μm, then the minimum feature size in the pattern will be of the same order of magnitude. To obtain a pattern with nanometer feature size we have to use so-called *deep-ultraviolet light* with a wavelength of the order of tens or hundreds of nanometers. In modern deep-ultraviolet photolithography a wavelength of $\lambda = 200$–400 nm is used. The sources of such a radiation are powerful Hg–Xe gas-discharge lamps and excimer lasers whose output power reaches hundreds of watts. The radiation wavelength of the excimer lasers depends on the composition of the active medium of the laser, for example, KrF ($\lambda = 248$ nm), ArF ($\lambda = 193$ nm), and F_2 ($\lambda = 157$ nm) are widely used. These lasers have a pulse duration of 5–20 ns with a repetition frequency of about 4 kHz and output power up to 50 W. Further increase of the photolithographic resolution can be achieved by using as a source of optical radiation excimer lasers with a wavelength of $\lambda = 13.5$ nm. In X-ray lithography soft (or low-energy) X-ray radiation with a wavelength of 0.5–5 nm is used for illumination.

In conclusion, let us note that the huge success of microelectronics in the last quarter of the twentieth century was due to the fast advances in the development of these processes. Many technological methods of microelectronics have the potential to be used also in nanoelectronics, with reduction of the characteristic feature size. In the transition from microelectronics to nanoelectronics special attention is paid to the self-organization processes which take place during epitaxial growth of nanostructures, atomic-force epitaxy in colloid solutions, etc. In particular, the rate of development of nanoelectronics in general depends entirely on the rate of development of industrial technologies for nanomaterials and nanostructures.

8.1.2 Epitaxy

Epitaxy is one of the methods of nanostructure fabrication using a bottom-up approach. Most of the fabrication methods using the assembly of nanoobjects from individual atoms are based on the phenomenon of *condensation*. By condensation we usually understand the transition of the substance from the gaseous state to a liquid or solid state as a result of its cooling or compression. Rain, snow, and dew all result from natural phenomena of condensation of water vapor from the atmosphere. The condensation of a vapor is possible only at temperatures below a critical temperature for a given substance. In a similar fashion we can condense atoms and molecules of other chemical elements. Condensation and the opposite process – evaporation – are examples of phase transitions of the substance. The process of phase transition of a gas into a liquid or of a liquid into a solid substance occurs during a certain time. During the initial stage nanoparticles are formed, with their further transformation into macroscopic objects occurring subsequently. We can fabricate nanoparticles by "freezing" phase transition at an earlier stage. As a result of condensation we can obtain fullerenes, carbon nanotubes, nanoclusters, and nanoparticles of various sizes.

Using the condensation method of fabrication of nanoparticles, we evaporate from macroscopic objects atoms that are assembled into nanoobjects. The evaporation can be done by thermal heating or by laser heating of the macroscopic material. The evaporated atoms have to be transferred into regions with lower temperatures, where the condensation of atoms into nanoparticles takes place. The controlled condensation of atoms on a crystal surface (*substrate*) is the foundation of *epitaxial technology*.

Epitaxy (the term is derived from the Greek words *epi*, which means *on*, and *táxis*, which means *arrangement*, or *order*) is an arranged growth of a substance during the process of condensation onto a substrate. Epitaxy of atoms on a crystalline surface can be done from the liquid phase as well as from the vapor phase. The process of epitaxy usually begins with the nucleation on the substrate of islands, which coalesce with each other, forming a continuous film. Modern epitaxial techniques allow one to control the growth with resolution up to a single atomic layer and to alternate layers with different physical and chemical properties.

One of the modern methods of epitaxial growth of nanostructures is *molecular-beam epitaxy* (MBE), which is based on the interaction of several molecular beams with a heated monocrystalline substrate. Molecular-beam epitaxy is an improved version of the technique of thermal sputtering under conditions of ultra-high vacuum. The pressure of the residual gases in the vacuum chamber of MBE is maintained at a level lower than 10^{-8} Pa. The fluxes of atoms (or very rarely entire molecules) are formed by the evaporation of liquids or sublimation of solid-state materials, which are located in the source – so-called *Knudsen effusion cells* or simply *effusion cells*. The effusion cell is a crucible of cylindrical or conical form of diameter 1–2 cm and length 5–10 cm. The outlet of the effusion cell is a round opening called a *diaphragm* of diameter 5–8 mm. The crucible is usually made of high-purity pyrolytic graphite or boron nitride (BN).

The fluxes of atoms (or molecules) of the necessary chemical elements are directed towards and subsequently deposited onto a substrate to form a substance of the desired composition. The number of effusion cells depends on the composition of the film and the dopants. To grow elementary semiconductors such as Si or Ge, we need only one source of main material and sources of dopants of n- and p-types. In the case of compound semiconductors (binary or ternary alloys) we need a separate source for each component of the material which is to be grown. The temperature of the effusion cells defines the magnitude of the flux of the particles deposited onto the substrate and it is strictly controlled. The control of flux is provided by so-called *shutters*, which shut off the various fluxes. The homogeneity of the grown material over the surface and its crystalline structure are defined by the homogeneity of the molecular beams. In some cases, to increase the homogeneity of the film the substrate with the forming film is constantly made to spin.

The epitaxial growth of semiconductor compounds involves a series of steps. The most important steps are (1) adsorption of atoms and molecules by the substrate, which leads to the nucleation and growth of the layer, and (2) migration and dissociation of the adsorbed particles. The growing material establishes a crystalline structure, which is defined by the crystalline properties of both the substrate and the deposited material. The atoms that are deposited onto the substrate are adsorbed by the surface. During the first stage, called *physisorption*, the physical adsorption is due to weak van der Waals and (or) electrostatic forces. During the second stage, called *chemisorption* (chemical adsorption), the molecules of the substance undergo a transition to a chemisorbed state during which electron transfer takes place, i.e., a chemical reaction between the surface atoms and the newly arrived atoms occurs. The binding energy of chemical adsorption is higher than of physical adsorption.

Figure 8.2 shows the main elements of the apparatus for the fabrication of semiconductor films of $Al_xGa_{1-x}As$ on GaAs substrate using MBE. Each heater has a crucible, which serves as a source of one of the components of the compound materials. The evaporated material is deposited onto the substrate with a relatively slow deposition rate under conditions of ultra-high vacuum. The heaters are arranged in such a way that the maxima of the distribution intensities of the beams are on the substrate. By selecting the temperatures of heaters and of a substrate we can fabricate structures and films with complex chemical compositions. For example, in the case of $Al_xGa_{1-x}As$, by controlling x, i.e., by controlling the fractions of Al and Ga, it is possible to grow a wide spectrum of materials, from

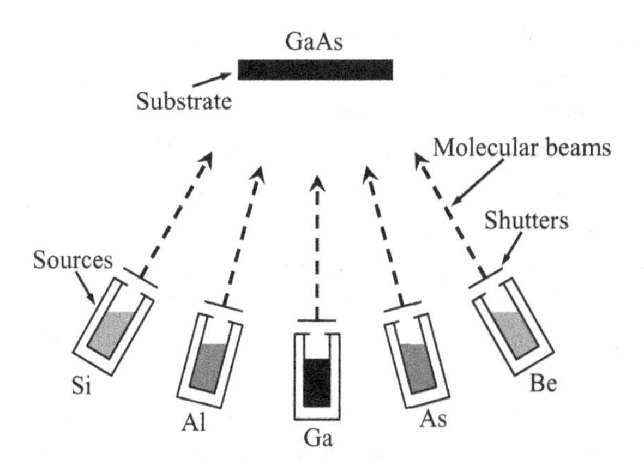

Figure 8.2 Molecular-beam epitaxial growth of GaAs/$Al_xGa_{1-x}As$ heterostructures.

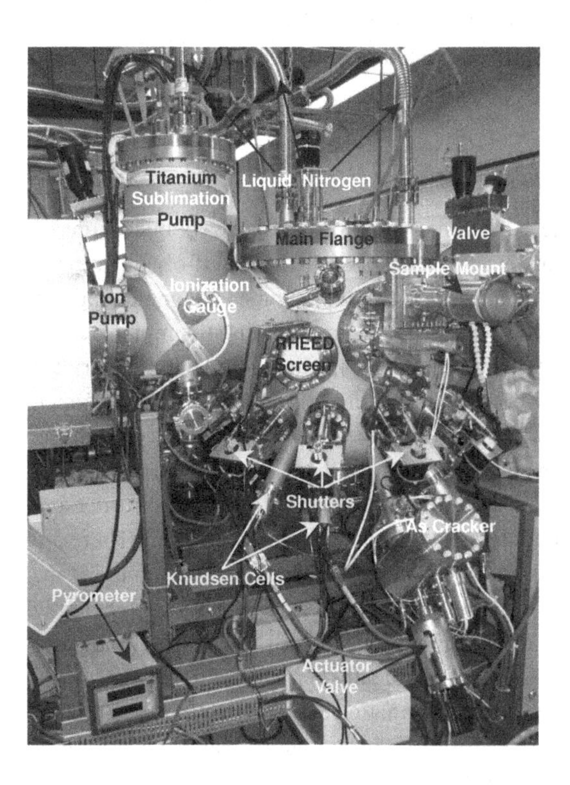

Figure 8.3 A photograph of a typical MBE chamber (University at Buffalo).

GaAs ($x = 0$) to AlAs ($x = 1$). Additionally, the growth process can be controlled by using shutters, which are positioned between the heaters and the substrate. Using these shutters, we can abruptly interrupt or resume the deposition onto the substrate of any molecular beam. For the discussed example of $Al_xGa_{1-x}As$, layers of materials with different x can be grown on top of each other. Figure 8.3 shows a picture of an actual MBE chamber used for research.

One of the main disadvantages of MBE is its low rate of growth – approximately one monatomic layer per second or about 1 μm per hour. By controlling the beams that are depositing material onto the substrate we can modulate layer by layer the doping of the thin-film structure with sufficient accuracy. The dopants can be of either p- or n-type. In such a way a superlattice with the desired number of layers and order can be fabricated.

There are many other technologies that use different forms of epitaxy, which can be realized from vapor, liquid or solid-state phases. Thus, for the growth of epitaxial layers of A_3B_5 compounds (compounds of elements from groups III and V of the Periodic Table of the elements, for example, GaAs) vapor-phase epitaxy from metallo-organic compounds is used. The methods of molecular-beam and metallo-organic epitaxy are widely used for the fabrication of certain types of structures in microelectronics (transistors, integrated circuits, and light-emitting diodes), in quantum electronics (multilayer semiconductor heterostructures and injection lasers), in the devices of integrated optics, in computer engineering, and so on.

8.1.3 Self-organization and self-assembly

As we have already mentioned, the process of formation of complex ordered structures from simple elements is called *self-organization*. Self-organization is one of the most fascinating and interesting natural phenomena, being connected with the formation of a more complex structure from an initial simpler one. In physics and chemistry self-organization is the transition from chaotic distributions of atoms and molecules to ordered structures. The field of science which studies self-organized systems is called *synergetics*, which is derived from the Greek word *synergos*, which means "working together." The main idea of synergetics is the fundamental possibility of the spontaneous formation of order and organization from disorder and chaos as a result of self-organization. There are many types of self-organizing system in nature. A characteristic example is the self-organization of the efforts of a large number of bees. Despite their apparently chaotic movements during the formation of honeycombs, the honeycombs take the form of a plane lattice consisting of practically perfect identical hexagons.

Among prospective approaches to the formation of nanostructures, those technologies which use the phenomena of self-organization to create nanostructures from individual atoms (the so-called *bottom-up technology*) are becoming more and more popular. One of the most difficult problems that nanotechnology has to resolve is how to make atoms and molecules group in a certain way, i.e., how to self-organize them in the most stable structures that can be used for the fabrication of new materials and devices. In the near future the self-organization of atoms and molecules will allow us to assemble materials from their constituents, and it will not be necessary to assemble a nanostructure by putting it together atom by atom manually. Manual assembly is not practical because it is a slow process and it requires disproportionate efforts. Therefore, nowadays the most natural method of nanostructure fabrication is based on self-organization.

Many of us know how the billiard balls are organized by a rack before the beginning of a game. In the closed space of the rack the balls spontaneously form an equilateral triangle. If we were to scatter them into a big box and shake them slightly, they would spontaneously take the form of a practically ideal box structure. In some cases atoms of the same type can be considered as homogeneous spheres, which we can always order in the same way in a closed volume as billiard balls do. In crystallography there is a term that defines such ordering – *close packing*. Analogously to billiard balls, nanoparticles can spontaneously pack on the surface of solid-state materials, forming regular geometrical structures. The main reasons behind the formation of such orderings of nanoparticles are forces that tend to decrease the overall surface area of nanoparticles and therefore their surface energy (we talk about interaction forces between atoms in Appendix C).

Ordered arrays of gold nanoparticles of size 4 nm were obtained by self-assembly for the first time in 1995. The same year structures of CdSe nanoparticles of size 5 nm were fabricated. Homogeneous initial nanoparticles result in regular arrays of nanoparticles. The size-homogeneous nanoparticles may be assembled in spatially ordered structures such as one-dimensional wires, two-dimensional closely packed layers, and three-dimensional arrays or small clusters. The type of organization of nanoparticles and the structure of the array formed depend on the conditions of synthesis, the diameter of the particles, and the nature of external forces acting on the structure. Two-dimensionally and three-dimensionally ordered arrays of Pt, Pd, Ag, Au, Fe, Co, FePt, Fe_3O_4, Co_3O_4, CoO, CdS, CdSe, CdTe, and PbSe nanocrystals have been synthesized to date. Orientationally ordered arrays have also successfully been synthesized from anisotropic nanoparticles.

Nowadays we know methods of self-assembly that allow us to fabricate useful ordered structures. To establish the special conditions required, the gravitational, electric or magnetic fields, capillary forces, mutual wettability or non-wettability of components of the system, and other factors must be taken into account. Self-assembly processes are actively being used in manufacturing processes – in particular, for the fabrication of a new generation of computer chips.

8.1.4 Fabrication of carbon nanostructures

One of the most studied types of carbon nanostructure is *graphene*, because of its promising applications. Long before the fabrication of graphene it was theoretically proved that it is impossible to obtain a free ideal two-dimensional film of monatomic thickness because of the instability caused by folding or twisting. Moreover, thermal fluctuations must lead to the melting of a two-dimensional crystal at any non-zero temperature. The first unsuccessful attempts at making graphene, which was attached to another material, were undertaken using ordinary pencil lead and AFM to mechanically remove layers of graphite. In 2004 graphene on a SiO_2 substrate was obtained, and stabilization of the two-dimensional film was achieved by bonding of the thin layer with SiO_2.

The main method of obtaining graphene is either mechanical splitting off from, or exfoliation of, graphite layers. The method of exfoliation is a relatively simple and flexible method, since it allows one to work with all crystals that have weakly coupled layers. This method allows one to obtain graphene samples of good enough quality, but it is not suitable for mass production of graphene since it is a manual procedure. The mass production of high-quality samples of graphene is still a very challenging problem. One of the prospective methods is the procedure of placing graphene oxide in pure hydrazine, which is a chemical compound of nitrogen and hydrogen. This procedure reduces graphene oxide to a graphene single layer. Another method of obtaining graphene, which is more suitable for mass production, is based on the thermal decomposition of SiC.

The ideal graphene consists exclusively of hexagons. The presence of pentagons or heptagons will lead to the existence of various defects. The presence of pentagons leads to the warping of the graphene plane into a cone. The presence of heptagons leads to the formation of a saddle-shaped flexing of the plane. The combination of such defects and hexagons may lead to the formation of various types of graphene surfaces. The presence of 12 hexagons leads to the formation of fullerene.

From the moment of discovery of fullerenes and carbon nanotubes, numerous methods for their fabrication were developed. The basis of some of these methods is evaporation of graphite by arc discharge or by laser ablation. Other methods are based on chemical transport reactions. An arc discharge is created in an evacuated reactor. After this, its volume is filled by an inert gas, with the pressure being about 0.5 bar. As a result of an arc discharge between two graphite electrodes (an electric current of $\sim$50 A is maintained) carbon atoms leave the electrode with the positive potential and are deposited onto the electrode with the negative potential and onto the walls of the reactor. As a result, tiny particles of soot are formed, which consist of arrays of carbon clusters differing in terms of size and composition. Among these molecular clusters, fullerenes, C_{60}, were discovered for the first time (see Fig. 1.9).

Carbon nanotubes can be obtained using the same methods: laser heating (or *laser ablation*), arc discharge, and chemical vapor deposition (in this case a catalyst is added). The method of laser heating suggests the existence of a graphite target containing particles of the catalyst. The graphite target has to be inserted into the reactor. Particles of Fe, Ni, and Co, and atoms of rare-earth elements, are used as catalysts. These particles become nucleation centers of carbon nanostructures. When the high-intensity beam of the pulsed laser is directed at the target, the carbon evaporates. The flow of argon gas carries carbon atoms from the high-temperature zone to a cooled collector, where the carbon nanotubes are formed. In this way single-walled carbon nanotubes of diameter

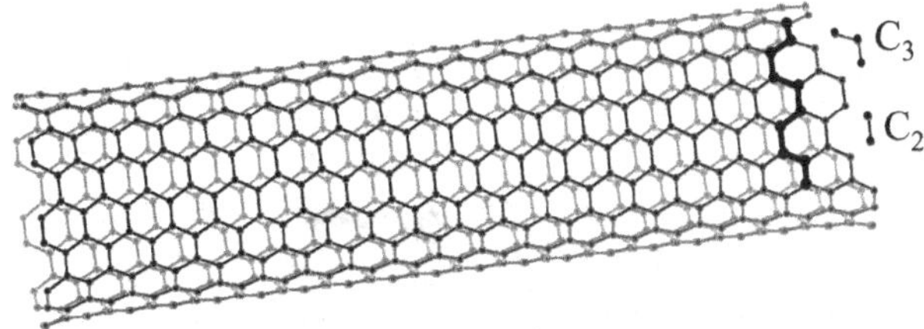

Figure 8.4 Armchair carbon-nanotube formation, showing carbon dimer, C_2, and trimer, C_3, units.

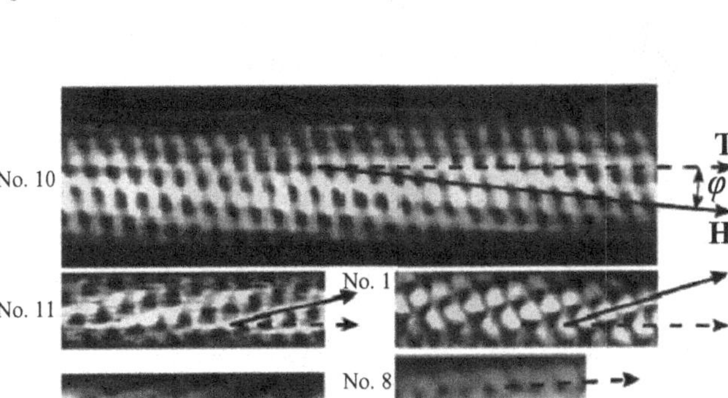

Figure 8.5 Atomically resolved STM images of single-walled carbon nanotubes with chirality. Reprinted with permission, from P. Moriarty, "Nanostructured materials," *Rep. Prog. Phys.*, **64**, 369 (2001).

about 1 nm and length up to several millimeters are formed. At the point of nucleation a carbon nanotube has a closed form, whereas from the side, where the carbon atoms are added, it is open-ended.

Figure 8.4 shows how carbon atoms are added to a single-walled armchair nanotube in the process of its growth. The sequential addition of C_2 dimers (clusters of two carbon atoms) leads to the continuous growth of nanotubes. In order to add hexagons and to eliminate pentagons, it is necessary also to use trimers C_3 (clusters of three carbon atoms). The inclusion of pentagons distorts the tube and makes the nanotube close, thus the growth of the nanotube ends. The inclusion of heptagons changes the size of the nanotube and its spatial orientation. Moreover, addition of "pentagon–heptagon" pairs brings about additional variations of nanotubes. Many properties of single-walled nanotubes are defined not only by their diameter but also by the so-called *chirality* or *torsion* (see Fig. 8.5, where the torsion (or chirality) of a carbon nanotube is shown).

To understand the notion of "chirality" or "chiral angle," we will start with a two-dimensional graphite sheet as shown in Fig. 8.6(a). The single sheet of graphite called *graphene* has the form of a hexagonal lattice. Let $\mathbf{a}_1$ and $\mathbf{a}_2$ be the graphene lattice vectors, and let n and m be integers. The diameter and helicity of the nanotube are uniquely characterized by the vector $\mathbf{C} = n\mathbf{a}_1 + m\mathbf{a}_2 \equiv (n, m)$ that connects crystallographically equivalent sites on a two-dimensional graphene sheet. Here $\mathbf{a}_{1,2}$ are in units of $a_0\sqrt{3}$, with a_0 being the carbon–carbon distance.

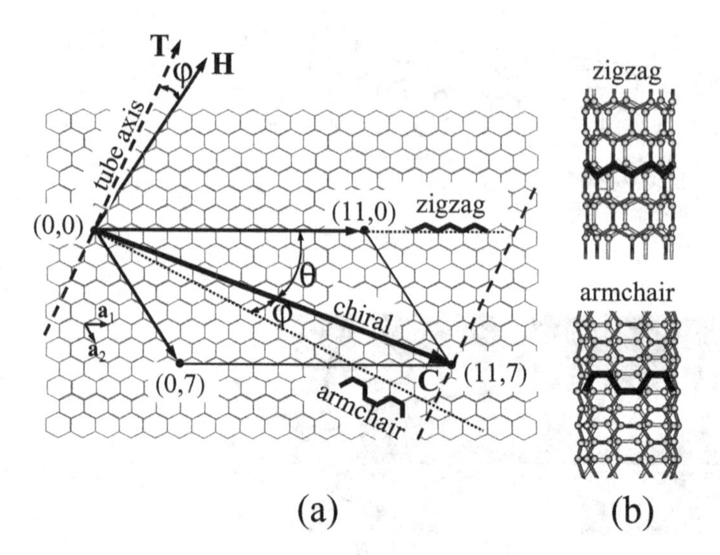

Figure 8.6 (a) The relation between the hexagonal carbon lattice and the chirality of carbon nanotubes. A carbon nanotube can be constructed from a single graphite sheet by rolling up the sheet along the wrapping vector **C**. (b) Fragments of "armchair" and "zigzag" carbon nanotubes.

By using the vector **C**, a carbon nanotube can be constructed by wrapping a graphene sheet in such a way that two equivalent sites of the hexagonal lattice coincide. The *wrapping vector* **C** defines the relative location of these two sites. In Fig. 8.6(a), the wrapping vector **C** connects the origin (0, 0) and the point with coordinates (11, 7). Thus, a nanotube denoted by indices (11, 7) is formed. A tube is called an *armchair* tube if n equals m, and a *zigzag* tube in the case of $m = 0$ (see the fragments of "zigzag" and "armchair" carbon nanotubes in Fig. 8.6(b)). Having wrapping vectors along the dotted lines leads to tubes that are of zigzag or armchair form. All other wrapping angles lead to chiral tubes whose wrapping angle is specified relative to either the zigzag direction θ or the armchair direction $\varphi = 30^\circ - \theta$. Both θ and the *wrapping angle* (*chiral angle*) φ are in the range $(0, 30^\circ)$ as a result of the hexagonal character of the two-dimensional lattice of the graphene. Dashed lines are perpendicular to **C** and run in the direction of the tube axis indicated by the vector **T**. The solid vector **H** is perpendicular to the armchair direction and specifies the direction of nearest-neighbor hexagon rows. The angle between **T** and **H** is the chiral angle φ. The unit cell of a nanotube can be constructed by finding the smallest lattice vector **T** which connects equivalent points of the lattice.

Depending on the magnitude of the diameter and the chiral angle (or chirality), nanotubes may have either metallic or semiconductor properties. In the process of fabrication two thirds of all nanotubes have semiconductor properties and the rest have metallic properties. Generally, nanotubes free of defects and without chirality have metallic properties. Such nanotubes consist of rows of hexagons parallel to the nanotube axis. If catalyst particles are not used, so-called *multiwalled* nanotubes are formed.

Example 8.1. A carbon nanotube of radius $r = 5$ nm is placed in a magnetic field with magnetic flux density $B = 1$ T, oriented in the direction of the nanotube axis, which is parallel to the surface of the nanotube. Find the orbital magnetic moment of an electron rotating around the nanotube axis and moving along the nanotube surface. Compare this magnitude with the magnetic moment of the electron in the hydrogen atom.
Reasoning. When an electron moves with the velocity v perpendicular to the magnetic field it experiences a magnetic force, F_m, which is equal to

$$F_\mathrm{m} = evB.$$

The electron undergoes centripetal acceleration and Newton's second law may be written as

$$m_\mathrm{e}\frac{v^2}{r} = evB. \tag{8.1}$$

Finding from the last equation the orbital velocity v, we can write the period of rotation, T, as

$$T = \frac{2\pi r}{v} = \frac{2\pi m_\mathrm{e}}{eB}. \tag{8.2}$$

The rotating electron creates a current, I, equal to

$$I = \frac{e}{T}. \tag{8.3}$$

Owing to the rotation, the electron acquires the following magnetic moment:

$$\mu = SI = \pi r^2 I = \frac{e^2 r^2 B}{2m_\mathrm{e}}. \tag{8.4}$$

For the electron motion around the nanotube axis the magnetic moment is equal to $\mu \approx 3.5 \times 10^{-25}$ A m^2. The electron magnetic moment in the hydrogen atom is equal to the Bohr magneton: $\mu_\mathrm{B} = 9.27 \times 10^{-24}$ A m^2. Therefore,

$$\frac{\mu_\mathrm{B}}{\mu} \approx 32. \tag{8.5}$$

8.2 Tools for characterization with nanoscale resolution

The minimum size of objects that the human eye can discriminate at the distance of their best visual acuity (about 25 cm) is about 0.1 mm. To study smaller objects, optical devices are used – from the simplest ones such as a magnifying glass to more complex ones consisting of several lenses, such as an optical microscope. Modern optical microscopes have magnifications up to 1500. This means that, with the help of these devices, we can discern objects of size about 10^{-7} m, i.e., objects of size about 100 nm. The increase of the resolution of optical microscopes is hindered by diffraction. If the objects are separated by a distance less than $d \approx \lambda/n$, then they are indistinguishable. Here, λ is the wavelength of the light and n is the index of refraction of a medium. The visible

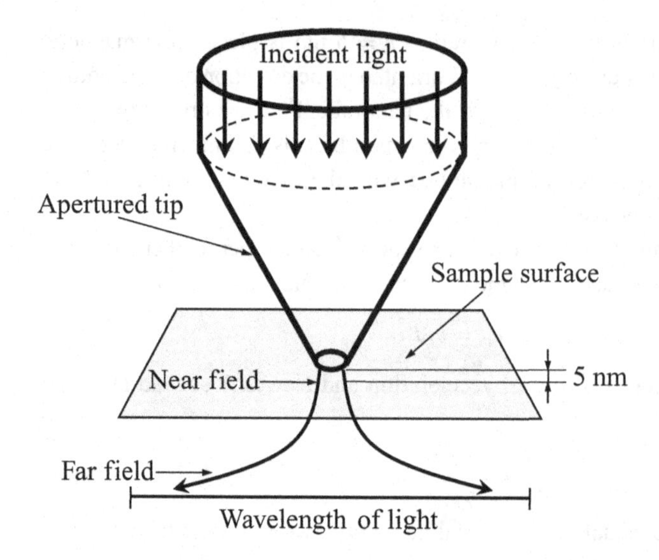

Figure 8.7 Near-field optical microscopy (schematic, not to scale).

part of the electromagnetic spectrum ranges from 400 to 800 nm. Therefore, even theoretically with the help of the most powerful optical microscope possible we cannot discern objects of size less than 200 nm. Some biological objects such as cells, whose size is of the order of hundreds and thousands of nanometers, can be studied using optical systems. Atoms, whose size is about one tenth of a nanometer, cannot be seen using optical microscopes. The simplest solution for this problem is to decrease the wavelength of the light. To discern atoms with the help of such a microscope, we have to decrease the wavelength of the radiation by a factor of 1000.

8.2.1 Near-field scanning optical microscopy

The *near-field scanning optical microscope* (NSOM) was introduced in 1972. This microscope is a special form of scanning-probe microscope, which utilizes visible light as its probe. The resolution of conventional optical microscopes, as we have noted, is limited by the diffraction limit of the light, and therefore it is limited by the wavelength of the light, which is about 0.5 μm. In the NSOM this limitation is overcome and the resolution is increased by one order of magnitude. Figure 8.7 illustrates how the diffraction limit of the light could be overcome in NSOM. The scanning tip used in an NSOM is called an *apertured tip*. The apertured tip is actually a "light funnel." The visible light emanates from the narrow side of the light funnel, which has a diameter of 10–30 nm. The light is registered by the detector either after its reflection from the sample or after transmission through the sample. The intensity of the optical signal is registered by the detector and the set of data from the scanned surface forms the NSOM image of the surface. With the help of an NSOM we can form the image of the scanned surface in visible light with a resolution of about 15 nm if the following condition is satisfied: the distance between the light source and the sample must be very small – about 5 nm. By making the scanning tip approach so close to the sample surface, we make the light interact with the surface *before its diffraction*, thus increasing the resolution of optical microscopy (see Fig. 8.7). The systems for placing the apertured tip and maintaining a constant distance between the sample and the light source are the most complicated parts of an NSOM. Usually, the apertured tip is made by heating an optical fiber and stretching it until its end has a small diameter, whereupon the other end of the fiber is chopped off. After this the optical fiber, except the end of the apertured tip, is covered with a metallic layer for better optical transmission. Another method for the fabrication of apertured tips is the following: a small hole is drilled in the sharp end of an atomic-force-microscope (AFM) cantilever and light is directed into this hole (we will consider AFM techniques in Section 8.2.3). The development of effective apertured tips is nowadays a field of intensive research efforts.

An NSOM must keep the distance between the sharp end of the scanning tip and the sample constant in order to obtain an optical image of the surface being studied. To do this, the traditional AFM methods for maintaining constant deflection of an AFM cantilever may be used. The distinctive feature of the NSOM compared with other scanning methods is that the image is formed directly in the optical range including visible light, but its resolution is much higher than the resolution of traditional optical microscopes.

8.2.2 Scanning-tunneling microscopy

As we have shown previously, the behavior of an electron in an atom is described by its wavefunction, the square of whose modulus defines the probability density of the electron being at a given point of space at a given time. The wavefunctions of the electrons which belong to an atom have non-zero values outside of the atom, i.e., at distances greater than the average size of the atom. Therefore, when we bring atoms that belong to different objects closer to each other, the wavefunctions of the valence electrons overlap before the forces of interatomic repulsion begin to manifest themselves. As a result there is a possibility of electron transfer from one atom to another, i.e., the exchange of electrons that belong to different objects is possible without mechanical contact. This effect may be realized if one of the objects has freely moving electrons, the other has vacant electron states, and the applied voltage between those objects is equal to or greater than the breakdown voltage of the air gap. If the voltage is much less than the breakdown voltage, the electric current occurring under these conditions is due to the *tunneling effect* and the current itself is called the *tunneling current*.

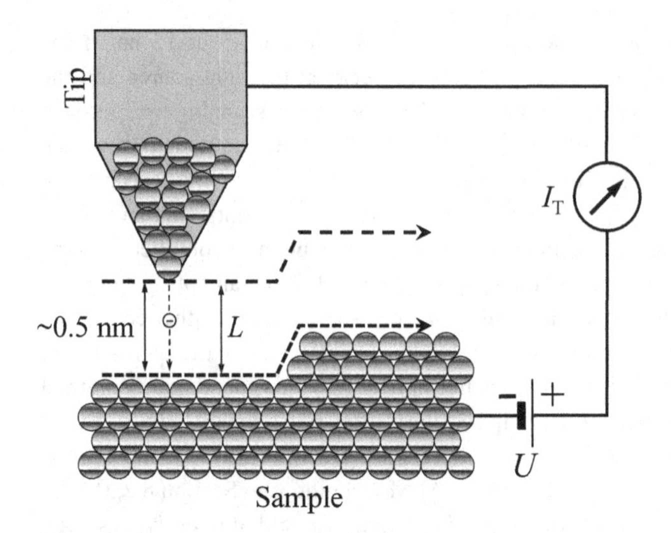

Figure 8.8 The scanning process in a scanning-tunneling microscope.

In tunneling microscopy, the tunneling current between the *probe* (or *tip*) and the conducting surface is measured at each point of the scanned surface (see Fig. 8.8). If we move the tip step by step (with a step of about the size of the atomic radius) while measuring the tunneling current, we will "feel" the individual atoms of the scanned surface. By *scanning* we mean the motion of the tip along the surface from one point to another. The picture of the distribution of the current measured over the sample surface is visualized on a computer display and allows us to study the surface topology on the atomic scale. The tunneling phenomenon was utilized in 1981 as the basic principle of the *scanning-tunneling microscope* (STM). One of the conductors had the form of a sharp needle, which was called the *tip*, and the other conductor was the surface of the object being studied.

In order to move electrons from the end of the tip to the studied surface, the conduction electrons must gain a certain energy. The energy depends on the distance, L, between the tip and the surface of the object, the voltage, U, between them, and the work functions, A_t and A_s, of the tip and the sample, respectively. On bringing the tip to the surface at a separation of $L \approx 0.5$ nm and applying a voltage $U \approx 0.1-1$ V a current, I_T, is produced due to the tunneling effect:

$$I_T \approx e n_0 v S. \tag{8.6}$$

Here, n_0 and v are the concentration and velocity of electrons, and S is the cross-section of the electron beam. The concentration of electrons in the beam is related to the concentration of electrons in the tip, n_0, by

$$n_0 = Dn, \tag{8.7}$$

where D is the probability of electron transmission through the air gap. In accordance with Eq. (3.156), this probability is defined as

$$D \approx \exp\left(-\frac{2L}{\hbar}\sqrt{2m_e U_{\text{eff}}}\right), \tag{8.8}$$

where U_{eff} is the magnitude of the effective potential barrier between the tip and the surface of the sample. For the majority of "tip–object" pairs the mean value of the potential barrier is equal to $U_{\text{eff}} \approx 4$–5 eV. To estimate the tunneling current, let us assume that the electron beam which emanates from the outermost atom of the tip has a diameter of about $d = 0.4$ nm to provide high scanning resolution. On substituting the values of the parameters $L = 0.5$ nm and $U_{\text{eff}} = 4$ eV into Eq. (8.8) we get $D \approx 10^{-5}$. Thus, taking into account that $n \approx 10^{28}$ m^{-3}, we get $n_0 \approx 10^{23}$ m^{-3}. By substituting the obtained value of n_0 and $v \approx 10^6$ m s^{-1} into Eq. (8.6) we can estimate the magnitude of the tunneling current as $I_T \approx 2 \times 10^{-9}$ A.

The magnitude of the tunneling current, I_T, depends exponentially on the distance, L, between the tip and the sample. Therefore, on increasing the distance only by 0.1 nm the magnitude of D and, correspondingly, the tunneling current I_T decrease by a factor of almost 10. Thus, small changes in the height of the surface relief induce substantial changes in the tunneling current, which provides the high resolution capabilities of the STM. Owing to this extremely high sensitivity, an STM may provide images of surfaces with a resolution of 0.01 nm in the vertical direction and with an accuracy of atomic size in the horizontal plane.

Usually two operation modes are used, for studying surfaces of different materials. The operating regime of *constant height* requires the motion of the tip in a strictly horizontal plane over the surface. The magnitude of the tunneling current will depend on the distance between the tip and the scanned surface. By registering the tunneling current, I_T, at each point of the surface and using computer processing, we can plot the image of the surface relief. In the operating regime of *constant current* a feedback system is used to maintain a constant tunneling current by changing the distance between the tip and the surface at each point of the scanned surface (see Fig. 8.9). The visualization of the surface relief is done using the data of the vertical displacements of the scanning device. As an example, a scanned surface of graphite is shown in Fig. 8.10.

Each of these two regimes has its advantages and drawbacks. The regime of constant height is fast enough because the system does not need to move the scanning device up and down. The drawback of this regime is that information may be obtained only from very flat and smooth surfaces. In the regime of constant current irregular surfaces may be scanned with high accuracy. The drawback of this regime is that the measurements take too much time. Let us note that the quality of the tip and the radius of its end define the resolution capabilities of the

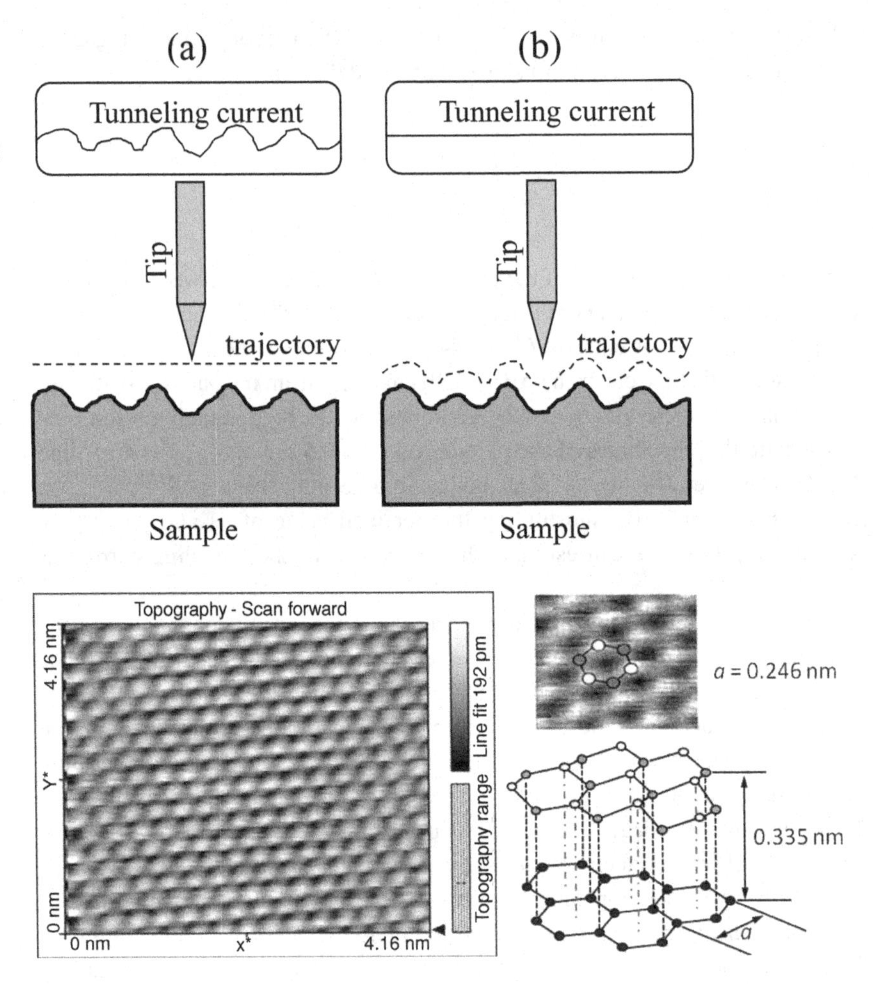

Figure 8.9 Tip motion over the scanned surface in the two operating regimes of an STM: (a) constant-height mode and (b) constant-current mode.

Figure 8.10 A picture of a graphite surface obtained by an undergraduate student carrying out an experiment from the lab course at the Undergraduate Nanoelectronics Laboratory at the University at Buffalo. There are two different positions of carbon atoms in a crystal lattice of graphite: one with a neighboring atom in the plane below (gray balls) and one without a neighbor in the plane below (white balls). It is clearly seen that the upper layer of graphite is not flat: the atoms that have neighboring atoms in the lower plane are more strongly attracted to each other than are the atoms depicted as white balls. Therefore, the atoms depicted as white balls are slightly above the plane while the atoms depicted as gray balls are slightly below the plane.

STM and the quality of the image. The end of the tips used in industry has the size of several atoms.

In the working regime of an STM the distance between the tip and the surface is about $L \approx 0.3$–1 nm, therefore the probability of air molecules being between them is very low at standard atmospheric pressure. Therefore, we may assume that passage of the tunneling current takes place under almost vacuum conditions.

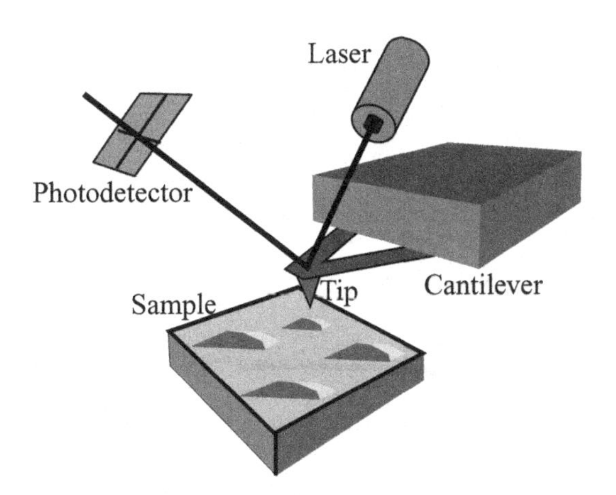

Figure 8.11 A schematic representation of the atomic-force microscope.

We can draw an important conclusion from the above for the practical operation of an STM: the operational regime of the STM in most cases does not require high vacuum, which is a necessary requirement, for example, for electron microscopes of other types.

The main drawback of an STM is the following: the object studied has to be a *conductor*, i.e., it must be either a metal or a doped semiconductor, otherwise the tunneling current would not flow. Therefore, we cannot study the surface structures of dielectrics, for example the surface structure of diamond, with the help of STM techniques.

8.2.3 Atomic-force microscopy

Five years after the invention of the STM in 1986, the *atomic-force microscope* (AFM) was developed. The measured parameter in an AFM is the interatomic interaction force between the tip and the surface of the sample. The operational principles of the AFM are the following. At the distance at which a tunneling current between the tip and the conducting surface occurs, there appears an appreciably large force between the tip and surface. This force, as in the case of tunneling current, significantly depends on the gap between the tip and the surface of the object studied. The force can be detected from the elastic deflection of the so-called *cantilever* which has the scanning tip at its end (see Fig. 8.11). A photodetector provides the detection of small deflections of the cantilever's end.

An AFM allows us to study the surface of any object regardless of its conductivity. However, new problems occur in this case: the dependence of the interaction force between the tip and the sample surface is complex and this force does not depend on the sample–tip distance in a monotonic way.

At large distances between the tip and the surface there is an attractive force, which at smaller distances becomes a repulsive force. The distance between the

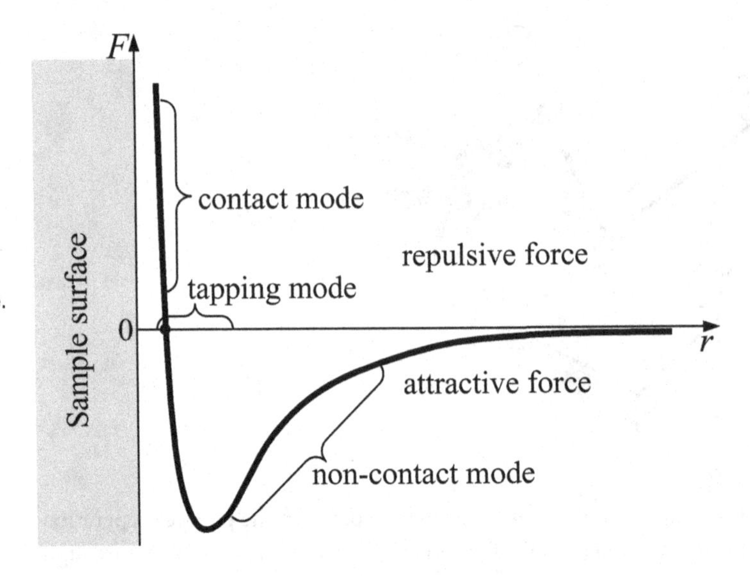

Figure 8.12 The dependence of the interatomic force, F, on the distance between the tip and the surface, r, showing the intervals of the three operational regimes: (1) contact mode, (2) tapping mode, and (3) non-contact mode.

cantilever's tip and the scanned surface at which the type of the force changes is approximately equal to 1 nm. Therefore, the distance range within which measurements can be made is very limited. Thus, it is very important to calibrate the measurement device carefully, otherwise the image obtained will be difficult to interpret.

Figure 8.12 shows schematically the dependence of the interaction force between the tip and the scanned surface on the distance between them. It also shows the types of operational regimes corresponding to different distances.

This new type of microscope allows us to overcome the principal limitation of the scanning-tunneling microscope: the AFM allows us to obtain images of conducting as well as non-conducting surfaces with atomic resolution under atmosphere conditions (see Fig. 8.13). An additional advantage of the AFM is the possibility of visualizing, together with the surface relief, its electric, magnetic, elastic, and other properties. Omitting the details, the operational principles of the AFM can be described as follows. The largest input into the interaction between the tip and the sample surface is from the force between an atom at the end of the tip and the nearest atoms of the sample surface. The closer the tip to the scanned surface, the more strongly atoms of the tip are attracted to the nearest atoms of the sample surface. The force of attraction will increase as these atoms approach each other, until the electron clouds of these atoms begin to repel each other. With a further decrease of the interatomic distance the electrostatic repulsion exponentially weakens the attraction force. At a distance between the atoms of the tip and the surface of about 1 nm the forces of attraction and repulsion become equal to each other.

Depending on the type of interaction, the AFM may operate in one of the following regimes (see Fig. 8.12). In the *contact mode* the tip comes very close

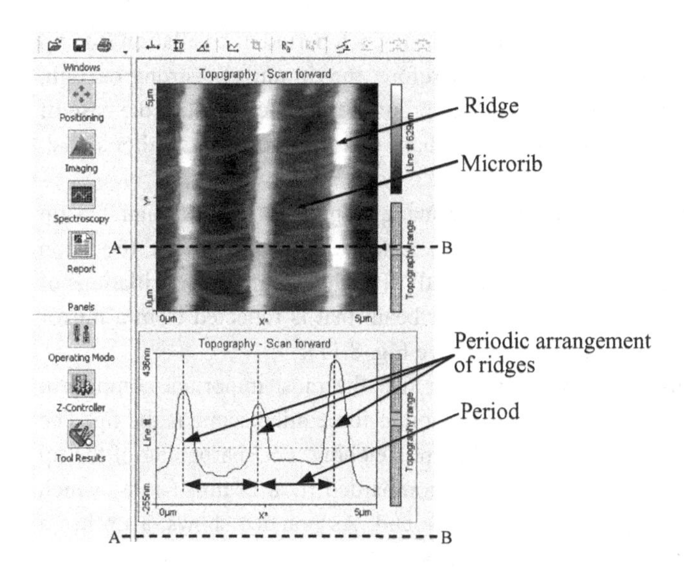

Figure 8.13 The structure of the scale of a Morpho butterfly wing obtained by an undergraduate student carrying out an experiment from the lab course at the Undergraduate Nanoelectronics Laboratory at the University at Buffalo. You can clearly see ridges and microribs that connect ridges. The lower picture shows a cross-section of the butterfly-wing scale along the horizontal line AB.

to the surface of the sample and the deflection of the cantilever is caused by the mutual repulsion of atoms on the end of the tip and on the sample surface. It is caused by the overlap of electron clouds and Coulomb repulsion of the atomic nuclei. In the *non-contact mode*, which corresponds to the region of attraction forces, the AFM registers the attractive van der Waals force acting between the scanning tip and the sample surface. The gap between the end of the tip and the surface in the non-contact mode is usually about 5–10 nm. The intermediate regime between contact and non-contact modes is the mode of periodic instant physical contacts of the tip with the surface during the scanning process. This mode is called the *tapping mode*. In this regime the cantilever with the tip oscillates with some fundamental resonance frequency having an amplitude of about 50–100 nm. The fundamental frequency of the cantilever with the tip is determined when the cantilever is far away from the surface. In operational mode the tip is moved closer to the surface and, at amplitudes of about 50–100 nm, the tip comes into brief physical contact with the sample surface when the cantilever has its maximum deflection down from the equilibrium position. This substantially changes the frequency, phase, and amplitude of the cantilever oscillations. All three of these parameters can be measured. The tapping mode is characterized by higher resolution in the horizontal plane than is possible in the contact-mode regime.

On comparing the operational principles of scanning-tunneling and atomic-force microscopes, we can see that they have much in common. The main difference between them is the design of the probe. In an STM the probe (the tip) must be rigidly clamped and must not touch the surface, and the sharper the tip the better. In an AFM the tip is mounted on the deflecting cantilever and can be operated in regimes in which the oscillating tip touches the surface of the material being studied for a short time during each period of oscillation (the tip literally "taps" the sample surface). Therefore, the tip must be strong enough, which requires a larger thickness of the end of the tip in contrast to the case of an STM tip. Moreover, a tip with a very sharp end will provide a smaller signal, which will be difficult to detect.

The first cantilevers for AFMs were made of gold foil with a diamond tip or of aluminum foil with a tungsten wire. Later developers switched to silicon cantilevers, which are widely used currently. The deflections and oscillations of the cantilever may be detected by a laser beam that is reflected from a mirror sputtered onto the top of the cantilever (see Fig. 8.11).

From the discussion above, it is clear that the most important component both of a scanning-tunneling and of an atomic-force microscope is the tip. The standard technology for sharpening the tip is etching. For sharpening of the tip also *ion sharpening* is used: the tip is bombarded by a beam of ions, which removes excessive material from the tip's end. As practice shows, a tip has a finite lifetime. During operation the distance between the tip and the sample is very small. Accidental contact with the scanned surface can damage the tip. Therefore, this component requires constant attention. When the properties of the tip deteriorate, the tip is usually replaced or it is sharpened without moving it away from the microscope.

8.2.4 Scanning-probe nanotechnology

The modification of scanning-probe microscopes allows us not only to characterize nanostructures, but also to fabricate nanostructures on the atomic level. When the voltage between the sample and the tip is slightly greater than that in the operational tunneling regime, the *atom* from the sample surface (precisely, the ion) can be transferred to the tip. By changing the polarity of the applied voltage we can force the ion to return to the sample surface. If during these two events the tip has been moved, then the ion will be placed onto a new position on the surface. So, it is possible to manipulate the positions of individual atoms with the help of a tip. For this kind of technique we need (a) a high-quality STM, which operates at low temperatures, in order for the atoms not to disperse as a result of thermal motion, (b) a proper tip, and (c) certain skills. Such a possibility of manipulation of atoms by use of an STM was demonstrated for the first time by the members of one of the IBM research centers. In 1985 they were granted a patent on the possibility of moving individual atoms from the end of the STM's tip onto the sample's surface. Moreover, in 1989 they were able to construct the logo of their company "IBM" on the surface of a gold film using 35 atoms of xenon. This event is considered the birth of *scanning-probe nanotechnology*.

Nowadays, there exist several methods for the displacement and assembly of nanostructures from individual atoms and molecules. The first method, as described above, consists of the capturing and displacing of atoms by applying a higher voltage. The sample must be at ultra-high vacuum during this procedure, otherwise the surface would quickly be covered by atoms from the environment. By filling the gap between the sample and the tip with inert gases we can obtain the same results as would be obtained in vacuum. So, the method of scanning-probe nanolithography in liquid and gaseous media was developed. By introducing into the gap between the tip and the sample's surface specifically selected substances and by changing the applied voltage, we can initiate chemical reactions at the point of the tip's contact with the surface. An example of such technology is the method of local anode oxidization of thin metallic films. This method allows one to obtain on the surface of a metal a pattern formed by its oxides with a width of several tens of nanometers. This method is used to fabricate electronic circuits of ultra-small size.

An AFM can also be used for the modification of a surface. The simplest method is to scratch the surface. With the help of direct contact of the end of the tip with the sample's surface, we can make pits and grooves or flatten the surface. For this purpose tips made from hard materials such as diamond are used. The capabilities of an AFM to modify a surface can be enhanced if the tip is made conductive. Being able to induce an electric current between the sample and the tip allows the operator to heat locally the surface, control the chemical reactions that can take place in the contact region, and transfer atoms and molecules from the tip to the sample and vice versa. The possibilities of modification of surfaces with the help of the AFM are similar to the possibilities that the STM has. At the same time, AFM techniques allow one to obtain images of fabricated structures even if their surfaces are not conducting.

Thus, the most important achievement of scanning-probe microscopy is that it is not only an instrument for the characterization of nanostructures but also an instrument for the fabrication of nanostructures.

The main problem of scanning-probe nanotechnology is its low production output. A single probe, even at maximum speed of operation, cannot provide mass production. This drawback can be overcome by the development of multiprobe devices. In such devices the nanostructures are produced simultaneously by several tens or even thousands of probes.

Currently STMs and AFMs are scientific instruments that are used widely. An entirely new industry that manufactures various components of microscopes, including tips and cantilevers, as well as complex research systems, has emerged.

Example 8.2. Estimate the pressure of an electron beam and the mechanical strain caused by the ponderomotive forces on the region of surface under the tip of the STM. The values of operational characteristics are the following: current $I = 10^{-2}$ A, applied voltage $U = 5$ V, radius of the tip $r = 20$ nm, and gap between tip and surface $L = 1$ nm.

Reasoning. The current density in the electron beam can be defined as

$$j = \frac{I}{S} = nev, \tag{8.9}$$

where n is the electron concentration in the beam, e the electron charge, and v the electron's velocity. The pressure of electrons in the beam is defined as

$$P = \frac{F}{S} = \frac{\Delta p}{S\,\Delta t} = \frac{(nvS\,\Delta t)m_e v}{S\,\Delta t} = nm_e v^2, \tag{8.10}$$

where S is the cross-sectional area of the electron beam, $nvS\,\Delta t$ is the number of electrons with mass m_e reaching the surface during the time period Δt at a given current I, and $m_e v$ is the momentum of an individual electron. The velocity of electrons in the beam can be estimated as

$$v = \sqrt{\frac{2eU}{m_e}} \approx 1.3 \times 10^6\ \mathrm{m\,s^{-1}}. \tag{8.11}$$

Taking into account the derived relationships, we obtain for the pressure of the electron beam, P, the following expression:

$$P = \frac{I}{Sev} m_e v^2 = \frac{m_e I}{eS}\sqrt{\frac{2eU}{m_e}} = \frac{I}{\pi r^2}\sqrt{\frac{2m_e U}{e}} \approx 6 \times 10^7\ \mathrm{Pa}. \tag{8.12}$$

The surface region under the tip experiences the action of the ponderomotive forces which are induced by the electric field created between the tip and the surface. Mechanical strain can be defined as negative pressure acting on the surface, i.e., in this case the surface particles are attracted to the tip. Considering the tip and the surface as two plates of a plane capacitor, we can estimate the magnitude of the force acting between these two plates:

$$F_E = QE_2 = S\sigma E_2 = S\sigma\frac{E}{2} = S\epsilon\epsilon_0\frac{E^2}{2}, \tag{8.13}$$

where Q is the charge of one of the plates, E_2 is the field induced by the other plate, σ is the surface charge density, $E = \sigma/(\epsilon\epsilon_0)$ is the electric field of the charged capacitor, and ϵ is the dielectric permittivity of the medium between the tip and the surface. As a result, the negative pressure on the studied region of the surface is equal to

$$P_E = \frac{F_E}{S} = \frac{\epsilon\epsilon_0 E^2}{2} = \frac{\epsilon\epsilon_0 U^2}{2L^2} \approx 10^8\ \mathrm{Pa}. \tag{8.14}$$

Thus, by controlling the applied voltage and the gap between the tip and the surface, we can achieve sufficiently large values of the positive and negative pressure, which will allow us to pull electrons away from the surface and to deform locally the sample surface.

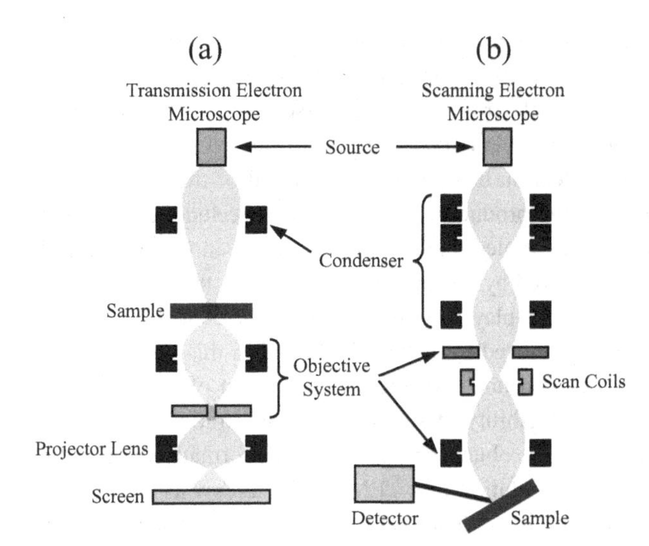

Figure 8.14 The TEM (a) and SEM (b).

8.2.5 Electron microscopy

The electron microscope, which uses an electron beam instead of a light beam for imaging, was developed in 1935. The invention of the electron microscope became possible because of the rapid development of quantum mechanics at the beginning of the twentieth century, when it was discovered that the electron has the properties of a wave as well as those of a particle. As we have already learned, the waves associated with electrons exhibit interference and diffraction, which are inherent to light waves. At the same time, electrons are charged particles and their motion can be controlled by electric and magnetic fields. Electron beams can be deflected by electric and magnetic fields just as light rays can be deflected by optical lenses. The focussing systems of electron beams are called *electron lenses*. The role of a light source in an optical microscope in electron microscopes is taken by the *electron gun* (usually a heated tungsten wire that serves as the source of electrons). The emitted electrons pass through the electron-condensing lens, which regulates the intensity of the flux and the area of the scanned surface subjected to exposure. After passing through the electron lens, which is used as an objective lens, the electron beam is directed onto a luminescent screen, which transforms the electron-distribution image into an image that can be either photographed or directly observed. There are two major types of electron microscopes: the *transmission-electron microscope* (TEM) and the *scanning-electron microscope* (SEM).

The TEM (see Fig. 8.14(a)) measures the intensity of the electron beam transmitted through the sample and it has the following principal features. Since the electron beam is strongly absorbed and/or scattered by atoms, the sample is placed in an evacuated chamber. For the same reasons the sample must be very thin (about 100 nm) and sample preparation is a rather difficult problem.

The SEM (see Fig. 8.14(b)) does not have restrictions on the thickness of the sample. The SEM design has much in common with the design of the TEM. The principal difference is that the electron beam is focussed onto a certain spot of the surface and the reflected or secondary electrons are studied. The process of interaction of the incident electron beam with the material of the object being studied results in the following products: (a) a beam of reflected electrons, (b) a beam of secondary and Auger electrons, and (c) radiation in the visible and X-ray range. These are registered by corresponding detectors, transformed into electric signals, amplified, and displayed. With the help of the deflection system, the incident electron beam is moved over the sample and in this way scans its surface. As a result, a magnified image of the studied region of the surface is formed on the display. The possibility of simultaneous registration of different types of radiation allows one to obtain comprehensive information about the structure being studied. The resolution of a TEM can be as low as 2 Å, which allows one to obtain images of individual atoms and molecules. The resolution of an SEM comes close to the resolution of a TEM and currently it reaches 5 Å. However, the SEM has a couple of advantages over the TEM. Since in an SEM the electrons which participate in the formation of an image do not penetrate the sample, there is no restriction on the thickness of the sample. The procedure of sample preparation is significantly simplified and there is no need for ultra-high vacuum as for the TEM, which simplifies the design of an SEM and reduces its cost.

There are also new devices that have been developed and are being developed. They use the principles described above for the measurement of electrical, optical, mechanical, and other properties of the sample surface. Currently the methods described above are widely used not only for the study of surface relief, but also for the control of the technological processes relating to the fabrication of nanostructures and nanomaterials. For example, the following system of two microscopes has been developed. Within an SEM system on its sample stage, an STM is mounted. First, with the help of the STM, a desired spot on the sample surface is identified. After this, the chosen spot is carefully studied by the SEM, which has much higher resolution (*subatomic resolution*) than that of the STM. Combinations of SEM and AFM techniques are also being developed.

8.3 Selected examples of nanodevices and systems

The tendency towards miniaturization in electronics is not only sustained but has intensified. Researchers have fabricated new types of electronic devices of ultra-small size, which have become the basis of nanoelectronics. Nanostructures are in turn used for the development of new devices, whose operational principles are fundamentally different from those of existing conventional semiconductor devices such as diodes and transistors.

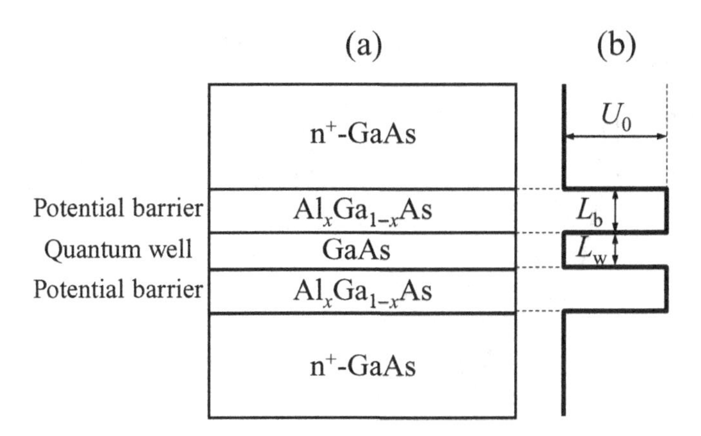

Figure 8.15 (a) The schematic structure of a GaAs/$Al_xGa_{1-x}As$ resonant-tunneling diode and (b) its potential profile. U_0 is the height of the potential barriers, L_b the thickness of the barriers, and L_w the width of the quantum well.

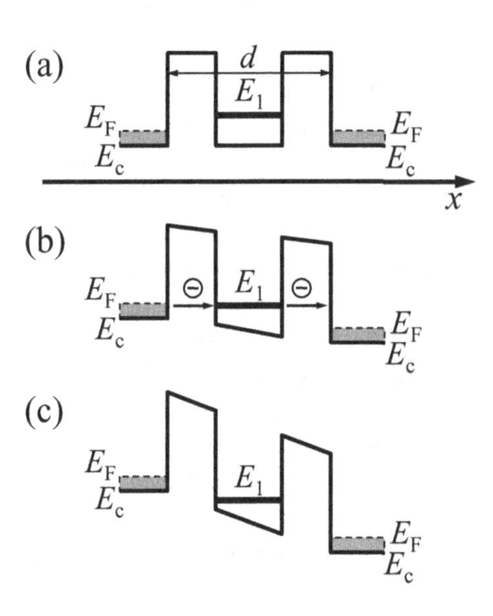

Figure 8.16 The system of energy levels in a resonant-tunneling diode: (a) under equilibrium conditions, (b) at resonance under an applied voltage, and (c) off resonance under an applied voltage. The electron potential energy and population of the energy band in contact n^+-regions are sketched in the left and right parts. E_c denotes the bottom of the conduction band.

8.3.1 Resonant-tunneling diodes

A nanoscale phenomenon called *resonant tunneling* can be described as follows. The current through a nanostructure consisting of two potential barriers with a quantum well between them (a so-called *double-barrier structure*), sharply increases when the Fermi energy of the electrode that supplies electrons becomes equal to the energy level in the quantum well. Such a double-barrier structure with two contacts is called a *resonant-tunneling diode* (RTD). The scheme of an RTD and its potential profile are shown in Fig. 8.15. The system of energy levels of such a simple structure is shown in Fig. 8.16. This structure consists of two barriers divided by a region with low potential energy (a quantum well with quantized electron levels). Let us assume that the quantum well has only

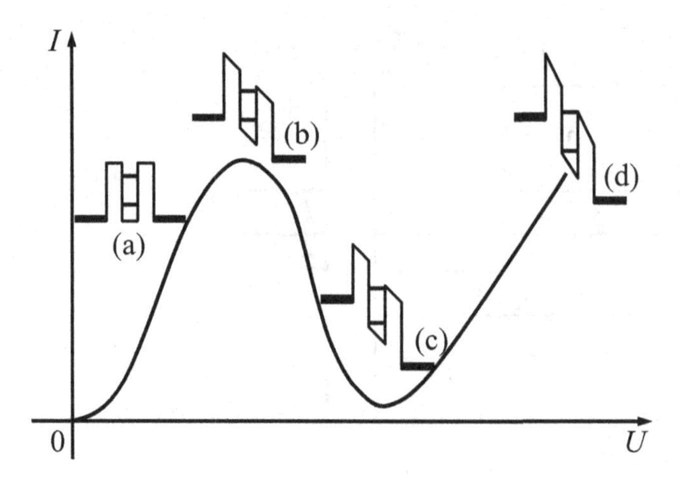

Figure 8.17 An energy diagram and current–voltage characteristic of the resonant-tunneling diode: the portions labeled (a), (b), and (c) correspond to the physical situations of Fig. 8.16. Part (d) shows what happens with the current when the second electron state enters into resonance with the Fermi level, E_F.

a single discrete energy level. The thickness of the barriers and width of the quantum well can be several nanometers. The regions in the right-hand and left-hand sides from the barriers serve as electron reservoirs, and these regions can be considered as the contacts to the structure. Such a double-barrier structure has one distinctive feature: its tunneling probability has a resonant character. Even if the transmission through each barrier is low, this does not mean that the probability of electron tunneling through the double-barrier structure is small. When the energy of electrons incident on the barrier is equal to the energy of one of the discrete levels of the quantum well between the two barriers, the tunneling probability for the double-barrier structure increases sharply. The effect of resonant tunneling can be understood in terms of *interference*. Indeed, due to the interference of the transmitted and reflected electron waves the wave that is reflected from the double barrier returns to the quantum well. Therefore, the wave incident on the double-barrier structure from the left has a high probability of being transmitted to the right.

Let us describe how the RTD works. The current that flows through the double-barrier structure depends on the applied voltage. Note that the voltage is mostly applied to the double-barrier region, since the regions from the left-hand and from the right-hand sides of the barriers have higher conductivity. If the applied voltage is small and the energy of electrons incident on the barrier from the left is lower than the energy of the discrete energy level in the quantum well, then the transmission through the barrier and, consequently, the current through the double-barrier structure will be small (see part (a) of the curve in Fig. 8.17 and Fig. 8.16(a)). The current reaches its maximum at the voltage at which the energy of electrons becomes equal to the energy of the discrete energy level (see part (b) of the curve in Fig. 8.17 and Fig. 8.16(b)). At higher voltages the energy of incident electrons becomes higher than the energy of a discrete energy level (see part (c) of the curve in Fig. 8.17 and Fig. 8.16(c)). As a result, the tunneling probability of the double-barrier structure decreases and consequently the current decreases. Thus, there appears a maximum of the current–voltage characteristic, which is followed by a region with *negative-differential resistance* (NDR), i.e., by a region where the current decreases with increasing voltage. With the further increase of the applied voltage, the current begins to increase again. If the potential well between two barriers has several discrete energy levels, then the current–voltage characteristic will have several maxima, each of them at a voltage at which the corresponding energy level comes into resonance with the Fermi energy, E_F, of the left-hand-side region (see part (d) of the curve in Fig. 8.17).

Having these regions, the RTD may be used in electronic circuits not only as a rectifier but also for performing various other functions. The existence of such an NDR region (or NDR regions) is promising for the development of multilevel logical elements, memory elements, and solid-state ultra-high-frequency generators. The addition of a control electrode in the central region of an RTD transforms it into a *resonant-tunneling transistor* and increases the number of its applications. Such transistors have turn-off and turn-on frequencies of about 10^{12} Hz. This is several orders of magnitude higher than the frequencies of the best silicon transistors in modern integrated circuits. Therefore, such transistors can be used in the next generation of integrated circuits.

8.3.2 Single-electron transistors

The operation of single-electron devices is based on the *effect of tunneling of a single electron*. Various structures and devices based on this effect have been realized experimentally. In the simplest version, this effect can be observed in a single-electron tunnel junction. The single-electron tunnel junction may be created in the form of two plates (we consider here the simplest case, in which both plates are made of the same metal) of small cross-sectional area, S, and a thin layer of dielectric of thickness d between them (see Fig. 8.18). The capacitance of this system is given by

$$C = \epsilon\epsilon_0 \frac{S}{d}, \tag{8.15}$$

where ϵ is the dielectric permittivity of the dielectric separating the metallic plates. The potential difference between the plates is determined by the charge accumulated on these plates:

$$U = \frac{Q}{C}. \tag{8.16}$$

If the charge, Q, is reduced by the charge of one electron, e, then the potential difference becomes

$$U = \frac{Q - e}{C}. \tag{8.17}$$

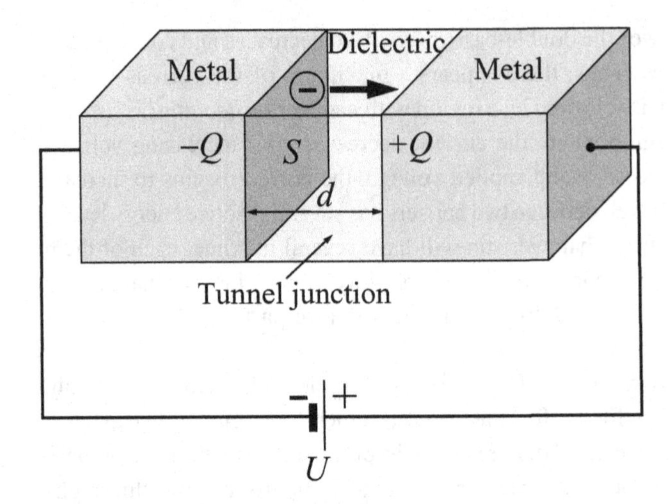

Figure 8.18 A capacitor with an induced charge and a single-electron tunnel junction.

As a result of transmission of a single electron from one electrode through the dielectric layer to the opposite electrode, the voltage at the tunnel junction changes by the following magnitude:

$$\Delta U = \frac{e}{C}. \tag{8.18}$$

The energy stored in the capacitor is defined as

$$E = \frac{Q^2}{2C}. \tag{8.19}$$

The change of the energy stored in the capacitor as a result of transfer of a single electron is equal to

$$\Delta E = \frac{Q^2}{2C} - \frac{(Q-e)^2}{2C} = \frac{e(Q-e/2)}{C}. \tag{8.20}$$

Electron tunneling occurs only if $\Delta E > 0$, i.e., if energy is decreased due to the transfer of a single electron. Thus, from Eq. (8.20) it follows that, for electron tunneling to occur, the initial charge on a metallic plate must satisfy the following condition:

$$Q \geq \frac{e}{2}. \tag{8.21}$$

This corresponds to the voltage between plates, U, being equal to

$$U \geq \frac{e}{2C}. \tag{8.22}$$

Thus, by applying a voltage higher than $e/(2C)$, we can make current flow. Since the current cannot flow if

$$U < \frac{e}{2C}, \tag{8.23}$$

the effect is known as *Coulomb blockade*. The current–voltage characteristic for a device with Coulomb blockade is shown in Fig. 8.19.

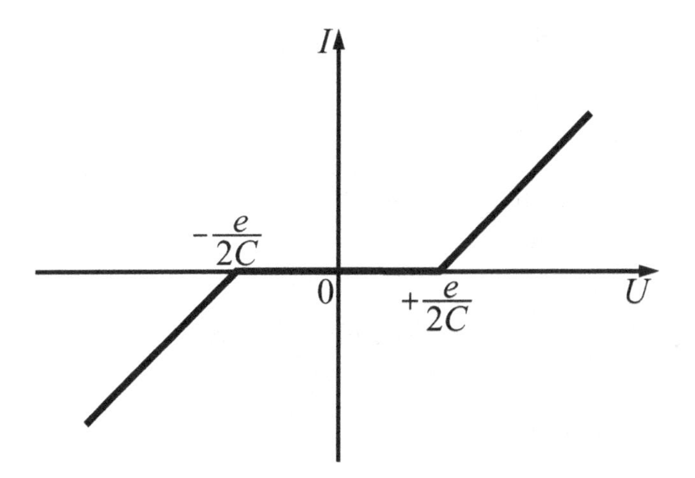

Figure 8.19 The current–voltage characteristic for a device with Coulomb blockade.

Because of the nanometer width of the tunnel junction, d, and small capacitance, C, each electron passing through the junction causes a substantial change in the voltage U (see Eq. (8.18)). If the condition (8.22) is not satisfied, the electron tunneling is blocked by a large blocking voltage, caused by the tunneling of a single electron. The energy which is necessary to charge the capacitor in order to satisfy the condition (8.22) is equal to

$$E_C = eU = \frac{e^2}{2C}. \tag{8.24}$$

This is the charging energy required to add a single electron with charge e to the capacitor. Note that for $Q = 0$ the expression (8.20) is reduced to Eq. (8.24).

Let us estimate the capacitance, C, and the energy, E_C, required to transfer a single electron in nanoscale capacitors. For example, for a capacitor with $S = 100$ nm^2, $d = 10$ nm, and $\epsilon = 1$ (we assume that the dielectric is vacuum) we get $C \approx 9 \times 10^{-18}$ F and $E_C \approx 0.9$ eV. The energy needed to transfer a single electron is equal to almost 1 eV! Let us compare the capacitance and charging energy of a nanoscale capacitor with those for a macroscopic capacitor with $S = 100$ mm^2, $d = 1$ mm, and $\epsilon = 1$. For these parameters we get $C \approx 9 \times 10^{-11}$ F and $E_C \approx 9 \times 10^{-8}$ eV. We see that the difference between the energy needed to transfer a single electron, E_C, in the case of a nanoscale capacitor and that for a macroscopic capacitor is enormous. In the case of a macroscopic capacitor, E_C is almost negligible. The charging energy of the macroscopic capacitor ($E_C = 9 \times 10^{-8}$ eV) is lower by six orders of magnitude than the thermal energy of an electron at room temperature ($E_{th} = 2.5 \times 10^{-2}$ eV)! Therefore, Coulomb blockade cannot be observed in macroscopic objects.

Let us now consider the Coulomb-blockade effect in a *quantum-dot single-electron transistor*, as shown in Fig. 8.20. Just like any conventional transistor, it has three electrodes: *source*, *drain*, and *gate*. A quantum dot is placed in

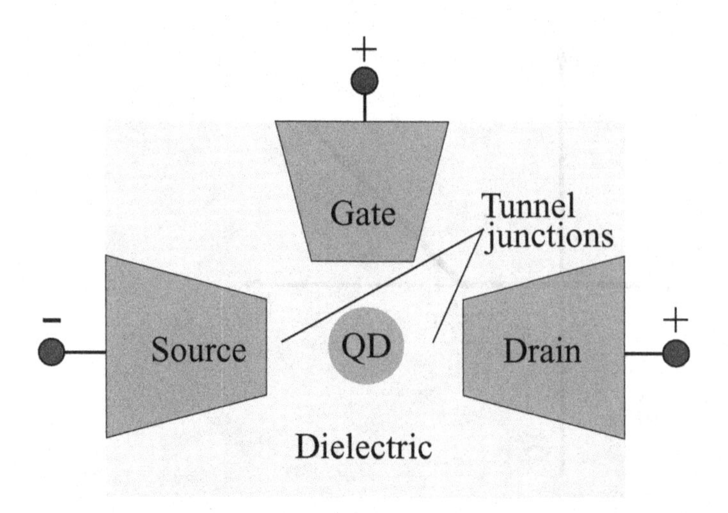

Figure 8.20 A schematic representation of a quantum-dot single-electron transistor: the quantum dot (QD) is isolated from the source, drain, and gate by dielectric.

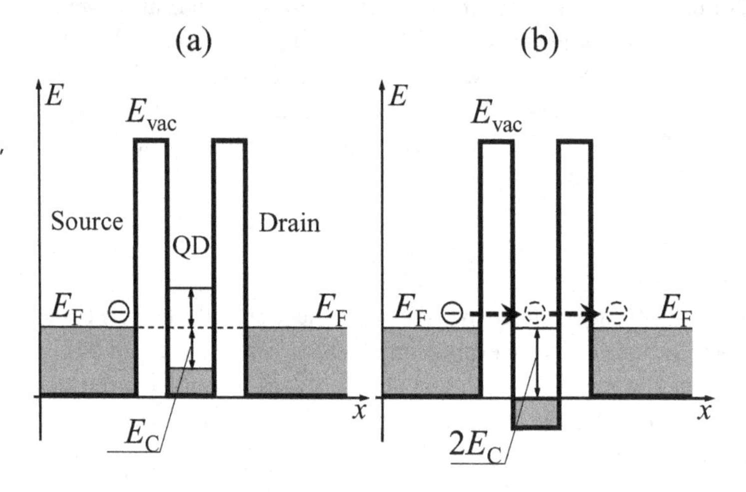

Figure 8.21 Band diagrams of a quantum-dot single-electron transistor with source–drain voltage, U_{SD}, equal to zero: (a) gate voltage $U_G = 0$ and (b) gate voltage $U_G > 0$. Here, $E_C = e^2/(2C)$.

the *channel*, i.e., in the region between the source and the drain. The quantum dot, which is often called the *island*, is either a metallic or a semiconductor nanoparticle, which is isolated from all three electrodes by a dielectric. Electron motion can take place only through the dielectric, under certain conditions.

Here we give you a very simplified version of the operational principles of a quantum-dot single-electron transistor. The band diagrams of a quantum-dot single-electron transistor without and with an applied positive gate voltage, U_G, are shown in Fig. 8.21 (the voltage between source and drain, U_{SD}, is equal to zero). The quantum dot (the island) in the channel of a single-electron transistor is a nanoparticle, which is made from the same metal as the source and the drain. As shown in Fig. 8.21(a), the electrons in the source and the drain have higher energy than the electrons have in a quantum dot. Therefore, it is energetically favorable for an electron from the source or the drain to tunnel to the quantum dot. However, because of Coulomb blockade, the electron must have energy greater than the charging energy E_C. At zero gate voltage ($U_G = 0$), as shown in Fig. 8.21(a), the electron in the source has energy lower than E_C and therefore tunneling of the electron from the source to the quantum dot cannot happen. If a positive gate voltage ($U_G > 0$) is applied to the gate, as shown in Fig. 8.21(b), the energy in the dot is decreased by E_C and the electron from the source will be able to tunnel to the quantum dot and then to the drain. Thus, by controlling the potential on the gate we can control the transmission of a single electron through the dielectric barrier and therefore control the current in a circuit. This is the essential feature of a transistor. If the electron that tunneled from the source stayed in the dot rather than going into the drain, then the energy in the dot would increase by E_C and tunneling of the next electron would be blocked.

In digital integrated circuits based on single-electron transistors one bit of information (two possible states, 0 or 1) may be presented as the presence or absence of a single electron in the quantum dot of the channel. A whole memory block with a capacity of 10^{11} bits (which is 100 times higher than that which modern chips have) could be placed on a crystal with a surface area of 1 cm^2. The practical realization of such devices is the goal of researchers all over the world.

8.3.3 Quantum-optoelectronic devices

Optoelectronics is a field of physics and technology that is related to the conversion of optical radiation into electrical signals and vice versa. The number of optoelectronic devices is large. The operational principles of these devices are based on quantum mechanics. We can distinguish two main types of optoelectronic devices: (1) light-emitting devices and (2) light-absorbing or photodetecting devices.

Light-emitting devices

These are the devices that convert electric current into light. They can be incandescent lamps, electroluminescent indicators, semiconductor light-emitting diodes (LEDs), and lasers. The most widely used devices developed with the help of nanotechnology are LEDs and lasers. We will talk about lasers – sources of coherent radiation – later. Now we will discuss the operational principles of the LEDs with a p–n-junction.

Light-emitting diodes are sources of non-coherent radiation. They emit light only when the current through the p–n-junction flows in the forward direction. The forward direction for the p–n-junction corresponds to the case when a negative potential is applied to the n-region. Electrons come into the junction region from the n-region and holes come from the p-region. After recombination, holes and electrons disappear and their energy in the form of a light quantum

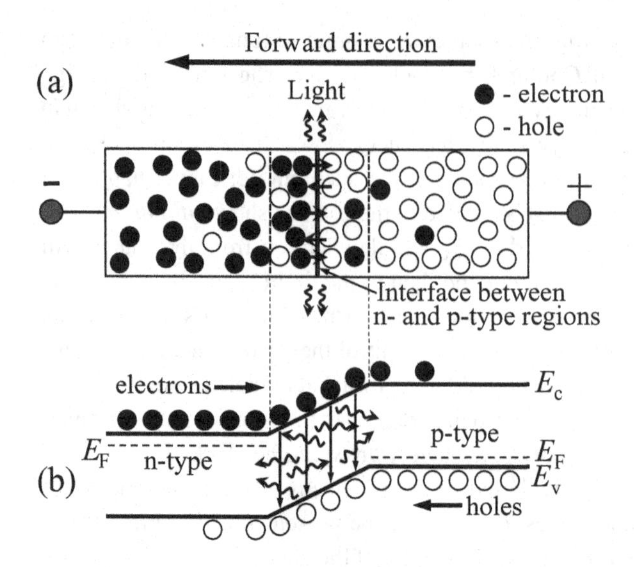

Figure 8.22 A schematic representation and band diagram of a semiconductor light-emitting diode.

is emitted, as shown in Fig. 8.22. The light intensity should be proportional to the current flowing through the LED. However, when the current increases the thermal heating of the LED also increases, leading to an increase of the non-radiative losses in the LED.

For a long time mass production of LEDs was limited to LEDs emitting only in the red and infrared regions of the electromagnetic spectrum. Green LEDs were made first from gallium phosphide (GaP) and blue LEDs from silicon carbide (SiC). However, these materials either emitted low-intensity radiation or quickly became overheated due to their low efficiency. Nitrides, materials based on elements from the third and fifth groups of the Periodic Table of the elements (AlN, GaN, and InN), have become practical for the fabrication of LEDs in the visible range of the electromagnetic spectrum. These materials emit throughout the entire range of the visible and ultraviolet regions of the electromagnetic spectrum from 240 to 640 nm. During the 1970s scientists from IBM fabricated violet and blue diodes based on epitaxial films of GaN. However, these diodes quickly overheated and thus became non-operational.

The development of nanotechnology has allowed the growth of heterostructures constructed on the basis of GaN, AlN, and InN that can be used for the fabrication of heterojunction LEDs. The junction is called a *heterojunction* when the interface is formed by two different materials. The junction is called a *homojunction* if both parts of the junction are made from the same material with different doping. The most frequently encountered type of homojunction is the p–n-junction, which has the same material on both sides of the junction, doped with donors (in the n-region) and acceptors (in the p-region) (the p–n-junction shown in Fig. 8.22 is a homojunction). A GaInN/GaAlN heterojunction is shown in Fig. 8.23.

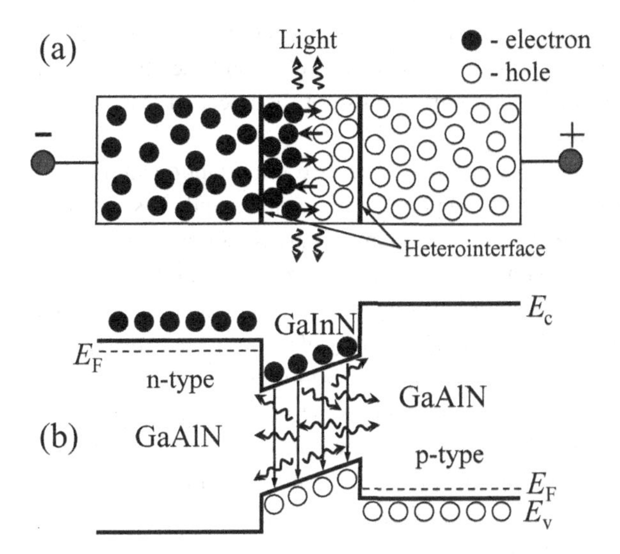

Figure 8.23 A schematic representation and band diagram of an GaInN/GaAlN light-emitting diode.

The advantage of a heterojunction lies in the possibility of controlling the doping in n- and p-regions independently of the size of the region where electrons recombine: in the case of the heterostructure shown in Fig. 8.23 the recombination region is controlled by the thickness of the GaInN layer. *Doping* is the addition of impurity atoms to a semiconductor, which defines its type of conductivity, p or n. The resistance of p- and n-regions significantly decreases when they are heavily doped. Therefore, when current flows through a more highly doped heterojunction there is less heat production in the highly doped n- and p-regions and lower currents may be used if the recombination region is made narrower than in the case of a homojunction. The concentration of electrons and holes in the narrow recombination region of the LED will be significantly higher than in the case of a homojunction and, therefore, the intensity of radiation will also be higher. The materials that can be grown by means of modern technology allow one to cover the entirety of the visible spectrum.

Lasers on quantum wells

A *laser*, which is the acronym of **l**ight **a**mplification by **s**timulated **e**mission of **r**adiation, is an *optical quantum generator*, i.e., it is a source of coherent electromagnetic radiation. Here, we will discuss lasers for the visible or close-to-visible range of the electromagnetic spectrum. Note that modern lasers cover all frequencies from X-rays (wavelength ~10 nm) up to 1 THz (wavelength ~300 μm). The operation of the laser is based on the utilization of the *stimulated emission* (or *induced emission*) of atoms, molecules, and crystalline structures. The coherence of laser radiation is due to the fact that light waves are emitted in a coordinated fashion: the emitted light waves have the same frequency, wavelength, and direction of propagation. The most compact, efficient, and reliable lasers are semiconductor lasers. Because of these properties they are used in CD and DVD players, laser printers, and computers. Telephony, the Internet, and optical and other types of cable communications got their second lease of life due to a wide use of semiconductor lasers.

The operation of any type of laser is based on two important quantum phenomena: (1) the existence of states with population inversion and (2) stimulated emission. A medium containing a system of quantum particles (atoms, molecules, or ions), whose quantum state with higher energy (*excited state*) can be populated significantly more than the state with lower energy, is called an *active medium*. The state of the active medium in which the majority of all particles of the system is in the excited state is called the *state with population inversion*. The excitation of a quantum system of particles is done by the *pump*, i.e., a constant or pulsed action on the active medium to maintain the state of population inversion of carriers. As an example, we consider here an active medium that consists of atoms rather than molecules or more complicated objects. In the active medium with population inversion photons, which either are injected into the medium or are created as a result of transitions of atoms from the excited to the non-excited state, are scattered more often by excited atoms than by non-excited atoms.

The emission radiated during the spontaneous transition of an electron in the excited atom from a higher energy level to the lower energy level is called *spontaneous emission*. Spontaneous emission, which occurs from a large number of atoms in the active medium, happens incoherently (i.e., in non-coordinated fashion), since quantum transitions of electrons with the emission of photons occur in each atom independently. However, the transition of an electron from a higher energy level, E_2, to a lower energy level, E_1, not only may be spontaneous, but also can happen under the influence of an external electromagnetic field, for example as a result of an atom's interaction with the photons which already exist in the system.

For such a type of transition the photon frequency, ω, must be equal to the frequency ω_{21}:

$$\omega = \omega_{21} = \frac{E_2 - E_1}{\hbar}. \quad (8.25)$$

Such radiation is called *stimulated* or *induced*. As a result of such an interaction of the excited atom with a photon whose frequency is equal to the frequency of the electron quantum transition from state E_2 to state E_1, two photons absolutely identical in energy, direction of propagation phase, and polarization occur (see Fig. 8.24). Thus, the electrons from the energy levels with population inversion emitting induced radiation amplify the original light beam, forming a light beam that is coherent, narrow, and polarized.

For the operation of any type of laser it is necessary to have an *optical resonator*, which is usually a system of two parallel mirrors. The role of the resonator is to confine electromagnetic radiation within the active medium and

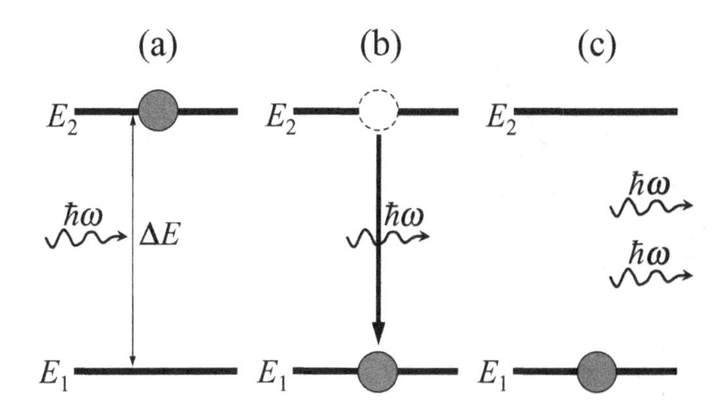

Figure 8.24 Stimulated emission. The incident photon with energy $\hbar\omega$ equal to $\Delta E = E_2 - E_1$ induces an electron from the excited level E_2 to emit a photon of the same energy $\hbar\omega$: (a) before the emission of a photon by the electron from the excited level, (b) during the emission, and (c) after the emission of a photon.

make it pass through it in the forward and reverse directions several times. The inverted state of the active medium is maintained by the pump.

To fabricate a laser on the basis of a planar semiconductor structure with quantum wells, it is necessary to fabricate two contacts through which electrons and holes are continuously injected into the active medium. Let, for example, electrons be injected through one of the contacts into the quantum well – a layer of narrow-bandgap semiconductor, surrounded by wide-bandgap semiconductor (for example, a narrow layer of InGaN in GaN). Electrons making quantum transitions from the conduction band to the valence band (i.e., electrons recombining with holes) emit photons. In accordance with the energy diagram shown in Fig. 8.25, the frequency of electron transition from the conduction band to the valence band is defined by the following condition:

$$\hbar\omega = E_{c1} + E_{v1} + E_g, \tag{8.26}$$

where E_{c1} and E_{v1} are the confinement energy levels of the electrons and holes in the quantum well, respectively, and E_g is the bandgap of the narrow-bandgap semiconductor. It is necessary to confine the radiation generated by the laser within the active medium of the device. Because of this, the refraction coefficients of the inner layers must be larger than those of the external layers. In the band diagram shown in Fig. 8.25 the external layers are the layers of the wide-bandgap semiconductor GaN. At the front and back facets of this waveguide, mirrors are deposited, which form the resonator.

Two competing processes take place in a laser during its operation. In one of them, when an electron recombines with a hole a quantum of light is emitted. In the second, some of the electrons in the ground state absorb light quanta, making the intensity and power of radiation smaller. As the current increases, the state with the population inversion is established and the process of emission begin to prevail over the absorption process. The use of lasers based on heterostructures resulted in a significant reduction of the minimum (or threshold) current at which a laser can emit coherent radiation since the injected electrons and holes

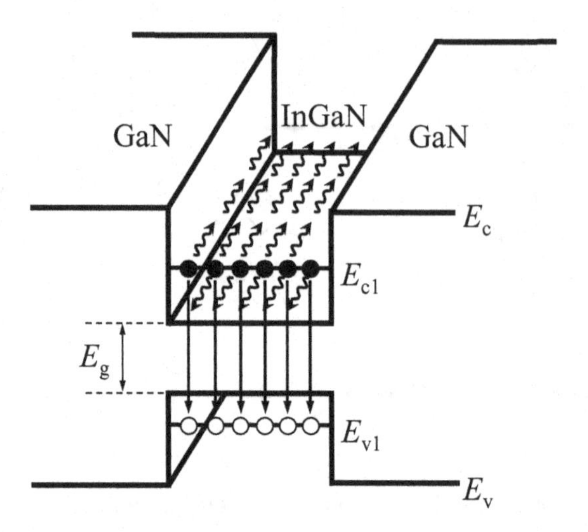

Figure 8.25 The band diagram of a semiconductor heterolaser on a quantum well.

are concentrated in the narrow region of the quantum well of the narrow-bandgap semiconductor.

Semiconductor lasers use multiple quantum wells, which form the active region, instead of one quantum well as is shown in Fig. 8.25. Lasers based on quantum wells allow one to tune the radiation frequency by controlling the energy states of quantum wells. Decreasing the width of the quantum wells results in an increase of E_{c1} and E_{v1} and consequently of the frequency of laser emission (see Eq. (8.26)). Therefore, lasers based on quantum structures are low-cost devices, use low currents, and are highly efficient at producing more light energy from the power supply. The most important parameter of any laser is its durability. Early semiconductor lasers working continuously operated for just several seconds, whereas modern lasers can operate for up to a million hours. Many groups are now developing lasers based on quantum structures with even lower dimensionality than quantum wells – lasers based on quantum dots.

Photodetectors based on quantum nanostructures

Detectors of electromagnetic signals, so-called *photodetectors*, convert light radiation into electric current. Photodetectors are used for the detection of radiation from the ultraviolet to terahertz frequency and for converting it into an electronic signal.

The operation of most photodetectors is based on the *internal photoelectric effect* – the phenomenon whereby mobile carriers (electron–hole pairs) are simultaneously generated when light is absorbed by a semiconductor (see Fig. 8.26). For light to be absorbed by the semiconductor, the energy of the incident radiation, $\hbar\omega$, must be greater than or equal to the energy bandgap of the semiconductor E_g. Following absorption of photons, the mobile electrons created in the conduction band of the semiconductor and holes created in the valence band increase the

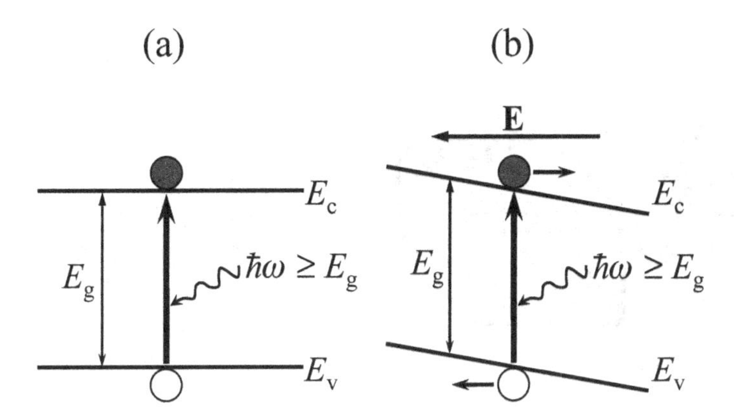

Figure 8.26 The internal photoelectric effect.

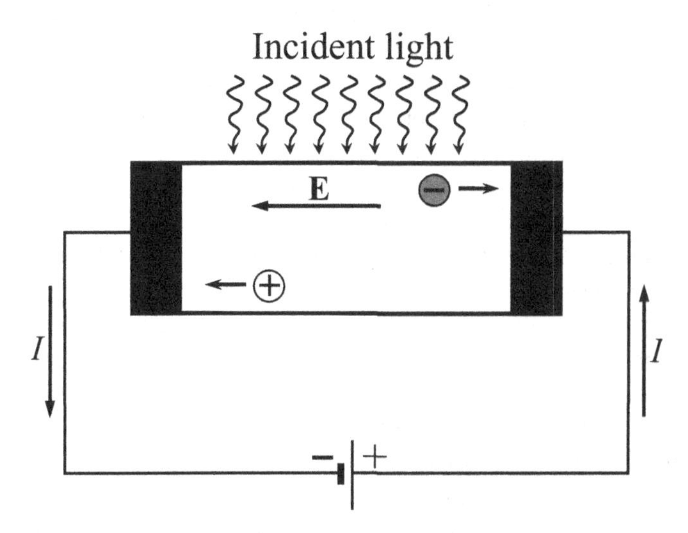

Figure 8.27 The operation of a photocell.

conductivity of the sample. The increase in conductivity after absorption of light is called *photoconductivity*. The device whose operation is based on the internal photoelectric effect is called a *photocell* (see Fig. 8.27).

Another type of photodetector is based on quantum heterostructures. They cover a wide range of the spectrum from the red into the infrared and far-infrared ranges. The operational principle of a heterostructure photodetector (for example, a photodetector based on quantum wells or quantum dots) is very simple: as a result of photo-ionization the carriers from a potential well are injected into the wide-bandgap semiconductor (potential barrier) (see Fig. 8.28). This increases the conductivity of the structure in the direction perpendicular to the layers of the heterostructure. The photoconductivity of the structure, as in the case of a simple photocell, is determined by the product of three factors: (1) the rate of optical generation, which in turn is proportional to the absorption coefficient; (2) the

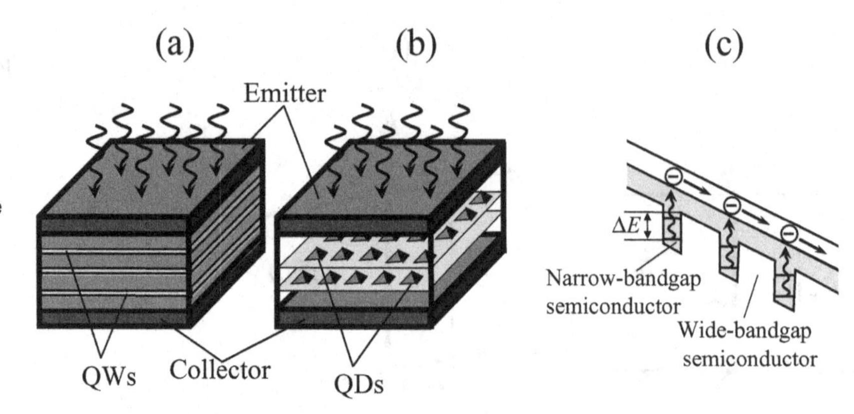

Figure 8.28 Photodetectors based on heterostructures: (a) a quantum-well photodetector, (b) a quantum-dot photodetector, and (c) the band diagram of heterostructure photodetectors. The incident light frequency, ω, must be greater than $\Delta E/\hbar$ in order to release confined electrons from the potential wells.

lifetime of carriers in the delocalized state; and (3) the effective electron mobility in the delocalized state. Heterostructure photodetectors are highly efficient. For example, photodetectors based on GaAs/GaAlAs quantum wells have quantum efficiencies close to 100%, i.e., one electron is released from the quantum well by each photon of the incident radiation.

One of the most important problems in the field of photodetectors is the development of multiwindow or multicolor photodetectors that can register radiation in different spectral ranges. The best suited devices for fabrication of multiwindow photodetectors are quantum heterostructures. In the case of photocells the only way to detect radiation with different wavelengths is to use different semiconductor materials with different bandgaps. The development of photodetectors based on heterostructures allows us to change the region of spectral sensitivity not just by changing the material of the structure as is done with more conventional photodetectors (photocells) but also by changing the width of the quantum wells or the size of quantum dots. This opens the possibility of using, for example, semiconductor heterostructures with different thicknesses of quantum wells to fabricate monolithic multicolor photodetecting elements that are sensitive for several specific spectral regions.

8.3.4 Quantum-dot cellular automata

The demand for the development of logic devices with maximum density of logic elements and minimum energy consumption for a single on/off switch led to the proposition to use in logical elements conducting islands of very small size – *quantum dots*. In such devices arrays of interacting quantum dots can be used for the implementation of logical Boolean functions. These new devices were called *quantum-dot cellular automata* (QCAs). The basic element of this device is a *cell*, which consists of four or five quantum dots.

As an example, Fig. 8.29 shows a cell consisting of five quantum dots: four dots are located at the four corners of a square and one dot is at the cell's center.

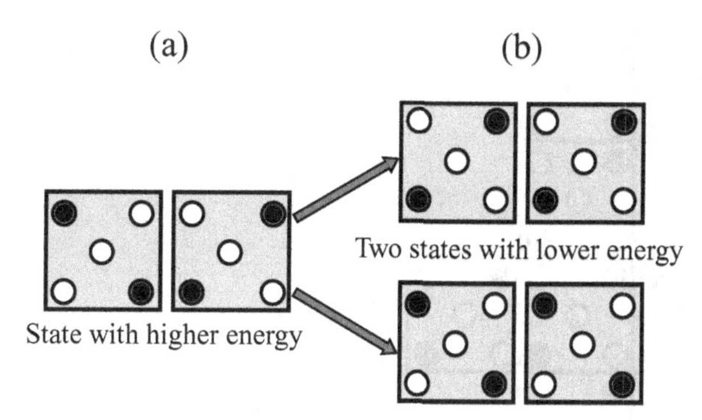

Figure 8.29 Quantum-dot cellular automata.

By applying a gate voltage to the cell, two electrons are injected into quantum dots, which makes the cell negatively charged. The quantum dots in the cell are arranged in such a way that tunneling through the central dot is possible. Because of the electrostatic repulsion between the two additional injected electrons, the whole system will have its minimum energy only in the case when the electrons are as far from each other as possible, i.e., when they are located at the corners along a diagonal. Since there are two such stable states (or two *polarizations*), one of them can be considered as a *logic "1" state* (see the upper part on the right-hand side of Fig. 8.29) and the other as a *logic "0" state* (see the lower part on the right-hand side of Fig. 8.29). When the system transfers from one stable state to another, the polarization of the system is changed and the distribution of electric fields around the cell is changed also. With the help of additional electrodes connected to the cell capacitatively, it is possible to switch the cell into the desired state – to put it into the logical state "1" or "0". If we place near the first cell a second cell, which also has two additional electrons, then the electrostatic field of the first cell will force the electrons of the second cell to move in order to minimize the electrostatic energy of the entire system (see Fig. 8.29).

By making combinations of different cells it is possible to realize various logic functions and carry out necessary logic transformations and computations. Figure 8.30 shows one of the examples of cell combinations, whose output state is defined by a majority of input states (this is the so-called "*majority*" logic function). Various combinations of cells have been suggested for the realization of different logic operations. On the basis of these logic elements the creation of a new "nanocomputer" is possible. It is important that the positional relationship of cells provides the transmission of a logic signal without charge transport along the chain of cells, i.e., only the polarization of the state is transmitted along the chain of cells.

The advantage of logic devices based on quantum-dot cellular automata is that they require significantly less volume of active region than do their counterparts based on field-effect transistors (FETs). For example, a summator on the basis of

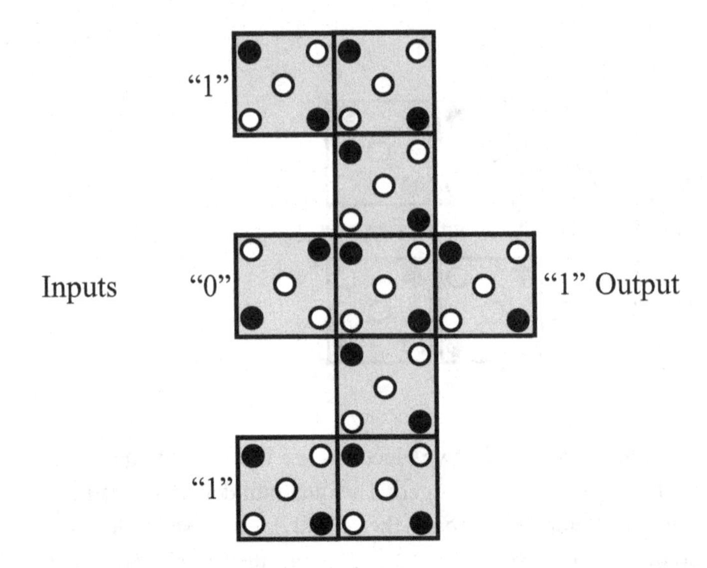

Figure 8.30 Combination of cells, when the output state is defined by the majority of input states: three inputs, 1, 0, and 1, and one output, 1.

cellular automata with the size of the quantum dot equal to 20 nm can be built on an area of 1 μm^2, whereas the same area is occupied practically by one FET. To build a summator on the basis of FETs, it is necessary to use approximately 40 transistors. If we take into account the area of connections between the transistors, which is even greater than that of the active devices, then the advantages of using cellular automata become apparent.

Computation in devices based on cellular automata is done by the transition of the whole array of cells to the state with minimum energy (or to the *ground state*). Since complex computational devices contain many cells, the state with minimum energy can be achieved in different ways. This may lead to errors in computations. Moreover, these systems are very sensitive to external factors and, therefore, they require very strict control of the external conditions. If the ambient temperature increases, the computational process can be disrupted. For a cell whose quantum-dot size is about 20 nm, the change of energy during quantum-dot recharging is about 1 meV, which corresponds to a temperature of about 12 K and is substantially lower than the 300 K of room temperature. As in the case of a single-electron transistor, the operation temperature of cellular automata can be increased by decreasing the cell's size (and correspondingly decreasing the size of each quantum dot). Unfortunately, there exists a significant roadblock, which has to be removed for successful operation of devices based on cellular automata. Since the electrostatic field of the cell affects the state of the neighboring cell in the output direction as well as in the input direction, signal propagation can, because of random influences, be not only from input to output, but also in the opposite direction. To get rid of this problem, the use of quantum-dot devices whose direction of signal propagation is controlled by the external field was suggested. Practical realization of these devices is in the initial stage and requires the solution of series of problems that are mainly technological.

Example 8.3. Find the temperature equivalent of the laser's active medium, which is responsible for the state with population inversion of energy levels: $N_2 > N_1$. Energy level E_2, which corresponds to the level population N_2, is higher than E_1, which corresponds to the level population N_1.
Reasoning. Under conditions of thermodynamical equilibrium the lowest energy level of atoms of the active medium, E_1, is more populated than the energy level E_2. In accordance with the classical law of the particles' distribution over the energy (i.e., Boltzmann's law) the number of particles on a particular energy level, E_i, is defined as

$$N_i = C\mathrm{e}^{-E_i/(k_\mathrm{B}T)}, \tag{8.27}$$

where C is a constant that depends on the number of particles per unit volume of active medium. In the inverted state the thermodynamical equilibrium of the active medium is violated. However, if we apply to this state Boltzmann's distribution law, Eq. (8.27), then the ratio of numbers of particles on corresponding levels will be defined by the following expression:

$$\frac{N_2}{N_1} = \exp\left(-\frac{E_2 - E_1}{k_\mathrm{B}T}\right) = \exp\left(-\frac{\hbar\omega_{21}}{k_\mathrm{B}T}\right), \tag{8.28}$$

where the frequency of transition $\omega_{21} > 0$. From the last expression we can find the temperature of the active medium in the inverted state:

$$T = -\frac{\hbar\omega_{21}}{k_\mathrm{B}\ln(N_2/N_1)}, \tag{8.29}$$

which in this case becomes negative. Note that the term "negative" can be applied only to non-equilibrium states, which is the case for a state with population inversion.

8.3.5 Josephson electronics

Quantum effects play a fundamental role in the operation of superconducting elements, which include the *Josephson junction*. The Josephson effect can be observed in the current–voltage characteristic of two superconductors separated by a narrow (about 1 nm) dielectric layer. When a sufficiently weak current flows through the Josephson contact there is no voltage at the contact, i.e., the current is purely superconducting. The existence of superconducting current is due to the existence of Cooper pairs, which are only partially destroyed when they pass through the thin non-superconducting layer. This phenomenon, called the *stationary Josephson effect*, was experimentally discovered in 1963. When the current through the contact is increased, an alternating voltage occurs between two superconducting films. This effect is called the *AC Josephson effect*. Small capacitance and the ability to switch from zero to non-zero voltage while exceeding the critical current allows one to use Josephson contacts as logical elements of computers. One or several Josephson contacts incorporated into a conventional electric circuit can provide automatic transformation from analog presentation of information to a discrete one. All free electrons in superconductors form *Cooper pairs*, which have double the electron charge. Since each Cooper pair has integer spin, the ensemble of such quasiparticles obeys Bose–Einstein statistics rather than the Fermi statistics for free electrons; therefore, all of the quasiparticles become correlated. The current created by them and the magnetic flux created by this current are quantized, i.e., within the ring made from Josephson's contacts and connected in parallel, only an integer number of electron wavelengths can be contained. Within such a ring only some multiple of an integer number of quanta of magnetic flux, equal to $\Phi_0 = \pi\hbar/e = 2.07 \times 10^{-15}$ Wb, can exist.

Elements of fast one-quantum logic, where the information unit is a quantum of magnetic flux, allow one to process signals with frequency higher than 100 GHz with a very low level of energy dissipation. It is especially important that such a structure is simultaneously a logical element and a memory cell. Since the volume of data transmitted through the Internet doubles every 3–4 months, even the best of semiconductor devices currently developed could not in the near future process such a huge data flow as will be needed. Three-dimensional structures built from Josephson electronic circuits stacked together seem to be the only alternative to planar semiconductor electronic circuits.

Nanostructure Josephson electronics is better suited as a physical medium for the construction of a quantum computer. On the basis of a two-dimensional net of Josephson contacts, a new type of computer memory can be developed. This memory is not based on the traditional logic but instead uses associative logic, distributed over the entire structure's memory in a manner similar to the neural networks of living organisms. Such a system will be able to recognize images, make quick decisions in multifactor situations (for example, in the economy, defense, and space exploration) in real time without considering all possible variants. Cryogenic electronics built on superconductors cannot compete with the traditional semiconductor electronics found in existing applications. The main purpose of cryogenic electronics is to provide the basis for a new generation of supercomputers and high-performance supporting telecommunication systems, whose development will be commercially profitable despite the necessity of cooling.

Many different types of Josephson elements and devices have been developed for use as logic elements and memory cells, devices for quantum encryption and data transfer, generators and detectors of millimeter and submillimeter waves, highly sensitive sensors of magnetic field, electric charge, voltage, current, heat flow, and so on. Josephson detectors for the registration of weak signals have a sensitivity of the level of the fundamental quantum limit, i.e., four orders of magnitude higher than that of traditional semiconductor devices. This allows one to use them for the development of non-contact devices for medical diagnostics, for example, magnetocardiographs and magnetoencephalographs. Today's agenda is the development of magnetic tomography, which will allow us to look in real time at the functioning of internal organs, the development of a fetus, and so on, by analyzing magnetic field data.

8.3.6 Spintronics

The term "spintronics" originates from the contraction of "spin electronics." Spintronics is a field of science and technology at the crossroads of nanoelectronics and nanomagnetics. Spintronics studies the properties of magnetic nanostructures and the peculiarities of the interaction of spins of electrons and nuclei with electromagnetic fields. Spintronics has produced devices used in magnetic memory, logic, and sensor systems with ultra-high sensitivity that use spin and spin properties for their operation. As we have already learned, the electron has a specific quantum property called *spin*, i.e., internal angular momentum. Its projection onto any chosen direction can have only two values: $+(1/2)\hbar$ and $-(1/2)\hbar$. If we place electrons in a magnetic field, then their spins will line up along the field. After the magnetic field has been switched off, all spins in a system of non-interacting electrons will return to their previous states, while the interacting electrons can preserve their states. The attractiveness of spintronic devices is based on the fact that the spin flip practically does not require energy. In between operations the device disconnects from the power supply. If we change the spin direction, the electron kinetic energy will not change. This means that practically energy will not be dissipated due to a change of spin direction. The speed of spin-position change is very high. Experiments have shown that the spin flip takes several picoseconds. Since electron spin projection has only two possible values, it is natural that engineers want to use these states for logic applications, especially because it is expected that silicon processors will reach the limits of their possibilities in the next 10–15 years.

Interest in spin electronics increased in 1988 with the discovery of the giant-magnetoresistance effect in multilayered magnetic Fe/Cr nanostructures, whose total thickness was about 100 nm. It was discovered that the resistance of the multilayer Fe/Cr structures drops by more than 50% in an external magnetic field. Since the decrease of the resistance was so large, the researchers called this effect *giant magnetoresistance (GMR)*.

The GMR effect is based on the phenomenon that electrons with different spin directions (and, respectively, with different internal magnetic moments) move differently in an external magnetic field. The discovery of GMR allowed the fabrication of high-precision magnetic-field sensors, detectors of angular rotation, and, most importantly, read heads of hard disks. The first GMR heads, with a density of writing of 2.69 Gb in^{-2}, were produced by IBM in 1997 and nowadays they are used in practically all hard disks. Let us consider the structure

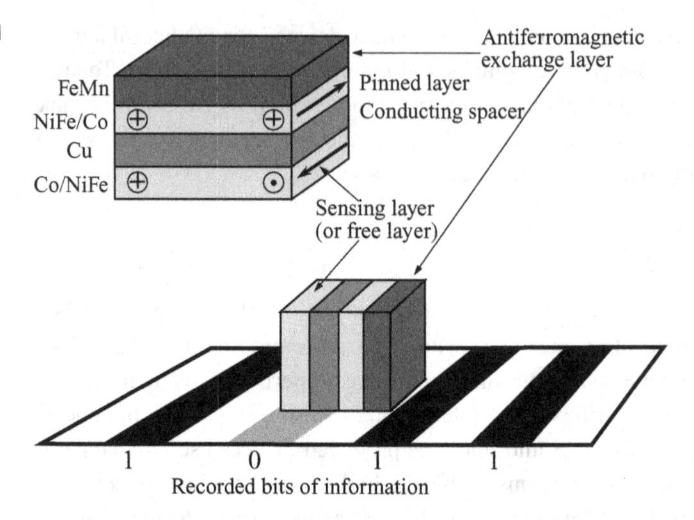

Figure 8.31 Layers of read magnetic head.

of a magnetic read head, which consists of four layers. The upper layer, which is antiferromagnetic, is called the *exchange layer* and is designed to fix the direction of the magnetic field of the second layer, which is called, therefore, the *pinned layer*. To have the necessary magnetic properties, the second layer is made of ferromagnetic material (alloys of nickel, iron, and cobalt). The magnetic field of the pinned layer is always directed in one direction defined by the exchange layer, as shown in Fig. 8.31. The third layer, which is called the *conducting layer*, is usually made of copper. The last layer, called the *sensing layer*, is made of ferromagnetic material. In contrast to that in the pinned layer, the direction of the magnetic field in the sensing layer is controlled by the external magnetic field created by the memory cell of the hard disk, which contains one bit of information. Each memory cell on the hard disk has two different orientations of the magnetic field that correspond to "0" and "1" states of the recorded information. The orientation of the magnetic field in the sensing layer changes depending on the state of magnetization of the cell.

If the orientations of the magnetic field in the sensing layer and in the pinned layer coincide, then the resistance of a sensor decreases to a minimum value. This is because electrons whose directions of spins coincide with the direction of magnetic field do not experience significant resistance and easily pass through all four layers. The electrons whose spins are directed against the magnetic field experience significant resistance in both ferromagnetic layers. If the orientations of the magnetic field in the sensing layer and in the pinned layer have opposite directions, then, regardless of spin orientation, electrons will experience significant resistance in one of the ferromagnetic layers. This effect has a purely quantum origin since its existence is directly connected with the existence of electron spin. The interaction of electron spins with a magnetic field provides

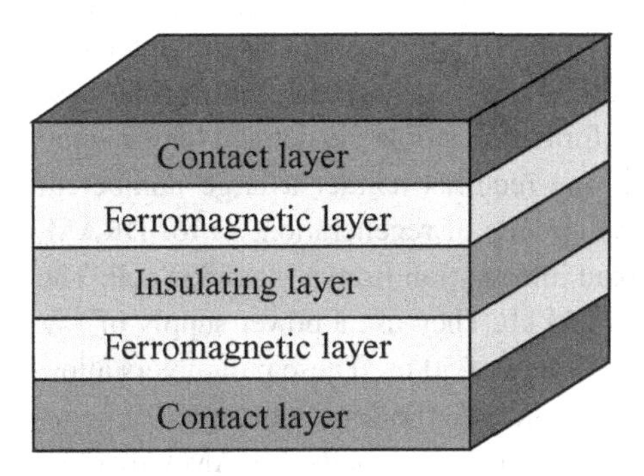

Figure 8.32 The structure of a magnetic tunnel junction (MTJ).

magnetic read heads whose operation is based on the GMR effect with very high sensitivity.

High-precision angle, position, and speed sensors based on the GMR are widely used in automobiles – for example, in an anti-block braking system (ABS), which provides straight-line motion of the vehicle while braking on a slippery pavement. Modern computer, television, and video technologies are hard to imagine without spintronic devices. Besides hard disks, spintronic devices can be found in high-definition television equipment and in DVD drives. Apart from in GMR devices, spin is used also in *magnetic tunneling junctions* (MTJs). The structure of a valve built on the basis of an MTJ consisting of two ferromagnetic layers separated by the insulator (for this purpose usually aluminum oxide is used) is shown in Fig. 8.32.

It looks as if electric current, which flows perpendicularly to the ferromagnetic layers, will not flow through such a structure at all. However, because of the tunneling effect electrons are able to move from one ferromagnetic layer, through the insulator, to another ferromagnetic layer. In MTJ valves, together with the tunneling effect, the fact that electrons with different spins move differently in a magnetic field is used. If the magnetic fields in the two ferromagnetic layers of the valve coincide directionally, then those electrons whose spin is directed along the magnetic field can easily tunnel from one ferromagnetic layer to another. Therefore, the resistance of the valve is very low in this case. If the magnetic fields in these layers are directed in opposite directions, then the tunneling transmission would be more difficult, since in one of the layers electrons will be moving against the magnetic field. Because of this, the resistance of the valve increases by 20%–30%.

A prospective application of spintronics known as *magnetic random-access memory* (MRAM) utilizes an MTJ. Its distinctive feature is that information is saved neither in the form of electric charges on the capacitor plates, as is done in dynamic memory DRAM, nor in the form of a trigger state as in static random-access memory (SRAM), but in the form of the magnetization of a ferromagnetic layer. The advantages of this type of memory are its low energy consumption, high speed of operation, and gigantic density of information storage. The maintenance of the magnetization state does not require energy consumption, which is the case with hard disks. The high density of information storage is defined by the small size of the memory cells, for which one requires neither a large number of transistors, as for SRAM, nor separate circuits of regeneration, as for DRAM. An MTJ can be used for MRAM to read information from a magnetic cell. The latest models of MRAM have reached 256 kb. They use a power supply of 3 V and provide a duration of the write/read cycle of 50 ns. These parameters allow MRAM to compete directly with flash-memory and offer the prospect of being able to compete with conventional dynamic memory. Finally, MRAM will allow the creation of computers whose dynamic memory data will not be lost when the power supply is turned off.

The potential of spintronics is not exhausted by the technologies which have already been developed. Despite the fact that research in this direction has been going on for more than 10 years, there are many unsolved scientific and technical problems. For example, to change the magnetization of some area of ferromagnetic material, first of all it is necessary to have a magnetic field. Since we can create a magnetic field by the use of an electric current (we do not take into account permanent magnets) there is a problem of localization of the magnetic field in the limited area. The smaller this area, the higher the density of information storage on the magnetic carrier which could be achieved. Recently, a particular experiment that showed the possibility of changing the magnetization with the help of spin-polarized electrons was carried out. With the help of photoemission from the semiconductor cathode, which is caused by polarized light, a beam of spin-polarized electrons was produced, which in its turn was transmitted through a magnetic film of width several nanometers. When electrons passed through the film, their spins changed, together with the spins of electrons in the magnetic film, which means that the magnetization of the film also changed. If the number of electrons transmitted through the film is comparable to the number of atoms of the material, then the change of magnetization of the film will be considerable. This effect can be used for writing information as well as for reading information with a lower intensity of the electron beam. Potentially this technology can provide a remagnetization speed (i.e., the speed of reading/writing of information) up to tens of GHz.

Another interesting effect is the formation of a pure spin flux of electrons without charge transport. In this experiment two colliding beams with oppositely directed spins were formed. Thus, the transfer of spin charge was achieved without an applied voltage. So far this phenomenon has been observed only at distances of about several nanometers, nevertheless research in this direction is continuing. One of the most important problems of spintronics is connected with the materials that are used. In spintronics you have to use ferromagnetics, whose magnetic properties give rise to various effects related to the spins of electrons. But ferromagnetic materials are metals, while modern electronics is based on semiconductors. It is precisely the properties of semiconductors which allow one to amplify current in transistors, whereas in metals this effect is impossible. Therefore, in order to create a device that uses electron spin as well as electron charge, it is necessary to have a ferromagnetic material that is simultaneously a semiconductor. Research in this direction is very active and new magnetic semiconductors that do not lose their magnetic properties even at room temperatures have been fabricated. One such material is titanium dioxide (TiO_2) with cobalt (Co) as impurity, which has been grown in the form of nanometer films by molecular-beam epitaxy. Under high-vacuum conditions the beams of atoms in certain proportions are directed onto a crystalline surface where they form the desired crystalline structure. Despite some problems with the growth of homogeneous films, it is a very tempting material for the development of new spintronic devices. Another similar material is an epitaxial film of alternating thin layers of GaSb and GaMn. The magnetic properties of this semiconductor are preserved up to 130 °C, which is quite high enough for the needs of modern electronics.

Using such materials, spintronics will allow us to process and store information on the same wafer, which will lead to an increase in the speed of operations and to a decrease in energy consumption. Integration of the achievements of electronics and spintronics will open new horizons in the development of traditional computers as well as quantum computers and may extend the time of validity of Moore's law.

8.3.7 Graphene and carbon nanotubes

The research group at Georgia Institute of Technology announced in 2006 that they had fabricated a field-effect transistor based on graphene and a device that uses the phenomenon of quantum interference. They stated that in the near future a new class of graphene nanoelectronics with a characterististic thickness of transistors equal to 10 nm will emerge. It is impossible to use graphene directly to create a field-effect transistor because graphene does not have a bandgap. This means that we cannot achieve a substantial difference in resistivity at any applied gate voltage, i.e., it is not possible to create two states for binary logic – conducting and non-conducting states. Therefore, for device fabrication it is necessary first to create a gap of substantial width at room temperature in order to reduce the input into the conductivity of the thermally excited electrons. One of the possible ways to do this is by the formation of narrow graphene ribbons with a width that provides quantum-dimensional effects in the plane of the sample. At the same time, the width of the bandgap, E_g, must be large enough to provide the device with a non-conducting state (turn-off state) at room temperature (the width of 20 nm possessed by the graphene nanoribbon corresponds to a bandgap width of about 30 meV). Owing to the high mobility of electrons in graphene, the operating speed of such a transistor may be substantially higher than that of the silicon transistors.

When depicting the band structure of graphene it is necessary to take into account the fact that the outer shell of the carbon atom contains four electrons, which form bonding with the neighboring atoms of the lattice by creating three sp^2-hybridized orbitals. The fourth electron is in the $2p_z$-state. It is precisely the $2p_z$-state which is responsible in *graphite* for the *formation of interlayer bondings*, whereas in *graphene* the $2p_z$-state is responsible for the *formation of energy bands*. The bonding energy of atoms in a crystalline lattice rapidly decreases with increasing interatomic separation. Therefore, the main input into the bonding energy of an atom with other lattice atoms and the formation of band structure arise from the atom's interaction with its nearest neighbors. In the electron tight-binding approximation, the energy spectrum of electrons in graphene has the form

$$E = \pm\sqrt{\gamma_0^2\left[1 + 4\cos^2(\pi k_y a) + 4\cos(\pi k_y a)\cos(\pi k_x \sqrt{3}a)\right]}, \tag{8.30}$$

where the $+$ sign corresponds to electrons and the $-$ sign to holes, and γ_0 is the overlap integral. This expression is obtained by taking into account interactions with the nearest neighboring atoms only.

From Eq. (8.30) it follows that near the points of intersection of the valence band and the conduction band the dispersion law for the carriers in graphene can be written in the following form:

$$E = \pm\hbar v_F k, \tag{8.31}$$

where v_F is the Fermi velocity (its experimental value is $v_F = 10^6$ m s^{-1}) and $k = \sqrt{k_x^2 + k_y^2}$ is the modulus of the wavevector in two-dimensional **k**-space. The photon, whose mass at rest is equal to zero, has a similar spectrum. Therefore, we can say that electrons and holes with energy defined by Eq. (8.31) have effective mass equal to zero too. The Fermi velocity, v_F, plays the role of the effective speed of light.

Transistors constructed on the basis of carbon nanostructures, as experts expect to exist in the very near future, will exceed the parameters of the conventional silicon devices. Up to now the efforts of researchers have been concentrated on the fabrication of single transistors on the basis of carbon nanotubes. Researchers from IBM recently announced that they had fabricated the first integrated circuit (IC) constructed on the basis of a single carbon nanotube. The importance of this achievement lies in the fact that the IC was fabricated using standard semiconductor-technology processes. This is an important step towards integration of the new technology with the existing electronic technologies. The researchers from IBM developed a ring oscillator – an IC with the help of which the developers can check the possibilities of new production processes or of new materials.

By fabricating an IC on a single carbon nanotube the researchers from IBM have achieved a speed of operation a million times higher than that of numerous ICs on nanotubes, which had been developed earlier. Despite the fact that this speed is still lower than that which the latest silicon devices have, these researchers from IBM are confident that new nanotechnologies will help to uncover the enormous possibilities of carbon nanotube electronics.

Lately an active development of memory modules constructed on the basis of carbon nanotubes has been undertaken. Such modules may become an alternative to traditional flash memory. Currently, energy-independent flash memory is used in various compact systems. In the main types of flash memory double-gate MOS transistors are used as elementary memory cells and it is expected that in future memory cells will be based on carbon nanotubes.

8.3.8 Nanoelectromechanical systems

Nanoelectromechanical systems (NEMSs) are integrated nanodevices or systems that combine electrical and mechanical components, and are fabricated using technologies that are compatible with the production of integrated circuits. To realize advanced scientific ideas in this field, it is necessary to utilize the achievements of various branches of nanotechnology.

An NEMS is a miniature system that has sensory, computational, and productive functions. *Sensory functionality* concerns the ability to define the existence and concentration of chemical or biological components in the environment. *Computational functionality* means that the NEMS has a processor, which carries out the necessary computations and prepares information for processing. *Productive functionality* means that the NEMS has nanoscale actuating mechanisms, so NEMSs can be used as controllers. Therefore, NEMSs usually contain a combination of two or more devices operating on the basis of electrical, mechanical, optical, chemical, biological, magnetic, and other properties, which are integrated on a single chip or on a multichip board. Work on NEMSs has rapidly become one of the leading fields of cutting-edge nanotechnology. Autonomous NEMSs, e.g., nanorobots, are actively used in manufacturing, medicine, and ecology.

Among the various elements of an NEMS, a very important role is played by the load-bearing elements, which can be made of carbon nanotubes. The controlled relative motion of layers in multiwalled nanotubes and their unique elastic properties makes it possible to use them as moving elements of an NEMS.

Devices based on the relative motion of carbon nanotubes have recently been suggested: rotational nanobearings, nanogears, nanoswitches, GHz oscillators, Brownian nanoengines, nanorelays, and also a nanonut–nanobolt pair. Moreover, there are experimentally developed nanomotors, which use as their axle and bush the layers of multiwalled carbon nanotubes.

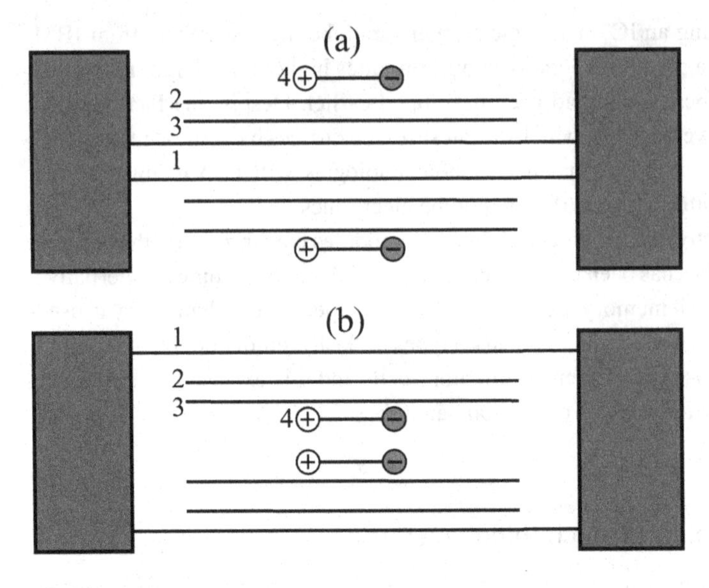

Figure 8.33 Schematic diagrams of nanoactuators: (a) with an inner layer as the stator and (b) with an external layer as the stator.

It has recently been shown that a two-walled carbon nanotube with non-zero chiral angle can be used as a nanonut–nanobolt pair (as can be seen from Fig. 8.5, the surface of a chiral nanotube is reminiscent of a thread on a bolt), i.e., the motion along the nanotube axis of the stator is transformed into rotation of the rotor (outer layers 2 and 4). This can be used for transformation of the force directed along the nanotube axis into the mechanical torque, which causes relative rotation of layers. Such a pair can be used in a so-called *nanoactuator* – a device that makes an NEMS able to move. The operational principle of such a device is the same as that of a humming-top toy. The principle of a nanoactuator is shown schematically in Fig. 8.33. In Fig. 8.33(a), 1 denotes the inner layer of a nanotube, whereas in Fig. 8.33(b) 1 denotes the outer layer of a nanotube. It serves as the stator of the nanomotor and is fixed. Other layers, labeled 2 and 3, serve as the rotor. The relative position of these layers has to be fixed. The stator and rotor serve as a rotational nanobearing. We are considering now the motor depicted in Fig. 8.33(a), but the operation of the motor shown in Fig. 8.33(b) is analogous. The external layers serve for transformation of the force applied to layer 4 and directed along the nanotube axis into torque, which rotates the rotor. Such a transformation is possible if layers 2 and 4 form a nanonut–nanobolt pair. The charges on the ends of layer 4 can be produced as a result of chemical adsorption and can be used for the control of a nanoactuator by an external electric field.

We can force the internal part of a two-walled nanotube to oscillate. In order to do this, the external layer has to be fixed (see Fig. 8.34). The inner layer has to be subjected to additional processing. At one end positive hydrogen ions are absorbed and at the other end negative fluorine ions. Under the influence of a

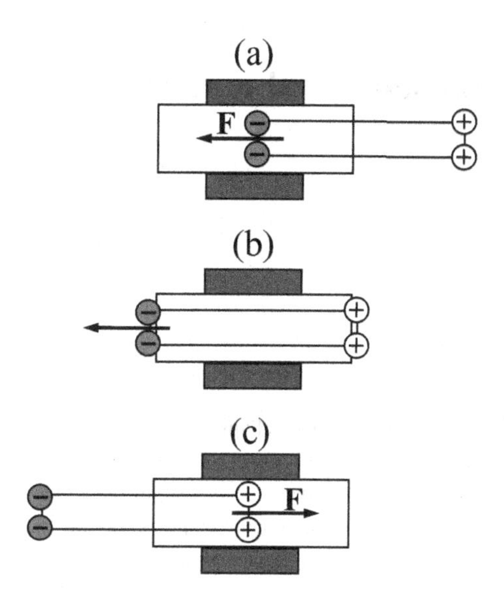

Figure 8.34 Subsequent positions of the two layers of a two-walled nanotube during the half-period of oscillations: (a) and (c) show the maximum amplitude of telescopic protrusion of the inner layer of the nanotube under the action of the force **F**.

non-uniform electric field the inner tube, which is a dipole, can move relative to the external tube, as shown in Fig. 8.34. This oscillatory motion can also be transformed into a rotational motion.

Currently there are numerous applications of nanotechnology. With its help, new anticorrosive coatings are made, nanostructures allow a significantly increased lifetime of steel-work, provide reliable operation of electricity-supply networks, reduce costs of equipment operation, and so on. According to some scientists, in the future nanotechnology could save mankind. When we have exhausted all the oil, coal, and gas on the Earth (which may be even sooner than we expect), people will be using nanotailored solar batteries that will be able to use solar energy for the production of hydrogen, oxygen, and electricity. In future biology and medicine nanoparticles will deliver medicine to the required parts of the human body. For example, nanoparticles mixed with blood will arrive at a tumor and then, by exposure to infrared radiation, the tumor will be eradicated because of overheating caused by absorption of radiation by the nanoparticles. Nanotechnology for space exploration has great prospects. New nanomaterials will make flights not only to the Moon but also to Mars a reality. One more perspective is the utilization of powdered nanotechnology for the development of a new generation of fuel cells that can produce electrical energy from any organic fuel, with an efficiency no less than 60%–70%. It is impossible to mention or foresee all the possible applications of nanotechnology. We can only mention that, according to some experts, during the next decade the world market of nanotechnology will reach the size of a trillion dollars and maybe will exceed the value of all markets related to electronics, medicine, chemistry, and energy altogether.

Appendix A
Classical dynamics of particles and waves

In this appendix we will study the basics of the classical mechanics of an individual particle as well as a system of particles. We will treat, here, a particle as a material point, i.e., an ideal object that has a mass and an infinitesimally small size. The importance of this chapter is defined by the fact that any macroscopic structure or medium can be considered as an array of material points interacting according to certain laws.

The famous three laws of classical dynamics formulated by Isaac Newton more than three centuries ago were for a long time considered as the basis for the description of motion of all objects – ranging from nanoparticles to cosmic objects. With the development of physics, it became clear that the region of validity of the classical description of a particle's motion is limited. Newton's mechanics cannot be applied for the description of motion with velocities close to the speed of light. For the description of such a motion, relativistic mechanics was introduced at the beginning of the twentieth century. The typical formulae of relativistic theory, in contrast to the corresponding formulae of classical mechanics, contain the radical $\sqrt{1-(v/c)^2}$, which contains the square of the ratio of the particle's velocity, v, to the speed of light in vacuum, c. For a wide group of problems the ratio $v/c \ll 1$; therefore, usage of the classical description at velocities that are lower than the speed of light by two orders of magnitude leads to an error that is significantly smaller than the error in measurements of the corresponding physical magnitude. However, Newton's mechanics cannot be applied for the description of elementary particles. The motion of electrons in atoms, molecules, nanostructures, semiconductors, and metals is not governed by the laws of classical mechanics. Nevertheless, the role of the laws of classical mechanics in the modern physics of the nanoworld is still significant. These laws define the motion of heavy particles, for example the motion of atoms or ions, and the wave motion of atoms or ions in crystalline structures.

In the current appendix we will also study free harmonic and damped oscillations of an individual particle, discuss their governing equations, and find their solutions. We will consider small-amplitude oscillations of a particle, with harmonic and anharmonic approximations for the description of oscillations for an arbitrary profile of the potential well.

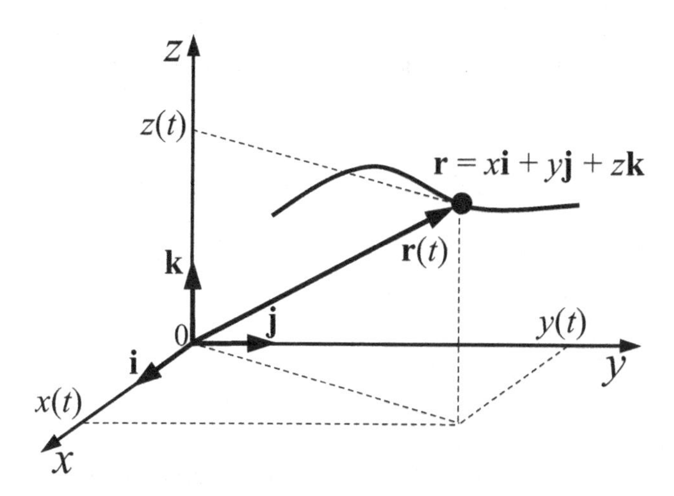

Figure A.1 The Cartesian system of coordinates and a particle's trajectory, $\mathbf{r}(t)$.

A.1 Classical dynamics of particles

A.1.1 Newton's second law of motion

According to the classical description of particle motion, a particle's location in space at any instant of time, t, is defined by three spatial coordinates, which depend on the choice of the system of coordinates. One such system – the Cartesian coordinate system – consists of three mutually orthogonal axes x, y, and z that intersect at the origin of coordinates ($x = y = z = 0$, with respect to which $x(t)$, $y(t)$, and $z(t)$ define the location of the particle at any given time t (see Fig. A.1)). For convenience the position of a particle can be written in the form of its radius vector, $\mathbf{r}(t)$:

$$\mathbf{r}(t) = \mathbf{i}x(t) + \mathbf{j}y(t) + \mathbf{k}z(t), \tag{A.1}$$

where $\mathbf{i}$, $\mathbf{j}$, and $\mathbf{k}$ are unit vectors directed along their respective coordinate axes. Since these vectors are perpendicular to each other, their scalar products are:

$$\mathbf{i}\cdot\mathbf{i} = \mathbf{j}\cdot\mathbf{j} = \mathbf{k}\cdot\mathbf{k} = 1 \quad \text{and} \quad \mathbf{i}\cdot\mathbf{j} = \mathbf{i}\cdot\mathbf{k} = \mathbf{j}\cdot\mathbf{k} = 0. \tag{A.2}$$

The array of points defined by the vector $\mathbf{r}(t)$ along a particle's motion defines the trajectory of the particle (Fig. A.1).

The change of a particle's state is described by Newton's second law, which in its general form can be written through the *momentum of a particle*, $\mathbf{p}$:

$$\mathbf{p} = m\mathbf{v}, \tag{A.3}$$

where

$$\mathbf{v} = \frac{\mathrm{d}\mathbf{r}}{\mathrm{d}t}. \tag{A.4}$$

Here, m is the mass and $\mathbf{v}$ is the velocity of the particle. According to Newton's second law of motion, the derivative of the momentum $\mathbf{p}$ with respect to time t is equal to the sum of all forces acting on the particle:

$$\frac{\mathrm{d}\mathbf{p}}{\mathrm{d}t} = \mathbf{F}, \tag{A.5}$$

where the *resultant force*, $\mathbf{F}$, is defined by a vector sum of all forces $\mathbf{F}_k$ acting on the particle:

$$\mathbf{F} = \mathbf{F}_1 + \mathbf{F}_2 + \cdots + \mathbf{F}_n = \sum_{k=1}^{n} \mathbf{F}_k, \tag{A.6}$$

where n is the total number of forces acting on the particle.

Now, let us discuss some of the forces existing in nature that can act on a particle, which has mass, m, and charge, q.

(a) Gravitational force, $\mathbf{F}_{\mathrm{gr}}$:

$$\mathbf{F}_{\mathrm{gr}} = m\mathbf{g}, \tag{A.7}$$

where $\mathbf{g}$, called the *acceleration due to gravity*, denotes the free-fall acceleration caused by the gravitational field at a point close to the Earth's surface where the particle is located. The acceleration due to gravity, $g = |\mathbf{g}|$, is approximately 9.8 m s^{-2} and is directed towards the center of the Earth.

(b) In electric and magnetic fields, the Lorentz force, $\mathbf{F}_{\mathrm{L}}$, is a sum of electric and magnetic forces applied to a particle of charge q:

$$\mathbf{F}_{\mathrm{L}} = q\left(\mathbf{E} + \mathbf{v} \times \mathbf{B}\right), \tag{A.8}$$

where $\mathbf{E}$ is the *electric field strength* and $\mathbf{B}$ is the *magnetic flux density*. The vector product, $\mathbf{v} \times \mathbf{B}$ (which is part of Eq. (A.8)), is given, according to the definition of the vector-product operation, by the third-order determinant:

$$\mathbf{v} \times \mathbf{B} = \begin{vmatrix} \mathbf{i} & \mathbf{j} & \mathbf{k} \\ v_x & v_y & v_z \\ B_x & B_y & B_z \end{vmatrix} = \mathbf{i}\left(v_y B_z - v_z B_y\right) + \mathbf{j}(v_z B_x - v_x B_z) + \mathbf{k}\left(v_x B_y - v_y B_x\right), \tag{A.9}$$

where we used the general rule for finding determinants. Let us note that because of the small masses of the electron and proton the gravitational force is several orders of magnitude smaller than the force of the electric field under laboratory conditions, even if the strength of the electric field is weak. For an electron these forces become of the same order of magnitude, i.e., $eE \approx F_{\mathrm{gr}}$, if the magnitude of the electric field, E, is about

$$E = \frac{m_{\mathrm{e}} g}{e} \approx 6 \times 10^{-11}\,\mathrm{V\,m^{-1}}, \tag{A.10}$$

where the mass of an electron is $m_e = 9.1 \times 10^{-31}$ kg and the charge of an electron is $e = 1.6 \times 10^{-19}$ C. This is a very small electric field and can be neglected in most practical situations.

For a non-relativistic particle, i.e., a particle that moves with a velocity v that is substantially smaller than the speed of light in vacuum, $c = 3 \times 10^8\ \mathrm{m\,s^{-1}}$, Newton's second law of Eq. (A.5) is often written, using Eq. (A.3), in the form

$$m\,\frac{\mathrm{d}\mathbf{v}}{\mathrm{d}t} = \mathbf{F}. \tag{A.11}$$

Here the mass of a particle, m, is considered a constant that does not depend on the velocity of the particle. In many practical cases, m depends on time and Eq. (A.11) becomes invalid. This is one of the reasons why we will use Eq. (A.5) rather than Eq. (A.11) throughout the remainder of the book.

One more standard expression for Newton's second law is

$$m\,\frac{\mathrm{d}^2\mathbf{r}}{\mathrm{d}t^2} = \mathbf{F}, \tag{A.12}$$

which combines Eq. (A.11) with Eq. (A.4). Equation (A.12) is a second-order differential equation and its solution has two constants of integration. Thus, to find the function $\mathbf{r}$ and its first derivative $\mathrm{d}\mathbf{r}/\mathrm{d}t$ as a function of time, t, it is necessary to have two initial conditions, for example, $\mathbf{r}(t_0) = \mathbf{r}_0$ and $\mathbf{v}(t_0) = \mathbf{v}_0$, or $\mathbf{p}(t_0) = \mathbf{p}_0$.

A.1.2 The dynamics of a system of particles

For n interacting particles the system of equations describing their behavior has the form

$$\frac{\mathrm{d}\mathbf{p}_i}{\mathrm{d}t} = \mathbf{F}_i + \sum_{k=1}^{n} \mathbf{F}_{ik}, \quad i = 1, 2, \ldots, n, \tag{A.13}$$

where $\mathbf{F}_i$ is the resultant of all external forces acting on the ith particle and $\mathbf{F}_{ik}$ are the internal forces acting between the ith and kth particles, while $\mathbf{F}_{kk} = 0$. On summing up all the equations of motion of particles in the system and taking into account that according to Newton's third law of motion $\mathbf{F}_{ik} = -\mathbf{F}_{ki}$ (therefore $\sum_{i=1}^{n} \sum_{k=1}^{n} \mathbf{F}_{ik} = 0$), we obtain the equation

$$\frac{\mathrm{d}\mathbf{P}}{\mathrm{d}t} = \mathbf{F}, \tag{A.14}$$

where

$$\mathbf{P} = \sum_{i=1}^{n} \mathbf{p}_i. \tag{A.15}$$

Equation (A.14) has exactly the same form as Eq. (A.5) but with two substantial differences. Here **P** is the sum of the momenta of all particles of the system and

$$\mathbf{F} = \sum_{i=1}^{n} \mathbf{F}_i \tag{A.16}$$

is the resultant force equal to the sum of all external forces $\mathbf{F}_i$ acting on the individual particle denoted by i. It is convenient to introduce other variables characterizing the system as a whole. The mass of the system is obviously equal to the sum of the masses of the individual particles that comprise the system:

$$M = \sum_{i=1}^{n} m_i. \tag{A.17}$$

By combining Eqs. (A.15) and (A.17), we can introduce the velocity $\mathbf{V}_c$ of the system using Eq. (A.3):

$$\mathbf{V}_c = \frac{\mathbf{P}}{M} \tag{A.18}$$

or

$$\mathbf{V}_c = \frac{\sum_{i=1}^{n} m_i \mathbf{v}_i}{\sum_{i=1}^{n} m_i}. \tag{A.19}$$

By continuing this analogy, we can introduce the coordinate $\mathbf{R}_c$ of a point:

$$\frac{d\mathbf{R}_c}{dt} = \mathbf{V}_c. \tag{A.20}$$

By substituting Eq. (A.19) into Eq. (A.20) we find that

$$\mathbf{R}_c = \frac{\sum_{i=1}^{n} m_i \mathbf{r}_i}{\sum_{i=1}^{n} m_i}. \tag{A.21}$$

The position vector $\mathbf{R}_c(t)$ describes the trajectory of the system as a whole. Therefore, it is called the *vector of the center of mass of the system*. Equations (A.14), (A.18), and (A.20) for the system of particles are the same as Eqs. (A.5), (A.3), and (A.4) for an individual particle. This means that we can treat any large system as a point particle with all the mass of the system, M_c, concentrated at the so-called *center of mass*, $\mathbf{R}_c$, as long as the positions of the individual particles of the system are not the subject of our interest and the size of a system is small relative to the coordinate space of interest.

For a so-called *closed system*, the sum of all external forces, **F** in Eq. (A.14) is equal to zero. Thus, the resultant momentum of the closed system, **P**, is constant, i.e.,

$$\frac{d\mathbf{P}}{dt} = 0, \tag{A.22}$$

$$\mathbf{P} = \sum_{i=1}^{n} \mathbf{p}_i = \text{constant}, \tag{A.23}$$

and

$$\mathbf{V}_c = \frac{\sum_{i=1}^{n} m_i \mathbf{v}_i}{\sum_{i=1}^{n} m_i} = \text{constant}. \tag{A.24}$$

Equations (A.22) and (A.23) describe one of the main principles of mechanics written in mathematical form – the *principle of conservation of momentum*, which implies that the *sum of the momenta of a closed system of particles that exert forces on each other is a constant.* From Eq. (A.24) it follows that, when the sum of all external forces acting on the system of interacting particles, **F**, is equal to zero, then the velocity of the system's center of mass, $\mathbf{V}_c$, is constant.

A.1.3 Rotational motion of a particle

Apart from the particle's momentum, **p**, there is another important characteristic of a moving particle. It is the particle's *angular momentum*, **L**, about a given point:

$$\mathbf{L} = \mathbf{r} \times \mathbf{p}, \tag{A.25}$$

where **r** is the position vector that defines the position of a particle with respect to a given point. The main equation that defines the change of angular momentum, **L**, of a particle can be derived from Eq. (A.25), if we take the derivatives of the left-hand and right-hand sides of this equation:

$$\frac{d\mathbf{L}}{dt} = \frac{d\mathbf{r}}{dt} \times \mathbf{p} + \mathbf{r} \times \frac{d\mathbf{p}}{dt}. \tag{A.26}$$

Since $d\mathbf{r}/dt = \mathbf{v}$ and $\mathbf{p} = m\mathbf{v}$,

$$\frac{d\mathbf{r}}{dt} \times \mathbf{p} = m(\mathbf{v} \times \mathbf{v}) = 0 \tag{A.27}$$

because the vector product $\mathbf{v} \times \mathbf{v}$ is equal to zero. Using Eq. (A.5) we arrive at the equation

$$\frac{d\mathbf{L}}{dt} = \boldsymbol{\tau}, \tag{A.28}$$

where

$$\boldsymbol{\tau} = \mathbf{r} \times \mathbf{F} \tag{A.29}$$

is the so-called *torque* with respect to the given point. If the resultant torque, $\boldsymbol{\tau}$, of all forces acting on the particle is equal to zero, then

$$\frac{\mathrm{d}\mathbf{L}}{\mathrm{d}t} = 0 \quad \text{and} \quad \mathbf{L} = \text{constant}. \tag{A.30}$$

The last equation reflects one of the main principles of mechanics – the *principle of conservation of angular momentum*: *the angular momentum is independent of time and it is conserved when the net torque is equal to zero.*

As an example let us consider the motion of a particle in a field of central forces (or *central field*). The vector of force, **F**, in the central field always coincides with the radius vector, **r**, directed from the central point to a point where the particle is located, and its magnitude depends only on the distance, r, between these two points, i.e.,

$$\mathbf{F}(\mathbf{r}) = F(r)\mathbf{e}_r, \tag{A.31}$$

where $\mathbf{e}_r = \mathbf{r}/r$ is a unit vector directed along the radius vector **r**. Note that in all cases of a particle's motion in the central field we have $F(r) < 0$ in Eq. (A.31). The torque of this force with respect to the central point is

$$\boldsymbol{\tau}_c = \mathbf{r} \times \mathbf{F} = F(r)(\mathbf{r} \times \mathbf{e}_r) = 0, \tag{A.32}$$

because the vectors **r** and $\mathbf{e}_r$ are collinear. This is why in the solar system the motion of planets around the Sun is along flat elliptic orbits – the angular momentum, **L**, of each of the planets is a constant vector.

A.1.4 Work and potential energy

A force, **F**, moving a particle by a certain distance, does a certain work. For the displacement d**r**, the work done, dW, is defined by the scalar product, $\mathbf{F} \cdot \mathrm{d}\mathbf{r}$:

$$\mathrm{d}W = \mathbf{F} \cdot \mathrm{d}\mathbf{r} = F_x\,\mathrm{d}x + F_y\,\mathrm{d}y + F_z\,\mathrm{d}z = F|\mathrm{d}\mathbf{r}|\cos\alpha = F_r\,\mathrm{d}r, \tag{A.33}$$

where α is the angle between the direction of force, **F**, and the direction of displacement, d**r**; $F_r = F\cos\alpha$ is the projection of the force, **F**, on the direction of displacement, d**r**; and $|\mathrm{d}\mathbf{r}| = \mathrm{d}r$. The total work done by the force, **F**, to move a particle from the point $\mathbf{r}_1$ to the point $\mathbf{r}_2$ is given by the following definite integral:

$$W_{12} = \int_{\mathbf{r}_1}^{\mathbf{r}_2} \mathbf{F} \cdot \mathrm{d}\mathbf{r} = \int_{r_1}^{r_2} F_r\,\mathrm{d}r. \tag{A.34}$$

Among all forces that can act on particles, we single out a wide class of forces whose work does not depend on the path followed between points $\mathbf{r}_1$ and $\mathbf{r}_2$ (i.e., the work done

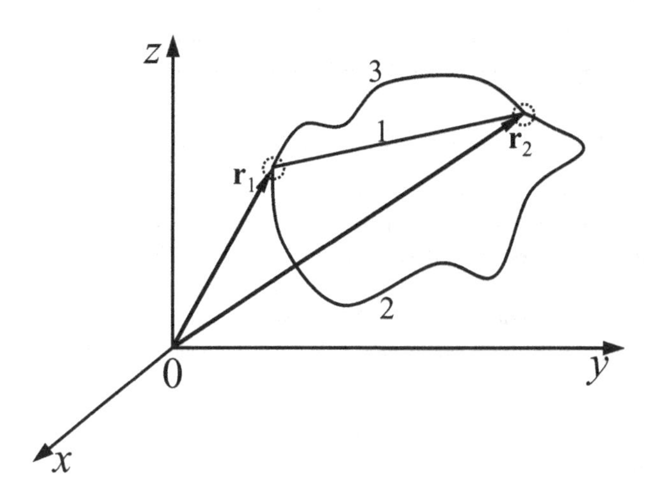

Figure A.2 The work done by the conservative forces around a closed path is equal to zero.

between points $\mathbf{r}_1$ and $\mathbf{r}_2$ is the same for paths 1, 2, and 3 in Fig. A.2), and for which the work, W_O, of moving a particle along a closed path is equal to zero:

$$W_O = \oint \mathbf{F} \cdot \mathrm{d}\mathbf{r} = 0. \tag{A.35}$$

Here we introduced the *symbol of integration around a closed path*, $\oint$. An example of a closed path is the transfer of a particle from point $\mathbf{r}_1$ to $\mathbf{r}_2$ along path 2 and then from point $\mathbf{r}_2$ to point $\mathbf{r}_1$ along path 3 (see Fig. A.2). The forces that satisfy Eq. (A.35) are called *conservative forces*, and they include, for example, gravitational and electrostatic forces. Note that for the forces of friction the work done depends on the path and therefore Eq. (A.35) is not valid for the forces of friction.

For the conservative force, $\mathbf{F}$, we can introduce a function, $U(\mathbf{r})$, that depends on the coordinate, $\mathbf{r}$, and relates to the force, $\mathbf{F}$, as

$$\mathbf{F} = -\nabla U(\mathbf{r}), \tag{A.36}$$

where we introduced the so-called *differential operator*, ∇:

$$\nabla = \mathbf{i}\frac{\partial}{\partial x} + \mathbf{j}\frac{\partial}{\partial y} + \mathbf{k}\frac{\partial}{\partial z}. \tag{A.37}$$

The function $U(\mathbf{r})$ is called *potential energy*. Taking into account Eq. (A.36) and substituting it into Eq. (A.34), we obtain the relationship between work W_{12} and potential energy $U(\mathbf{r})$:

$$W_{12} = -\int_{\mathbf{r}_1}^{\mathbf{r}_2} \nabla U(\mathbf{r}) \cdot \mathrm{d}\mathbf{r} = -\int_{(1)}^{(2)} \left(\frac{\partial U}{\partial x}\mathrm{d}x + \frac{\partial U}{\partial y}\mathrm{d}y + \frac{\partial U}{\partial z}\mathrm{d}z \right), \tag{A.38}$$

where the function under the integral sign is the so-called *total differential*, $dU(x, y, z)$, of the function $U(x, y, z)$. The expression for the work done takes the form

$$W_{12} = -\int_{(1)}^{(2)} dU(x, y, z) = -\int_{(1)}^{(2)} dU(\mathbf{r}) = -(U_2 - U_1) = -[U(\mathbf{r}_2) - U(\mathbf{r}_1)]. \tag{A.39}$$

Thus, the work of conservative forces does not depend on the path, but depends on the initial and final coordinates, $\mathbf{r}_1$ and $\mathbf{r}_2$, only. Moreover, the work of these forces is equal to the difference of potential energies at the initial and final positions of the particle, and hence does not depend on the choice of the origin of the coordinate system for $U(r)$. Let us consider as examples of potential fields homogeneous and non-homogeneous electric fields.

A.1.5 Kinetic and total mechanical energy

The total mechanical energy, E, of a particle is defined as the sum of its potential energy, $U(\mathbf{r})$, and kinetic energy, K:

$$E = K + U. \tag{A.40}$$

The expression for a particle's kinetic energy, K, can be derived from the equation of motion, Eq. (A.11), after multiplying its right-hand and left-hand sides by the displacement vector, $d\mathbf{r} = \mathbf{v}\, dt$:

$$m\mathbf{v} \cdot d\mathbf{v} = \mathbf{F} \cdot d\mathbf{r}. \tag{A.41}$$

This expression can be transformed into

$$d\left(\frac{mv^2}{2}\right) = \mathbf{F} \cdot d\mathbf{r}, \tag{A.42}$$

because $d(v^2) = 2\mathbf{v} \cdot d\mathbf{v}$. The quantity

$$K = \frac{mv^2}{2} = \frac{p^2}{2m} \tag{A.43}$$

is called the *kinetic energy* of a particle. If we integrate Eq. (A.42) from $\mathbf{r}_1$ to $\mathbf{r}_2$, we will obtain

$$K_2 - K_1 = \frac{mv_2^2}{2} - \frac{mv_1^2}{2} = \int_{\mathbf{r}_1}^{\mathbf{r}_2} \mathbf{F} \cdot d\mathbf{r}. \tag{A.44}$$

Here, $K_i = (mv_i^2)/2$ and $v_i = v(r_i)$ is the velocity of the particle at point i, where $i = 1, 2$.

The force $\mathbf{F}$ under the integral sign is the sum of the conservative, $\mathbf{F}_c$, and non-conservative (for example, friction), $\mathbf{F}_{fr}$, forces:

$$\mathbf{F} = \mathbf{F}_c + \mathbf{F}_{fr}. \tag{A.45}$$

The work of the conservative forces can be defined as the difference of potential energies of the particle. In accordance with Eq. (A.39),

$$\int_{\mathbf{r}_1}^{\mathbf{r}_2} \mathbf{F}_c \cdot d\mathbf{r} = U(\mathbf{r}_1) - U(\mathbf{r}_2), \tag{A.46}$$

where $U(\mathbf{r})$ is the potential energy of the conservative force. Therefore, Eq. (A.44) can be rewritten as

$$K_2 + U_2 - K_1 - U_1 = \int_{\mathbf{r}_1}^{\mathbf{r}_2} \mathbf{F}_{fr} \cdot d\mathbf{r}. \tag{A.47}$$

Equation (A.47) is called the *work–energy theorem*. The change in total mechanical energy of a particle, E, is equal to the work of friction forces along the entire path of a particle's motion. In the absence of non-conservative forces, the total mechanical energy, $E = K + U$, is conserved, i.e.,

$$K_1 + U_1 = K_2 + U_2. \tag{A.48}$$

During the process of a particle's motion in a potential field the kinetic energy of a particle is transformed into its potential energy or vice versa:

$$K_2 - K_1 = -(U_2 - U_1). \tag{A.49}$$

A.1.6 Equilibrium conditions for a particle

The conditions of equilibrium play a very important role in the analysis of the motion of a particle in a potential field. The mechanical equilibrium of a body assumes satisfaction of two vector conditions:

$$\sum_{i=1}^{n} \mathbf{F}_i = 0, \tag{A.50}$$

$$\sum_{i=1}^{n} \boldsymbol{\tau}_i = 0. \tag{A.51}$$

The sum of all forces and the sum of all torques acting on a particle must be equal to zero. The first of the above-mentioned conditions is required for physical bodies that can be considered as particles (or material points). With Eq. (A.36) taken into account, the equilibrium condition for a particle in the potential field can be rewritten in the form

$$\nabla U(x, y, z) = 0. \tag{A.52}$$

From the last expression, we can write the following expressions for the potential energy:

$$\frac{\partial U}{\partial x} = 0, \quad \frac{\partial U}{\partial y} = 0, \quad \frac{\partial U}{\partial z} = 0, \tag{A.53}$$

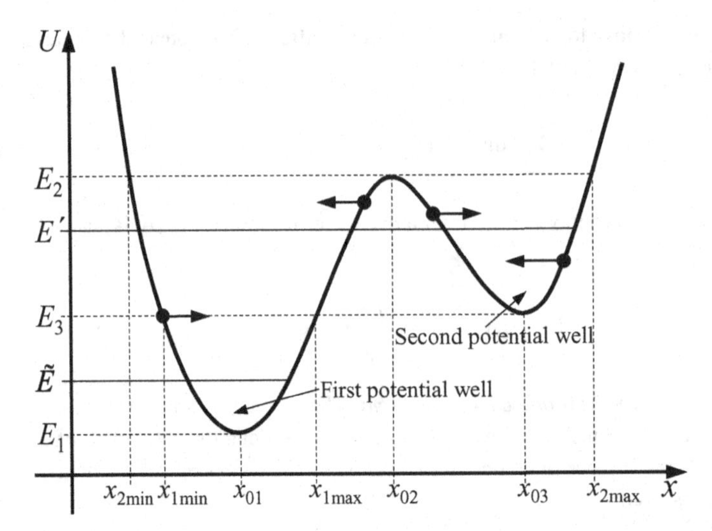

Figure A.3 A potential profile with two stable (x_{01} and x_{03}) equilibrium positions and one unstable (x_{02}) equilibrium position.

which have to be satisfied at the position of equilibrium. For convenience the position of equilibrium is often denoted by the coordinates x_0, y_0, and z_0.

The position of equilibrium can be either stable or unstable. In the case of a stable equilibrium position, when the particle is displaced out of the equilibrium, the force, $\mathbf{F} = -\nabla U(\mathbf{r})$, that appears due to the dependence of $U(\mathbf{r})$ on $\mathbf{r}$ will return the particle to the state of equilibrium. In the case of an unstable equilibrium, when the body is displaced out of the equilibrium, the force $\mathbf{F} = -\nabla U(\mathbf{r})$ will take the particle away from the state of equilibrium. When the force depends only on a single coordinate (let us choose it to be the x-coordinate) it is defined as

$$\mathbf{F} = -\mathbf{i}\,\frac{\mathrm{d}U(x)}{\mathrm{d}x}. \tag{A.54}$$

From Eq. (A.54) it follows that for $\mathrm{d}U(x)/\mathrm{d}x > 0$ the force $\mathbf{F}$ is directed against the vector $\mathbf{i}$, and, in the case when $\mathrm{d}U(x)/\mathrm{d}x < 0$, $\mathbf{F}$ is parallel to the vector $\mathbf{i}$.

In the one-dimensional case we can easily depict the function $U(x)$ near its equilibrium positions. In Fig. A.3 two positions of stable equilibrium are defined by the coordinates x_{01} and x_{03}, and near these positions the function $U(x)$ has the form of a "potential well." At the position x_{02} the state of equilibrium is unstable. The directions of the forces that appear upon displacement of a particle from the respective equilibrium positions are depicted by arrows. It is clearly seen that the minimum of potential energy corresponds to the position of stable equilibrium. The conditions of a stable equilibrium state at an arbitrary point $x = x_0$ in the one-dimensional case have the form

$$\left.\frac{\mathrm{d}U}{\mathrm{d}x}\right|_{x=x_0} = 0 \quad \text{and} \quad \left.\frac{\mathrm{d}^2U}{\mathrm{d}x^2}\right|_{x=x_0} > 0. \tag{A.55}$$

An unstable equilibrium state is defined by the conditions

$$\left.\frac{\mathrm{d}U}{\mathrm{d}x}\right|_{x=x_0} = 0 \quad \text{and} \quad \left.\frac{\mathrm{d}^2U}{\mathrm{d}x^2}\right|_{x=x_0} < 0. \tag{A.56}$$

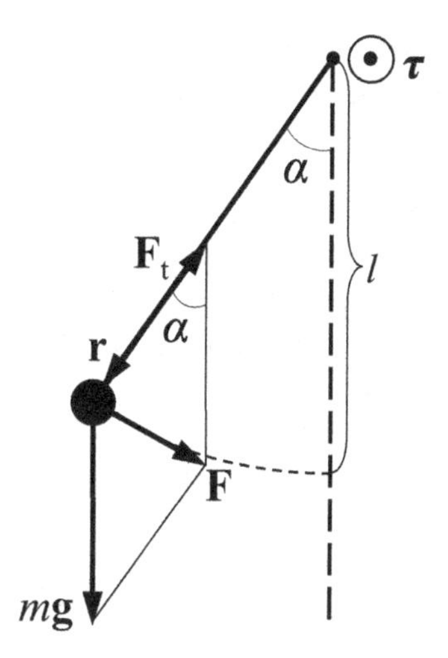

Figure A.4 A simple pendulum.

If the particle's total energy is equal to its minimal possible energy, $E_1 = U(x_{01})$, the particle can only be at the position of equilibrium x_{01}. For a given profile of the potential energy, the total energy of a particle, E, cannot be less than E_1; otherwise its kinetic energy, K, would be negative. With an increase of energy, a particle can oscillate near its equilibrium position. For example, for the energy $E = \tilde{E}$, where $E_1 < \tilde{E} < E_3$, the particle will move in the region of positive kinetic energy between points $x_{1\min}$ and $x_{1\max}$.

At $E = E_2$ the particle is in the position of unstable equilibrium, x_{02}. After leaving the position of equilibrium x_{02}, the particle may travel between the first and second potential wells in the region from $x_{2\min}$ to $x_{2\max}$. A further increase of energy above E_2 leads to an increase of the interval of allowed coordinates, x. At energies $E = E'$, where $E_3 < E' < E_2$, the "classical" particle can be located only in one of two potential wells near their equilibrium positions, x_{01} or x_{03}. In order to transfer the particle from the first potential well to the second well, it is necessary to supply the energy deficit, i.e., the energy difference between E_2 and E'. As will be shown later, for "quantum" particles this ban is lifted.

A.2 Oscillatory motion of a particle

A.2.1 Simple harmonic motion

If a particle follows a motion that is periodic, i.e., repetitive in time, then

$$\mathbf{r}(t + nT) = \mathbf{r}(t), \tag{A.57}$$

where T is a period and n is an integer number. Along each of the Cartesian coordinates the particle undergoes a periodic motion, which is called an *oscillatory motion*.

The most famous example of oscillatory motion is the oscillation of a *simple pendulum*, i.e., an oscillation of a point mass on a long massless inextensible string with length l (Fig. A.4). Let the mass of the oscillating particle be m and let us derive the equation that describes oscillations of such a pendulum. At an arbitrary displacement of a pendulum at an angle α, two forces act on a particle – the weight, $m\mathbf{g}$, and the tension of the string, $\mathbf{F}_t$.

Since the pendulum undergoes rotational motion with respect to the point of suspension O, we can write the main equation of rotational motion (Eq. (A.28)) in the form of a projection onto the axis of rotation:

$$\frac{\mathrm{d}L}{\mathrm{d}t} = \tau, \tag{A.58}$$

where L and τ are projections of the angular momentum and torque of an oscillating particle onto the axis perpendicular to the plane of the pendulum's oscillations. Since the line along which the tension force of the string, $\mathbf{F}_t$, is directed goes through the point of suspension, this force does not create the torque. The torque that returns the particle to the equilibrium position is created only by the particle's weight, $m\mathbf{g}$. The magnitude of this force's torque, τ, is equal to

$$\tau = -mgl\sin\alpha, \tag{A.59}$$

where the negative sign occurs because at any angle of displacement, α, the torque τ is restoring, i.e., torque pushes the particle back to the equilibrium position. For the pendulum shown in Fig. A.4, the position angle α is positive. Note that the displacement angle, α, around a fixed axis is considered as a vector directed along the axis of rotation according to the right-hand rule (see Fig. A.4). The angular momentum according to Eq. (A.25) is

$$L = mvl = ml^2\Omega = ml^2\frac{\mathrm{d}\alpha}{\mathrm{d}t}, \tag{A.60}$$

where the linear velocity of a particle, v, along the arc of a circle of radius l is equal to

$$v = \Omega l \tag{A.61}$$

and the angular velocity is defined as

$$\Omega = \frac{\mathrm{d}\alpha}{\mathrm{d}t}. \tag{A.62}$$

After substitution of L and τ into Eq. (A.58), the last equation becomes

$$ml^2\frac{\mathrm{d}^2\alpha}{\mathrm{d}t^2} = -mgl\sin\alpha. \tag{A.63}$$

Let us rewrite Eq. (A.63) in the standard form of the equation for a simple pendulum's oscillatory motion,

$$\frac{\mathrm{d}^2\alpha}{\mathrm{d}t^2} + \omega^2\sin\alpha = 0, \tag{A.64}$$

where

$$\omega = \sqrt{\frac{g}{l}} \tag{A.65}$$

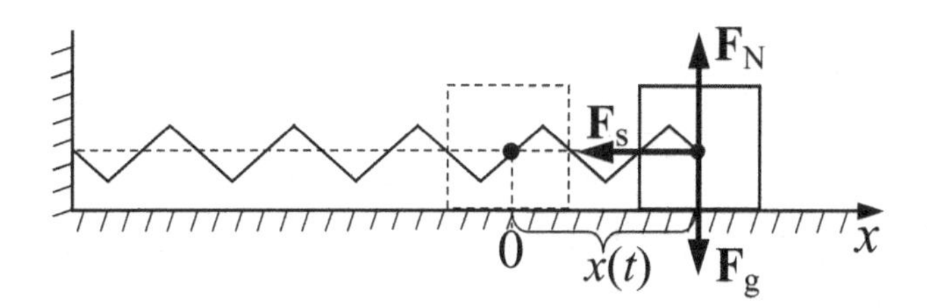

Figure A.5 The oscillations of a particle connected to a spring along a smooth surface.

is the angular frequency of the pendulum's oscillations. Equation (A.64) is valid at any displacement of the pendulum from the position of equilibrium. However, its solution at arbitrary values of the angle α cannot be described by elementary functions. Therefore, let us consider oscillations with small amplitude when the angle α is much less than 1 ($\alpha \ll 1$) and $\sin\alpha \approx \alpha$. In this case, Eq. (A.64) becomes a standard form of equation for small displacements of a simple pendulum:

$$\frac{\mathrm{d}^2\alpha}{\mathrm{d}t^2} + \omega^2\alpha = 0. \tag{A.66}$$

The following function is the solution of Eq. (A.66):

$$\alpha(t) = \alpha_0 \cos(\omega t + \phi), \tag{A.67}$$

where α_0 and ϕ are the amplitude and phase of oscillations, respectively. This function defines the angular position of the pendulum at any given time. Oscillatory motion defined by a function of this type is called *simple harmonic motion*. These oscillations have the most simple form, from the mathematical point of view, because they can be described by an elementary function of Eq. (A.67).

A.2.2 One-dimensional harmonic oscillator

Another example of harmonic motion is that of a particle connected to a spring (see Fig. A.5). Three forces act on the particle: the gravitational force, $\mathbf{F}_\mathrm{g}$, the normal force, $\mathbf{F}_\mathrm{N}$, and the spring force, $\mathbf{F}_\mathrm{s}$. The net force is equal to the spring force because the forces $\mathbf{F}_\mathrm{N}$ and $\mathbf{F}_\mathrm{g}$ exactly compensate for each other. Here we neglect the force of friction (see the next section). According to Hooke's law the spring force $\mathbf{F}_\mathrm{s}$ depends linearly on the displacement $x(t)$ from the equilibrium position $x = 0$ (we chose the x-axis to be in the direction of the force $\mathbf{F}_\mathrm{s}$) as shown in Fig. A.5:

$$F_\mathrm{s} = -kx(t), \tag{A.68}$$

where k is a spring constant. Then the equation of motion can be written as

$$m\frac{\mathrm{d}^2x}{\mathrm{d}t^2} = F_\mathrm{s}, \tag{A.69}$$

or

$$m\,\frac{\mathrm{d}^2x}{\mathrm{d}t^2} = -kx. \tag{A.70}$$

Introducing the angular frequency of oscillations,

$$\omega = \sqrt{\frac{k}{m}}, \tag{A.71}$$

we arrive at the following differential equation:

$$\frac{\mathrm{d}^2x}{\mathrm{d}t^2} + \omega^2 x = 0, \tag{A.72}$$

which is the same as Eq. (A.66) for small-amplitude oscillations of a simple pendulum, which are harmonic oscillations.

The general solution of Eq. (A.72) can be found in the form of any of the following expressions:

$$x(t) = \begin{cases} A_1 \mathrm{e}^{\mathrm{i}\omega t} + A_2 \mathrm{e}^{-\mathrm{i}\omega t}, \\ B_1 \cos(\omega t) + B_2 \sin(\omega t), \\ C \cos(\omega t + \phi), \end{cases} \tag{A.73}$$

where the constants $A_{1,2}$, $B_{1,2}$, C, and ϕ are constants of integration. Since all three solutions satisfy the same equation, these constants are functionally related and these relations can easily be established using well-known trigonometric relations.

Let us find first the relationship between the coefficients $A_{1,2}$ and $B_{1,2}$. By substituting the formulae

$$\cos(\omega t) = \frac{1}{2}\left(\mathrm{e}^{\mathrm{i}\omega t} + \mathrm{e}^{-\mathrm{i}\omega t}\right) \quad \text{and} \quad \sin(\omega t) = \frac{1}{2\mathrm{i}}\left(\mathrm{e}^{\mathrm{i}\omega t} - \mathrm{e}^{-\mathrm{i}\omega t}\right) \tag{A.74}$$

into the second expression of Eqs. (A.73) and regrouping terms with the same exponents, we obtain

$$x(t) = \left(\frac{B_1}{2} + \frac{B_2}{2\mathrm{i}}\right)\mathrm{e}^{\mathrm{i}\omega t} + \left(\frac{B_1}{2} - \frac{B_2}{2\mathrm{i}}\right)\mathrm{e}^{-\mathrm{i}\omega t}. \tag{A.75}$$

On comparing the last equation with the first expression of Eq. (A.73), we get

$$A_1 = \frac{1}{2}(B_1 - \mathrm{i}B_2) \quad \text{and} \quad A_2 = \frac{1}{2}(B_1 + \mathrm{i}B_2). \tag{A.76}$$

Inversely,

$$B_1 = A_1 + A_2 \quad \text{and} \quad B_2 = \mathrm{i}(A_1 - A_2). \tag{A.77}$$

Let us define the relation of coefficients C and ϕ with $A_{1,2}$ by rewriting the third expression of Eq. (A.73) in terms of the exponents:

$$C\cos(\omega t+\phi)=\frac{C}{2}\left(\mathrm{e}^{\mathrm{i}(\omega t+\phi)}+\mathrm{e}^{-\mathrm{i}(\omega t+\phi)}\right)=\frac{C}{2}\mathrm{e}^{\mathrm{i}\phi}\mathrm{e}^{\mathrm{i}\omega t}+\frac{C}{2}\mathrm{e}^{-\mathrm{i}\phi}\mathrm{e}^{-\mathrm{i}\omega t}. \tag{A.78}$$

On comparing this equation with the first expression of Eq. (A.73), we get

$$A_1=\frac{C}{2}\mathrm{e}^{\mathrm{i}\phi}\quad\text{and}\quad A_2=\frac{C}{2}\mathrm{e}^{-\mathrm{i}\phi}. \tag{A.79}$$

Inversely,

$$C=2\sqrt{A_1A_2}\quad\text{and}\quad \phi=\frac{1}{2\mathrm{i}}\ln\left(\frac{A_1}{A_2}\right). \tag{A.80}$$

By substituting Eq. (A.76) into Eq. (A.80) we can find the relation of the coefficients C and ϕ with $B_{1,2}$. The constants themselves are defined from the initial conditions.

A.2.3 The energy of harmonic oscillations

For a particle that follows simple harmonic oscillatory motion, its velocity and acceleration are also given by harmonic functions. Choosing the dependence of a particle's displacement, x, from its equilibrium position as

$$x(t)=C\cos(\omega t+\phi), \tag{A.81}$$

we obtain for the velocity, v,

$$v(t)=\frac{\mathrm{d}x}{\mathrm{d}t}=-\omega C\sin(\omega t+\phi), \tag{A.82}$$

and for the acceleration, a,

$$a(t)=\frac{\mathrm{d}v}{\mathrm{d}t}=-\omega^2C\cos(\omega t+\phi)=-\omega^2x(t). \tag{A.83}$$

From the last equation it follows that, for harmonic oscillations, the acceleration, a, is directly proportional to the displacement, x, and its proportionality constant is always equal to the square of the angular frequency with a negative sign, $-\omega^2$.

Let us define now the total mechanical energy, E, of the oscillating particle. A particle's kinetic energy is defined by the expression

$$K(t)=\frac{mv^2(t)}{2}=\frac{m\omega^2}{2}C^2\sin^2(\omega t+\phi). \tag{A.84}$$

The potential energy is related to the force by Eq. (A.36), which for the one-dimensional case has the form

$$F_x = -\frac{\mathrm{d}U}{\mathrm{d}x} = -kx. \tag{A.85}$$

From Eq. (A.85) we obtain the quadratic dependence of the potential energy, U, on the displacement, x:

$$U(x) = \frac{kx^2}{2}. \tag{A.86}$$

By substituting the displacement $x(t)$, defined by Eq. (A.81), into Eq. (A.86) we obtain for the potential energy of an oscillating particle the relation

$$U(x) = \frac{kx^2(t)}{2} = \frac{m\omega^2}{2}C^2\cos^2(\omega t + \phi). \tag{A.87}$$

We see that the kinetic as well as the potential energy varies in time with doubled frequency (since $2\sin^2\beta = 1 - \cos(2\beta)$ and $2\cos^2\beta = 1 + \cos(2\beta)$), whereas the total mechanical energy, E, does not depend on time, t, and stays constant:

$$E = K(t) + U(t) = \frac{m\omega^2 C^2}{2}. \tag{A.88}$$

A.2.4 Damped oscillatory motion

The oscillations of a particle become damped if in the process of the oscillatory motion a particle loses its energy. For a number of problems, e.g., the motion of a particle in viscous media, the occurrence of damping in an oscillatory system can be taken into account by introducing into Eq. (A.69) the *damping force*, which is proportional to the velocity of a particle, $v = \mathrm{d}x/\mathrm{d}t$:

$$m\,\frac{\mathrm{d}^2x}{\mathrm{d}t^2} = -kx - b\,\frac{\mathrm{d}x}{\mathrm{d}t}. \tag{A.89}$$

Let us introduce new parameters, $b/m = 2\beta$ and $k/m = \omega^2$, where β is the damping factor and ω is the frequency of the particle's oscillations at $\beta = 0$. As a result, Eq. (A.89) is reduced to a standard form of the equation with damped oscillations

$$\frac{\mathrm{d}^2x}{\mathrm{d}t^2} + 2\beta\,\frac{\mathrm{d}x}{\mathrm{d}t} + \omega^2 x = 0. \tag{A.90}$$

Let us seek the solution of this equation in the form

$$x(t) = A\mathrm{e}^{\lambda t}. \tag{A.91}$$

After substitution of $x(t)$ into Eq. (A.90), we obtain the characteristic equation for finding the parameter λ:

$$\lambda^2 + 2\beta\lambda + \omega^2 = 0. \tag{A.92}$$

This equation has two roots:

$$\lambda_{1,2} = -\beta \pm \sqrt{\beta^2 - \omega^2}. \tag{A.93}$$

Depending on the magnitudes of ω and β, we can distinguish two different types of damped oscillatory motion: *aperiodic*,

$$\omega \leq \beta, \tag{A.94}$$

and *oscillatory*,

$$\omega > \beta. \tag{A.95}$$

In the first case both roots λ_1 and λ_2 are real numbers and both are negative. Such a case corresponds to the strong damping when $x(t)$ exponentially decreases in time (Eq. (A.91)). The second case, when $\omega > \beta$, is more interesting for us. Moreover, for a sufficiently distinctive oscillatory behavior an even stronger inequality has to be fulfilled: $\omega \gg \beta$. In this case both roots are complex numbers:

$$\lambda_{1,2} = -\beta \pm \mathrm{i}\sqrt{\omega^2 - \beta^2}. \tag{A.96}$$

For these two roots of the characteristic equation, we can write the general solution of Eq. (A.90) in the form

$$\begin{aligned} x(t) &= A_1 \mathrm{e}^{\lambda_1 t} + A_2 \mathrm{e}^{\lambda_2 t} = \mathrm{e}^{-\beta t}\left(A_1 \mathrm{e}^{\mathrm{i}\Omega t} + A_2 \mathrm{e}^{-\mathrm{i}\Omega t}\right) = C\mathrm{e}^{-\beta t}\cos(\Omega t + \phi) \\ &= C(t)\cos(\Omega t + \phi), \end{aligned} \tag{A.97}$$

where $\Omega = \sqrt{\omega^2 - \beta^2}$ is the angular frequency of damped oscillations and $C(t) = C\mathrm{e}^{-\beta t}$. Since $\Omega < \omega$, the period of damped oscillations, $T = 2\pi/\Omega$, is always greater than the period of free oscillations, $T_0 = 2\pi/\omega$.

In Fig. A.6 the graph of the time-dependent displacement, $x(t)$, of a particle from its position of equilibrium, $x = x(0) = x_0$, is shown for the case of damped oscillations. At $t = 0$, the displacement of a particle from the position of equilibrium is equal to

$$x(0) = x_0 = C\cos\phi. \tag{A.98}$$

To characterize the rate of damping, it is convenient to introduce a number of quantities apart from the coefficient of damping, β, such as the *time of relaxation*, τ, and the *logarithmic decrement of damping*, δ. The time of relaxation is equal to

$$\tau = \frac{1}{\beta}, \tag{A.99}$$

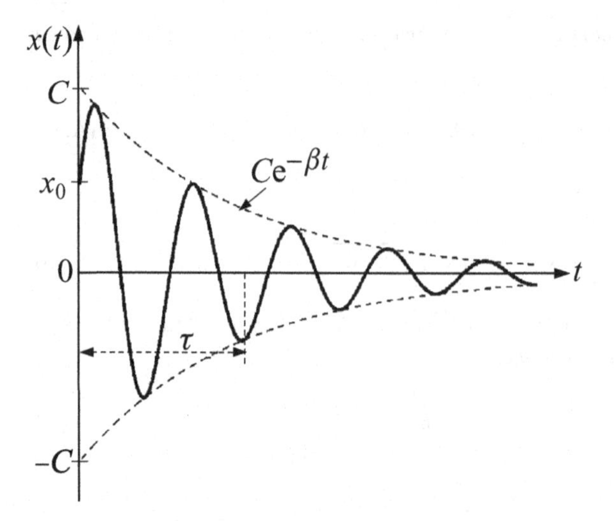

Figure A.6 Damped oscillations.

and it defines the time interval during which the amplitude of oscillations decreases by a factor of e (e ≈ 2.718). The logarithmic decrement of damping, δ, is introduced in the following way:

$$\delta = \ln\left(\frac{C(t)}{C(t+T)}\right) = \ln\left(\frac{C\mathrm{e}^{-\beta t}}{C\mathrm{e}^{-\beta(t+T)}}\right) = \ln \mathrm{e}^{\beta T} = \beta T. \tag{A.100}$$

Taking into account that $\beta = 1/\tau$, we obtain

$$\delta = \frac{T}{\tau} = \frac{1}{N_\tau}, \tag{A.101}$$

where $N_\tau = \tau/T$ is the number of full periods for which the amplitude of oscillations decreases by a factor of e. The smaller the logarithmic decrement of damping, δ, the larger is the number of oscillations that a particle makes during time τ, and the more evident the oscillatory character of a particle's motion.

The velocity of a particle that is oscillating with a damping can be found by taking the first derivative with respect to time of $x(t)$ as defined by Eq. (A.97):

$$v(t) = \frac{\mathrm{d}x}{\mathrm{d}t} = -C\mathrm{e}^{-\beta t}\left[\Omega \sin(\Omega t + \phi) + \beta \cos(\Omega t + \phi)\right]. \tag{A.102}$$

If at $t = 0$ the particle's velocity is equal to zero, then it follows from the last equation that the initial phase of oscillations, ϕ, is equal to

$$\phi = -\arctan\left(\frac{\beta}{\Omega}\right). \tag{A.103}$$

The total energy of an oscillating particle, $E = K + U$, decreases after each period of oscillations by the value of the work done by friction forces, A_T:

$$A_T = -b \int_{x(t)}^{x(t+T)} v \, \mathrm{d}x = -2\beta m \int_t^{t+T} v^2(t) \mathrm{d}t. \tag{A.104}$$

This energy is transformed into the internal energy of the medium, i.e., into the thermal energy of the medium – heat.

If the damping factor is large, $\beta \geq \omega$, the roots of the characteristic equation Eq. (A.92) are real, and the solution of Eq. (A.90) can be written in the form

$$x(t) = \mathrm{e}^{-\beta t} \left[A_1 \mathrm{e}^{\sqrt{\beta^2 - \omega^2} t} + A_2 \mathrm{e}^{-\sqrt{\beta^2 - \omega^2} t} \right]. \tag{A.105}$$

The motion of a particle described by this solution is not oscillatory and there is only an exponential approach of the particle to the equilibrium position. The velocity and energy of a particle exponentially drop with time, t.

A.2.5 Small-amplitude oscillations

If a particle executes oscillatory motion in an arbitrary potential field, then, near the position of equilibrium, x_0, the potential energy of a particle, $U(x)$, can be expanded in a Taylor series with respect to the displacement from the position of equilibrium:

$$U(x) = U_0 + \left(\frac{\mathrm{d}U}{\mathrm{d}x} \right)_{x=x_0} (x - x_0) + \frac{1}{2} \left(\frac{\mathrm{d}^2 U}{\mathrm{d}x^2} \right)_{x=x_0} (x - x_0)^2 + \frac{1}{6} \left(\frac{\mathrm{d}^3 U}{\mathrm{d}x^3} \right)_{x=x_0} (x - x_0)^3 + \cdots. \tag{A.106}$$

Here, $U_0 = U(x_0)$ is the value of the potential energy at the position of equilibrium. Since, at the position of equilibrium, the sum of all forces acting on a particle must be equal to zero, the coefficient $(\mathrm{d}U/\mathrm{d}x)_{x=x_0}$, which defines the force acting on the particle at the equilibrium position, is equal to zero. The coefficient $(\mathrm{d}^2U/\mathrm{d}x^2)_{x=x_0}$ in the expression for the potential energy, $U(x)$, is a coefficient of proportionality for the quadratic displacement of a particle from the equilibrium position. Therefore, this coefficient is actually the spring constant k, i.e.,

$$\left(\frac{\mathrm{d}^2 U}{\mathrm{d}x^2} \right)_{x=x_0} = k. \tag{A.107}$$

In the case of small-amplitude oscillations we can neglect in Eq. (A.106) the terms of expansion with powers greater than two. Then, the expression for the potential energy will take the form

$$U(x) = U(x_0) + \frac{1}{2} k (x - x_0)^2. \tag{A.108}$$

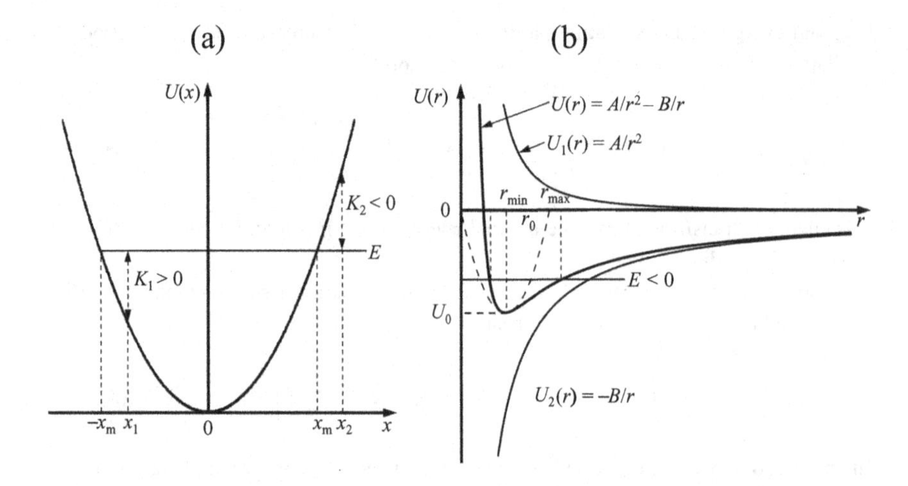

Figure A.7 Small-amplitude oscillations in a parabolic potential (a) and in the potential defined by Eq. (A.113)(b).

Such a representation of a potential energy is called the *harmonic approximation*. If we choose the position of equilibrium at the origin of coordinates, $x_0 = 0$, and put the potential energy at this point equal to zero ($U(x_0) = U(0) = 0$), the expression for the potential energy, Eq. (A.108), becomes equivalent to Eq. (A.86). We can conclude that, in the case of an arbitrary dependence of U on x for small displacements of a particle from a stable equilibrium position, the motion of the particle is always harmonic. The terms of the expansion proportional to $(x - x_0)^3$ and higher powers of the particle's displacement define the anharmonic effects in the oscillatory motion of a particle.

Since the potential energy of a particle in a field of elastic forces is quadratic with respect to the displacement of the particle from the equilibrium position, we can say that its motion takes place in a *parabolic potential* (see Fig. A.7(a)). From Fig. A.7(a) you can see that the motion of a particle at given total energy, E, takes place within the following range of displacements:

$$-x_{\mathrm{m}} \leq x(t) \leq x_{\mathrm{m}}, \tag{A.109}$$

i.e., inside of the potential well where the kinetic energy is equal to

$$K = E - U \geq 0. \tag{A.110}$$

The particle cannot move outside of this potential well because at $x > x_{\mathrm{m}}$ or at $x < -x_{\mathrm{m}}$ the kinetic energy of the particle would have to be negative, which is impossible in classical physics. Since at the point x_1 the inequality $U(x) < E$ holds, we have $K(x_1) > 0$. At the point x_2 the value of $U(x_2)$ is greater than E ($U(x_2) > E$), so $K(x_2) < 0$. At the points of maximum displacement, $x = \pm x_{\mathrm{m}}$, the particle's velocity becomes equal to zero and $K(\pm x_{\mathrm{m}}) = 0$. At these points the particle's velocity changes its sign to the opposite. For this reason these points are called *turning points*. The turning points of the maximum displacement are found from the condition

$$U(x) = E. \tag{A.111}$$

Then, according to Eq. (A.86), the turning points are defined as

$$x_0 = \sqrt{\frac{2E}{k}}. \tag{A.112}$$

The point $x = 0$ is the particle's position of stable equilibrium.

In real situations we very often have to solve the problem of finding the period of small-amplitude oscillations in a given potential field. Since the force acting on the particle in such a field is defined by Eq. (A.36), by using the corresponding expansion in series of the given function $U(x)$ we can write the particle's equation of motion. By limiting the expansion of $U(x)$ to the quadratic term we can solve the particle's equation of motion in the limit of small-amplitude oscillations and find the period of such oscillations.

Example A.1. The potential energy of a particle with mass m is given by the function

$$U(r) = \frac{A}{r^2} - \frac{B}{r}, \tag{A.113}$$

see Fig. A.7(b). Find the position of equilibrium, r_0, the energy of the particle, U_0, at the position of equilibrium, the angular frequency, ω, and the period, T, of the small-amplitude oscillations of the particle.

Reasoning. The potential energy, $U(r)$, given above defines the oscillatory motion of ions in metals. The coordinate $r = r_0$ as shown in Fig. A.7 is the position of equilibrium of a particle. The first term in the expression for $U(r)$ (Eq. (A.113)) is related to the forces of repulsion that define the particle's behavior at small distances from $r = 0$, i.e., to the left of $r = r_0$. The second term is related to the forces of attraction that are significant at large distances, i.e., at $r \gg r_0$. At the position of equilibrium, $r = r_0$, the potential energy, $U(r)$, reaches its minimum, U_0 (at the position of equilibrium the sum of repulsion and attraction forces is equal to zero). Thus,

$$\left(\frac{\mathrm{d}U}{\mathrm{d}r}\right)_{r=r_0} = -\frac{2A}{r_0^3} + \frac{B}{r_0^2} = 0. \tag{A.114}$$

From the last equation we get

$$r_0 = \frac{2A}{B} \quad \text{and} \quad U(0) = U(r)_{r=r_0} = \frac{A}{r_0^2} - \frac{B}{r_0} = -\frac{B^2}{4A}. \tag{A.115}$$

To define the frequency of oscillations, we find the coefficient k, using Eq. (A.107):

$$k = \left(\frac{\mathrm{d}^2U}{\mathrm{d}r^2}\right)_{r=r_0} = \frac{2}{r^3}\left(\frac{3A}{r} - B\right)_{r=r_0} = \frac{B^4}{8A^3}. \tag{A.116}$$

The potential energy, $U(r)$, near the equilibrium position r_0 can be written in the form

$$U(r) = U(r)_{r=r_0} + \frac{1}{2}\left(\frac{\mathrm{d}^2U}{\mathrm{d}r^2}\right)_{r=r_0}(r - r_0)^2 = -\frac{B^2}{4A} + \frac{B^4}{16A^3}(r - r_0)^2. \tag{A.117}$$

The equation of the particle's oscillatory motion in such a field can be written in the following form, if we take into account the relation $F = -\mathrm{d}U/\mathrm{d}r$:

$$m\,\frac{\mathrm{d}^2 r}{\mathrm{d}t^2} = -\frac{B^4}{8A^3}\,(r - r_0)\,. \tag{A.118}$$

The right-hand side of Eq. (A.118) was obtained by differentiating Eq. (A.117) with respect to r. To find the solution of Eq. (A.118), let us make the change of variable $r - r_0 = R$. Then, the last equation can be rewritten as

$$m\,\frac{\mathrm{d}^2 R}{\mathrm{d}t^2} = -kR \tag{A.119}$$

or

$$\frac{\mathrm{d}^2 R}{\mathrm{d}t^2} + \omega^2 R = 0, \tag{A.120}$$

where the angular frequency of oscillations

$$\omega = \sqrt{\frac{k}{m}} = \frac{B^2}{\sqrt{8A^3 m}}$$

and the period of oscillations

$$T = \frac{4\pi\sqrt{2A^3 m}}{B^2}.$$

A.2.6 Probability and probability density of oscillatory motion

Let us consider a particle oscillating harmonically as $x(t) = x_0 \sin(\omega t)$ in the given region of one-dimensional space and analyze the probability of finding the particle at a specific location, x. Since the period of oscillations is

$$T = \frac{2\pi}{\omega}$$

and the time the particle remains within the given interval $\mathrm{d}x$ is equal to $2\,\mathrm{d}t$ (one $\mathrm{d}t$ for each direction of motion during the period T), the corresponding probability is the ratio of the time which the particle spends in the given time interval to the period:

$$\mathrm{d}P(x) = \frac{2\,\mathrm{d}t(x)}{T} = \frac{\mathrm{d}x}{\pi\sqrt{x_0^2 - x^2}}. \tag{A.121}$$

To obtain this expression, we carried out the following transformations:

$$\frac{\mathrm{d}x}{\mathrm{d}t} = \omega x_0 \cos(\omega t) = \omega\sqrt{x_0^2 - x_0^2 \sin^2(\omega t)} = \omega\sqrt{x_0^2 - x^2} \tag{A.122}$$

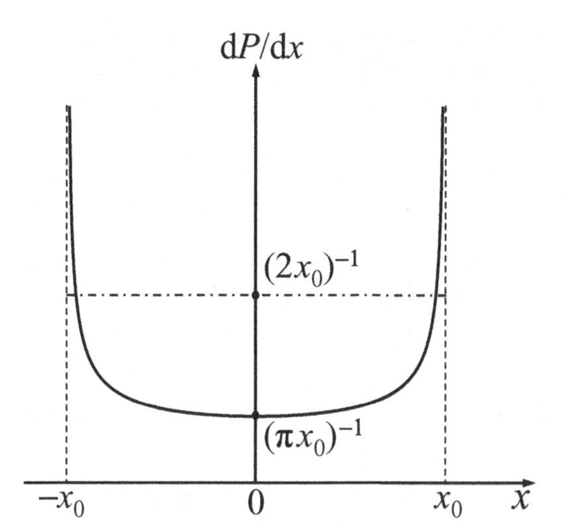

Figure A.8 The probability density, $\mathrm{d}P/\mathrm{d}x$, of finding a particle at a given point x.

and

$$\mathrm{d}t = \frac{\mathrm{d}x}{\omega\sqrt{x_0^2 - x^2}}. \tag{A.123}$$

Let us introduce the *probability density*, the entity that is defined by the ratio of the probability of a particle being in the given interval, $\mathrm{d}P$, to the width of this interval, $\mathrm{d}x$:

$$\frac{\mathrm{d}P}{\mathrm{d}x} = \frac{1}{\pi\sqrt{x_0^2 - x^2}}. \tag{A.124}$$

From Eq. (A.124) it follows that the probability of finding a particle near the equilibrium position, $x = 0$, is minimal, whereas when the particle approaches the turning points, $\pm x_0$, where its velocity approaches zero, the probability density tends to infinity (Fig. A.8). The probability of finding the particle outside of the interval $[-x_0, x_0]$ is equal to zero: $\mathrm{d}P/\mathrm{d}x = 0$. At the same time the total probability of finding the particle in the interval from $-x_0$ to x_0 is equal to unity. Indeed, the probability of finding the particle in the interval $[-x_0, x_0]$ is equal to

$$P(-x_0, x_0) = \frac{1}{\pi}\int_{-x_0}^{x_0} \frac{\mathrm{d}x}{\sqrt{x_0^2 - x^2}} = 1, \tag{A.125}$$

where we took into account that

$$\int_{-1}^{1} \frac{\mathrm{d}\xi}{\sqrt{1 - \xi^2}} = \pi. \tag{A.126}$$

Let us compare the above-considered oscillations with the case when a particle oscillates between two vertical parallel walls with distance $2x_0$ between them, from which the particle is elastically reflected when it hits them perpendicularly (we neglect the influence of the gravitational force). Let us consider the velocity of a particle equal to the average velocity of a harmonically oscillating particle, i.e.,

$$v = \frac{4x_0}{T} = \frac{2x_0\omega}{\pi}. \tag{A.127}$$

Here we took into account that during the period T the particle covers the distance $4x_0$. Since during elastic reflections the magnitude of a particle's velocity does not change, the probability of finding it within the given interval does not depend on the coordinate, i.e.,

$$\mathrm{d}P = \frac{2\,\mathrm{d}x}{vT} = \frac{\mathrm{d}x}{2x_0}. \tag{A.128}$$

Thus, the probability density is constant (the dash–dotted line in Fig. A.8):

$$\frac{\mathrm{d}P}{\mathrm{d}x} = \frac{1}{2x_0}. \tag{A.129}$$

A.3 Summary

1. Newton's second law of motion establishes the relationship between the acceleration of a particle and the sum of forces acting on the particle. Presented in the form of a differential equation of first or second order, it is one of the main equations of classical mechanics, with a wide range of applicability – from macroobjects to microparticles and from the macroscale to the microscale.
2. By force in physics we understand a quantitative measure of the intensity of interaction of two particles (bodies) or a particle and a field. Among all the forces that affect bodies and particles, we can distinguish a wide class of conservative forces. For them the work done does not depend on the form of the trajectory and the work done along a closed contour is equal to zero. The gravitational, electrostatic, and elastic forces are conservative.
3. For a conservative force we can always introduce a function $U(\mathbf{r})$, which is called *potential energy*, and is connected with force as $\mathbf{F} = \nabla(\mathbf{r})$. The force that acts on a particle and the work that is done to displace the particle are real physical magnitudes, whereas potential energy is defined to within the precision of some constant.
4. The most fundamental laws of physics are the three laws of conservation of momentum, angular momentum, and energy. The first two laws are valid for closed systems, whereas the law of conservation of energy applies for conservative systems.
5. The total mechanical energy, E, is defined as the sum of the potential energy, $U(\mathbf{r})$, and kinetic energy, K, which is a quadratic function of the momentum or velocity of a particle.
6. At mechanical equilibrium of a system of particles two vector conditions must be satisfied: the sum of all forces and the sum of torques must be equal to zero. For a system that can be considered a material point it is sufficient if only the first condition is satisfied and the equilibrium condition can be written as $\nabla U(\mathbf{r}) = 0$.

7. The equilibrium position can be either stable or unstable. In the first case the body is returned to the stable equilibrium position by the forces which occur during the displacement, whereas in the second case the body is taken away from the unstable equilibrium position by the same forces.
8. Oscillations that are harmonic are caused by the influence of an elastic or quasielastic force. By a quasielastic force we understand a force that is not connected with elastic interaction, but depends linearly on the displacement, as is the case for elastic force.
9. In harmonic oscillations the kinetic and potential energy separately change with time at double the frequency; however, the energy of oscillations does not depend on time.
10. If oscillations have small amplitude then terms in the expansion of the potential energy over the displacement of a particle from the equilibrium position of higher than second order can be neglected (the harmonic approximation). The terms in the expansion which are of higher than second order define the anharmonic effects of oscillatory motion. If the potential energy of a particle is quadratic with respect to the displacement from the equilibrium position then it is assumed that its motion takes place in a "parabolic potential."
11. The probability of finding an oscillating particle near the equilibrium position is the lowest. The case of a particle approaching the turning points, where the particle has velocity equal to zero, has a probability density tending to infinity.

A.4 Problems

Problem A.4.1. A ball of mass m is thrown at point $x = y = 0$ at initial time $t = 0$ with initial momentum $\mathbf{p}_0 = m\mathbf{v}_0$, at an angle $\alpha = 45^\circ$ (see Fig. A.9). Find the dependence of the magnitude of the ball's momentum on time during the ball's motion, the minimal magnitude of the ball's momentum, and the magnitude of the momentum at the time when the ball hits the ground. Write the expression for the ball's trajectory. Disregard the friction of air.

Problem A.4.2. A particle with mass m and initial velocity $\mathbf{v}_0$ enters a fluid, where a retarding force, $\mathbf{F}_{\mathrm{rt}} = -b\mathbf{v}$, is applied to it. Find the time during which the particle's velocity drops to $\mathbf{v}_0/n$ (n is an integer number greater than 1), and find the distance traveled by the particle during this time. Ignore the gravitational force acting on the particle.

Problem A.4.3. A particle with mass m begins its motion from the state of rest under the influence of an oscillating force $F(t) = F_0 \sin(\gamma t)$. Find the particle's velocity and the distance traveled at time t_1, when the force reaches its first maximum, and at time t_2, when the force becomes equal to zero.

Problem A.4.4. Two particles, which have masses m_1 and m_2 and velocities before the collision v_1 and v_2, undergo a perfect inelastic collision. Find the velocity of the particle formed after the collision and find the proportion of the mechanical energy that has been transformed into internal energy.

Problem A.4.5. Find the velocities of two particles after their perfectly elastic central collision. The initial state of the two particles before the collision is given by the

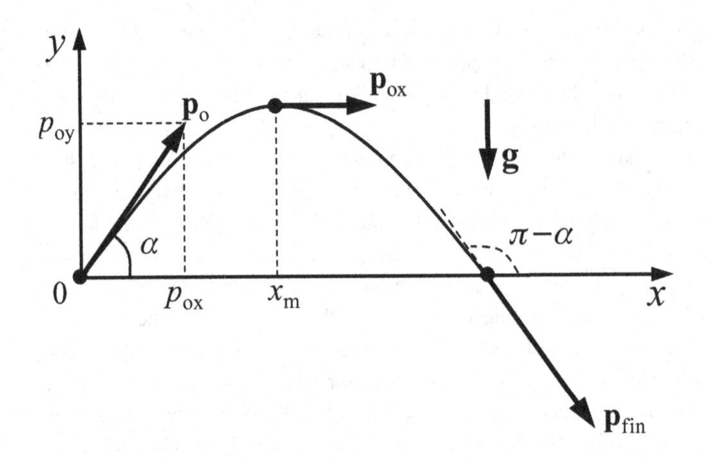

Figure A.9 The trajectory of a ball with initial momentum $\mathbf{p}_0$ and initial angle α.

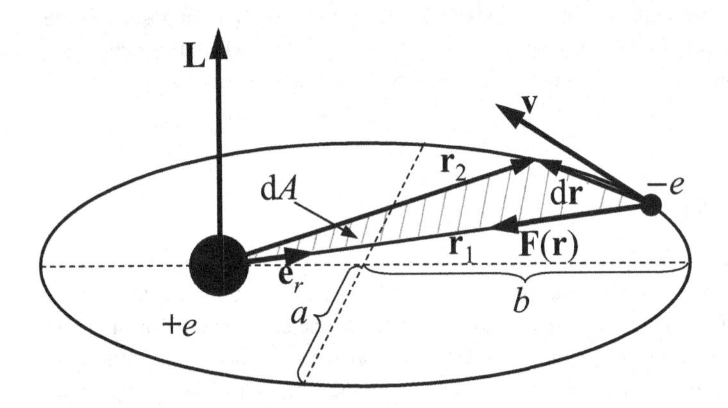

Figure A.10 The elliptic orbit of an electron in the field of a central force $F(r)\mathbf{e}_r$.

parameters m_1, $\mathbf{v}_1$, m_2, and $\mathbf{v}_2$. A *central collision* is defined as a *collision that happens along the line connecting the centers of the particles.*

Problem A.4.6. According to one of the first models of the hydrogen atom (Thomson's model), an electron (a particle with mass $m_e = 9.1 \times 10^{-31}$ kg and negative charge $-e = -1.6 \times 10^{-19}$ C) is located at the center of the atom and all the rest of the atomic space is uniformly filled by a spherical cloud of positive charge, the total charge of which is equal to $+e$, which does not hinder the motion of an electron. Considering that the hydrogen atom's radius is known ($a \approx 0.5 \times 10^{-10}$ m), estimate the oscillation frequency of an electron. (Answer: $\omega_e \approx 4.5 \times 10^{16}$ s^{-1}.)

Problem A.4.7. In Sommerfeld's model of the hydrogen atom the electron rotates around the proton along stationary elliptic orbits (see Fig. A.10). Find the electron's angular momentum in terms of the parameters of the orbit (the semimajor and semiminor axes of the ellipse are a and b, respectively) and the period of the orbital rotation, T. Find the electron's angular momentum for a circular orbit.

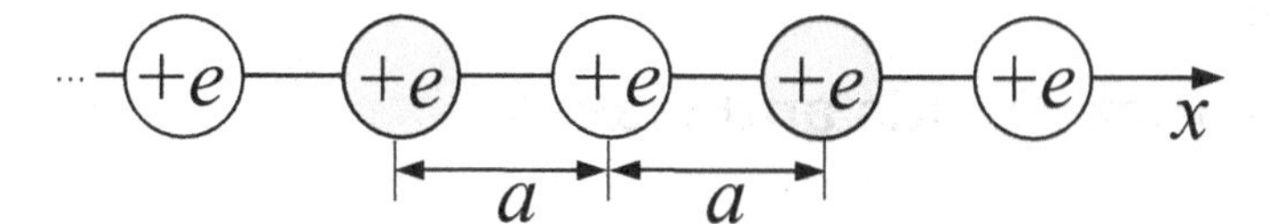

Figure A.11 A linear chain of ions.

Problem A.4.8. An electron with velocity **v** enters a uniform magnetic field **B**, which is perpendicular to the electron's velocity. Find the radius, r, of the circle along which the electron is moving, its angular momentum, L, and its magnetic moment, μ. The magnetic moment of an electron can be calculated according to the formula

$$\mu = -IS\mathbf{e}_B, \tag{A.130}$$

where I is the circular current induced by the electron's motion along a circle, S is the cross-sectional area of a current loop, and $\mathbf{e}_B$ is a unit vector along vector **B**.

Problem A.4.9. A particle with positive charge q, mass m, and velocity v_1 enters a region with an electric field. Associated with the electric force there is an electric potential ϕ. While moving along the electric field lines from a point with electrostatic potential ϕ_1 to a point with electrostatic potential ϕ_2, the particle accelerates. Find its final velocity v_2.

Problem A.4.10. Show that the force acting on a particle in a homogeneous electrostatic field is conservative and that the field itself is a potential field.

Problem A.4.11. Show that the kinetic energy of a charged particle in a magnetic field does not change.

Problem A.4.12. Consider an infinite linear chain of ions with mass m and charge e. The distance between the neighboring ions is a (Fig. A.11). Find the period of linear oscillations of one of the ions, considering that all other ions are motionless. Estimate the frequency, ω, of a chain of copper ions having $e = 1.6 \times 10^{-19}$ C, $m = 1.06 \times 10^{-25}$ kg, and $a = 3 \times 10^{-10}$ m. (Answer: $\omega \approx 2 \times 10^{14}$ s^{-1}.)

Problem A.4.13. Consider a linear chain of ions that consists of ions with opposite charges. The distance between ions is equal to a. Find the energy of the electrostatic interaction of a separate ion with all of the others, considering the number of ions, N, to be large. Find the energy of the whole chain of ions.

Appendix B
Electromagnetic fields and waves

In Appendix A, Section A.2, we considered the oscillations of particles using Newton's laws of classical mechanics. The distinctive feature of Newton's mechanics, when considering interactions between particles (for example, the gravitational interaction), is the instantaneous transmission of the interaction between particles, i.e., transmission occurs with infinite speed. The interaction between charged particles is realized through an electromagnetic field, which possesses energy and momentum and is carried through space with finite speed. Electromagnetic waves can exist without any charges in a space in which there is no substance, i.e., in vacuum. This is substantiated by the fact that the equations of classical electrodynamics allow solutions in the form of electromagnetic waves for such conditions.

The main equations of classical electrodynamics are Maxwell's equations, which were formulated after analyzing numerous experimental data. In this sense they are analogous to Newton's equations of classical mechanics. Maxwell's equations are the basis for electrical and radio engineering, television and radiolocation, integrated and fiber optics, and numerous phenomena and processes that take place in materials placed in an electromagnetic field. Together with Newton's equations, Maxwell's equations are the fundamental equations of classical physics. Just like Newton's equations, Maxwell's equations have their limits of applicability. For example, they do not sufficiently well describe the state of an electromagnetic field in a medium at frequencies higher than 10^{14}–10^{15} Hz. This limitation is due to the manifestation at high frequencies of the quantum nature of the interaction of electromagnetic radiation with materials.

In this appendix we will consider Maxwell's equations, the wave equations derived from them, and their solutions, which describe electromagnetic waves. In contrast to elastic waves, in which the material particles take part in the wave movement, in an electromagnetic wave oscillations of the electric and magnetic vectors take place. Together with the solution of the boundary problem and the definition of coefficients of reflection and transmission through the interface between two media, we will consider the most characteristic wave phenomena – interference and diffraction – which prove the wave nature of electromagnetic radiation.

B.1 Equations of an electromagnetic field

B.1.1 Characteristics of electric and magnetic fields

The electric, **E**, and magnetic, **B**, fields are two vector fields that describe an electromagnetic field. Sometimes **B**, which is measured in tesla ([T]), is called the *magnetic* *flux density* and $\mathbf{E}$, which is measured in volt/meter ([V m^{-1}]), is called the *electric field intensity*. Charged particles are the sources of electric and magnetic fields. An isolated particle with charge q creates around itself a field, the electric component of which is

$$\mathbf{E} = \frac{k_e q}{r^3}\mathbf{r}, \tag{B.1}$$

and the magnetic component of which is

$$\mathbf{B} = \frac{k_m q}{r^3}\mathbf{v} \times \mathbf{r}, \tag{B.2}$$

where $\mathbf{r}$ is the vector directed from the position of charge q to the point where the field is measured, and $\mathbf{v}$ is the velocity of the particle. The coefficients k_e and k_m are defined as

$$k_e = \frac{1}{4\pi\epsilon_0} \tag{B.3}$$

and

$$k_m = \frac{\mu_0}{4\pi}, \tag{B.4}$$

where $\epsilon_0 = 8.85 \times 10^{-12}$ F m^{-1} (farad/meter) is known as the *permittivity of free space* and $\mu_0 = 4\pi \times 10^{-7}$ H m^{-1} (henry/meter) is known as the *permeability of free space*. It follows from Eq. (B.1) that the electric field is created by a charge and it does not matter whether the charge is moving or not. A magnetic field can be created only by a *moving* (in the chosen frame of reference) *charge*. Stationary charges do not generate magnetic fields.

If there is a system of n charged particles, the field at an arbitrary point $\mathbf{r}$ is defined as a vector sum of the fields created by particles, $\mathbf{E}_i$ and $\mathbf{B}_i$:

$$\mathbf{E} = \sum_{i=1}^{n} \mathbf{E}_i \tag{B.5}$$

and

$$\mathbf{B} = \sum_{i=1}^{n} \mathbf{B}_i. \tag{B.6}$$

Equations (B.5) and (B.6) describe the *superposition principle* of the fields in electromagnetism (note that electromagnetism is the field of physics that studies electrical and magnetic phenomena). The physical quantities $\mathbf{E}$ and $\mathbf{B}$ are the *force-field characteristics of electric and magnetic fields* because they define the electrical, $\mathbf{F}_e = q\mathbf{E}$, and magnetic, $\mathbf{F}_m = q\mathbf{v}_0 \times \mathbf{B}$, components of the total force that acts on a point charge, q, placed in the electromagnetic field:

$$\mathbf{F} = \mathbf{F}_e + \mathbf{F}_m = q(\mathbf{E} + \mathbf{v} \times \mathbf{B}), \tag{B.7}$$

where $\mathbf{v}$ is the velocity of the charge q. From Eq. (B.7) it follows that the magnetic force, $\mathbf{F}_m$, in contrast to the electric force, $\mathbf{F}_e$, acts only on a moving charged particle. The directional movement of charged particles creates an electric current, **I**, whose density, **j**, is given by the expression

$$\mathbf{j} = qn\mathbf{v}, \tag{B.8}$$

where **j** is measured in ampere/meter^2 ([A m^{-2}]), q is the charge of the carriers measured in coulombs ([C]), and n is the concentration of charged carriers (or number of charged carriers in the volume unit, measured in 1/meter^3 ([m^{-3}])). The directional movement of electrons in the conductor creates an electric current, I, whose magnitude in the general case is given by the expression

$$I = \int_S \mathbf{j} \cdot \mathrm{d}\mathbf{S}, \tag{B.9}$$

where I is measured in amperes ([A]), S is the cross-section of the conductor, measured in meter^2 ([m^2]). In the case of a uniform distribution across the cross-section, S, of the current density, j, Eq. (B.9) takes the form

$$I = jS. \tag{B.10}$$

Since the magnetic field, **B**, according to Eq. (B.2) is created by moving charges, it follows that it can be generated by currents. In accordance with the Biot–Savart law,

$$\mathrm{d}\mathbf{B} = \frac{k_{\mathrm{m}} I}{r^3} \, \mathrm{d}\mathbf{l} \times \mathbf{r}, \tag{B.11}$$

where **dl** is the vector element of a linear conductor, the direction of which coincides with the direction of a current, and **r** is the vector directed from the element with the current to the point where the magnetic field element, **dB**, is measured.

When considering the electromagnetic field in a medium, it is necessary to take into account the fact that fields can be created by external charges and currents as well as by bound charges (in dielectrics) and currents of bound charges (in magnetic materials). The vectors **E** and **B** are the characteristics of the total field of external and bound charges and currents. For convenience, quantities that define only external sources will be introduced. For an electric field such a vector is the *electric displacement* **D**, which is measured in coulomb/meter^2 ([C m^{-2}]), and for a magnetic field it is the *magnetic field intensity* **H**, which is measured in ampere/meter ([A m^{-1}]). In the case of an isotropic medium the above-mentioned quantities are defined as

$$\mathbf{D} = \epsilon_0 \epsilon \mathbf{E}, \tag{B.12}$$

$$\mathbf{H} = \frac{\mathbf{B}}{\mu_0 \mu}, \tag{B.13}$$

where we introduced the *relative permittivity*, ϵ, and *relative permeability*, μ, of a medium, which define the ratio of fields in a medium and in vacuum:

$$\mathbf{E}_{\mathrm{medium}} = \frac{1}{\epsilon} \mathbf{E}_{\mathrm{vacuum}}, \tag{B.14}$$

$$\mathbf{B}_{\mathrm{medium}} = \mu \mathbf{B}_{\mathrm{vacuum}}. \tag{B.15}$$

From Eqs. (B.14) and (B.15) it follows that ϵ and μ are dimensionless quantities. Using these quantities, an important classification of different media can be made.

All materials can be divided into three broad classes with respect to their magnetic properties:

(a) *diamagnetic materials, for which the static value of* μ *is less than* 1,
(b) *paramagnetic materials with* $\mu \geq 1$, and
(c) *strongly magnetic materials (ferromagnetic, ferrimagnetic, and antiferromagnetic materials) with* $\mu \gg 1$.

A similar classification can be made on the basis of ϵ. Materials with $\epsilon \gg 1$ are called *ferroelectrics*, whereas materials with $\epsilon \geq 1$ are *paraelectrics*.

Lately, the so-called "*left-handed media*," for which ϵ and μ are less than zero simultaneously, have attracted great attention. The optical properties of such media are very different from the properties of conventional media with $\epsilon > 0$ and $\mu > 0$.

B.1.2 Maxwell's equations

Maxwell's equations, which relate electromagnetic fields to charges and currents, can be written in the form of the following differential equations:

$$\nabla \times \mathbf{H} = \frac{\partial \mathbf{D}}{\partial t} + \mathbf{j}, \tag{B.16}$$

$$\nabla \times \mathbf{E} = -\frac{\partial \mathbf{B}}{\partial t}, \tag{B.17}$$

$$\nabla \cdot \mathbf{B} = 0, \tag{B.18}$$

$$\nabla \cdot \mathbf{D} = \rho, \tag{B.19}$$

where ρ is the charge density, measured in coulomb/meter^3 ([C m^{-3}]), and ∇ is the gradient operator. From Eqs. (B.16)–(B.19) it follows that (a) not only electric charges but also time-varying magnetic fields can be the sources of electric fields; and (b) the sources of magnetic fields are moving electric charges (currents) as well as time-varying electric fields. The Cartesian coordinates of **E**, **B**, **H**, and **D** can be written in the following form:

$$\begin{aligned}
\frac{\partial H_z}{\partial y} - \frac{\partial H_y}{\partial z} &= \frac{\partial D_x}{\partial t} + j_x, \\
\frac{\partial H_x}{\partial z} - \frac{\partial H_z}{\partial x} &= \frac{\partial D_y}{\partial t} + j_y, \\
\frac{\partial H_y}{\partial x} - \frac{\partial H_x}{\partial y} &= \frac{\partial D_z}{\partial t} + j_z, \\
\frac{\partial E_z}{\partial y} - \frac{\partial E_y}{\partial z} &= -\frac{\partial B_x}{\partial t}, \\
\frac{\partial E_x}{\partial z} - \frac{\partial E_z}{\partial x} &= -\frac{\partial B_y}{\partial t},
\end{aligned} \tag{B.20}$$

$$\frac{\partial E_y}{\partial x} - \frac{\partial E_x}{\partial y} = -\frac{\partial B_z}{\partial t}, \tag{B.21}$$

$$\frac{\partial B_x}{\partial x} + \frac{\partial B_y}{\partial y} + \frac{\partial B_z}{\partial z} = 0, \tag{B.22}$$

$$\frac{\partial D_x}{\partial x} + \frac{\partial D_y}{\partial y} + \frac{\partial D_z}{\partial z} = \rho. \tag{B.23}$$

We see that Maxwell's equations have the form of a system of linear first-order partial differential equations. Equations (B.12) and (B.13) may be used to reduce the number of unknown functions in the system of Maxwell's equations. Let us note that in textbooks and in the scientific literature the vector product $\nabla \times \mathbf{A}$ is sometimes denoted as rot **A** (rotation of vector **A**) and the scalar product $\nabla \cdot \mathbf{A}$ as div **A** (divergence of **A**), i.e.,

$$\nabla \times \mathbf{A} = \operatorname{rot} \mathbf{A}, \tag{B.24}$$

$$\nabla \cdot \mathbf{A} = \operatorname{div} \mathbf{A}. \tag{B.25}$$

If there is an interface that separates two media, then the following boundary conditions for the introduced characteristics of electric and magnetic fields have to be satisfied:

$$(\mathbf{D}_2 - \mathbf{D}_1) \cdot \mathbf{n} = \sigma, \tag{B.26}$$

$$(\mathbf{E}_2 - \mathbf{E}_1) \cdot \boldsymbol{\tau} = 0, \tag{B.27}$$

$$(\mathbf{B}_2 - \mathbf{B}_1) \cdot \mathbf{n} = 0, \tag{B.28}$$

$$\mathbf{n} \times (\mathbf{H}_2 - \mathbf{H}_1) = \mathbf{j}_s. \tag{B.29}$$

Here **n** is the unit vector in the direction perpendicular to the interface of the two media, which is directed from medium 1 to medium 2; $\boldsymbol{\tau}$ is a unit vector oriented along the interface; σ is the charge surface density with dimension [C m^{-2}]; and $\mathbf{j}_s$ is the vector of current surface density ([A m^{-1}]) at the interface. From Eqs. (B.26)–(B.29) it follows that the tangential component, $\mathbf{E}_t$, of the vector **E** is continuous across the interface, i.e., $\mathbf{E}_t$ undergoes no change at the interface: $\mathbf{E}_{t1} = \mathbf{E}_{t2}$. The same is true for the normal component, $\mathbf{B}_n$, of the vector **B**, i.e., $\mathbf{B}_{n1} = \mathbf{B}_{n2}$. At an interface without any free charges and currents, the normal components of the vector **D** and the tangential components of the vector **H** are also continuous. If the interface is charged, then the normal component of the vector **D** has a discontinuity, and if there is a current along the interface, then the tangential component of the vector **H** has a discontinuity.

B.1.3 Energy and energy density

Electric and magnetic fields have the ability to store energy. The total energy can be defined through the energy density of the respective fields at each point of the volume occupied by the field:

$$W = \int_V (w_e + w_m) \mathrm{d}V, \tag{B.30}$$

where the energy density of the electric field is

$$w_e = \frac{1}{2}\mathbf{E}\cdot\mathbf{D} = \frac{\epsilon\epsilon_0 E^2}{2} \tag{B.31}$$

and the energy density of the magnetic field is

$$w_m = \frac{1}{2}\mathbf{B}\cdot\mathbf{H} = \frac{B^2}{2\mu\mu_0}. \tag{B.32}$$

Using the dependences of field characteristics on coordinates, we can find with the help of Eq. (B.30) the total energy of an electromagnetic field in any arbitrary volume.

Example B.1. Let us consider a sphere of radius R, which is uniformly charged with the total charge Q. Find the energy of the electric field inside the sphere and outside it, considering that the field intensity distribution in such a sphere is known:

$$E(r) = \begin{cases} \dfrac{k_e Q}{r^2}, & r \geq R, \\ \dfrac{k_e Q}{R^2}\dfrac{r}{R}, & r \leq R. \end{cases} \tag{B.33}$$

Outside of the sphere the electric field is equal to the electric field of a point charge Q placed at the center of the sphere. Inside of the sphere the field at a distance r from its center is defined also as a field of a point charge

$$Q_{in} = Q\frac{r^3}{R^3}, \tag{B.34}$$

which is placed into the center of the sphere and is equal to a charge inside of a sphere of radius r.

The distribution of the intensity of the electric field, $E(r)$, defined by Eq. (B.33) is shown in Fig. B.1.

Reasoning. Let us substitute Eq. (B.33) into the expression for the energy density of the electric field, Eq. (B.31):

$$w_e(r) = \begin{cases} \dfrac{k_e Q^2}{8\pi}\dfrac{1}{r^4}, & r \geq R, \\ \dfrac{k_e Q^2}{8\pi}\dfrac{r^2}{R^6}, & r \leq R. \end{cases} \tag{B.35}$$

Let us calculate with the help of Eq. (B.30) the energy of the electric field, W_1, concentrated outside of the sphere, and the energy W_2 inside of the sphere. Taking into account that in the case of spherical symmetry the differential volume is defined as

$$dV = 4\pi r^2\, dr, \tag{B.36}$$

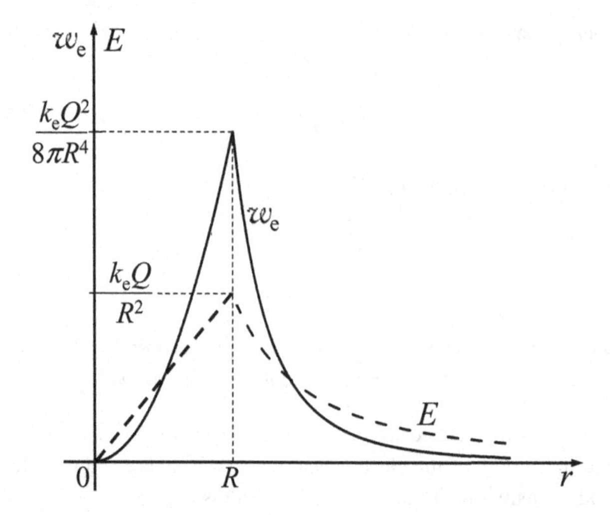

Figure B.1 Distributions of the electric field, $E(r)$ (dashed line), and density of electric field energy, $w_e(r)$ (solid line), in the case of an electrically charged sphere.

for the above-mentioned energies we obtain the following expressions:

$$W_1 = \int_{V_{out}} \frac{k_e Q^2}{8\pi} \frac{1}{r^4} \, dV = \frac{k_e Q^2}{8\pi} \int_R^\infty \frac{4\pi r^2}{r^4} \, dr = \frac{k_e Q^2}{2R}, \tag{B.37}$$

$$W_2 = \int_{V_{in}} \frac{k_e Q^2}{8\pi} \frac{r^2}{R^6} \, dV = \frac{k_e Q^2}{8\pi} \int_0^R \frac{4\pi r^2 r^2}{R^6} \, dr = \frac{k_e Q^2}{10R}. \tag{B.38}$$

The ratio of total energies of the electric field outside and inside of the sphere does not depend on the radius, R, and is equal to

$$\frac{W_1}{W_2} = 5. \tag{B.39}$$

Example B.2. The equation that defines the law of charge conservation and establishes at each point of space the relation between the density of charge, $\rho(\mathbf{r})$, and the density of current, $\mathbf{j}(\mathbf{r})$, can be obtained from Maxwell's equations. Derive the charge-conservation equation.

Reasoning. Let us perform the scalar multiplication of the left-hand and right-hand sides of Eq. (B.16) by the operator ∇ and, in the first term on the right-hand side of the equation obtained, let us change the order of taking time and spatial derivatives:

$$\nabla \cdot (\nabla \times \mathbf{H}) = \frac{\partial \nabla \cdot \mathbf{D}}{\partial t} + \nabla \cdot \mathbf{j}. \tag{B.40}$$

In Eq. (B.40) the left-hand side is identically equal to zero:

$$\nabla \cdot (\nabla \times \mathbf{H}) = (\nabla \times \nabla) \cdot \mathbf{H} = 0, \tag{B.41}$$

because $\nabla \times \nabla = 0$. On the right-hand side of Eq. (B.40) let us replace $\nabla \cdot \mathbf{D}$ by ρ, taking into account Eq. (B.19). As a result we obtain the equation

$$\frac{\partial \rho}{\partial t} + \nabla \cdot \mathbf{j} = 0, \tag{B.42}$$

which is called the *continuity equation* (there is an analogous equation in hydrodynamics, where ρ is the density of liquid and $\mathbf{j} = \rho \mathbf{v}$ is the density flux of liquid).

Let us choose an arbitrary point $\mathbf{r}$, and encircle it by the closed surface S. Then let us integrate Eq. (B.42) over the volume V which this surface encloses:

$$\frac{\partial}{\partial t} \int_V \rho(\mathbf{r}) \mathrm{d}V = - \int_V (\nabla \cdot \mathbf{j}) \mathrm{d}V. \tag{B.43}$$

The integral on the left-hand side of Eq. (B.43) is equal to the total charge Q in the volume V that is enclosed by the surface S. The integral over the volume, V, on the right-hand side of Eq. (B.43) can be transformed into an integral over the closed surface, S:

$$\frac{\mathrm{d}Q}{\mathrm{d}t} = - \oint_S \mathbf{j} \cdot \mathrm{d}\mathbf{S} = -I. \tag{B.44}$$

This equation represents the fact that the change of the charge inside of the closed surface is defined by the current through this surface.

B.2 Electromagnetic waves

B.2.1 Wave equations

It follows from Eqs. (B.1) and (B.2) that electric and magnetic fields can exist in a medium as well as in vacuum. These fields can exist near charges and currents, as well as far from them. Moreover, it follows from Maxwell's equations, (B.16)–(B.19) that time-varying fields can exist without any charges or currents. The change in time of the flux, D, leads to the creation of the solenoidal magnetic field, H, and the changing magnetic flux density, B, induces the solenoidal electric field, E. The above-mentioned changes (disturbances), which are called *electromagnetic waves*, propagate in a medium as well as in vacuum.

The wave character of electromagnetic field propagation is built into Maxwell's equations. To check this statement, let us consider a uniform and isotropic medium from which free charges and currents are absent ($\rho = 0$ and $\mathbf{j} = 0$). Let us extract from the system of equations (B.16)–(B.19) two equations with rotors and let us replace the field flux densities by field intensities, assuming that ϵ and μ do not depend on time, t:

$$\nabla \times \mathbf{H} = \epsilon_0 \epsilon \frac{\partial \mathbf{E}}{\partial t}, \tag{B.45}$$

$$\nabla \times \mathbf{E} = -\mu_0 \mu \frac{\partial \mathbf{H}}{\partial t}. \tag{B.46}$$

Let us take the time derivative of both sides of Eq. (B.45):

$$\frac{\partial}{\partial t}\nabla \times \mathbf{H} = \nabla \times \frac{\partial \mathbf{H}}{\partial t} = \epsilon_0 \epsilon \frac{\partial^2 \mathbf{E}}{\partial t^2}, \tag{B.47}$$

where we changed the order of time and spatial derivatives. Let us substitute into Eq. (B.47) $\partial \mathbf{H}/\partial t$ from Eq. (B.46):

$$-\frac{1}{\mu_0 \mu}\nabla \times (\nabla \times \mathbf{E}) = \epsilon_0 \epsilon \frac{\partial^2 \mathbf{E}}{\partial t^2}. \tag{B.48}$$

The rule of opening of a double vector product of any three vectors **A**, **B**, and **C** is the following:

$$\mathbf{A} \times (\mathbf{B} \times \mathbf{C}) = \mathbf{B} \cdot (\mathbf{A} \cdot \mathbf{C}) - \mathbf{C} \cdot (\mathbf{A} \cdot \mathbf{B}). \tag{B.49}$$

Using this rule, we can carry out the following transformations:

$$\nabla \times (\nabla \times \mathbf{E}) = \nabla \cdot (\nabla \cdot \mathbf{E}) - \nabla^2 \mathbf{E} = -\nabla^2 \mathbf{E}. \tag{B.50}$$

Here we took into account that, according to Eq. (B.19), for the case considered here (ϵ = constant and $\rho = 0$) the following equality has to be fulfilled:

$$\nabla \cdot \mathbf{D} = \nabla \cdot (\epsilon \mathbf{E}) = \epsilon \nabla \cdot \mathbf{E} = 0, \tag{B.51}$$

i.e., $\nabla \cdot \mathbf{E} = 0$. By substituting Eq. (B.50) into Eq. (B.48) we obtain the wave equation for the electric field, **E**:

$$\nabla^2 \mathbf{E} = \frac{\epsilon \mu}{c^2}\frac{\partial^2 \mathbf{E}}{\partial t^2}. \tag{B.52}$$

Analogously, we can obtain the wave equation for the magnetic field intensity, **H**:

$$\nabla^2 \mathbf{H} = \frac{\epsilon \mu}{c^2}\frac{\partial^2 \mathbf{H}}{\partial t^2}. \tag{B.53}$$

Here we introduced the quantity c:

$$c = \frac{1}{\sqrt{\epsilon_0 \mu_0}} = 3 \times 10^8 \text{ m s}^{-1}, \tag{B.54}$$

the magnitude and dimension of which coincide with those of the speed of light in vacuum. The parameter

$$v = \frac{c}{\sqrt{\epsilon \mu}} \tag{B.55}$$

is the phase velocity of the electromagnetic field propagating in a medium. For vacuum $\epsilon = \mu = 1$, and the phase velocity, v, of the electromagnetic wave coincides with the speed of light, c.

To define the fields associated with an electromagnetic wave, let us write the solutions for Eqs. (B.52) and (B.53) and for the field intensities of electric and magnetic fields in the form of plane monochromatic *traveling* waves:

$$\mathbf{E} = \mathbf{E}_0 \cos(\omega t - \mathbf{k} \cdot \mathbf{r}), \tag{B.56}$$

$$\mathbf{H} = \mathbf{H}_0 \cos(\omega t - \mathbf{k} \cdot \mathbf{r}), \tag{B.57}$$

where $\mathbf{E}_0$ and $\mathbf{H}_0$ are the amplitudes of the corresponding fields, ω is the frequency of the oscillations of field intensities $\mathbf{E}$ and $\mathbf{H}$ in the wave, and $\mathbf{k}$ is the wavevector.

The magnitude of the wavevector, k, is equal to

$$k = \frac{\omega}{v} = \frac{2\pi}{\lambda}, \tag{B.58}$$

where $\lambda = vT$ is the wavelength, which is equal to the distance covered by the wave with phase velocity v during the period of oscillations $T = 2\pi/\omega$.

The direction of wave propagation coincides with the direction of the wavevector $\mathbf{k}$. Indeed, if we assume that the argument in Eqs. (B.56) and (B.57) is constant, it leads to the equation

$$\mathbf{k} \cdot \mathbf{r} - \omega t = \text{constant}, \tag{B.59}$$

which defines the space planes of equal wave phases propagating with velocity $\mathbf{v}$ in the direction of the wavevector $\mathbf{k}$. The scalar product $\mathbf{k} \cdot \mathbf{r}$ can be written in terms of the projections on the Cartesian coordinates

$$\mathbf{k} \cdot \mathbf{r} = k_x x + k_y y + k_z z, \tag{B.60}$$

where $k_x = k \cos\alpha$, $k_y = k \cos\beta$, and $k_z = k \cos\gamma$; α, β, and γ are the angles between the wavevector, $\mathbf{k}$, and the corresponding coordinate axis. The constants $\cos\alpha$, $\cos\beta$, and $\cos\gamma$ are called the *directional cosines of the vector* $\mathbf{k}$ (see Fig. B.2). Thus, the spatial change of the wave's phase takes place in the direction of the vector $\mathbf{k}$, namely this vector defines the direction of wave propagation in an isotropic medium.

Following substitution of solutions (B.56) and (B.57) into Eqs. (B.52) and (B.53), we can obtain the relation between the frequency, ω, and the wavenumber, k. As a result we will obtain the dispersion equation for the electromagnetic wave in an isotropic non-absorbing medium,

$$\omega^2 = \frac{c^2}{\epsilon\mu} k^2, \tag{B.61}$$

or

$$\omega = vk. \tag{B.62}$$

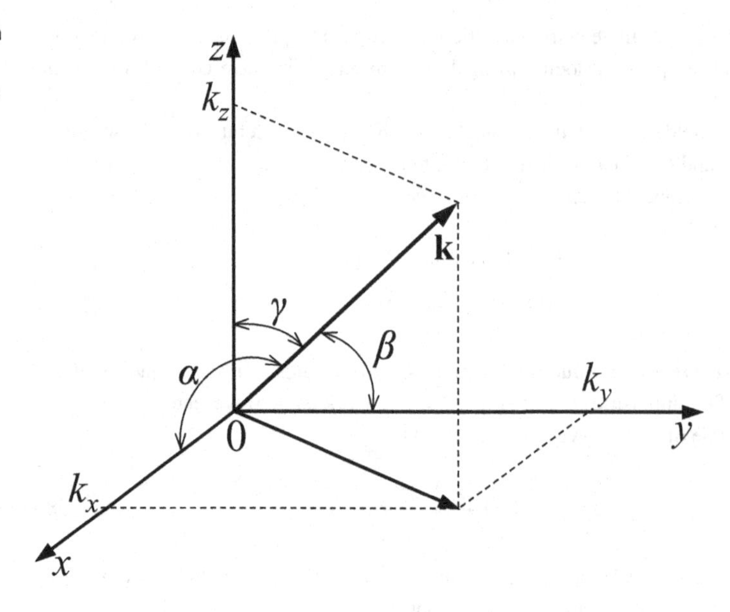

Figure B.2 The orientation of the wavevector **k** in terms of the angles of the directional cosines α, β, and γ.

If the parameters of the medium ϵ and μ depend on the frequency, then the phase velocity, v, also depends on the frequency, i.e., such a medium is dispersive. If ϵ and μ do not depend on the frequency (this happens for real media in certain frequency ranges), then such a medium is not dispersive (in this case the phase and group velocities coincide). Let us substitute Eqs. (B.56) and (B.57) into Eqs. (B.45) and (B.46), then we get

$$\mathbf{k} \times \mathbf{H} = -\epsilon_0 \epsilon \omega \mathbf{E}, \tag{B.63}$$

$$\mathbf{k} \times \mathbf{E} = \mu_0 \mu \omega \mathbf{H}. \tag{B.64}$$

According to the definition, the directions of the three vectors that compose the scalar triple product, **E**, **H**, and **k**, are perpendicular to each other and constitute in this order a right-handed system.

Vectors **E** and **H** in a monochromatic wave oscillate in orthogonal planes; the vector, **k**, is directed along the line of intersection of these planes in the direction of wave propagation (see Fig. B.3). Such a wave is called *linearly polarized*. There exist waves for which vectors **E** and **H**, while staying perpendicular to each other during wave propagation, rotate clockwise or counter-clockwise. These waves are called *circularly polarized waves* or *elliptically polarized waves*, depending on the trajectory of the end of the vector **E**.

B.2.2 Energy flux and wave momentum

Let us scalarly multiply Eq. (B.63) by **E** and Eq. (B.64) by **H**, and then subtract the first result from the second:

$$\mathbf{H} \cdot (\mathbf{k} \times \mathbf{E}) - \mathbf{E} \cdot (\mathbf{k} \times \mathbf{H}) = \omega(\mu_0 \mu H^2 + \epsilon_0 \epsilon E^2). \tag{B.65}$$

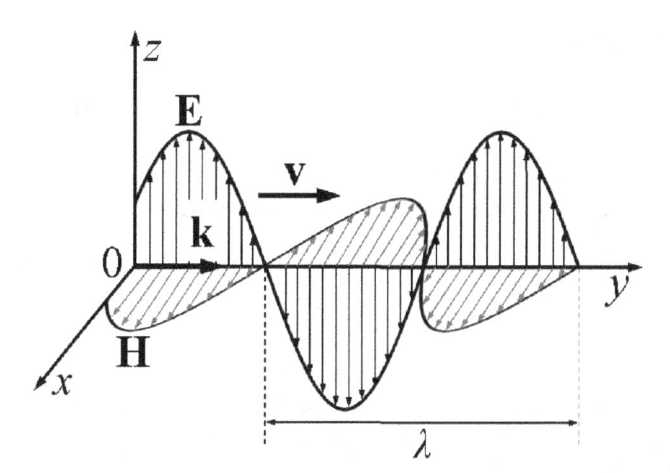

Figure B.3 The propagation of an electromagnetic wave with wavelength λ; **E**, **H**, and **k** constitute a right-handed system.

The expression in parentheses on the right-hand side of Eq. (B.65) is equal to twice the energy density of the electromagnetic field, i.e.,

$$\mu_0\mu H^2 + \epsilon_0\epsilon E^2 = 2(w_\text{m} + w_\text{e}) = 2w. \tag{B.66}$$

The left-hand side of Eq. (B.65), according to *cyclic permutation* for any scalar triple product

$$\mathbf{A}\cdot(\mathbf{B}\times\mathbf{C}) = \mathbf{C}\cdot(\mathbf{A}\times\mathbf{B}) = \mathbf{B}\cdot(\mathbf{C}\times\mathbf{A}), \tag{B.67}$$

will transform to

$$\mathbf{H}\cdot(\mathbf{k}\times\mathbf{E}) - \mathbf{E}\cdot(\mathbf{k}\times\mathbf{H}) = \mathbf{k}\cdot(\mathbf{E}\times\mathbf{H}) - \mathbf{k}\cdot(\mathbf{H}\times\mathbf{E}) = 2\mathbf{k}\cdot(\mathbf{E}\times\mathbf{H}). \tag{B.68}$$

Let us introduce the vector **P**:

$$\mathbf{P} = \mathbf{E}\times\mathbf{H}. \tag{B.69}$$

P is known as the *Poynting vector*. This vector defines the density of energy flux of the electromagnetic field and has the dimension [J m^{-2} s^{-1}]. In an isotropic medium this vector is parallel to the vector **k**. Now, taking into account Eqs. (B.61) and (B.65), Eq. (B.69) can be rewritten as follows:

$$\mathbf{P} = vw\mathbf{e}_k, \tag{B.70}$$

or

$$P = vw. \tag{B.71}$$

Here, $\mathbf{e}_k$ is a unit vector oriented in the direction of the vector **k**. Taking into account the orthogonality of vectors **E**, **H**, and **k**, and using Eqs. (B.63) and (B.64), we can write the relation between the moduli of vectors **E** and **H**:

$$\sqrt{\mu_0\mu}H = \sqrt{\epsilon_0\epsilon}E. \tag{B.72}$$

Taking into account this relation, it can be shown that, in traveling waves described by Eqs. (B.56) and (B.57), the energy density of the electric field is equal to the energy density of the magnetic field, and the total energy density is

$$w = \epsilon_0\epsilon E^2 = \mu_0\mu H^2 = \sqrt{\epsilon_0\mu_0\epsilon\mu}EH. \tag{B.73}$$

The magnitude of the energy transferred by the electromagnetic wave is defined by the *intensity of the electromagnetic wave*, $\mathcal{I}$, which is equal, according to the definition, to the average value of the density of the energy flux during a period T. For the monochromatic waves described by Eqs. (B.56) and (B.57), taking into account that $P = vw$, we obtain

$$\mathcal{I} = \langle P\rangle = \frac{1}{T}\int_0^T P(t)\mathrm{d}t = \frac{c}{\sqrt{\epsilon\mu}}\epsilon_0\epsilon E_0^2\left\langle\cos^2(\omega t - \mathbf{kr})\right\rangle = \sqrt{\frac{\epsilon_0\epsilon}{\mu_0\mu}}\frac{E_0^2}{2}, \tag{B.74}$$

which shows that the intensity of the wave is proportional to the square of its field's amplitude. The dimension of intensity is the same as that for the density of energy flux, namely [J s^{-1} m^{-2}] =[W m^{-2}].

An electromagnetic wave, in addition to energy, also carries linear momentum. According to Einstein's relativistic theory, an *object that moves with the speed of light must have mass at rest equal to zero*. At the same time the momentum, p, of such an object is equal to

$$p = \frac{W}{c}, \tag{B.75}$$

where W is the object's energy. The object in the case considered here is the electromagnetic wave (later we will show that an electromagnetic wave can be considered as a flux of photons moving with the speed of light and having mass at rest equal to zero). For the momentum density of an electromagnetic wave, π, we can write

$$\pi = \frac{p}{V} = \frac{W}{Vc} = \frac{w}{c}. \tag{B.76}$$

If we take into account the density of energy flux in vacuum (Eq. (B.71)), then Eq. (B.76) can be rewritten in vector form in terms of the Poynting vector, **P**:

$$\pi = \frac{\mathbf{P}}{c^2} = \frac{1}{c^2}\mathbf{E}\times\mathbf{H}. \tag{B.77}$$

Because the electromagnetic wave has momentum, when it is incident upon a surface it exerts pressure on it. Let a plane wave be incident perpendicularly upon the plane surface of a body, which absorbs this wave. Let us calculate the momentum that the wave transmits to the part of the body's surface S during the time Δt. It must be equal to the volume of

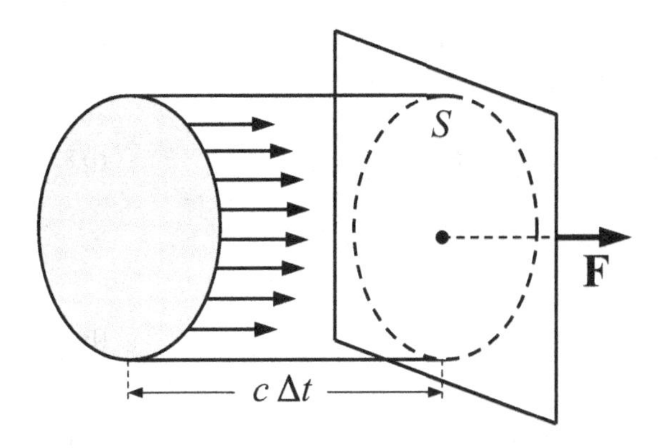

Figure B.4 The pressure, $\mathcal{P} = F/S = \pi c$, exerted by an electromagnetic wave on a plane surface of cross-section S. The momentum density of the electromagnetic wave is π.

the cylinder, $Sc\,\Delta t$ (see Fig. B.4) multiplied by the momentum density π:

$$\Delta p = \pi\, Sc\,\Delta t. \tag{B.78}$$

The pressure, $\mathcal{P}$, is defined by the force, F, acting on the surface, divided by the cross-section S, i.e.,

$$\mathcal{P} = \frac{F}{S} = \frac{1}{S}\frac{\Delta p}{\Delta t} = \pi c = w. \tag{B.79}$$

In the case of the surface of a mirror that completely reflects the incident wave, the momentum transmitted to the surface during the same time is twice as large, and thus the pressure in this case is

$$\mathcal{P} = 2w. \tag{B.80}$$

Let us estimate the magnitude of this pressure for a laser beam, for which the electromagnetic wave has an amplitude of the electric field intensity of $E_0 = 10^3\ \mathrm{V\,m^{-1}}$:

$$\mathcal{P} = 2\epsilon_0 E_0^2 = 2 \times 8.85 \times 10^{-12} \times 10^6\ \mathrm{F\,m^{-1}} \cdot \mathrm{V^2\,m^{-2}} \simeq 1.8 \times 10^{-5}\ \mathrm{Pa},$$

which is 10^{10} times smaller than standard atmospheric pressure ($\mathcal{P}_{\mathrm{atm}} \approx 10^5$ Pa). Nevertheless, the pressure of an electromagnetic wave can nowadays easily be detected and measured with high precision.

Example B.3. For wave fields as well as for static fields the superposition principle (Eq. (B.5)) is valid. Using this principle, estimate the intensity of the total wave field created by two linearly polarized plane waves

$$\begin{aligned} \mathbf{E}_1 &= \mathbf{e}_1 E_0 \cos(\omega t - \mathbf{k}\cdot\mathbf{r}), \\ \mathbf{E}_2 &= \mathbf{e}_2 E_0 \cos(\omega t - \mathbf{k}\cdot\mathbf{r}), \end{aligned} \tag{B.81}$$

propagating in vacuum in the same direction. The unit vectors $\mathbf{e}_1$ and $\mathbf{e}_2$ define the directions of the wave polarizations, which in general are different.

Reasoning. In accordance with the superposition principle, the total electric field $\mathbf{E}$ is equal to the sum of two fields $\mathbf{E}_1$ and $\mathbf{E}_2$:

$$\mathbf{E} = \mathbf{E}_1 + \mathbf{E}_2. \tag{B.82}$$

Then, the intensity, $\mathcal{I}$, of the total electric field is defined as

$$\mathcal{I} = \sqrt{\frac{\epsilon_0}{\mu_0}} \left\langle E^2 \right\rangle = \sqrt{\frac{\epsilon_0}{\mu_0}} \left\langle E_1^2 + E_2^2 + 2\mathbf{E}_1 \cdot \mathbf{E}_2 \right\rangle. \tag{B.83}$$

Because $\left\langle E_1^2 \right\rangle = E_0^2/2$, $\left\langle E_2^2 \right\rangle = E_0^2/2$, and $2\langle \mathbf{E}_1 \cdot \mathbf{E}_2 \rangle = E_0^2(\mathbf{e}_1 \cdot \mathbf{e}_2)$, after averaging and summation in Eq. (B.83) we obtain

$$\mathcal{I} = \sqrt{\frac{\epsilon_0}{\mu_0}} E_0^2(1 + \mathbf{e}_1 \cdot \mathbf{e}_2) = 2\mathcal{I}_0(1 + \mathbf{e}_1 \cdot \mathbf{e}_2), \tag{B.84}$$

where

$$\mathcal{I}_0 = \sqrt{\frac{\epsilon_0}{\mu_0}} \frac{E_0^2}{2} \tag{B.85}$$

is the intensity of an individual wave. The maximal magnitude of the total intensity, $\mathcal{I}_{\text{max}} = 4\mathcal{I}_0$, will occur in the case of parallel vectors $\mathbf{E}_1$ and $\mathbf{E}_2$, i.e., when $\mathbf{e}_1 \cdot \mathbf{e}_2 = 1$. In this case the amplitudes of the fields are added to each other, so we can say that the waves reinforce each other. The minimal magnitude of the intensity of the total wave, $\mathcal{I}_{\text{min}} = 0$, occurs in the case of anti-parallel vectors $\mathbf{E}_1$ and $\mathbf{E}_2$, i.e., when $\mathbf{e}_1 \cdot \mathbf{e}_2 = -1$ and the waves cancel each other out. If the polarizations of two waves are orthogonal to each other, i.e., $\mathbf{e}_1 \cdot \mathbf{e}_2 = 0$, then $\mathcal{I} = 2\mathcal{I}_0$, and the waves propagate independently from each other. The example considered here illustrates the superposition principle with respect to the wave fields.

Example B.4. Consider the harmonic-in-time wave process when only the averaged magnitude of the density of energy flux $\mathbf{P}$ over the time period has physical meaning. Find $\langle \mathbf{P} \rangle$ by presenting fields $\mathbf{E}$ and $\mathbf{H}$ in complex form.
Reasoning. Using the complex form of fields $\mathbf{E}$ and $\mathbf{H}$

$$\begin{aligned} \mathbf{E} &= \mathbf{E}_0 \mathrm{e}^{\mathrm{i}(\omega t - \mathbf{k}\cdot\mathbf{r})}, \\ \mathbf{H} &= \mathbf{H}_0 \mathrm{e}^{\mathrm{i}(\omega t - \mathbf{k}\cdot\mathbf{r})}, \end{aligned} \tag{B.86}$$

it is convenient to substitute into the expressions for the density of energy flux real values of fields, i.e.,

$$\frac{\mathbf{E} + \mathbf{E}^*}{2} \quad \text{and} \quad \frac{\mathbf{H} + \mathbf{H}^*}{2}.$$

On substituting these magnitudes into Eq. (B.69) and averaging over time, we obtain

$$\langle \mathbf{P} \rangle = \frac{1}{4} \langle (\mathbf{E} + \mathbf{E}^*) \times (\mathbf{H} + \mathbf{H}^*) \rangle = \frac{1}{4} \langle \mathbf{E} \times \mathbf{H} + \mathbf{E}^* \times \mathbf{H}^* + \mathbf{E}^* \times \mathbf{H} + \mathbf{E} \times \mathbf{H}^* \rangle . \tag{B.87}$$

Taking into account that the dependence on time of the fields from Eq. (B.87) is proportional to $e^{\pm i\omega t}$, we obtain for two terms

$$\langle \mathbf{E} \times \mathbf{H} \rangle = \langle \mathbf{E}^* \times \mathbf{H}^* \rangle = 0. \tag{B.88}$$

Let us prove that the first term is equal to zero (for the second term the calculations are similar):

$$\langle \mathbf{E} \times \mathbf{H} \rangle = \mathbf{E}_0 \times \mathbf{H}_0 \frac{1}{T} e^{-2i\mathbf{k}\cdot\mathbf{r}} \int_0^T e^{2i\omega t}\, dt = \frac{\mathbf{E}_0 \times \mathbf{H}_0}{T} e^{-2i\mathbf{k}\cdot\mathbf{r}} \frac{1}{2i\omega} \left(e^{2i\omega T} - 1 \right) = 0, \tag{B.89}$$

since $\omega = 2\pi/T$ and

$$e^{2i\pi} = \cos(2\pi) + i\sin(2\pi) = 1. \tag{B.90}$$

As a result, we obtain for the average density of flux

$$\langle \mathbf{P} \rangle = \frac{1}{4} \langle \mathbf{E}^* \times \mathbf{H} + \mathbf{E} \times \mathbf{H}^* \rangle . \tag{B.91}$$

We have to leave only the real parts in the last equation. Let us show that the real parts of both terms in Eq. (B.91) are the same. For this purpose, let us present the complex-conjugate magnitudes we have,

$$\begin{aligned} \mathbf{E}^* &= \mathbf{E}_0 e^{-i(\omega t - \mathbf{k}\cdot\mathbf{r})}, \\ \mathbf{H}^* &= \mathbf{H}_0 e^{-i(\omega t - \mathbf{k}\cdot\mathbf{r})}. \end{aligned} \tag{B.92}$$

The results of multiplication of the fields from Eqs. (B.86) and (B.92) show that both products in Eq. (B.91) are real and that they are the same. Thus, the average density of flux can be written as follows:

$$\langle \mathbf{P} \rangle = \frac{1}{2} \mathrm{Re}(\mathbf{E} \times \mathbf{H}^*). \tag{B.93}$$

B.3 Reflection of a plane wave from the interface between two media

B.3.1 Plane waves in absorbing media

We have considered so far waves that are not damped while propagating. This is possible only in vacuum or in a non-absorbing medium. All real media absorb waves to some extent.

The magnitude of such absorption greatly depends on the frequency of the propagating wave. For some regions of the electromagnetic spectrum this absorption can be so weak that we can neglect it, but for other regions of the spectrum the absorption can be very significant. Let us consider the peculiarities of the propagation of plane waves in an absorbing medium.

Absorption in the medium has to be taken into account if we are dealing with a medium that has a conductivity, σ, that is not equal to zero. Since the current density, **j**, of the free charges is related to the intensity of the electric field, **E**, by Ohm's law (in differential form) as

$$\mathbf{j} = \sigma \mathbf{E}, \tag{B.94}$$

Maxwell's equation (B.16) that contains **j** can be rewritten in the following form:

$$\nabla \times \mathbf{H} = \epsilon\epsilon_0 \frac{\partial \mathbf{E}}{\partial t} + \sigma \mathbf{E}. \tag{B.95}$$

Using Maxwell's equation (B.17),

$$\nabla \times \mathbf{E} = -\mu\mu_0 \frac{\partial \mathbf{H}}{\partial t}, \tag{B.96}$$

we can obtain the wave equation for the case considered here (in the same way as we obtained Eq. (B.52)):

$$\nabla^2 \mathbf{E} = \frac{\epsilon\mu}{c^2} \frac{\partial^2 \mathbf{E}}{\partial t^2} + \mu\mu_0\sigma \frac{\partial \mathbf{E}}{\partial t}. \tag{B.97}$$

The corresponding wave equation can be obtained for the intensity of the magnetic field, **H**:

$$\nabla^2 \mathbf{H} = \frac{\epsilon\mu}{c^2} \frac{\partial^2 \mathbf{H}}{\partial t^2} + \mu\mu_0\sigma \frac{\partial \mathbf{H}}{\partial t}. \tag{B.98}$$

These wave equations, in contrast to Eqs. (B.52) and (B.53), contain terms with the time derivative in the first order, which are responsible for the damping of the wave in a medium.

We will seek solutions of Eqs. (B.97) and (B.98) in the form of monochromatic plane waves, which for convenience can be presented in the exponential form,

$$\mathbf{E}(t, \mathbf{r}) = \mathbf{E}_0 \mathrm{e}^{\mathrm{i}(\omega t - \mathbf{k}\cdot\mathbf{r})}, \tag{B.99}$$

$$\mathbf{H}(t, \mathbf{r}) = \mathbf{H}_0 \mathrm{e}^{\mathrm{i}(\omega t - \mathbf{k}\cdot\mathbf{r})}, \tag{B.100}$$

where $\mathbf{E}_0$ and $\mathbf{H}_0$ are the amplitudes of the electric and magnetic fields of the wave. This form of presentation of wave equations is generally used in electromagnetism, where the fields **E** and **H** are defined as complex functions, while the real wave processes are described by the real parts of Eqs. (B.99) and (B.100).

By substitution of Eq. (B.99) into Eq. (B.97), we arrive at the equation

$$\left[k^2 - \frac{\omega^2}{c^2}\mu\left(\epsilon - \mathrm{i}\frac{\sigma}{\epsilon_0\omega}\right)\right]\mathbf{E} = 0, \tag{B.101}$$

from which we obtain the dispersion relation that connects the wavenumber, k, with the wave frequency, ω, and the medium parameters:

$$k^2 = k_0^2\mu\left(\epsilon - \mathrm{i}\frac{\sigma}{\epsilon_0\omega}\right) \quad \text{or} \quad k = k_0\sqrt{\mu\left(\epsilon - \mathrm{i}\frac{\sigma}{\epsilon_0\omega}\right)}. \tag{B.102}$$

Here, we introduced the wavenumber, $k_0 = \omega/c$, for the wave propagating in vacuum. From the expressions obtained it follows that the wavenumber, k, for a wave propagating in a conducting medium is a complex number, i.e.,

$$k = k^{'} - \mathrm{i}k^{''}. \tag{B.103}$$

The sign before the imaginary part of the wavenumber is chosen to be negative to make the wave that propagates in the positive direction in an absorbing medium damped instead of amplifying it. Let us substitute Eq. (B.103) into Eq. (B.102), and let us separate the real and imaginary parts. As a result we obtain two equations from which to find $k^{'}$ and $k^{''}$:

$$(k^{'})^2 - (k^{''})^2 = k_0^2\epsilon\mu, \tag{B.104}$$

$$2k^{'}k^{''} = k_0^2\frac{\mu\sigma}{\epsilon_0\omega}. \tag{B.105}$$

By solving the system of (B.104) and (B.105), we obtain the following expressions for the real, $k^{'}$, and the complex, $k^{''}$, parts of the complex wavenumber, k:

$$k^{'} = k_0\sqrt{\frac{\epsilon\mu}{2}\left(\sqrt{1 + \frac{\sigma^2}{\epsilon_0^2\epsilon^2\omega^2}} + 1\right)}, \tag{B.106}$$

$$k^{''} = k_0\sqrt{\frac{\epsilon\mu}{2}\left(\sqrt{1 + \frac{\sigma^2}{\epsilon_0^2\epsilon^2\omega^2}} - 1\right)}. \tag{B.107}$$

The above relations for the real, $k^{'}$, and imaginary, $k^{''}$, parts of the wavenumber, k, are written in the general form and are valid for weakly absorbing media as well as for strongly absorbing media over a wide frequency range (from $\omega = 0$ to $\omega = 10^{14}$ Hz, where quantum effects become significant). Depending on the media and on the frequency ranges, the magnitude of $\sigma/(\epsilon_0\epsilon\omega)$ can be either large or small, which can be used for simplification of the above expressions.

B.3.2 The skin effect

Let us assume that a plane wave, whose electric and magnetic fields are described by Eqs. (B.99) and (B.100), is propagating in the y-direction. Then, the partial derivatives with respect to x and z are equal to zero and in Eqs. (B.97) and (B.98) the operator ∇^2 is simplified to the form

$$\nabla^2 = \frac{\partial^2}{\partial y^2}. \tag{B.108}$$

Taking into account that the wavenumber of Eq. (B.103) is a complex variable, the solutions (B.99) and (B.100) can be rewritten as

$$\mathbf{E}(t, y) = \mathbf{E}_0 \mathrm{e}^{-k'' y} \mathrm{e}^{\mathrm{i}(\omega t - k' y)}, \tag{B.109}$$

$$\mathbf{H}(t, y) = \mathbf{H}_0 \mathrm{e}^{-k'' y} \mathrm{e}^{\mathrm{i}(\omega t - k' y)}, \tag{B.110}$$

where $\mathbf{E}_0$ and $\mathbf{H}_0$ are the amplitudes of the wave fields in the plane $y = 0$. It follows from Eqs. (B.109) and (B.110) that the amplitudes of wave fields decrease exponentially in the direction of wave propagation (the y-direction):

$$\mathbf{E}(y) = \mathbf{E}_0 \mathrm{e}^{-k'' y}, \tag{B.111}$$

$$\mathbf{H}(y) = \mathbf{H}_0 \mathrm{e}^{-k'' y}. \tag{B.112}$$

The imaginary part of the wavenumber, k'', defines the rate of the decrease in amplitude along the direction of wave propagation. Its reciprocal, δ, which is called the *skin depth,*

$$\delta = \frac{1}{k''}, \tag{B.113}$$

defines the penetration depth of the field in the absorbing medium. At the distance δ the amplitude of the wave has decreased by a factor of e:

$$\mathbf{E}(\delta) = \mathbf{E}_0 \mathrm{e}^{-k'' \delta} = \mathbf{E}_0 \mathrm{e}^{-1}. \tag{B.114}$$

Figure B.5 shows the distribution of the electric field of the wave in a medium with damping for two time instants. If a wave with an amplitude of $\mathbf{E}_0$ penetrates from vacuum into a medium with non-zero conductivity, then it practically dampens at a depth of several δ. Since the intensity of the wave $\mathcal{I}$ is proportional to E^2, at the distance $y = \delta$ the intensity of the wave will have decreased by a factor of $\mathcal{I}_0/\mathcal{I} = \mathrm{e}^2 \cong 7.4$, and at the distance $y = 2\delta$ the intensity will have decreased by a factor of $\mathcal{I}_0/\mathcal{I} = \mathrm{e}^4 \cong 55$. We can conclude that the energy of the transmitted wave in an absorbing medium is located in a layer of depth several times δ. For media with high conductivity the depth of such a layer can be relatively small (for example, for copper a wave of frequency $f = \omega/(2\pi) = 1$ MHz has a penetration depth of $\delta \approx 0.07$ mm). This is why the near-surface region of depth δ, where most of the penetrated electromagnetic wave is concentrated, is called the *skin layer*, the magnitude δ is called the *skin depth*, and the effect itself is called the *skin effect*. Let us consider two

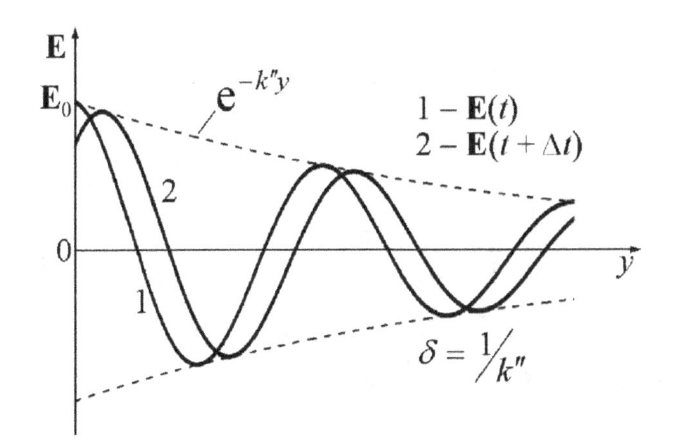

Figure B.5 The electric field of an electromagnetic wave in an absorbing medium at two instants of time, t and $t + \Delta t$. The magnitude of the skin depth, δ, is indicated.

limiting cases for Eqs. (B.106) and (B.107) for low and high conductivities.

(a) Low conductivity (or the high-frequency limit), $\sigma \ll \epsilon_0 \epsilon \omega$. In this case

$$\delta = \frac{1}{k''} = \frac{2\epsilon_0 c}{\sigma}\sqrt{\frac{\epsilon}{\mu}} = \frac{2}{\sigma}\sqrt{\frac{\epsilon_0 \epsilon}{\mu_0 \mu}}, \tag{B.115}$$

i.e., the depth of wave penetration into the medium does not depend on its frequency. With decreasing σ the wave field penetrates further into the medium. In the case of $\sigma \to 0$ the medium becomes non-conducting and the penetration depth for the electromagnetic field tends to infinity, $\delta \to \infty$.

(b) High conductivity (or the low-frequency limit), $\sigma \gg \epsilon_0 \epsilon \omega$. In this case the skin depth, δ, is defined as

$$\delta = c\sqrt{\frac{2\epsilon_0}{\mu \sigma \omega}} = \sqrt{\frac{2}{\mu_0 \mu \sigma \omega}}, \tag{B.116}$$

i.e., with increasing conductivity, σ, and frequency, ω, the penetration depth decreases. At $\omega \to 0$ the depth of penetration into the medium tends to infinity, $\delta \to \infty$. This is a feature of static fields, both electric and magnetic.

B.3.3 Solution of the boundary problem

Real wave-guiding structures always have finite dimensions. Therefore, there is partial reflection from boundary surfaces and transmission into the neighboring medium during wave propagation in such structures. Depending on the type of material, the geometry of the reflecting surfaces, the structure of the incident wave, and the angle of incidence, the problems of finding fields of reflected and transmitted waves seem to be problems of different levels of complexity. Let us consider the simplest problem of the reflection and transmission of a monochromatic plane wave through a plane interface at normal incidence.

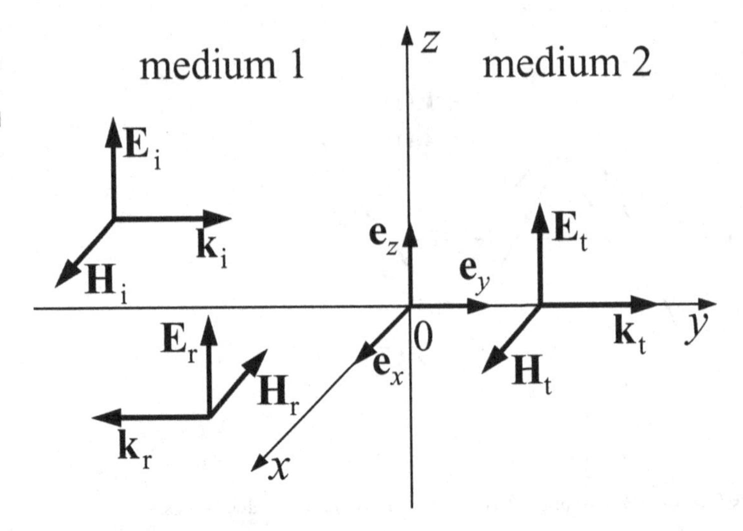

Figure B.6 The interface between two semi-infinite homogeneous media: medium 1 with ϵ_1, μ_1, and σ_1; and medium 2 with ϵ_2, μ_2, and σ_2.

Let the interface between two semi-infinite homogeneous media coincide with the plane $y = 0$ (see Fig. B.6). The medium located to the left of the interface ($y < 0$) is characterized by the parameters ϵ_1, μ_1, and σ_1, whereas the medium located to the right of the interface ($y > 0$) is characterized by the parameters ϵ_2, μ_2, and σ_2. A plane wave of frequency ω and wavevector $\mathbf{k}_i = \mathbf{e}_y k_1$ is normally incident (along the y-axis). The reflected wave propagates along the negative y-axis, and the wave transmitted into the second medium propagates in the positive direction of the y-axis (see Fig. B.6).

To find the amplitude of the reflected and transmitted waves, let us solve the boundary problem, which includes sewing fields at the interface using the boundary conditions (B.26)–(B.29). In writing the expressions for the fields we will drop the time factor $e^{i\omega t}$ because it is present in each expression for the fields of each wave.

The field of the incident wave can be written in the following form:

$$\mathbf{E}_i(y) = \mathbf{e}_z E_i e^{-ik_1 y}, \tag{B.117}$$

$$\mathbf{H}_i(y) = \mathbf{e}_x H_i e^{-ik_1 y}, \tag{B.118}$$

where $\mathbf{e}_x$ and $\mathbf{e}_z$ are the unit vectors of the corresponding coordinate axes, and wavenumber k_1 is defined by Eqs. (B.106) and (B.107). The relation between the amplitudes of these waves is given by the expression

$$E_i = Z_1 H_i, \tag{B.119}$$

where the wave resistance Z_1 of the first medium is

$$Z_1 = \sqrt{\frac{\mu_0 \mu_1 \omega}{\epsilon_0 \epsilon_1 \omega - i\sigma_1}}. \tag{B.120}$$

The fields of the reflected wave propagating in the opposite direction have to be written in the following form:

$$\mathbf{E}_r(y) = \mathbf{e}_z E_r e^{ik_1 y}, \tag{B.121}$$

$$\mathbf{H}_r(y) = -\mathbf{e}_x H_r e^{ik_1 y}, \tag{B.122}$$

where $E_r = Z_1 H_r$, and the negative sign appears in the expression for $\mathbf{H}_r(y)$ because $\mathbf{E}_r$, $\mathbf{H}_r$, and $\mathbf{k}_r$ constitute a right-handed system.

For the wave transmitted across the interface we can write

$$\mathbf{E}_t(y) = \mathbf{e}_z E_t e^{-ik_2 y}, \tag{B.123}$$

$$\mathbf{H}_t(y) = \mathbf{e}_x H_t e^{-ik_2 y}, \tag{B.124}$$

where $E_t = Z_2 H_t$; Z_2 is defined analogously to Z_1:

$$Z_2 = \sqrt{\frac{\mu_0 \mu_2 \omega}{\epsilon_0 \epsilon_2 \omega - i\sigma_2}}. \tag{B.125}$$

At the interface, which does not have any charges and any currents along it, according to Eqs. (B.26)–(B.29) the tangential components of the field intensities of the electric and magnetic fields must be continuous. On the left-hand side of the interface ($y < 0$) the field is presented as a sum of the incident and reflected waves, and on the right-hand side of the interface ($y > 0$) it is given solely by the transmitted wave. So at the interface ($y = 0$) we obtain the following system of equations:

$$\mathbf{E}_i(0) + \mathbf{E}_r(0) = \mathbf{E}_t(0), \tag{B.126}$$

$$\mathbf{H}_i(0) + \mathbf{H}_r(0) = \mathbf{H}_t(0). \tag{B.127}$$

Let us substitute into this system of equations the expressions for the fields of the incident, reflected, and transmitted waves, and let us drop the common factors. As a result, we will obtain the following system of equations for the amplitudes of the corresponding fields:

$$E_i + E_r = E_t, \tag{B.128}$$

$$H_i - H_r = H_t. \tag{B.129}$$

On proceeding in the second equation with the amplitudes of the electric fields, we obtain

$$E_i - E_r = \frac{Z_1}{Z_2} E_t. \tag{B.130}$$

By solving this equation self-consistently with Eq. (B.128) we obtain

$$E_r = \frac{Z_2 - Z_1}{Z_2 + Z_1} E_i, \tag{B.131}$$

$$E_t = \frac{2Z_2}{Z_2 + Z_1} E_i. \tag{B.132}$$

From these relations we can find the coefficients of reflection, R, and transmission, T:

$$R = \frac{E_r}{E_i} = \frac{Z_2 - Z_1}{Z_2 + Z_1} = |R| e^{i\varphi_r}, \tag{B.133}$$

$$T = \frac{E_t}{E_i} = \frac{2Z_2}{Z_2 + Z_1} = |T| e^{i\varphi_t}. \tag{B.134}$$

In the general case of absorbing media the coefficients R and T are complex variables. This means that the phases of reflected and transmitted waves do not coincide with the phase of the incident wave. The corresponding phase shifts φ_r and φ_t can be found from Eqs. (B.133) and (B.134).

B.3.4 Analysis of the reflection and transmission coefficients

Let us consider several important special cases for several interfaces.

(1) Let the reflection happen from the interface of two transparent non-magnetic dielectrics, for which $\mu_1 = \mu_2 = 1$ and $\sigma_1 = \sigma_2 = 0$. For such media the impedances are real:

$$Z_1 = \sqrt{\frac{\mu_0}{\epsilon_0 \epsilon_1}}, \tag{B.135}$$

$$Z_2 = \sqrt{\frac{\mu_0}{\epsilon_0 \epsilon_2}}. \tag{B.136}$$

Then, the coefficients of reflection, R, and transmission, T, we obtain from Eqs. (B.133) and (B.134),

$$R = \frac{\sqrt{\epsilon_1} - \sqrt{\epsilon_2}}{\sqrt{\epsilon_1} + \sqrt{\epsilon_2}}, \tag{B.137}$$

$$T = \frac{2\sqrt{\epsilon_1}}{\sqrt{\epsilon_1} + \sqrt{\epsilon_2}}, \tag{B.138}$$

have real values in this case. There is no phase shift between incident and transmitted waves (i.e., $\varphi_t = 0$) regardless of the magnitudes of ϵ_1 and ϵ_2. The total electric field in the region $y < 0$ has the form

$$\mathbf{E}_1(y, t) = \mathbf{e}_z E_0[\cos(\omega t - k_1 y) + R \cos(\omega t + k_1 y)]. \tag{B.139}$$

The phase shift between the incident and the reflected wave depends on the relation between ϵ_1 and ϵ_2. If $\epsilon_1 < \epsilon_2$, there is a phase shift equal to π ($\varphi_r = \pi$, see Fig. B.7)

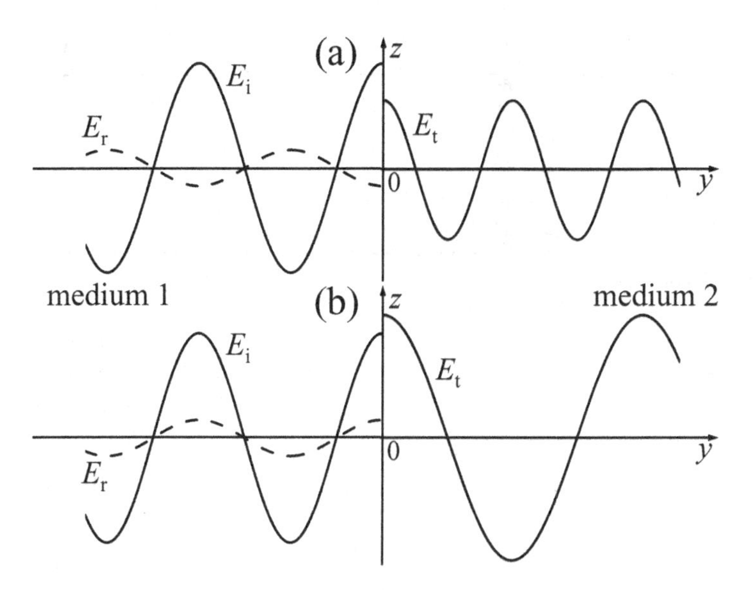

Figure B.7 The incident, E_i, reflected, E_r, and transmitted, E_t, waves for media with (a) $\epsilon_1 < \epsilon_2$ and (b) $\epsilon_1 > \epsilon_2$.

between incident and reflected waves. When the wave is incident from the optically more dense medium to the optically less dense medium (i.e., $\epsilon_1 > \epsilon_2$) there is no phase shift between the incident and reflected waves (see Fig. B.7).

(2) Let the wave be incident from the non-absorbing medium ($\sigma_1 = 0$) to the interface with an ideal conductor ($\sigma_2 \to \infty$). In this case $Z_2 = 0$, $R = -1$, and $T = 0$, i.e., the wave does not penetrate into the second medium and the wavenumber k_1 is real. Therefore, a standing wave is formed in the first medium, and, as we found from Eq. (B.139) for $R = -1$,

$$\mathbf{E}_1(y, t) = 2\mathbf{e}_z E_i \sin(k_1 y)\sin(\omega t). \tag{B.140}$$

The expressions for the magnetic field can be obtained analogously:

$$\mathbf{H}_1(y, t) = \frac{1}{Z_1} 2\mathbf{e}_x E_i \cos(k_1 y)\cos(\omega t). \tag{B.141}$$

From this expression we see that the wave formed as a result of the overlapping of incident and reflected waves with phase shift $\varphi = \pi$ is a standing wave. The magnetic field of this wave has a phase shift equal to $\pi/2$ in space as well as in time relative to the electric field.

Example B.5. For real media the material parameters are functions of frequency. Find the reflection coefficient of an electromagnetic wave from a medium that has for a given frequency the dielectric constant $\epsilon_2(\omega) < 0$ and $\sigma_2(\omega) = 0$. The parameters of the medium in which the wave is initially propagating are $\epsilon_1(\omega) > 0$ and $\sigma_1(\omega) = 0$. Media 1 and 2 are non-magnetic media with $\mu_1 = \mu_2 = 1$.

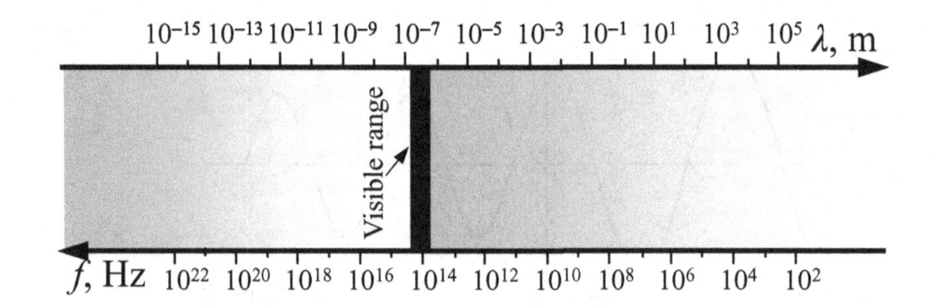

Figure B.8 The visible range of the electromagnetic spectrum.

Reasoning. According to Eq. (B.120) the expressions for the wave resistance of each of the media have the forms

$$Z_1 = \sqrt{\frac{\mu_0}{\epsilon_0 \epsilon_1}}, \tag{B.142}$$

$$Z_2 = \sqrt{\frac{\mu_0}{\epsilon_0 \epsilon_2}} = \mathrm{i}\sqrt{\frac{\mu_0}{\epsilon_0 |\epsilon_2|}}. \tag{B.143}$$

On substituting these expressions into Eq. (B.133), which defines the reflection coefficient, we find

$$R = \frac{Z_2 - Z_1}{Z_2 + Z_1} = \frac{\mathrm{i}\sqrt{\epsilon_1} - \sqrt{|\epsilon_2|}}{\mathrm{i}\sqrt{\epsilon_1} + \sqrt{|\epsilon_2|}} = \frac{\sqrt{\epsilon_1} + \mathrm{i}\sqrt{|\epsilon_2|}}{\sqrt{\epsilon_1} - \mathrm{i}\sqrt{|\epsilon_2|}}. \tag{B.144}$$

The expression for the reflection coefficient, R, can be rewritten in the form

$$R = |R| \mathrm{e}^{\mathrm{i}\varphi_{\mathrm{r}}}, \tag{B.145}$$

where

$$\tan \varphi_{\mathrm{r}} = \frac{2\sqrt{\epsilon_1 |\epsilon_2|}}{\epsilon_1 + |\epsilon_2|} \quad \text{and} \quad |R| = 1. \tag{B.146}$$

Therefore, in the case considered here the modulus of the reflection coefficient $|R| = 1$. The non-absorbing medium with $\epsilon_2 < 0$ completely reflects the wave, shifting its phase by φ_{r}. In this case it is said that we observe the effect of "*metallic reflection.*" Media with $\epsilon(\omega) < 0$ in the corresponding frequency range are often used in devices in which complete reflection of a wave with frequency ω is needed. This is important when for any reason you cannot use metallic layers in the device because of their high conductivity.

B.4 Light and its wave properties

The electromagnetic radiation that we observe in nature occupies a very wide range of wavelengths: from 10^{-15} m to hundreds of kilometers (see Fig. B.8). Nowadays we understand light to be electromagnetic radiation in the range from ultraviolet to infrared waves, with corresponding wavelengths from 0.01 μm to 100 μm, although these boundaries are not exact, whereas the light visible to human beings has much narrower boundaries and lies in the interval of wavelengths from 0.35 μm to 0.67 μm. It is quite obvious that all of the properties of electromagnetic waves, such as interference, polarization, diffraction, and dispersion, apply to light. Let us briefly discuss interference and the last two phenomena with a specific focus on light.

B.4.1 Interference

During the simultaneous propagation of several waves in a medium, overlapping of the waves occurs. From the linearity of Maxwell's equations with respect to the field vectors **E**, **D**, **B**, and **H**, and the lack of dependence of the material constants ϵ, μ, and σ on these fields, it follows that the wave processes described by Maxwell's equations satisfy the superposition principle. This means that the oscillations of the field vectors at each point are the vector sum of oscillations of the field vectors of the individual waves, which propagate separately. Media for which the superposition principle is valid are called *linear media*. A linear medium represents an idealized model. We can use this model for the description of waves propagating through a real medium only if the amplitudes of propagating waves, E_{0i}, are much smaller than the internal electric field, E_{int}, acting on the valence electrons of the constituent atoms. An estimate of the magnitude of this field is

$$E_{\text{int}} \sim \frac{k_e e}{a^2} \sim 10^{11}\ \text{V m}^{-1}, \tag{B.147}$$

where a is the radius of the valence electron's orbit ($\sim 10^{-10}$ m). Such magnitudes are much larger than most wave fields. For example, the field intensity in the beam of a continuous-wave laser with power of about 1 W and beam cross-sectional area 1 mm^2 attains magnitudes of $E_0 \sim 10^4$ V m^{-1}, which is many orders of magnitude less than the internal field intensity of Eq. (B.147). Thus, the medium for such a relatively powerful laser beam is linear. Vacuum is always a linear medium for electromagnetic fields of any intensity. Its dielectric and magnetic permittivities, ϵ_{vac} and μ_{vac}, depend neither on the magnitude of the field nor on its frequency: $\epsilon_{\text{vac}} = \mu_{\text{vac}} = 1$, and the conductivity $\sigma = 0$.

Let us consider the superposition of two plane harmonic waves of the same polarization, propagating in a non-absorbing medium along two arbitrary directions. At any given point the magnitudes $\mathbf{E}_j$ and $\mathbf{H}_j$ of these two waves undergo harmonic oscillations. The vectors of the electric and magnetic fields of these oscillations at a point, **r**, are

$$\mathbf{E}_j = \mathbf{e}_z A_j \cos(\omega_j t - \Delta_j) = \frac{\mathbf{e}_z}{2} A_j \left[e^{i(\omega_j t - \Delta_j)} + e^{-i(\omega_j t - \Delta_j)} \right], \tag{B.148}$$

$$\mathbf{H}_j = \mathbf{e}_x \gamma A_j \cos(\omega_j t - \Delta_j) = \frac{\mathbf{e}_x}{2} \gamma A_j \left[e^{i(\omega_j t - \Delta_j)} + e^{-i(\omega_j t - \Delta_j)} \right], \tag{B.149}$$

where $j = 1, 2$; $\Delta_j = \mathbf{k}_j \cdot \mathbf{r}$; and $\gamma = \sqrt{\epsilon\epsilon_0/(\mu\mu_0)}$. The wavenumber, k_j, of each wave is related to the frequency, ω_j, and wavelength, λ_j, by

$$k_j = \frac{\omega_j}{v} = \frac{2\pi}{\lambda_j}, \quad j = 1, 2, \tag{B.150}$$

where the speed of both waves is defined by Eq. (B.55) ($v = c/\sqrt{\epsilon\mu}$). The sums of the electric and magnetic fields of both waves will depend only on time, t, and will change from point to point, i.e.,

$$\mathbf{E}(\mathbf{r}, t) = \mathbf{E}_1(\mathbf{r}, t) + \mathbf{E}_2(\mathbf{r}, t), \tag{B.151}$$

$$\mathbf{H}(\mathbf{r}, t) = \mathbf{H}_1(\mathbf{r}, t) + \mathbf{H}_2(\mathbf{r}, t). \tag{B.152}$$

Let us calculate the total intensity $\mathcal{I} = \langle P \rangle$. Using the definitions (B.74) and (B.69), we obtain

$$\mathcal{I} = \langle P \rangle = \frac{\gamma}{2}\left[A_1^2 + A_2^2 + 2A_1A_2\frac{1}{T}\int_0^T \mathrm{d}t \cos[(\omega_1 - \omega_2)t - (\Delta_1 - \Delta_2)]\right], \tag{B.153}$$

where the period, T, of total oscillations is defined in terms of the periods of individual oscillations, T_j, as

$$T = \frac{2\pi}{\omega_1 - \omega_2} = \frac{T_1T_2}{T_1 - T_2}. \tag{B.154}$$

If the frequencies, ω_1 and ω_2, of the added waves are different, then the integral in Eq. (B.153) becomes equal to zero, and the intensity of the superposition of the two wave fields is equal to the sum of the intensities of the individual waves. If the two individual oscillations are not correlated with each other and the total phase difference of oscillations of added wave fields

$$\Delta = (\omega_1 - \omega_2)t + (\Delta_2 - \Delta_1) \tag{B.155}$$

changes with time, then such oscillations are called *incoherent*. If the phase difference does not change with time (this occurs when the frequencies of the two waves are the same), then such oscillations (and, correspondingly, waves) are called *coherent*. In the case of incoherent waves their superposition does not produce a so-called *interference pattern*, i.e., an increase of the intensity at some points and a decrease at other points. The intensity of the total field at a particular point for coherent waves ($\omega_1 = \omega_2$), taking into account Eq. (B.153), is given by the expression

$$\mathcal{I}(r, t) = \frac{\gamma}{2}\left[A_1^2 + A_2^2 + 2A_1A_2\cos(\Delta_2 - \Delta_1)\right]. \tag{B.156}$$

Since the intensities of the individual waves are equal to

$$\mathcal{I}_j = \frac{\gamma}{2}A_j^2, \tag{B.157}$$

Eq. (B.156) can be rewritten as

$$\mathcal{I} = \mathcal{I}_1 + \mathcal{I}_2 + 2\sqrt{\mathcal{I}_1\mathcal{I}_2}\cos\Delta, \quad \text{(B.158)}$$

where we introduced the phase difference of the two waves,

$$\Delta = \Delta_2 - \Delta_1. \quad \text{(B.159)}$$

The last term in Eq. (B.158),

$$\mathcal{I}_{\text{int}} = 2\sqrt{\mathcal{I}_1\mathcal{I}_2}\cos\Delta, \quad \text{(B.160)}$$

is called the *interference term*. The intensity in the interference pattern greatly depends on the magnitude of the interference term.

At the points of space where $\cos\Delta > 0$ the total intensity is $\mathcal{I} > \mathcal{I}_1 + \mathcal{I}_2$. At the points where $\cos\Delta < 0$ the total intensity $\mathcal{I} < \mathcal{I}_1 + \mathcal{I}_2$. As a result of the superposition of coherent waves the intensity of waves is redistributed in space – at some points the intensity increases and at others it decreases. This phenomenon is called the *interference of waves*. At the points of space where $\Delta = 2m\pi$ and $\cos\Delta = 1$ the intensity is maximal (these are the so-called points of *constructive interference*),

$$\mathcal{I}_{\max} = \left(\sqrt{\mathcal{I}_1} + \sqrt{\mathcal{I}_2}\right)^2, \quad \text{(B.161)}$$

and at the points where $\Delta = (2m - 1)\pi$ and $\cos\Delta = -1$, the intensity is minimal (these are the so-called points of *destructive interference*),

$$\mathcal{I}_{\min} = \left(\sqrt{\mathcal{I}_1} - \sqrt{\mathcal{I}_2}\right)^2. \quad \text{(B.162)}$$

The interference effect becomes especially apparent when the amplitudes of the interfering waves are equal to each other, $\mathcal{I}_1 = \mathcal{I}_2$. In this case, at the points with maximal intensity $\mathcal{I}_{\max} = 4\mathcal{I}$, and at the points with minimal intensity $\mathcal{I}_{\min} = 0$. For incoherent waves in this case the intensity is the same everywhere, and, according to Eq. (B.158), is given by $\mathcal{I} = 2\mathcal{I}_1$.

Interference is a natural phenomenon in which wave properties become strikingly apparent. Interference is characteristic for waves of any nature and it is relatively easily observed for acoustic (sound) waves or for waves on the surface of water. Interference of electromagnetic waves, in particular light waves, can be observed only under certain conditions. Conventional sources (i.e., not laser sources) do not radiate monochromatic light waves. Non-laser sources of light have a wide spectrum of radiation. Light waves that originate from such sources are not coherent and do not create a stable interference pattern. The simplest way to obtain two coherent waves is the following. A wave from one source (with spectral width $\Delta\lambda$ and mean wavelength λ) is divided into two waves, which then are brought together at a screen. The two waves formed after the division can be considered as waves coming from two individual points or linear sources.

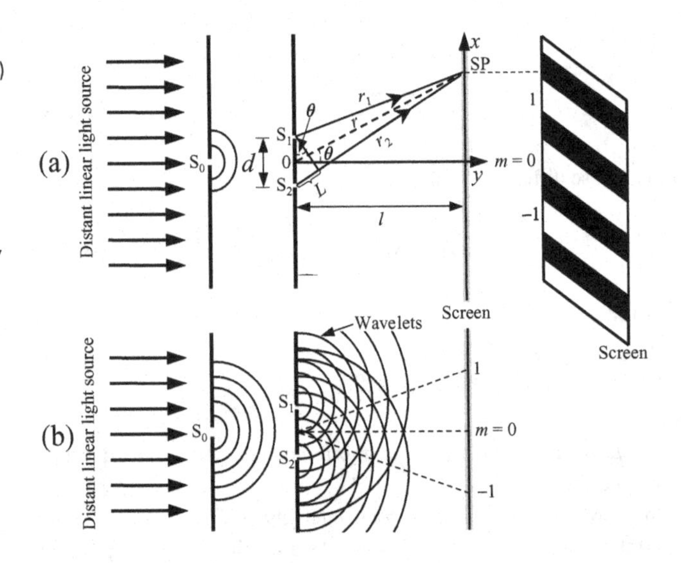

Figure B.9 Young's double-slit experiment. (a) White stripes at the screen correspond to maxima of intensity and black stripes to minima. (b) Plane waves from the distant linear source (e.g., the Sun) are transformed into concentric wavelets, which originate from the slits S_0, S_1, and S_2.

There are many other ways to bring about such division of one wave into two. In particular, it can be achieved by using two narrow slits, S_1 and S_2, separated by a small distance, d, which are illuminated by a distant linear source (e.g., the Sun). This method is called *Young's scheme*. The wavelets coming from the slits overlap, establishing a system of alternating maxima and minima at the screen (see Fig. B.9). Let us introduce the so-called *path-length difference*, L, of two waves propagating from corresponding slits to the point SP at the screen:

$$L = r_2 - r_1. \tag{B.163}$$

If the path-length difference, L, contains an integer number of wavelengths (in the medium being considered)

$$L = \frac{m\lambda_0}{\sqrt{\epsilon\mu}} = m\lambda, \tag{B.164}$$

then the oscillations of the field vectors of these waves at the given point are in phase, which results in interference maxima. Note that, if the path-length difference, L, of these waves propagating from the source to the point of the screen SP is not greater than a characteristic length $l_{\text{coh}} = \lambda^2/\Delta\lambda$, which is called the *coherence length*, then the random changes of the amplitude and phase of the fields of the two waves happen coherently, and a stable interference pattern is established. The points where the path-length difference is equal to an integer multiple of half-wavelengths correspond to the interference minima.

The formation of a standing wave by two monochromatic waves with the same frequencies and amplitudes, which are propagating in opposite directions, is an important particular case of interference. For a wave propagating in the positive direction of the y-axis the field intensities of the electric, $\mathbf{E}^+$, and magnetic, $\mathbf{H}^+$, fields are defined by the expressions

$$\mathbf{E}^+ = \mathbf{e}_z E_0 \cos(\omega t - ky), \tag{B.165}$$

$$\mathbf{H}^+ = \mathbf{e}_x H_0 \cos(\omega t - ky). \tag{B.166}$$

The equations for the wave propagating in the opposite direction can be written as

$$\mathbf{E}^- = \mathbf{e}_z E_0 \cos(\omega t + ky + \varphi), \tag{B.167}$$

$$\mathbf{H}^- = -\mathbf{e}_x H_0 \cos(\omega t + ky + \varphi), \tag{B.168}$$

where φ is the phase shift between the two waves traveling in opposite directions. In writing the expression for $\mathbf{H}^-$ it is necessary to take into account that the vectors $\mathbf{E}^-$, $\mathbf{H}^-$, and $\mathbf{k}^-$ form a right-handed system. Therefore, the polarization of the field $\mathbf{H}^-$ must be directed opposite to the x-axis. As a result of overlapping of fields with opposite directions we obtain a wave whose electric and magnetic fields are given by the following expressions:

$$\mathbf{E} = \mathbf{E}^+ + \mathbf{E}^- = 2\mathbf{e}_z E_0 \cos\left(ky + \frac{\varphi}{2}\right) \cos\left(\omega t + \frac{\varphi}{2}\right), \tag{B.169}$$

$$\mathbf{H} = \mathbf{H}^+ + \mathbf{H}^- = 2\mathbf{e}_x H_0 \sin\left(ky + \frac{\varphi}{2}\right) \sin\left(\omega t + \frac{\varphi}{2}\right). \tag{B.170}$$

Each of these equations represents a so-called *standing wave*, because for such a wave, in contrast to the case of a *traveling wave*, there is no displacement of the wavefront in space. Equations (B.169) and (B.170) describe the oscillations of the field intensities with the frequency ω and amplitudes, A_e and A_m, for electric and magnetic fields that change from point to point according to the relations

$$A_\mathrm{e} = 2E_0 \left|\cos\left(ky + \frac{\varphi}{2}\right)\right|, \tag{B.171}$$

$$A_\mathrm{m} = 2H_0 \left|\sin\left(ky + \frac{\varphi}{2}\right)\right|. \tag{B.172}$$

For each field separately we can indicate the points where the amplitude is maximal (*anti-nodes*) or equal to zero (*nodes*). For the electric field the anti-nodes are located at the points with coordinates

$$y_\text{anti-node} = \frac{m\pi - \varphi/2}{k}, \tag{B.173}$$

and the nodes are located at the points with coordinates

$$y_\text{node} = \frac{\left(m + \frac{1}{2}\right)\pi - \varphi/2}{k}, \tag{B.174}$$

where m is an arbitrary integer number. For the magnetic field all anti-nodes and nodes are shifted along the y-coordinate by $\Delta y = \pi/(2k)$. Thus, at the points where the electric field has anti-nodes, the magnetic field has nodes, and vice versa. The distance between adjacent nodes and anti-nodes for each of the fields is equal to the quarter wavelength of the traveling wave: $\Delta y_{n+1,n} = \pi/k = \lambda/4$. Between two adjacent nodes of the field it oscillates in phase. After crossing the node the phase of the field oscillations changes by π, i.e., the symmetric field oscillations on the two sides of the node are in anti-phase. The magnitude of the phase difference, φ, of the waves propagating in opposite directions depends on the conditions of their formation. Control of this magnitude allows us to shift all the nodes and anti-nodes along the space coordinate.

There is a shift of a quarter of the period between the oscillations of electric and magnetic fields not only in space but also in time. If at some instant in time at all points $|E|$ attains its maximum and H is equal to zero, then after a time $\Delta t = T/4$ at all points $E = 0$ and $|H|$ attains its maximum. Thus, during the oscillations in the standing wave the electric field transforms into the magnetic field and vice versa. The standing wave does not transfer energy because for each of the fields the energy is concentrated between the nodes of the field. At the same time, after each quarter of a period, the energy of a standing wave transforms from electric to magnetic energy, and vice versa. The amplitudes of the fields satisfy Eq. (B.72), i.e., $\sqrt{\epsilon\epsilon_0}E_0 = \sqrt{\mu\mu_0}H_0$. Such a process of energy transformation in a standing wave is analogous to the processes in harmonic oscillators, whereby the energy is transformed from one type into another. For a spring–mass system and the simple pendulum, the energy of oscillation is transformed from kinetic to potential energy. Thus, a standing wave can be considered as a wave harmonic oscillator from the point of view of energy.

B.4.2 Diffraction

We understand diffraction of light to be the *deviation of light waves from direct propagation in the presence of obstacles in their path, wave bending, and penetration into the region of geometrical shadow.* To observe the effect of diffraction, various types of obstacles that limit the propagation of a part of the wavefront can be placed in the path of light waves. If we put a screen behind the obstacle, then under certain conditions a diffraction picture in the form of a pattern of intensity maxima and minima of the light field appears on the screen. *The interference of the infinite number of coherent spherical wave pulses (**wavelets**) emitted by every point of the wavefront is the physical origin of diffraction.*

One of the most important examples of electromagnetic wave diffraction is the diffraction of X-rays on crystalline lattices, by means of which the geometries of many crystalline structures are investigated. For observation of X-ray diffraction it is necessary to have the period of the diffraction grating comparable to the wavelength of the X-rays. Natural diffraction gratings for such radiation are crystalline lattices with period about $0.1-1$ nm. If a monochromatic plane wave is incident on a crystalline lattice with basis vectors $\mathbf{a}_1$, $\mathbf{a}_2$, and $\mathbf{a}_3$, at angles α_0, β_0, and γ_0 with respect to the crystallographic axes, then the directions of the diffraction maxima, defined by the angles α, β, and γ, satisfy the so-called *Laue equations*:

$$\begin{aligned} a_1(\cos\alpha - \cos\alpha_0) &= m_1\lambda, \\ a_2(\cos\beta - \cos\beta_0) &= m_2\lambda, \\ a_3(\cos\gamma - \cos\gamma_0) &= m_3\lambda, \end{aligned} \tag{B.175}$$

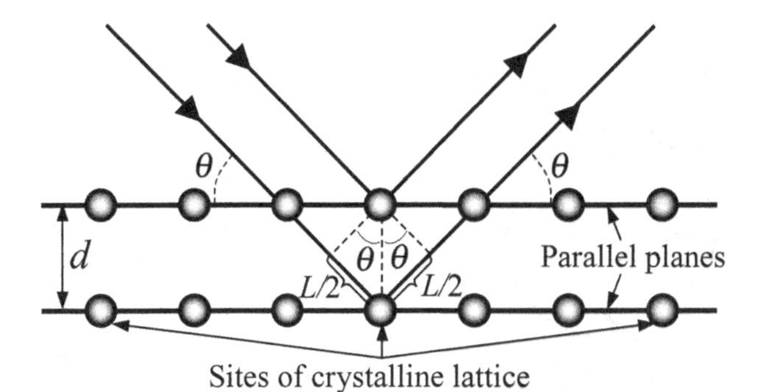

Figure B.10 Diffraction of X-rays from crystalline planes.

where $m_{1,2,3}$ are integer numbers that define the order of the interference maximum.

The conditions (B.175) can be formulated in a different way. The diffraction in a three-dimensional crystalline structure can be considered as a result of mirror reflections from a system of parallel planes, on which the sites of the crystalline lattice are located (see Fig. B.10). The wavelets reflected from the different atomic planes are coherent and may interfere with each other. The path-length difference of two waves that are mirror reflected from neighboring parallel planes, as in Fig. B.10, is equal to

$$L = 2d \sin\theta, \tag{B.176}$$

where d is the distance between two adjacent planes and θ is the angle between the incident ray and the plane. The directions for which the diffraction maxima are developed are defined by Bragg's law:

$$2d \sin\theta_m = m\lambda, \tag{B.177}$$

where $m = 1, 2, 3, \ldots$ We can plot many sets of crystalline planes with different distances between two adjacent planes. Each set of crystalline planes gives its own diffraction maxima in the directions defined by Eq. (B.177). However, the diffraction maxima that occur near the planes with the highest density of atoms have the largest magnitudes of intensity.

B.4.3 Dispersion

The dependence of the propagation speed, v, of a monochromatic light wave in a medium on its frequency, ω, is called *dispersion*. Since the speed of the electromagnetic wave in a medium, according to Eq. (B.55), is defined by the expression

$$v(\omega) = \frac{c}{\sqrt{\mu(\omega)\epsilon(\omega)}}, \tag{B.178}$$

the dispersion is actually defined by the dependence of the medium's material parameters, μ and ϵ, on the frequency, ω. All substances are to some extent dispersive, but the frequency dependence of the dielectric constant, ϵ, and magnetic permeability, μ, for

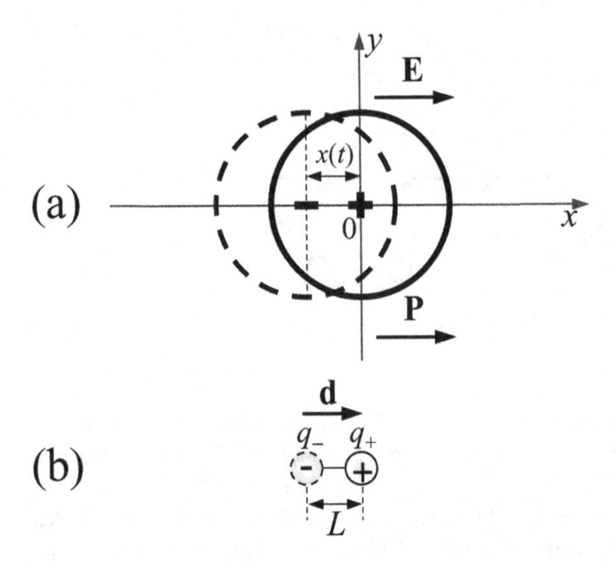

Figure B.11 (a) The polarization of an atom. (b) The dipole moment **d**.

most media becomes apparent only in relatively narrow regions of the spectrum. For the dielectric constant, ϵ, such a dispersive region usually lies in the visible and infrared regions. In many weakly magnetic materials the magnetic permeability practically does not depend on frequency and is close to unity throughout the entire frequency range. Only in strongly magnetic materials (such as ferromagnetic materials and ferrites) is the frequency dependence, $\mu(\omega)$, substantial in the ultra-high-frequency region of the spectrum ($\omega \sim 10^9 - 10^{11}$ s^{-1}). Therefore, let us concentrate on the analysis of the frequency dependence of the dielectric constant, $\epsilon(\omega)$, which can be considered qualitatively in the framework of classical concepts of the behavior of materials in electromagnetic fields.

Let us consider an *isotropic dielectric medium*, which consists of non-interacting atoms. Let us suppose that the atoms of the medium do not have a dipole moment in the absence of an external electric field, i.e., the centers of positive charges of nuclei coincide with the centers of negative charges of electrons. In the presence of an electric field the centers of valence electrons' orbits shift with respect to the nucleus, atoms acquire a dipole moment oriented along the direction of the electric field, and the medium in general acquires polarization (see Fig. B.11(a)). The dipole moment, **d**, is defined as the product of two physical magnitudes: the charge $q = |q_-| = q_+$ and the distance, L, between charges q_- and q_+. The dipole moment, **d**, is a vector, which is directed from the negative charge q_- to the positive charge q_+ (see Fig. B.11(b)):

$$\mathbf{d} = q\mathbf{L}\,. \tag{B.179}$$

Considering this, let us take into account that in the electric field of the electromagnetic wave the center of the valence electron's orbit, coupled with the atom, undergoes an oscillatory motion. The motion of the electron can be described by Newton's second law:

$$m_e \frac{\mathrm{d}^2\mathbf{x}}{\mathrm{d}t^2} = -k\mathbf{x} - b\frac{\mathrm{d}\mathbf{x}}{\mathrm{d}t} - e\mathbf{E}(t). \tag{B.180}$$

Here, the first term on the right-hand side of Eq. (B.180) defines the restoring force exerted on the electron by an atom, whereas the second term is the friction force. The third term describes the periodic driving force of the wave's electric field, $\mathbf{E}(t)$:

$$\mathbf{E}(t) = \mathbf{E}_0 \mathrm{e}^{\mathrm{i}\omega t} = E_0 \mathbf{e}_x \mathrm{e}^{\mathrm{i}\omega t}. \tag{B.181}$$

The steady-state solution of Eq. (B.180) has the form

$$\mathbf{x}(t) = \mathbf{e}_x x(t), \tag{B.182}$$

with

$$x(t) = x_0 \mathrm{e}^{\mathrm{i}\omega t}. \tag{B.183}$$

Here we introduced the complex amplitude of the electron's oscillations, x_0,

$$x_0 = \frac{\omega^2 - \omega_0^2 + 2\mathrm{i}\beta\omega}{\left(\omega^2 - \omega_0^2\right)^2 + 4\beta^2\omega^2} \frac{eE_0}{m_\mathrm{e}} \tag{B.184}$$

and the standard parameters $\omega_0^2 = k/m_\mathrm{e}$ and $2\beta = b/m_\mathrm{e}$. Finally, the dipole moment, $\mathbf{d}$, that the atom acquires is defined as

$$\begin{aligned}\mathbf{d} = -e\mathbf{x}(t) = -ex(t)\mathbf{e}_x &= -e\frac{\omega^2 - \omega_0^2 + 2\mathrm{i}\beta\omega}{\left(\omega^2 - \omega_0^2\right)^2 + 4\beta^2\omega^2} \frac{eE_0}{m_\mathrm{e}} \mathrm{e}^{\mathrm{i}\omega t}\mathbf{e}_x \\ &= -\frac{\omega^2 - \omega_0^2 + 2\mathrm{i}\beta\omega}{\left(\omega^2 - \omega_0^2\right)^2 + 4\beta^2\omega^2} \frac{e^2}{m_\mathrm{e}} \mathbf{E}(t).\end{aligned} \tag{B.185}$$

Then, the dipole moment for unit volume of a medium, which is called the *polarization vector*, $\mathbf{P}$, is defined as

$$\mathbf{P} = n\mathbf{d} = -n\frac{\omega^2 - \omega_0^2 + 2\mathrm{i}\beta\omega}{\left(\omega^2 - \omega_0^2\right)^2 + 4\beta^2\omega^2} \frac{e^2}{m_\mathrm{e}} \mathbf{E}(t). \tag{B.186}$$

Here, e is the charge of the electron and n is the number of atoms in the unit volume. The electric flux density, $\mathbf{D}$, in the medium consists of the flux density of the external electric field, $\epsilon_0 \mathbf{E}$, and the internal flux density, which is equal to the polarization vector, $\mathbf{P}$:

$$\mathbf{D} = \epsilon_0 \mathbf{E} + \mathbf{P}. \tag{B.187}$$

For the so-called *linear media* the polarization vector, $\mathbf{P}$, is linearly proportional to the external electric field, $\mathbf{E}$:

$$\mathbf{P} = \kappa \epsilon_0 \mathbf{E}, \tag{B.188}$$

where we introduced the so-called *electric susceptibility*, κ, of the material. Thus,

$$\mathbf{D} = \epsilon_0 \mathbf{E} + \kappa \epsilon_0 \mathbf{E} = \epsilon \epsilon_0 \mathbf{E}, \tag{B.189}$$

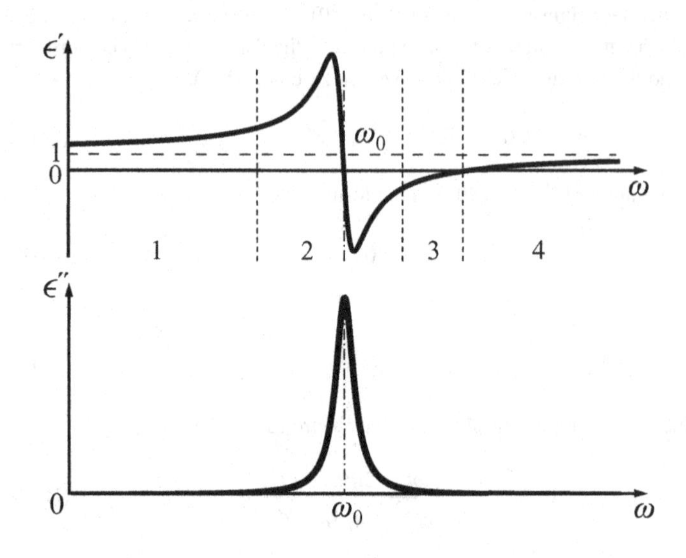

Figure B.12 The dependences of ϵ' and ϵ'' on the frequency, ω.

where

$$\epsilon = 1 + \kappa. \tag{B.190}$$

Finally, the polarization vector, **P**, can also be presented in the form

$$\mathbf{P} = (\epsilon - 1)\epsilon_0 \mathbf{E}. \tag{B.191}$$

By comparing Eqs. (B.186) and (B.191), we find the complex dielectric constant, ϵ:

$$\epsilon = 1 - \frac{n}{\epsilon_0} \frac{\omega^2 - \omega_0^2 + 2\mathrm{i}\beta\omega}{\left(\omega^2 - \omega_0^2\right)^2 + 4\beta^2\omega^2} \frac{e^2}{m_\mathrm{e}}. \tag{B.192}$$

Now, we can find the expressions for the real and imaginary parts of the complex dielectric constant, $\epsilon = \epsilon' - \mathrm{i}\epsilon''$:

$$\epsilon' = 1 - \frac{ne^2}{\epsilon_0 m_\mathrm{e}} \frac{\omega^2 - \omega_0^2}{\left(\omega^2 - \omega_0^2\right)^2 + 4\beta^2\omega^2}, \tag{B.193}$$

$$\epsilon'' = \frac{ne^2}{\epsilon_0 m_\mathrm{e}} \frac{2\beta\omega}{\left(\omega^2 - \omega_0^2\right)^2 + 4\beta^2\omega^2}. \tag{B.194}$$

The phase velocity of the wave in the medium, $v_{\mathrm{ph}} = c/\sqrt{\mu\epsilon'}$, is directly related to the real part of the dielectric constant, ϵ'. The damping of a wave in a medium, which according to Eqs. (B.104) and (B.105) is proportional to ϵ'', is related to the imaginary part of the dielectric constant, ϵ''.

Let us analyze the obtained frequency dependences of the magnitudes ϵ' and ϵ'', which are shown in Fig. B.12. The entire frequency interval of change of ϵ' and ϵ'' can be divided into four regions. In region 1 there is no absorption because ϵ'' is practically equal to zero. This region is called the *optical-transparency window*. In this region $\epsilon' > 1$ and thus the wave speed in the medium is less than the speed of light in vacuum. Region 2 is called the *optical-absorption window* because in this region ϵ'' has sufficiently large magnitude. In this region the medium becomes non-transparent and waves with the corresponding frequencies quickly fade in the medium. Note that any dielectric medium, even if it does not have free carriers of charge, always has a frequency interval within which you can see strong absorption. This is caused by quantum transitions in atoms. Region 3 is called the *metallic-reflection window*. It begins from frequencies at which there is practically no absorption, and it ends at frequencies for which $\epsilon' \to 0$. In this region $\epsilon' < 0$ and $\epsilon'' = 0$. Thus, the coefficient of wave reflection from the interface is practically equal to unity, which is characteristic of metals. Region 4 also is a region with optical transparency, since at these frequencies $\epsilon'' = 0$. In this region $0 < \epsilon' < 1$, thus the wave's phase velocity is greater than the speed of light in vacuum.

As we mentioned before, monochromatic waves are only an idealization and real waves propagate in the form of wave packets transferring energy with the group velocity, v_{gr}:

$$v_{gr} = \frac{d\omega}{dk}. \tag{B.195}$$

There is a relation between phase and group velocities:

$$v_{gr} = v_{ph} - \lambda \frac{dv_{ph}}{d\lambda}. \tag{B.196}$$

The region where $dv_{ph}/d\lambda > 0$ is the region of so-called *normal dispersion* and in this region $v_{gr} < v_{ph}$. In the region of so-called *anomalous dispersion* we have $dv_{ph}/d\lambda < 0$ and $v_{gr} > v_{ph}$. In the absence of dispersion, when the phase velocity does not depend on the wavelength, $v_{gr} = v_{ph}$ and the wave packet propagates without spreading. Vacuum is a medium of this kind.

Example B.6. Estimate the real and imaginary parts of the dielectric constant, ϵ, for graphite at the X-ray wavelength of $\lambda = 0.05$ nm. The density of graphite is $\rho = 1.6 \times 10^3$ kg m^{-3}.

Reasoning. Note that the carbon atom has four electrons in its outer shell (the electron configuration of carbon is $1s^2 2s^2 2p^2$, see Table 6.7). Hard X-rays ionize these electrons and make them free, i.e., uncouple them from their nuclei. Thus, the returning force from the core of the atom in Eq. (B.180) is equal to zero, i.e., $\kappa = 0$, so the frequency ω_0 can also be considered equal to zero. As a result Eqs. (B.193) and (B.194) take the forms

$$\epsilon'(\omega) = 1 - \frac{ne^2}{\epsilon_0 m_e} \frac{1}{\omega^2 + 4\beta^2} = 1 - \frac{\omega_p^2}{\omega^2 + 4\beta^2}, \tag{B.197}$$

$$\epsilon''(\omega) = \frac{ne^2}{\epsilon_0 m_e} \frac{2\beta}{\omega(\omega^2 + 4\beta^2)} = \frac{2\omega_p^2 \beta}{\omega(\omega^2 + 4\beta^2)}, \tag{B.198}$$

where $\omega_p = \left[ne^2/(\epsilon_0 m_e)\right]^{1/2}$ is the so-called *plasma frequency*. For its calculation, let us estimate the concentration of free electrons created by X-rays. The number of carbon atoms in the unit volume, n_{C}, is

$$n_{\mathrm{C}} = \frac{\rho N_{\mathrm{A}}}{A}, \tag{B.199}$$

where $A = 12 \times 10^{-3}$ kg mol^{-1} and $N_{\mathrm{A}} = 6.02 \times 10^{23}$ mol^{-1}. After carrying out the requisite calculations, we obtain

$$n = 4n_{\mathrm{C}} = 3.2 \times 10^{29}\ \mathrm{m}^{-3},$$

and the plasma frequency is $\omega_{\mathrm{p}} \approx 3.2 \times 10^{16}$ s^{-1}. The wavelength $\lambda = 0.05$ nm corresponds to the frequency $\omega = 2\pi c/\lambda = 3.8 \times 10^{19}$ s^{-1}, which is three orders of magnitude higher than ω_{p}. The parameter β plays the role of the scattering frequency and it has the following order: $\beta \sim 10^{14}\text{–}10^{15}$ s^{-1}. Therefore, ϵ'' is equal to zero with high precision, which indicates that graphite is transparent for X-rays. The magnitude of the real part of the dielectric constant is practically equal to unity because $1 - \epsilon' \sim 10^{-6}$. In accordance with Eqs. (B.137) and (B.138), the coefficients of reflection, R, and transmission, T, of X-rays incident from vacuum on graphite are equal to

$$R = \frac{1 - \sqrt{\epsilon'}}{1 + \sqrt{\epsilon'}} \approx 0, \tag{B.200}$$

$$T = \frac{2}{1 + \sqrt{\epsilon'}} \approx 1. \tag{B.201}$$

B.5 Summary

1. Maxwell's equations are the main equations of classical electrodynamics, which sufficiently correctly describe the processes of propagation and interaction of electromagnetic waves with various media up to frequencies of $10^{14}-10^{15}$ Hz. At higher frequencies the quantum character of the interaction of electromagnetic radiation with a substance comes to dominate.
2. Maxwell's equations are a system of linear partial differential equations of first order, whose unknowns are the field intensities, **E** and **H**, and flux densities **D** and **B** of the electric and magnetic fields, respectively. For these magnitudes the superposition principle is valid.
3. Electric and magnetic fields possess energy. The volume energy density at each point of an electromagnetic field is equal to

$$w = \frac{\mathbf{E}\cdot\mathbf{D} + \mathbf{H}\cdot\mathbf{B}}{2} = \frac{\epsilon\epsilon_0 E^2 + \mu\mu_0 H^2}{2}.$$

4. The vectors **E**, **H**, and **k** of an electromagnetic wave in an isotropic medium are perpendicular to each other and in this order form a right-handed system of vectors. For a monochromatic wave the oscillations of **E** and **H** are harmonic.
5. The Poynting vector, $\mathbf{P} = \mathbf{E} \times \mathbf{H}$, defines the flux energy density which is carried by an electromagnetic wave, which has the dimensionality [W m^{-2}]. In an isotropic medium this vector is parallel to the wavevector **k**.

6. In absorbing media the amplitude of an electromagnetic wave exponentially decreases during its propagation. The effect of localization of electromagnetic wave energy in the subsurface layer of an absorbing sample is called the skin effect and it can most clearly be observed in metals.
7. At the interface of two media, which is not charged, and there is no current flowing along this interface, the tangential components of the field intensity and normal components of flux density of electric and magnetic fields must be continuous. To find the amplitudes of transmitted and reflected waves we have to solve a boundary problem, which requires sewing fields at the interface taking into account the above-mentioned boundary conditions.
8. Interference, polarization, diffraction, and dispersion, which occur widely in nature and are observed throughout the entire electromagnetic spectrum, are explained as arising solely because of the wave nature of electromagnetic radiation.

B.6 Problems

Problem B.6.1. Maxwell's equation (B.16) can be rewritten in the following form:

$$\nabla \times \mathbf{H} = \mathbf{j}_{\text{dis}} + \mathbf{j}, \tag{B.202}$$

where we introduced the density, $\mathbf{j}_{\text{dis}}$, of the displacement current, which is measured in [A m^{-2}]:

$$\mathbf{j}_{\text{dis}} = \frac{\partial \mathbf{D}}{\partial t}. \tag{B.203}$$

We see from Eq. (B.202) that it is precisely the displacement currents which are responsible for the appearance of vortex magnetic fields in the case when there is no current, $\mathbf{j}$, of external charges. Let a monochromatic plane wave described by Eq. (B.56), with an amplitude of the electric field of $E_0 = 10^3$ V m^{-1} and frequency $\omega = 3 \times 10^{15}$ s^{-1} (red light), propagate in vacuum. Find the amplitudes of the displacement current, $j_{\text{dis},0}$, and magnetic field intensity, H_0, of the wave.

Problem B.6.2. The point of interference, SP, of two monochromatic plane waves, propagating in vacuum, is defined by the radius vector $\mathbf{r}$, which forms angles φ_1 and φ_2 with the wavevectors, $\mathbf{k}_1$ and $\mathbf{k}_2$, of the two waves (Fig. B.13). Find the frequency of the wave at the point SP which corresponds to the intensity maximum of the wave field.

Problem B.6.3. Derive the equation for the location of maxima and minima in the interference pattern. Assume that the setup of Young's scheme (the two-slit experiment; see Fig. B.9) is placed in water to perform the same experiment as was done in the air.

Problem B.6.4. The field intensities of electric and magnetic fields in a non-absorbing medium oscillate in phase. Show that, for a wave in an absorbing medium, the oscillations of the vectors $\mathbf{E}$ and $\mathbf{H}$ do not coincide in phase. Find the relation of the phase shift of these two vectors to the conductivity of the medium, σ.

Problem B.6.5. The conductivity of sea water is $\sigma_{\text{s.w.}} \approx 5\ \Omega^{-1}$ m^{-1} and that of copper is $\sigma_{\text{Cu}} \approx 6 \times 10^7\ \Omega^{-1}$ m^{-1}. Compare the values of skin depth for these two media at

Table B.6.1. *The skin depth, δ, of sea water and copper*

	Frequency (Hz)			
Skin depth (m)	10	10^4	10^7	10^{10}
Sea water	71.4	2.26	7.1×10^{-2}	2.3×10^{-3}
Copper	2.1×10^{-2}	6.6×10^{-4}	2.1×10^{-5}	6.6×10^{-7}

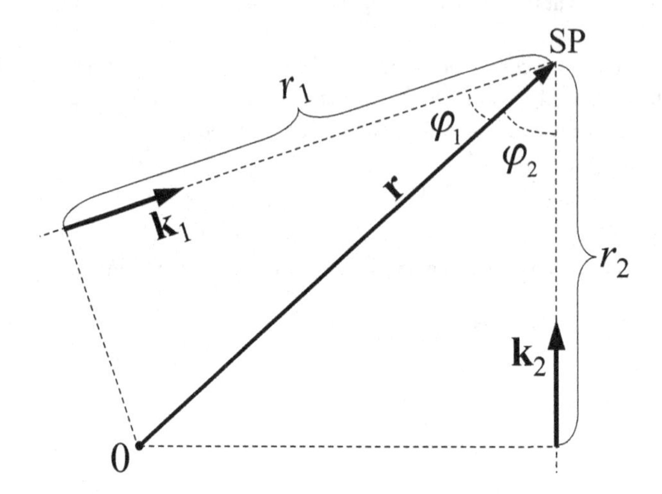

Figure B.13 The interference of two plane waves with wavevectors $\mathbf{k}_1$ and $\mathbf{k}_2$. Here $r_1 = r\cos\varphi_1$ and $r_2 = r\cos\varphi_2$.

frequencies $f = 10$, 10^4, 10^7, and 10^{10} Hz, considering for simplicity that the dielectric constant, ϵ, and magnetic permeability, μ, equal unity for both media. (Answer: Table B.6.1.)

Problem B.6.6. Let us consider an optical micro-resonator that consists of two parallel-plate mirrors applied from the opposite sides of a narrow layer of optical glass with width $L = 10$ μm. Find the number, N, of possible types of electromagnetic field oscillations in the region between the mirrors (i.e., standing waves with different wavelengths) that can be excited in the visible range of wavelengths. Take the dielectric constant of glass as $\epsilon = 1.55$ and its magnetic permeability as $\mu = 1$. (Answer: $N = 27$.)

Problem B.6.7. The density of charge distribution in the electron cloud of a hydrogen atom in the ground state is defined by the following expression:

$$\rho(r) = -\frac{e}{\pi a^3}\mathrm{e}^{-2r/r_1}, \tag{B.204}$$

where r_1 is the radius of the first Bohr orbit and e is the electron charge. Find the energy of the electrostatic Coulomb interaction of the electron cloud with the nucleus of the hydrogen atom (i.e., with the proton).

Problem B.6.8. Find the screening radius (Debye radius), r_D, and potential, φ, of a point defect with charge q in an n-type semiconductor at room temperature, $T = 300$ K.

Assume that positively charged ions of univalent impurities are located at the points of the lattice and that the free electrons are in thermal equilibrium. The energy distribution of electrons is described by Boltzmann's classical function. The density of impurities is $n_0 = 10^{25}\ \mathrm{m}^{-3}$ and the dielectric constant of the crystal is $\epsilon = 10$. (Answer: $r_\mathrm{D} \approx 1.2\ \mathrm{nm}$.)

Problem B.6.9. A plane electromagnetic wave from a medium with index of refraction n_1 is obliquely incident on an interface with a medium with index of refraction n_2. Find the condition of total internal reflection, the wave field in the second medium, and the depth of penetration into this medium.

Problem B.6.10. Find the energy, W, which is transmitted in vacuum by a plane electromagnetic wave propagating in the x-direction through a plane contour with area $S = 1\ \mathrm{m}^2$ perpendicular to this direction during one second. Assume that the amplitude of the electric field intensity is equal to $E_0 = 10^{-4}\ \mathrm{V\ m}^{-1}$ and that the frequency is equal to $f = 10^6$ Hz. (Answer: $W \approx 1.3 \times 10^{-15}$ J.)

Problem B.6.11. A linearly polarized wave is normally incident on a slab of thickness L. The dielectric constant of the slab is ϵ, its conductivity is $\sigma = 0$, and its permeability is $\mu = 1$. Find the amplitudes of the waves reflected from and transmitted through the slab and of the wave inside of the slab. Find the energy coefficients of reflection and transmission. Find the angle of minimal wave reflection from the slab.

Appendix C
Crystals as atomic lattices

Three-dimensional quantum-dot superlattices can be considered as nanocrystals. Spherical nanoparticles consisting of a big enough number (from 10 to 1000) of atoms or ions, which are connected with each other and are ordered in a certain fashion, can be considered as the structural units of such nanocrystals. Examples of nanocrystals that are of natural origin are the crystalline modifications of boron and carbon which have as their structural units the molecules B_{12} and C_{60}. The boron molecule B_{12} consists of 12 boron atoms, and the carbon molecule C_{60}, which is called *fullerene*, consists of 60 carbon atoms. The fullerene molecule resembles a soccer ball, i.e., it consists of 12 pentagons and 20 hexagons, with carbon atoms at their corners. These nanoparticles form face-centered superlattices with a period of about 1–10 nm. At these distances between molecules of C_{60} weak molecular forces, which provide the crystalline state of fullerene, act.

In addition to nanocrystals of natural origin, numerous artificial three-dimensional superlattices consisting of various types of nanoparticles have been fabricated. The variety of nanocrystalline structures as well as of conventional crystals is defined by the differences in the distribution of electrons over the quantum states of atoms. The most significant role in the formation of individual nanoparticles as well as of crystals is played by the electrons in the outer shells of atoms. First of all, the energy minimum of a nanoparticle itself is predominantly defined by the interaction of valence electrons. Configurations of nanoparticles in which electrons completely fill the energy shells have stable structures consisting of a certain number of atoms. For this type of structure there exist so-called *magic numbers*, which define the total number of atoms or ions in a nanoparticle. For one group of metals (Ag, Au, Cu, and others) nanoparticles are formed on the basis of a face-centered cubic elementary cell built around the central atom and have the following series of magic numbers: $N = 1, 13, 55, 147, 309, 561, \ldots$ For another group of metals (Mg, Zn, and others) nanoparticles form on the basis of a hexagonal densely packed elementary cell and have the following magic numbers: $N = 1, 13, 57, 163, 321, 581, \ldots$

Owing to the exchange of valence electrons or their collectivization, and due to the spatial distribution of their wavefunctions, arrays of atoms, ions, or nanoparticles will, under certain conditions, form regular crystalline structures of various configurations. The stable state in such structures is almost completely provided by the forces of electrostatic attraction involving the subsystem of valence electrons in ion shells. In addition to attraction forces at large distances, significant forces of repulsion can occur at small distances between atoms and ions because of overlap of electron shells. There are significant differences in behavior of electrons among metals, dielectrics, and semiconductors that are defined mainly by their different band-energy spectra. We will consider some of the features of the formation of the above-mentioned structures and their energy spectra.

C.1 Crystalline structures

C.1.1 Spatial periodicity

All objects consist of a collection of atoms, molecules, or ions. The average distance between them is defined by their interaction with each other and by temperature. In the crystalline state the particles of a medium are arranged in an ordered fashion, and the type of ordering depends on the chemical species as well as on the details of the interaction among the particles.

In 1912 the German physicist Max von Laue discovered the diffraction of X-rays from crystals. Since X-rays are electromagnetic waves, diffraction can take place only in spatially-periodic structures consisting of atoms or ions with interatomic distances comparable to the wavelength of X-rays. Modern experimental methods such as scanning tunneling microscopy (STM) allow us to see directly the regular spatial distribution of atoms in a crystal.

The main property of crystalline structures is their so-called *translational symmetry*. The crystal can be thought of as being generated by the repetition of a basic atomic group over a fixed vector, which is called the *period of a crystalline lattice*. In such a case it is said that there is *long-range order in a crystalline structure*, i.e., the particles are distributed in an orderly fashion throughout the entire crystal.

In addition to the translational symmetry in ideal crystals, there is so-called *point symmetry*. A crystal that has point symmetry coincides with itself after an operation of *rotation* and/or *reflection* has been carried out. The symmetry of a crystalline lattice depends on the distribution of its structural elements in space. Symmetry defines the physical properties of a crystal and the directions along which these properties become apparent. For example, after a rotation of a cube around one of its three axes drawn through the centers of the opposite cube plane, the cube will coincide with itself after each 90° rotation. If we divide a cube by a plane going through the diagonals of the opposite bases and then reflect it with respect to this plane, then the cube will coincide with itself. The existence of such elements of symmetry can be demonstrated on the crystalline structure of sodium chloride (NaCl). It is represented by a cubic lattice with alternating ions of sodium (Na) and chlorine (Cl) in the sites of the lattice. Each Na^+ ion is surrounded by six Cl^- ions, and in its turn each Cl^- ion is surrounded by six Na^+ ions (see Fig. C.1). The minimal distance between ions is 2.81 Å. Let us note that on a macroscopic scale we are dealing with a very large number of crystalline elements. For example, if the linear dimensions of a crystal are about 1 cm, the crystal contains of the order of 10^{22} ions.

How can one build an infinite lattice of a *simple* (consisting of only one type of atoms or ions) crystal? Let us choose the origin of coordinates in any arbitrary place, and let us put an atom (or ion) at this point. Let us draw three mutually-orthogonal vectors $\mathbf{a}_1$, $\mathbf{a}_2$, and $\mathbf{a}_3$ (in general they are not required to be mutually orthogonal), whose length is equal to the distance between the nearest atoms (or ions). The parallelepiped whose edges are the vectors $\mathbf{a}_1$, $\mathbf{a}_2$, and $\mathbf{a}_3$ is called a *unit cell*. The three vectors $\mathbf{a}_1$, $\mathbf{a}_2$, and $\mathbf{a}_3$ are also called *basis vectors*, because, using them, one can build an infinite spatial lattice. In order

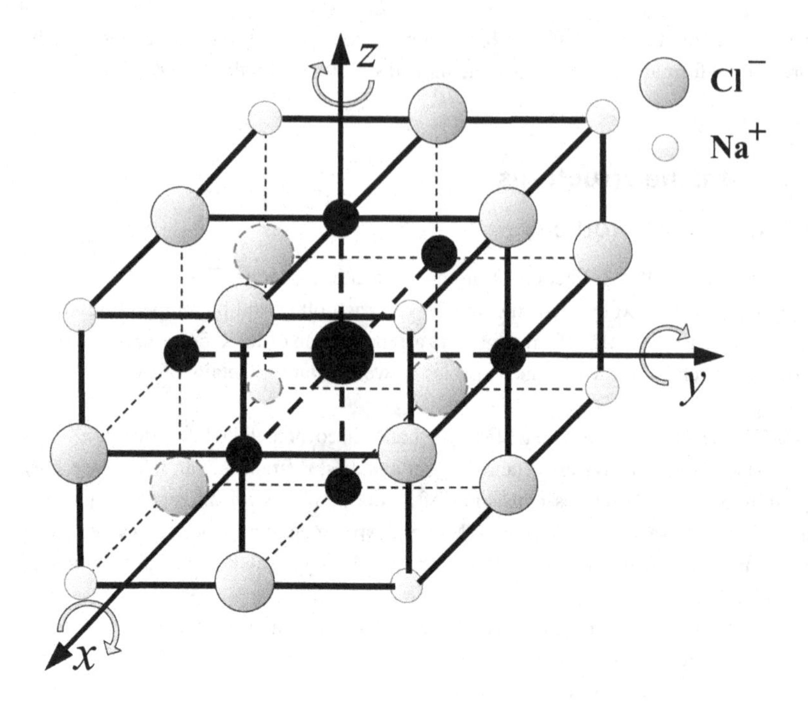

Figure C.1 The unit cell of the NaCl crystalline structure.

to do this it is necessary to move all atoms (or ions) of a lattice from their initial positions by distances equal to the translational vectors $\mathbf{a}_n$:

$$\mathbf{a}_n = n_1\mathbf{a}_1 + n_2\mathbf{a}_2 + n_3\mathbf{a}_3, \tag{C.1}$$

where n_1, n_2, and n_3 are arbitrary integer numbers. We can choose unit cells with other basis vectors and, by translation of these unit cells, we may obtain the same spatial crystalline lattice. A unit cell of minimal volume is called a *primitive cell* (see Fig. C.2 for examples of translational symmetry).

There are many physical phenomena in which the atomic structure of the material is not manifested directly. While studying these phenomena the material can be considered as a continuum, disregarding its internal structure. Such phenomena include, for example, thermal expansion of bodies, their deformation under the influence of external forces, dielectric permeability, and many optical phenomena. These material properties are characteristic for a continuum and are called *macroscopic properties*. The indicated properties significantly depend for most crystals on the direction along which the measurements are carried out. The origin of this dependence is connected with the structure of the crystal and its symmetry. For example, stretching a cubic crystal in the direction parallel to the edges of cubic cells of a lattice will not be the same as stretching along the diagonals of the cells, because the coupling energy between atoms depends on the distance between them. The dependence of physical properties on the direction is called *anisotropy*. Anisotropy is the characteristic property of crystals and in this sense they are fundamentally different from isotropic media such as liquids and gases, whose properties are the same in all directions.

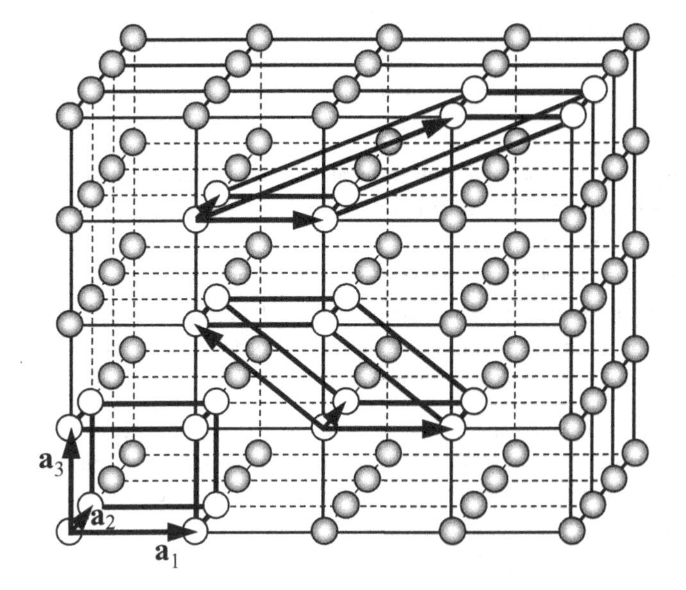

Figure C.2 Translational symmetry of a three-dimensional lattice.

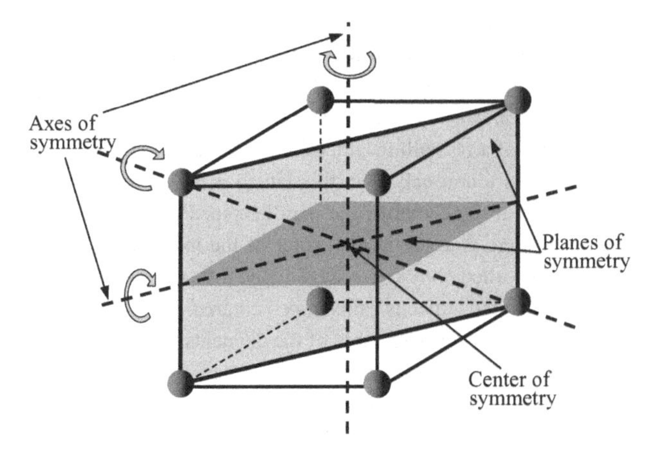

Figure C.3 Axes, planes, and the center of symmetry in a simple cubic lattice.

C.1.2 Symmetry elements and types of crystalline lattices

What are the main symmetry elements of an object? Let us briefly describe them.

Plane of symmetry. The presence of a plane of symmetry means that one part of the figure coincides with the other if we move all its points equal distances to the other side of a plane in a direction perpendicular to it. Thus the plane of symmetry can be treated as a mirror, through which one part of the figure is reflected (see Fig. C.3).

Axis of symmetry. An axis of symmetry is a straight line around which rotation of the figure by a certain angle makes the figure coincide with itself. The order, n, of the axis of symmetry is defined by the equation

$$n = \frac{360}{\alpha}, \tag{C.2}$$

where α is the minimal angle of rotation at which the figure coincides with itself. Only five axes of symmetry can exist in crystals: of first, second, third, fourth, and sixth orders (see Fig. C.3).

Center of symmetry. A crystal has a center of symmetry if any straight line drawn through it at the opposite sides of a crystal goes through identical points. Thus, there are equal planes, edges, and angles at the opposite sides of the crystal with respect to the center of symmetry (see Fig. C.3).

There are 32 possible combinations of planes, axes, and centers of symmetry in crystals. In general, a crystal does not possess only one element of symmetry. The full set of symmetry elements is called the *symmetry group*. Why is the symmetry group so important for the physics of crystals? It turns out that the symmetry group of the crystal very often defines the physical properties of the crystal. Depending on the relation of magnitudes and on the self-orientation of edges of the unit cell, there are 14 types of crystalline lattices, which are called *Bravais lattices* (see Fig. C.4). The above-mentioned 14 types of lattices constitute 7 different systems: *triclinic*, *monoclinic*, *orthorhombic*, *tetragonal*, *rhombohedral*, *hexagonal*, and *cubic*. Each of the systems is characterized by the ratio of sides of the unit cells and the angles α, β, and γ between them. Many important materials have *simple* (or *primitive*), *base-centered*, *body-centered*, and *face-centered* Bravais lattices. If the sites of a crystalline lattice are located only at the vertices of a parallelepiped, which represents a unit cell, then this lattice is called *primitive* or *simple*. If there are additional sites at the center of the parallelepiped's base, then this lattice is called *base-centered*. If there is a site at the center of the intersection of the spatial diagonals, then this lattice is called *body-centered*. Finally, if there are sites at the centers of all of the lateral faces, then the lattice is called *face-centered*.

Almost half of the elements from the Periodic Table of the elements form crystals of cubic or hexagonal symmetry, which we will study now in more detail.

Three lattices are possible for the crystals of cubic systems: simple, body-centered, and face-centered (see Fig. C.5). In the cubic system all the angles of the unit cell are equal to 90° and all the edges are equal to each other. A right prism with a rhombus of angles 60° and 120° as the base represents the unit cell of the hexagonal system. The two angles between the unit cell's axes are right angles and one of them is equal to 120° (see Fig. C.4).

In most cases we consider a crystal as a system of hard spheres that are in contact with each other. A structure in which the spheres are as densely packed as possible corresponds to the minimal energy of the structure. Let us compare three possible cubic structures in such a model. In a simple cubic structure atoms are present only at the sites of a cube. In this case one atom corresponds to one primitive cell. In a face-centered cubic lattice atoms are present not only at the sites of a lattice, but also at the centers of six faces (NaCl has such a structure). In a body-centered cubic lattice the atoms are at the sites of the cube and in addition there is one at the center of the cube. The least densely packed is the simple cubic structure, and the chemical elements prefer not to crystallize in such structures,

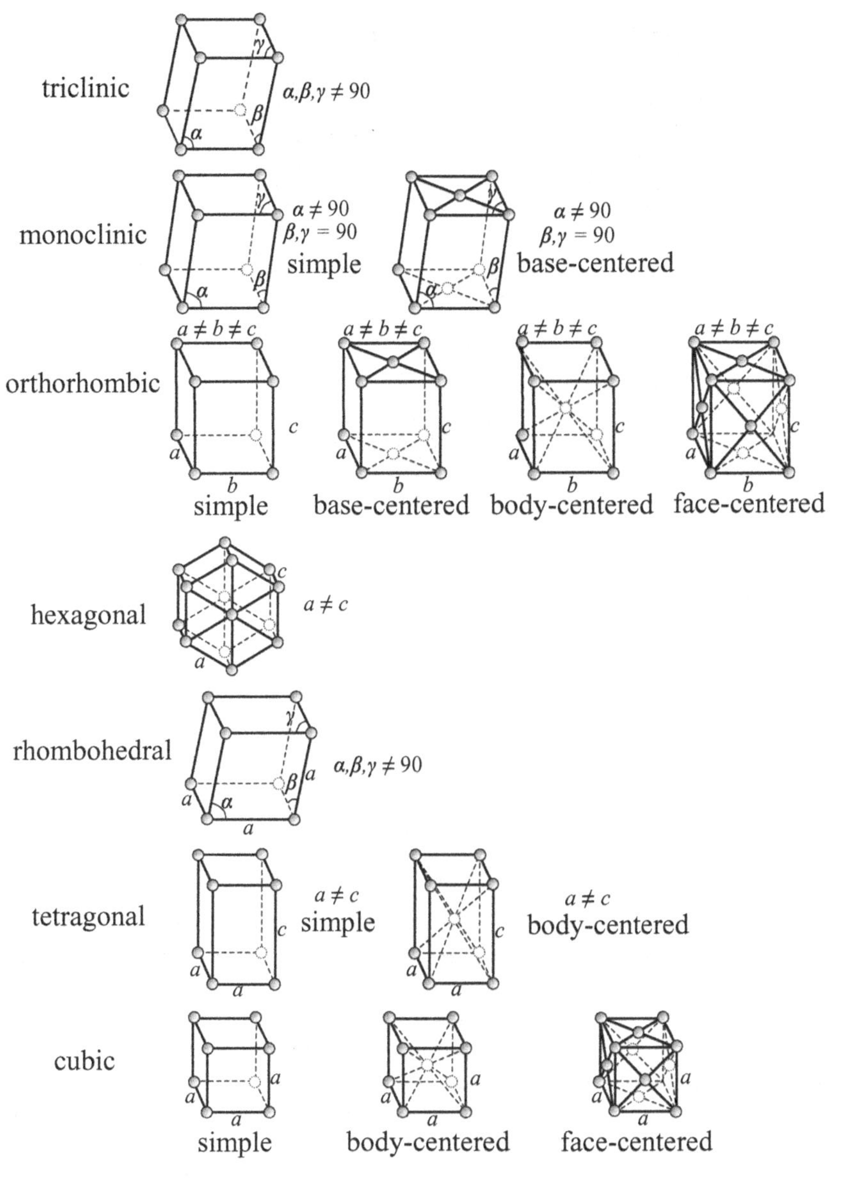

Figure C.4 The 14 Bravais lattices.

Figure C.5 Unit cells of the cubic lattices:
(a) simple,
(b) body-centered, and
(c) face-centered.

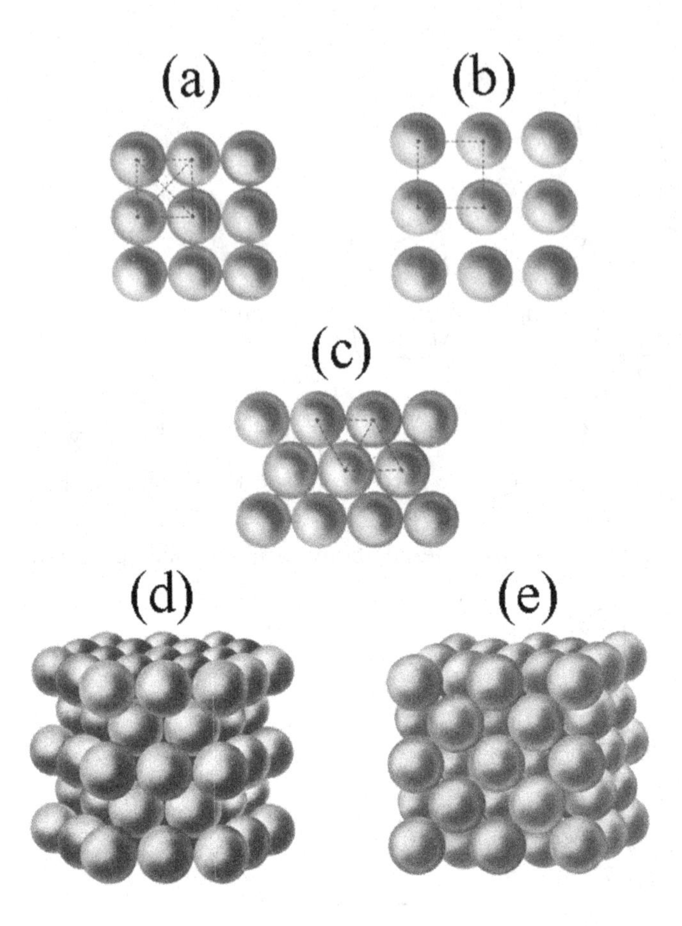

Figure C.6 Types of packing: (a) dense square-packed layer, (b) loose square-packed layer, (c) dense hexagonally packed layer, (d) dense-packed hexagonal crystalline lattice, and (e) dense-packed face-centered cubic crystalline lattice.

although many substances do in their crystalline states have the structure of a simple cube (e.g., CsCl or CuPd). The most densely packed is the *face-centered structure*, which for this reason is also called a *cubic structure with the most dense packing*. However, it is possible to have the same packing density if we were to arrange in space identical hard spheres composing a hexagonal dense packing. In this structure the density of packing is the same as in a face-centered cubic lattice. This is why most metals at certain temperatures very easily change their structure from face-centered cubic to hexagonal dense packing and vice versa. The first hexagonal plane of the skeleton of the hexagonal dense-packing lattice corresponds to the dense packing of hard spheres (see Fig. C.6). The next atomic plane is packed in a similar way, but it is shifted in such a manner that its atoms are in between the atoms of the first plane; the third plane is packed in the same way, but its atoms are exactly over the atoms of the first plane; the fourth plane has its atoms located as in the second one, and so on.

Figure C.6 shows the two most densely packed crystalline lattices: hexagonal and face-centered cubic lattices with dense packing.

Classification by crystal system gives us an idea about the geometrical characteristics of a crystal, but it does not tell us about the nature of the forces that keep atoms (ions or molecules) at certain sites within a crystal lattice. Classification of the crystals can also be done in terms of the physical nature of forces acting between the particles of a crystal (we will talk about it later). In this case we will have four main types of crystalline lattices: *ionic*, *atomic*, *metallic*, and *molecular*.

C.2 The nature of attraction and repulsion forces

C.2.1 The potential energy of a crystal

It is well known that a large enough number of neutral atoms when brought close together will transform into a stable crystalline state of matter. This fact indicates the existence of attraction forces between them that at atomic distances (about 0.1 nm) are counterbalanced by repulsion forces. In order for the atoms or ions to form a crystal, their interaction has to be defined by a potential energy of the following form:

$$U(r) = \frac{A}{r^m} - \frac{B}{r^n}, \tag{C.3}$$

where A, B, m, and n are positive parameters defined by a certain type of interaction. The first term in Eq. (C.3) describes the repulsion forces and the second term concerns the attraction forces. In order for repulsion forces to be greater than attraction forces at small distances ($r < r_0$), it is necessary to satisfy the condition $m > n$. The equilibrium state between atoms when forces of repulsion and attraction compensate for each other is defined by the following condition:

$$\left.\frac{\mathrm{d}U}{\mathrm{d}r}\right|_{r=r_0} = 0. \tag{C.4}$$

By equating the first derivative of the expression (C.3) to zero we obtain

$$r_0 = \left(\frac{nB}{mA}\right)^{\frac{1}{m-n}}. \tag{C.5}$$

At this distance the potential interaction energy reaches its minimum:

$$U_0 = -\frac{A}{r_0^m}\left(\frac{m}{n} - 1\right). \tag{C.6}$$

The true theory of atomic interaction in a certain crystalline structure must be based on quantum-mechanical consideration of the electron motion. As a rule, in such an approach atomic nuclei are considered motionless (the so-called *adiabatic approximation*) since their mass is three to four orders of magnitude larger than the mass of electrons The main (and close to each other in strength) types of bonding in crystals are ionic, metallic, and covalent. Other types of bonding (intermolecular interactions and hydrogen bonding) are one to two orders weaker than the above three types. By the *bonding energy*,

Table C.1. *Bonding energies in kJ mol^{-1} for crystals with various types of bonding*

LiF	1013	V	510	C	711
NaCl	762	Fe	414	Si	448
KI	623	Cu	339	Ge	372

U_b, we understand the energy which is required for separation of one mole of crystal to individual neutral atoms. In Table C.1 the values of bonding energy for crystals with ionic, metallic, and covalent bondings, which we will define, are shown. For comparison, the characteristic energy of intermolecular bonding, which is called van der Waals bonding, is about 1 kJ mol^{-1}. It is most clearly manifested in crystals of the inert gases, which crystallize at temperatures of about 10–100 K, and in molecular crystals formed from individual molecules (for example, crystals of fullerene). This bonding is caused by interaction of dipole moments of atoms, which occur as a result of fluctuations in the distribution of charge of atoms. Even in a spherically-symmetric state the instantaneous position of the center of the electron cloud need not coincide with the center of the atom, which may lead to the occurrence of a non-zero dipole moment at that instant. An instantaneous dipole moment of one atom induces instantaneous moments of neighboring atoms of such direction that it causes attraction. For this type of bonding, in the potential energy arising from the interaction of neighboring atoms (Eq. (C.3)) the parameter $n = 6$. Let us briefly consider the main types of bonding.

C.2.2 Ionic and metallic bonding

The easiest type of bonding to interpret from the classical point of view is the bonding in ionic crystals, which is the result of Coulomb attraction of adjacent ions. Ionic bonding manifests itself especially clearly in such compounds as NaCl. In Example C.1 later we will obtain the expression for the potential energy of an ion in a one-dimensional periodic chain of ions with alternating signs:

$$U(r) = -\alpha \frac{k_e e^2}{r}, \tag{C.7}$$

where r is the distance between ions with charge $\pm e$ and α is Madelung's constant, which is equal in the case of a one-dimensional periodic chain of atoms to $\alpha = 2 \ln 2 = 1.386$. In the case of a three-dimensional crystal with a cubic lattice Madelung's constant is equal to $\alpha = 1.748$. By comparing Eq. (C.7) with Eq. (C.3), we obtain $B = \alpha k_e e^2$ and $n = 1$.

To estimate the energy of ionic-type bonding we have to multiply Eq. (C.7) by the Avogadro number, N_A:

$$U_{ib}(r) = -\alpha N_A \frac{k_e e^2}{r}. \tag{C.8}$$

After substituting into Eq. (C.8) the values of the physical magnitudes, we obtain for the bonding energy $U_{ib} = 900\,\text{kJ}\,\text{mol}^{-1}$, which is in good agreement with tabulated values for the bonding energy of ionic crystals.

Typical metals (Cu, Ag, Fe, and others) have relatively high conductivity, optical reflectance, and plasticity. These properties unambiguously tell us that in metals there is a large number of free electrons, which even under weak electric fields can move around over large distances within crystals. The simplest model of a metal suggested by Drude is an array of charged ions located at the points of the lattice and an ideal gas of electrons freely moving between ions throughout the crystal. This simplified model qualitatively explains most of the experimental data. If we take into account that an ideal electron gas obeys quantum statistics, then the model becomes valid for the description of a wider range of phenomena taking place in metals.

Let us consider the simplest case of a univalent metal, for which to each univalent ion there corresponds one free electron. In the stationary state the concentration maxima of electrons will be at positions in the middle between the ions. From the electrostatic point of view this model can be treated as an ionic crystal with non-localized negative ions. The potential energy of Coulomb interaction for each ion and bonding energy can be written in the form of Eqs. (C.8) and (C.7), with substitution of α by the effective Madelung constant, α_{eff}:

$$U(r) = -\alpha_{\text{eff}}\frac{k_{\text{e}}e^2}{r}, \qquad U_{\text{ib}}(r) = -\alpha_{\text{eff}}N_{\text{A}}\frac{k_{\text{e}}e^2}{r}, \tag{C.9}$$

where $r = a/2$ is the distance between "opposite ions," which can be estimated as half of the lattice constant. Thus, for the metal in Eq. (C.3), the parameters B and n are the same as for an ionic crystal. The spreading in space of negative charge is taken into account by the effective Madelung constant, α_{eff}, which, because of this spreading of charge, is less than the constant α and significantly depends on the distribution of the electron wavefunction in the given metal.

C.2.3 Covalent bonding

This type of bonding is of special importance because it is the most important not only in molecules but also in crystals. As was shown for the hydrogen molecule, this bonding occurs between neutral atoms and is characterized by saturation and distinctive orientation. The saturation of bonding in the case of hydrogen atoms is manifested by the impossibility of forming a hydrogen molecule from more than two atoms. This is because the attraction occurs only between atoms whose electrons are able to form pairs with oppositely directed spins. The orientation of covalent bonding is defined by the orientation (absence of spherical symmetry) of the wavefunctions of the valence electrons of the majority of atoms that contribute to make bound states.

Thus, the nitrogen atom, N, has electron configuration $1s^2 2s^2 2p^3$. It is well known that the spins of the three 2p-electrons are parallel, i.e., they are not spin-saturated. Therefore, a nitrogen atom can form three covalent bonds and is a *trivalent* atom. For example, in the ammonia molecule, NH_3, the lobes of the wavefunctions of the three 2p-electrons extend along three mutually orthogonal directions. The energy gain as a result of formation of covalent bonds is significantly defined by the overlapping of the electron wavefunctions of the corresponding pair with anti-parallel spins. Therefore, the hydrogen atoms in the ammonia molecule take positions along these three directions where

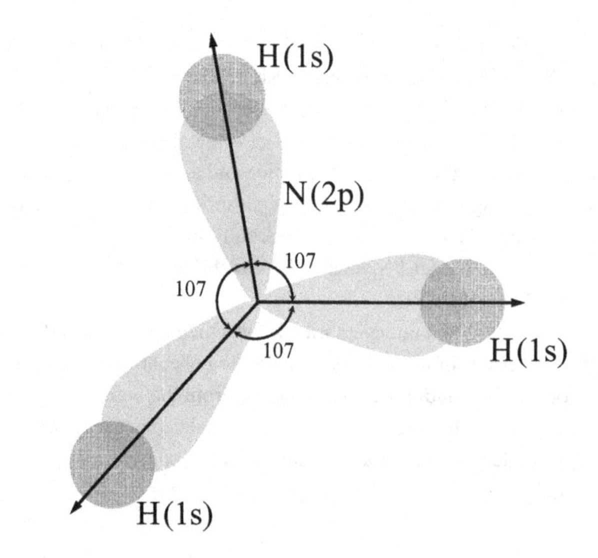

Figure C.7 The ammonia molecule, NH_3.

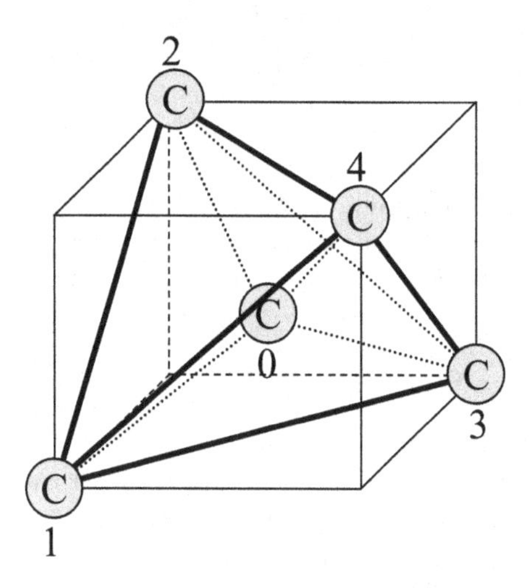

Figure C.8 The diamond structure.

partial overlapping of the wavefunction lobes of nitrogen p-electrons and the spherically-symmetric wavefunctions of hydrogen 1s-electrons take place (Fig. C.7). Because of the mutual repulsion of the hydrogen atoms, the angle between bond directions is equal to 107°, not 90°.

The carbon atom, C, has electron configuration $1s^2 2s^2 2p^2$. Therefore, it must be *bivalent*. However, carbon is *quadrivalent* in the majority of cases, since it participates in chemical compounds not in its ground state but in the excited state $1s^2 2s^1 2p^3$. In this case the spins of all four electrons in 2s- and 2p-states are parallel (not saturated) and may form four covalent bonds. Thus, in diamond the directions of these bonds coincide with directions 01, 02, 03, and 04 of the tetrahedron whose vertexes and center contain the carbon atom (see Fig. C.8).

Crystals of the semiconductors silicon, Si, and germanium, Ge, have the same structure. The silicon atom, Si, has four electrons in its M-shell with $3s^2 2p^2$ states, and its K- and L-shells are completely filled and are spin-saturated. Therefore, in forming crystalline bonds the silicon atom behaves similarly to the carbon atom. Germanium, Ge, which has four electrons in its N-shell in $4s^2 4p^2$ states, has similar properties.

Chemical compounds such as A_3B_5, i.e., compounds of elements from groups III and V of the Periodic Table of the elements, share most of the properties of crystalline structures of elements of group IV (silicon and germanium). For example, InSb and GaAs belong to this group of compounds. Thus, indium, In, has in its O-shell three electrons with configuration $5s^2 5p^1$ and antimony, Sb, has five electrons in the states $5s^2 5p^3$. During the formation of covalent bonds one p-electron of Sb is transferred to In and as a result the electron configurations of the ions formed become similar to the electron configurations of Ge and Si. The same takes place in the case of GaAs.

Covalent bonding is very strong bonding. For example, in the diamond crystal the energy of bonding between two carbon atoms is 7.3 eV, which is comparable to the bonding energy of ionic crystals. The magnitude of such bonding is defined mainly by the overlapping of adjacent atoms' wavefunctions. Therefore, it decreases exponentially with increasing distance between atoms. If this distance is greater than several interatomic distances, the covalent bonding becomes negligibly small.

Crystals with ionic and covalent types of bonding can be considered as extreme cases. In between these two types of crystals there are crystals with mixed types of bonding. Atoms with almost-filled electron shells (Li, F, Na, Cl, K, and I) have predominantly ionic bonding. Atoms from groups III, IV, and V of the Periodic Table of the elements (Ga, In, Si, Ge, As, and Sb) form crystals with predominantly covalent bonding. Thus, in crystals of Si and Ge the bonding is completely covalent, whereas in crystals of InSb and GaAs the contribution of ionic bonding to the total binding energy amounts to 32%.

C.2.4 The nature of repulsion forces

Repulsion between atoms (ions) at small distances ($r \leq 0.1$ nm) has a purely *quantum origin*. This is because, at uniform pressure of a crystalline specimen, electron shells penetrate into each other and, thus, increase the specimen's electron concentration, n. In addition, the kinetic energy of electrons is increased, which can be ascribed to Heisenberg's uncertainty principle. Indeed, when the specimen's volume decreases, the uncertainty in coordinate position for each electron decreases, and as a result of this the uncertainty in electron momentum increases. The increase in momentum uncertainty is possible only with an increase of momentum and, therefore, with an increase of the average kinetic energy of electrons $\langle E \rangle$. To explain qualitatively the effect of repulsion, let us consider "free" electrons in a metal. The array of these electrons constitutes a *degenerate electron gas*. Taking into account the expression for the average energy, $\langle E \rangle$, of a degenerate electron gas (see Eqs. (C.32) and (C.38) in Section C.3.2 of this appendix), we can write the expression for the density of kinetic energy of the electron gas, w_e:

$$w_e = \langle E(n) \rangle \, n = \gamma n^{5/3}, \tag{C.10}$$

where

$$\gamma = \frac{3^{5/3} \pi^{4/3} \hbar^2}{10 m_e}. \tag{C.11}$$

The internal pressure of free electrons in metals, $\mathcal{P}(n)$, is related to the density of electron kinetic energy, $w_e(n)$, in metals by

$$\mathcal{P}(n) = \frac{2}{3} w_e(n) = \frac{2}{3} \langle E(n) \rangle \, n. \tag{C.12}$$

By applying uniform pressure, we decrease the volume, and as a result we increase the electron density in the specimen. Since

$$n = \frac{N}{V} \approx \frac{N}{N\beta r^3} = \frac{1}{\beta r^3}, \tag{C.13}$$

where β is a factor that depends on the geometry of the lattice and is approximately equal to unity, and r is the average distance between electrons, we have

$$\mathrm{d}n = -\frac{3}{\beta r^4}\,\mathrm{d}r = -3n\frac{\mathrm{d}r}{r} \tag{C.14}$$

(under compression $\mathrm{d}r < 0$, therefore $\mathrm{d}n > 0$). From Eq. (C.12) it follows that an increase of electron density, $\mathrm{d}n$, leads to an increase of the electron-gas pressure:

$$\mathrm{d}\mathcal{P} = \frac{2}{3}\frac{\mathrm{d}w_e}{\mathrm{d}n}\,\mathrm{d}n = -2n\left(\frac{\mathrm{d}w_e}{\mathrm{d}n}\right)\frac{\mathrm{d}r}{r}. \tag{C.15}$$

The estimate of this quantity for copper with electron density $n \approx 8 \times 10^{28}\ \mathrm{m}^{-3}$ in the case of decreasing the distance between electrons (due to uniform compression) by 10%, i.e., at $|\mathrm{d}r| = 0.1r$, is

$$\mathrm{d}\mathcal{P} \approx 4 \times 10^9\ \mathrm{Pa} \approx 4 \times 10^4\ \mathrm{atm}. \tag{C.16}$$

Here we used the dependence w_e on n for a degenerate electron gas. Such overpressure is due solely to the high forces of repulsion.

Thus, when ions in a metal approach each other, the lattice constant decreases. Therefore, the density and pressure of the electron gas increase, which leads to the emergence of significant repulsion forces. Let us substitute the average distance between electrons,

$$r \approx n^{-1/3}, \tag{C.17}$$

into the expression for the average energy,

$$\langle E \rangle = \gamma n^{2/3} = \frac{\gamma}{r^2}. \tag{C.18}$$

This quantity can be interpreted as the potential energy of repulsion of neighboring ions in a metal, i.e.,

$$U(r) = \frac{A}{r^2}, \qquad A = \gamma. \tag{C.19}$$

As the distance between atoms and ions decreases, the overlapping of electron wave-functions not only of outer shells but also of inner shells increases. As a result, the forces of repulsion sharply increase. The quantum-mechanical theory gives the following expression for the potential energy of repulsion:

$$U(r) = A'\mathrm{e}^{-r/a}, \tag{C.20}$$

which slightly differs from the first term in Eq. (C.3), A/r^m, when the parameter m takes the range of values $m \approx 9-11$. Because of overlapping of electron shells, forces of repulsion occur in dielectric crystals. Note that in dielectrics the parameter m has large values, whereas in metals this parameter is small ($m = 2$). This is why most dielectric crystals are fragile and metals are ductile.

Example C.1. Estimate the energy of van der Waals bonding of atoms located at distance $r \approx 3 \times 10^{-10}$ m from each other.

Reasoning. Let us assume that the instantaneous dipole moment of one of the atoms, which occurs as a result of fluctuation of charge, is equal to $\mathbf{p}_1$. At the distance r from this atom, at the center of another atom, which is positioned on the same axis as the first atom, an electric field, $\mathbf{E}$, is created:

$$\mathbf{E} = \frac{2k_e\mathbf{p}_1}{r^3}. \tag{C.21}$$

This field induces an instantaneous dipole moment of the second atom:

$$\mathbf{p}_2 = \alpha\mathbf{E} = \frac{2k_e\alpha\mathbf{p}_1}{r^3}, \tag{C.22}$$

where α is the atom's polarizability. Now, the energy of interaction of two atoms can be defined as the energy of interaction of two electric dipoles. The general expression for the energy of such interaction has the form

$$U(r) = k_e\left[\frac{\mathbf{p}_1\cdot\mathbf{p}_2}{r^3} - \frac{3(\mathbf{p}_1\cdot\mathbf{r})(\mathbf{p}_2\cdot\mathbf{r})}{r^5}\right]. \tag{C.23}$$

Since the dipole moments of atoms $\mathbf{p}_1$ and $\mathbf{p}_2$ are parallel to $\mathbf{r}$, Eq. (C.23) takes the form

$$U(r) = -\frac{2k_e p_1 p_2}{r^3} = -\frac{4k_e^2\alpha p_1^2}{r^6}. \tag{C.24}$$

From Eq. (C.24) it follows that these two atoms are attracted to each other because of instantaneous dipole–dipole interaction. Electron polarizability has the dimensionality [m^3], and for its estimation we can use the following approximate expression:

$$\alpha \approx a^3, \tag{C.25}$$

where a is the atomic radius. To estimate the dipole moment of an atom we can use the following expression:

$$p_1 \approx ea. \tag{C.26}$$

As a result, for the interaction energy of two atoms we obtain the following expression:

$$U(r) \approx -\frac{4k_e e^2 a^5}{r^6}. \tag{C.27}$$

Assuming $a \approx 10^{-10}$ m and $r \approx 4 \times 10^{-10}$ m, we get $U(r) \approx 2.5 \times 10^{-21}$ J ≈ 0.02 eV. Recalculating for one mole, this energy is equal to 1.5 kJ mol^{-1}, which is of the same order of magnitude as the real van der Waals binding energy.

C.3 Degenerate electron gas

C.3.1 The quantum statistics of electrons

Electrical, optical, and other properties of metals mostly are defined by the state of conduction electrons in metals. The conduction electrons are called *free electrons* because they are not bound to any particular atom of the crystalline lattice. As we have already noted, free electrons in metals obey quantum statistics. Therefore a gas of free electrons is called degenerate. In classical electron theory at $T = 0$ K the energy of all free electrons is equal to zero, i.e., all electrons are in one lowest-energy state. Quantum statistics does not allow such a state of the ensemble of electrons. Because of Pauli's principle, even at zero temperature the electrons must be in different energy states. Beginning from the lowest state, electrons fill one after another (with two electrons with spins having opposite directions on a single level) discrete energy levels of the allowed valence band up to an energy level called the *Fermi energy*, which is denoted as E_F. The filling of energy levels by electrons is defined by the Fermi function, $f(E)$:

$$f(E) = \frac{1}{e^{(E-E_F)/(k_B T)} + 1}, \tag{C.28}$$

which defines the probability of filling by electrons of the corresponding level with energy E under conditions of thermodynamic equilibrium of the electrons in crystal. Let us find out how electrons fill the energy levels in a metal at zero and non-zero temperatures.

If the temperature of a metal tends to absolute zero, then, according to Eq. (C.28), for electrons with $E > E_F$ the function $f(E) \to 0$, and for electrons with $E < E_F$ we have $f(E) = 1$. It follows from the above that energy levels lower than the Fermi level are occupied by electrons, since the probability of filling these levels is equal to unity. Energy levels higher than the Fermi level are unoccupied, since the probability of their filling is equal to zero. Figure 7.3 shows a plot of the function $f(E)$ at absolute zero. Thus, the Fermi energy is the energy of the highest level occupied by electrons in a metal at $T = 0$ K.

In the case of non-zero temperature, i.e., at $T > 0$ K, the probability of occupation by electrons of an energy level with $E = E_F$ is equal to 1/2. If $E < E_F$, then the probability of occupation of levels with such energy is greater than 1/2. If $E > E_F$, then $f(E) < 1/2$. Therefore, at $T > 0$ K levels that are lower than the Fermi level have higher probabilities of being occupied than do the levels higher than the Fermi level. The Fermi function, $f(E)$, is practically equal to unity for energies lower than the Fermi energy by several $k_B T$, whereas $f(E)$ is practically equal to zero for energies higher than the Fermi energy by several $k_B T$. Figure 7.3 shows the Fermi function for two temperatures. The higher the temperature, the wider is the energy interval over which the Fermi function changes from $f(E) = 1$ to $f(E) = 0$. The width of this interval is about (4–6) $k_B T$ (at room temperature $k_B T \approx 0.025$ eV). The width of the plot along the energy axis depends on the magnitude of the Fermi energy, E_F, which for metals is about several electron-volts. Thus, the graphical plot of the Fermi function in Fig. 7.3 is merely indicative since it does not take into account the real width of the function $f(E)$ along the energy axis.

C.3.2 The Fermi energy and average energy of a degenerate electron gas

In order to define E_F let us take into account that this is the energy of the highest level occupied by electrons at $T = 0$ K. All N conduction electrons have energy within the interval from 0 to E_F. Therefore,

$$N = \int_0^{E_F} g(E) \mathrm{d}E. \tag{C.29}$$

Here, the function $g(E)$ is the density of states of a three-dimensional crystal, which takes into account that each energy level can be occupied by two electrons. Let us substitute into Eq. (C.29) the equation obtained in Section 7.1, namely Eq. (7.16), for the density of quantum states for three-dimensional crystals:

$$N = \frac{\sqrt{2} V m_e^{3/2}}{\pi^2 \hbar^3} \int_0^{E_F} \sqrt{E}\, \mathrm{d}E = \frac{V (2 m_e E_F)^{3/2}}{3\pi^2 \hbar^3}. \tag{C.30}$$

From this expression we get the relationship that connects the concentration of free electrons, $n = N/V$, and the Fermi energy, E_F:

$$n = \frac{(2 m_e E_F)^{3/2}}{3\pi^2 \hbar^3}. \tag{C.31}$$

Equation (C.31) gives an important definition of the Fermi energy, E_F, in terms of the concentration of free electrons, n:

$$E_F = \frac{\hbar^2}{2 m_e} \left(3\pi^2 n\right)^{2/3}. \tag{C.32}$$

As an example, let us find the Fermi energy for free electrons in copper, Cu. Since each copper atom provides one conduction electron (valence electron), the concentration of

Table C.2. *Fermi-energy values for some metals*

Metal	Cs	K	Na	Li	Ag	Al	Be	Cu
E_F (eV)	1.53	2.14	3.12	4.72	5.5	11.9	14.6	6.9

free electrons in the Cu crystal is equal to the concentration of Cu atoms:

$$n = \frac{N_A \rho}{A} = \frac{6.02 \times 10^{26} \times 8.9 \times 10^3}{64} \approx 8.4 \times 10^{28}\ \mathrm{m}^{-3}, \tag{C.33}$$

where N_A is the Avogadro number, ρ is the density of copper, and A is the mass of one kilomole of copper. On substituting the obtained concentration of free electrons into Eq. (C.33), we obtain

$$E_F = \frac{(1.05 \times 10^{-34})^2}{2 \times 9.1 \times 10^{-31}} \left(3 \times \pi^2 \times 8.4 \times 10^{28}\right)^{2/3} \approx 1.1 \times 10^{-18}\ \mathrm{J} \approx 6.9\ \mathrm{eV}. \tag{C.34}$$

In Table C.2 we present the Fermi energy of degenerate electron gases in some metals.

Thus, at absolute zero temperature in the allowed energy band in which all of the valence electrons of a metal are present (this band is called the *conduction band*), all levels up to the Fermi level are occupied. At non-zero temperatures the energy of lattice thermal oscillations is two orders of magnitude less than the Fermi energy. Therefore, when the temperature of the metal increases, only a small number of electrons that obtained additional energy from lattice thermal oscillations (as a result of electron scattering on phonons) will move to higher unoccupied energy levels of the conduction band. Only electrons that are close to the Fermi level (i.e., electrons that are separated from unoccupied states by several $k_B T$) can make such a transition. Thus, the change in temperature affects the energy only of those electrons in the conduction band of the metal which are close to the Fermi level.

Only a small number of electrons that occupy levels lower than the Fermi level will move to the unoccupied levels that are higher than the Fermi level. The number of thermally excited electrons can be estimated with high precision as

$$N_T \approx N \frac{2 k_B T}{E_F}, \tag{C.35}$$

where N is the total number of electrons in the conduction band.

The main indication of degeneracy of an electron gas is the fact that the energy of the whole ensemble and of the individual electrons practically does not depend on temperature. The electron gas stays degenerate until each of its electrons could exchange energy with the crystalline lattice. This is possible only when the average energy of lattice thermal oscillations, $k_B T$, is comparable to the Fermi energy, E_F. The temperature

$$T_F = \frac{E_F}{k_B}, \tag{C.36}$$

at which an electron gas transforms from a non-degenerate state to a degenerate state is called the *degeneracy temperature*. In metals at all practically possible temperatures (for which the metal stays in a condensed state) the electron gas is degenerate. The electron gas stays non-degenerate only in semiconductors with a concentration of electrons less than $n = 10^{19}\ \text{cm}^{-3}$.

To find the average energy of an electron in a degenerate electron gas at $T = 0$ K, let us first find the total energy of the electron gas:

$$E_N = \int_0^{E_F} Eg(E)\text{d}E = \frac{\sqrt{2}Vm_e^{3/2}}{\pi^2\hbar^3}\int_0^{E_F} E\sqrt{E}\,\text{d}E = \frac{V(2m_e)^{3/2}}{\pi^2\hbar^3}(E_F)^{5/2}. \qquad \text{(C.37)}$$

Let us divide this energy by the number of electrons in the ensemble and take into account the expression for the Fermi energy (C.32). As a result, we find the following important relation for the average energy of a single electron:

$$\langle E\rangle = \frac{E_N}{N} = \frac{(2m_e)^{3/2}}{5\pi^2 n\hbar^3}(E_F)^{5/2} = \frac{3}{5}E_F. \qquad \text{(C.38)}$$

From Eq. (C.38) it follows that an electron gas in equilibrium has more electrons with high energy than electrons with low energy. According to Eq. (C.38), the average energy of electrons in Cu is close to 4.15 eV.

More accurate consideration shows that the Fermi energy and average electron energy in a metal change with temperature. With increasing temperature the Fermi energy negligibly decreases and the average electron energy slightly increases. Thus, when the temperature increases from 0 K to 1000 K, the Fermi energy of silver, which is equal to 5.5 eV at T = 0 K, decreases by only 0.01 eV.

Example C.2. Find the molar heat capacity of an electron gas and the total heat capacity of a metallic crystal.

Reasoning. According to classical theory, all conduction electrons in a metal must have input to the heat capacity of the electron gas, as in the case of a monatomic ideal gas. For one mole of monatomic ideal gas (the number of particles is equal to the Avogadro number, N_A), which is at temperature T, the total thermal energy is equal to

$$E_\mu = \frac{3}{2}k_B T N_A = \frac{3}{2}RT, \qquad \text{(C.39)}$$

where R is the universal gas constant. The molar heat capacity of such a gas is defined as

$$c_\mu = \frac{\text{d}E_\mu}{\text{d}T} = \frac{3}{2}R. \qquad \text{(C.40)}$$

If conduction electrons in metals obey classical statistics, the total molar heat capacity of a univalent metallic crystal (taking into account the heat capacity of the crystalline lattice $c_{cr} = 3R$) must be equal to

$$c_{met} = c_{cr} + c_{el} = 3R + \frac{3}{2}R = \frac{9}{2}R. \qquad \text{(C.41)}$$

However, at room temperature, for most metals the Dulong–Petit law is valid with very high precision, i.e., the molar heat capacity of a metal (or non-metal) is equal to $c_{\mathrm{met}} = 3R$. This fact shows that the electron gas in metals is not classical, i.e., this gas does not obey classical statistics, namely Maxwell–Boltzmann statistics.

As we have already mentioned, in metals a change of temperature leads to a change in energy only of those electrons which are on levels that differ from the Fermi energy by several $k_{\mathrm{B}}T$. Therefore, most of the conduction electrons will not have input to the heat capacity since their energy stays constant. We can estimate the magnitude of the electron heat capacity if we take into account that the number of "active" electrons is defined by N_T and their energy at thermal excitation increases approximately by $2k_{\mathrm{B}}T$. The energy of one mole of electron gas can be presented as

$$E_{\mu} = \frac{3}{5} E_{\mathrm{F}} N_{\mathrm{A}} + \frac{3}{2} k_{\mathrm{B}} T \left(\frac{2k_{\mathrm{B}}T}{E_{\mathrm{F}}} \right) N_{\mathrm{A}}. \tag{C.42}$$

As a result, an approximate expression for the molar heat capacity of a degenerate electron gas, $c_{\mathrm{el}} = \mathrm{d}E_{\mu}/\mathrm{d}T$, has the following form:

$$c_{\mathrm{el}} = 6R \left(\frac{k_{\mathrm{B}}T}{E_{\mathrm{F}}} \right). \tag{C.43}$$

If each atom of metal has not one but Z valence electrons, then we have

$$c_{\mathrm{el}} = 6ZR \left(\frac{k_{\mathrm{B}}T}{E_{\mathrm{F}}} \right). \tag{C.44}$$

More accurate calculations of the molar heat capacity of a degenerate electron gas instead of the coefficient 6 give $\pi^2/2$.

From the expressions written above it follows that the input from conduction electrons to the total heat capacity of metals at room temperatures is less than 1% of the input from the lattice heat capacity, which explains the validity of the Dulong–Petit law for metals. This input becomes noticeable only at very low temperatures, for which the heat capacity of the crystalline lattice, which is proportional to T^3, becomes smaller than the electron heat capacity, which is proportional to T. This temperature range is at liquid-helium temperatures for most metals.

C.4 Waves in a crystalline lattice and normal coordinates

The kinetic and potential energy of oscillatory motion of atoms in a crystalline lattice are functions of the displacement of each atom of a crystal. Since the displacements of all atoms are coupled, the total energy of a lattice, which is the sum of its kinetic and potential energies, depends on the coupled displacements of the lattice atoms. It is very important that, for the theoretical analysis and explanation of processes taking place in a crystal, the total energy of oscillations of atoms during wave propagation is presented in the form of a sum of energies of independent waves. Such a representation is possible as a result of transition from the usual representation of displacements of atoms to the so-called *normal coordinates* and to the corresponding *generalized momenta*. Let us introduce them using the example of a one-dimensional (linear) chain of atoms.

Each nth atom of a linear chain participates in an infinite number of displacements $u_n(q)$ with different wavenumbers q and corresponding ω_q and A_q:

$$u_n(q) = A_q e^{i(qan-\omega_q t)}. \tag{C.45}$$

Arbitrary motion can be presented in the form of the linear superposition of all possible waves of the form (C.45) with different wavenumbers q and corresponding frequencies ω_q and amplitudes A_q:

$$u_n = \sum_q \left[A_q e^{i(qan-\omega_q t)} + A_q^* e^{-i(qan-\omega_q t)}\right]. \tag{C.46}$$

Let us introduce a set of variables that can be convenient for further analysis:

$$a_q = \sqrt{N} A_q e^{-i\omega_q t}, \tag{C.47}$$

through which Eq. (C.46) can be rewritten in the form

$$u_n = \frac{1}{\sqrt{N}} \sum_q \left(a_q e^{iqan} + a_q^* e^{-iqan}\right). \tag{C.48}$$

Since the lattice contains a finite number of atoms, N, and from the cyclic boundary conditions it follows that q has N discrete values, $q = [2\pi/(aN)]g$, where $g = 1, 2, \ldots, N$, the summation in Eq. (C.48) can be done over all mentioned values of wavenumber, q.

The kinetic and potential energies of a linear chain of interacting atoms are given by the expressions

$$K = \frac{m}{2} \sum_{n=1}^{N} \left(\frac{du_n}{dt}\right)^2, \tag{C.49}$$

$$U = \frac{\beta}{2} \sum_{n=1}^{N} \left(u_n - u_{n-1}\right)^2. \tag{C.50}$$

From the expression for the potential energy (C.50) we can obtain the force acting on the nth atom from the neighboring atoms. Indeed,

$$F_n = -\frac{\partial U}{\partial u_n} = -\beta\left(2u_n - u_{n-1} - u_{n+1}\right). \tag{C.51}$$

On substituting Eq. (C.48) into the expressions for energies of Eqs. (C.49) and (C.50) and taking into account that

$$\frac{da_q}{dt} = -i\omega a_q, \qquad \frac{da_q^*}{dt} = i\omega a_q^*, \tag{C.52}$$

we come, after some cumbersome transformations, which we will omit here (see Example C.3), to the following expressions:

$$K = \frac{m}{2} \sum_q \omega_q^2 \left(2a_q a_q^* - a_q a_{-q} - a_q^* a_{-q}^*\right), \tag{C.53}$$

$$U = \frac{m}{2} \sum_q \omega_q^2 \left(2a_q a_q^* + a_q a_{-q} + a_q^* a_{-q}^*\right). \tag{C.54}$$

The total energy of a linear chain reduces to

$$E = K + U = 2m \sum_q \omega_q^2 a_q a_q^*. \tag{C.55}$$

Let us define now the normal coordinates of the lattice,

$$x_q = a_q + a_q^*, \tag{C.56}$$

and the corresponding momenta, which, taking into account Eq. (C.52), have the form

$$p_q = m \frac{\mathrm{d}x_q}{\mathrm{d}t} = \frac{m\omega_q}{\mathrm{i}} \left(a_q - a_q^*\right). \tag{C.57}$$

Both of the quantities x_q and p_q are real. Let us evaluate the parameters a_q and a_q^*:

$$a_q = \frac{1}{2}\left(x_q + \mathrm{i}\frac{p_q}{m\omega_q}\right), \qquad a_q^* = \frac{1}{2}\left(x_q - \mathrm{i}\frac{p_q}{m\omega_q}\right), \tag{C.58}$$

and let us substitute them into the expression for the total energy (C.55). As a result we obtain

$$E = \sum_q \left(\frac{p_q^2}{2m} + \frac{m\omega_q^2 x_q^2}{2}\right). \tag{C.59}$$

It follows that, represented in normal coordinates and corresponding momenta, the total energy of the wave motion of atoms in a linear chain is the sum of energies of independent harmonic oscillators, $\hbar\omega_q$. This representation, as will be shown in the next sections, allows us to describe the interaction of electrons in a crystal with lattice vibrations using the language of crystal quasiparticles. In addition, the elementary lattice excitations represented in terms of the normal coordinates (Eq. (C.59)) are called *phonons*. Phonons are one of the main types of quasiparticle in a crystal.

Example C.3. Obtain the expression for the kinetic energy of oscillations of atoms in a linear chain using the representation of an atom's displacement Eq. (C.48) expressed in terms of the variables a_q introduced by the relationship (C.47).
Reasoning. Let us substitute Eq. (C.48) into the expression for the kinetic energy (C.49):

$$K = \frac{m}{2} \sum_{n=1}^{N} \left(\frac{\mathrm{d}u_n}{\mathrm{d}t} \frac{\mathrm{d}u_n}{\mathrm{d}t}\right). \tag{C.60}$$

Taking into account Eq. (C.48),

$$K = \frac{m}{2N} \sum_{n=1}^{N} \frac{\mathrm{d}}{\mathrm{d}t} \left(\sum_q a_q \mathrm{e}^{\mathrm{i}qan} + a_q^* \mathrm{e}^{-\mathrm{i}qan}\right) \frac{\mathrm{d}}{\mathrm{d}t} \left(\sum_{q'} a_{q'} \mathrm{e}^{\mathrm{i}q'an} + a_{q'}^* \mathrm{e}^{-\mathrm{i}q'an}\right). \tag{C.61}$$

In the above equations only a_q and $a_{q'}$ are functions of t; thus, taking into account Eq. (C.52), we get

$$K = \frac{m}{2N} \sum_{qq'} \sum_{n=1}^{N} \omega_q \omega_{q'} \times \left[a_q a_{q'} \mathrm{e}^{\mathrm{i}(q+q')an} - a_q a_{q'}^* \mathrm{e}^{\mathrm{i}(q-q')an} - a_q^* a_{q'} \mathrm{e}^{-\mathrm{i}(q-q')an} + a_q^* a_{q'}^* \mathrm{e}^{-\mathrm{i}(q+q')an} \right]. \quad \text{(C.62)}$$

Let us do the summation in the last expression over the index n first, using the fact that the sums over this index are geometric progressions. Indeed,

$$\sum_{n=1}^{N} \mathrm{e}^{\mathrm{i}(q \pm q')an} = \sum_{n=1}^{N} \mathrm{e}^{\mathrm{i}\frac{2\pi}{N}gn} = \mathrm{e}^{\mathrm{i}\frac{2\pi}{N}g} + \mathrm{e}^{\mathrm{i}\frac{2\pi}{N}2g} + \cdots + \mathrm{e}^{\mathrm{i}\frac{2\pi}{N}Ng}$$
$$= \frac{\mathrm{e}^{\mathrm{i}\frac{2\pi}{N}g}\left(1 - \mathrm{e}^{\mathrm{i}2\pi g}\right)}{1 - \mathrm{e}^{\mathrm{i}\frac{2\pi}{N}g}} = \begin{cases} 0, & g \neq 0, q \pm q' \neq 0, \\ N, & g = 0, q \pm q' = 0, \end{cases} \quad \text{(C.63)}$$

where we must take into account that the wavenumber $q + q'$, reduced to the first Brillouin zone, takes N values because the number $g = 1, 2, \ldots, N$. Thus, the above-mentioned sum plays the role of Kronecker's symbol, i.e.,

$$\frac{1}{N} \sum_{n=1}^{N} \mathrm{e}^{\mathrm{i}(q \pm q')an} = \delta_{q \pm q'} = \begin{cases} 0, & q \pm q' \neq 0, \\ 1, & q \pm q' = 0, \end{cases} \quad \text{(C.64)}$$

which allows us to do the summation over the index n in Eq. (C.62):

$$K = \frac{m}{2} \sum_{qq'} \omega_q \omega_{q'} \left[a_q a_{q'} \delta_{q+q'} - a_q a_{q'}^* \delta_{q-q'} - a_q^* a_{q'} \delta_{q-q'} + a_q^* a_{q'}^* \delta_{q+q'} \right]. \quad \text{(C.65)}$$

Taking into account that $\omega_{-q} = \omega_q$ as a result of summation over the index q', we come to Eq. (C.53).

Example C.4. Find the molar heat capacity of a crystalline lattice consisting of atoms of the same type, using the classical description of the energy of thermal vibrations.
Reasoning. The crystalline lattice contains N atoms, which are in a state of thermodynamic equilibrium at a given temperature T. Atoms oscillate near their positions of equilibrium. For an arbitrary atom of the three-dimensional crystal there are three degrees of freedom of oscillatory motion (oscillatory motion in three independent directions). Considering a crystal as a system of non-interacting three-dimensional oscillators, we can write the crystal's energy of thermal motion as

$$E(T) = 3N \langle \varepsilon(T) \rangle, \quad \text{(C.66)}$$

where $\langle \varepsilon(T) \rangle$ is the mean thermal energy of a single linear oscillator that depends on the temperature. To find it, let us take into account that the mechanical energy of a harmonic oscillator averaged over a single period (or over a big time interval) is, according to Eq. (A.40), equal to

$$\varepsilon = K + U = 2\langle K\rangle. \tag{C.67}$$

Here we took into account that the mean potential and kinetic energies of an oscillator are equal to each other. Thus,

$$E(T) = 6N\,\langle K(T)\rangle\,. \tag{C.68}$$

In accordance with the well-known theorem of classical physics, in thermodynamical equilibrium each degree of freedom of a particle's translational motion corresponds to the mean kinetic energy of the crystal's thermal vibrations: $\langle K(T)\rangle = k_B T/2$. Thus, for the total energy of the crystal's thermal vibrations we obtain

$$E(T) = 3Nk_B T. \tag{C.69}$$

If the number of atoms in a crystal is equal to the Avogadro number, N_A, i.e., if we take one mole of a substance, its energy will be

$$E(T) = 3N_A k_B T = 3RT, \tag{C.70}$$

where $R = N_A k_B = 8.31\,\mathrm{J\,mol^{-1}K^{-1}}$ is the so-called *universal gas constant.* Since the energy found is an internal energy of a crystal, then, by definition, the molar heat capacity is equal to

$$c = \frac{\mathrm{d}E}{\mathrm{d}T} = 3R. \tag{C.71}$$

This relation is called the *Dulong–Petit law*. It works sufficiently well at high temperatures ($T > T_D$, where T_D is the so-called *Debye temperature* of the given substance), but it is not valid for low temperatures. At $T \to 0$ the molar heat capacity of all solids tends to zero. This fact can be explained only with the help of the quantum-mechanical description. If the substance is complex and in a unit cell there are s different atoms or ions, then its molar heat capacity is defined by the relation

$$c = 3sR. \tag{C.72}$$

C.5 The energy spectrum of an electron in a crystal

As we have already mentioned, ordinary crystals, which have as a structural unit individual atoms or ions and distances between them of several ångström units, are superlattices with various types of ordering of their structural units. The correct description of the electron dynamics in such a superlattice, i.e., in a periodic field of atomic nuclei, requires a quantum-mechanical approach based on the solution of the Schrödinger equation. The behavior of the entire system of electrons and nuclei is defined by its wavefunction. In our case, even without taking into account spin, the wavefunction of such a system will depend on the $3N$ spatial coordinates of the nuclei and the $3ZN$ coordinates of the electrons (here Z is the number of electrons per atom). Since for a crystal of macroscopic size N is approximately equal to 10^{23}, finding such a wavefunction and its analysis must be understood to be unrealistic. Because of this, for the solution of problems involving electrons in crystals certain well-established approximations, which allow one to calculate physical quantities observed in experiments, are used. The main approximations that are used in the electron theory of crystals are the *adiabatic* and *one-electron* approximations.

In the adiabatic approximation, which is based on the smallness of the ratio of the electron mass to the mass of a nucleus, the quantum-mechanical problem of the behavior of the system of electrons and nuclei splits into two simpler problems. The first problem is connected with finding the state of the electrons in the field of motionless nuclei, which are located at the vertexes of the crystalline lattice, i.e., at their equilibrium positions. The second problem is connected with the consideration of small-amplitude oscillations of nuclei near their positions of equilibrium as a result of interaction with the system of electrons whose states were defined in the first stage. This problem is reduced on the classical level to studying elastic oscillations and waves in crystalline structures. The correctness of the classical approach for the consideration of lattice vibrations is justified by the smallness of the de Broglie wavelength of the atoms and ions in comparison with the lattice constant ($\lambda_{\mathrm{Br}} \ll a$). In the case of thermal oscillations of a crystalline lattice the velocity of oscillating atoms may be estimated as follows:

$$v_T \approx \sqrt{\frac{3k_{\mathrm{B}}T}{m}}. \tag{C.73}$$

Using Eq. (C.73), we can find the de Broglie wavelength of these atoms:

$$\lambda_{\mathrm{Br}} = \frac{2\pi\hbar}{mv_T} \approx \frac{2\pi\hbar}{\sqrt{3mk_{\mathrm{B}}T}}. \tag{C.74}$$

Since for Si atoms $m \approx 4.7 \times 10^{-26}$ kg, their de Broglie wavelength is equal to $\lambda_{\mathrm{Br}} \approx 3 \times 10^{-11}$ m, which is one order of magnitude smaller than the lattice constant, a.

The one-electron approximation allows significant simplification of the complex problem of description of the motion of a system of electrons interacting with each other and with the motionless nuclei. This problem is reduced to the simpler problem of independent motion of each electron in a self-consistent periodic field, $U(\mathbf{r})$, which is created by all other particles of the crystalline structure. The form of this field is defined by the symmetry properties of the crystal. Even though the coordinate dependence of $U(\mathbf{r})$ is almost the same for different crystals, there are sharp differences among the behaviors of electrons in metals, dielectrics, and semiconductors. To understand the origin of these differences, let us analyze the behavior of electron systems in crystals using the above-mentioned approximations.

C.5.1 The wavefunction of an electron in a crystal

Considering hydrogen molecules in Section 6.6, we showed that electron atomic states become non-stationary when atoms form a molecule. Owing to the tunneling phenomenon, electrons in a molecule constantly tunnel from one atom to another, thus becoming

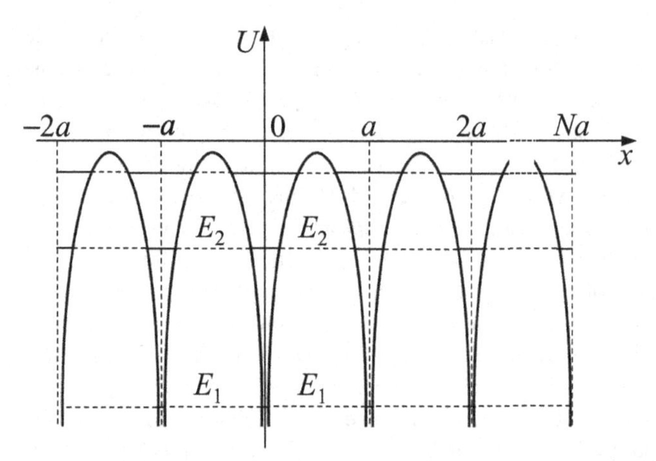

Figure C.9
Inner-crystalline potential.

collectivized for both atoms. This collectivization provides strong covalent bonding. An analogous situation takes place in a crystal, whose potential energy has the form of regularly placed potential wells separated by potential barriers. Figure C.9 shows a one-dimensional case of such a distribution of potential energy. Each electron of any atom in a crystal has a limited probability of tunneling through the potential barrier from one atom to another. The maximum probability will be for transitions between the same states of electrons in atoms. Since the probability of tunneling through the barrier decreases exponentially with increasing width and height of the potential barrier, electrons from the outermost shells have a higher probability of tunneling than do the internal electrons.

When two hydrogen atoms unite to form a hydrogen molecule, as a result of the Pauli exclusion principle each atomic level, E_q (the index q defines the set of all electron quantum numbers in an atom) splits into two close sublevels (see Section 6.6). The distance between two sublevels is defined as

$$E_q^{(2)} - E_q^{(1)} \approx 2A_q \ll E_q, \tag{C.75}$$

where A_q is the exchange integral for corresponding atomic states. An increase of the number of atoms binding increases the number of sublevels into which each initial atomic level is split. As a result, because of the exchange interaction, each atomic level, defined by a set of quantum numbers $\{q\}$, transforms into a band of closely placed sublevels whose width is proportional to $2A_q$. For deep-lying atomic levels, which correspond to internal electrons, this splitting is small. Therefore, these deep-lying bands are narrow.

For levels that correspond to the electrons from the outermost shells the splitting is considerable and the width of bands may be comparable to the separation between atomic levels. The electron quantum states of an individual atom are not in general stationary for a crystal. Each electron, as a result of tunneling from atom, to atom, belongs not to an individual atom but to the entire crystal. Only the internal electrons can be considered as belonging to their corresponding atoms. To describe such electrons in a crystal, their wavefunctions, taking into account the degeneracy of states, can be presented as a superposition of atomic wavefunctions $\psi_{0q}(\mathbf{r} - \mathbf{a}_n)$, corresponding to the vertex $\mathbf{a}_n$

and the atomic state with the set of quantum numbers $\{q\}$:

$$\psi_q(\mathbf{r}) = \sum_n C_n \psi_{0q}(\mathbf{r} - \mathbf{a}_n). \tag{C.76}$$

Here, $\mathbf{a}_n$ is the translation vector of the crystalline lattice, and the coefficients C_n define the probability amplitudes of finding an electron at a corresponding atom. Let us note that the electron wavefunction in a periodic inner-crystal potential has, according to Bloch's theorem, in the general case the form

$$\psi_q(\mathbf{r}) = G_k(\mathbf{r})\mathrm{e}^{\mathrm{i}\mathbf{k}\cdot\mathbf{r}}, \tag{C.77}$$

where $\mathbf{k}$ is the electron wavevector and the function $G_k(\mathbf{r})$ is a periodic function. The translation by a lattice vector $\mathbf{a}_n$ leads to multiplication of the electron wavefunction in the crystal by the exponential factor corresponding to this shift:

$$\psi_q(\mathbf{r} + \mathbf{a}_n) = \mathrm{e}^{\mathrm{i}\mathbf{k}\cdot\mathbf{a}_n} \psi_q(\mathbf{r}). \tag{C.78}$$

From this expression it follows that the expansion coefficients, C_n, in Eq. (C.76) must have the following form:

$$C_n = \mathrm{e}^{\mathrm{i}\mathbf{k}\cdot\mathbf{a}_n}. \tag{C.79}$$

The form of the Bloch electron wavefunction in a crystal (Eq. (C.77)) indicates the commonality with free-electron character of its behavior. The factor $G_k(\mathbf{r})$ modulates the wavefunction of a free electron $\mathrm{e}^{\mathrm{i}\mathbf{k}\cdot\mathbf{r}}$ with the periodicity of the crystalline lattice potential. Taking into account the temporal dependence, the electron wavefunction in a crystal can be presented in the following form:

$$\psi_q(\mathbf{r}, t) = G_k(\mathbf{r})\mathrm{e}^{\mathrm{i}(\mathbf{k}\cdot\mathbf{r} - E_q t/\hbar)}. \tag{C.80}$$

C.5.2 Energy bands in crystals

The one-electron wavefunction (C.76) is a solution of the Schrödinger equation with a self-consistent inner-crystal potential, $U_{\mathrm{eff}}(\mathbf{r})$, i.e.,

$$\hat{H}\psi_q(\mathbf{r}) = E_q \psi_q(\mathbf{r}), \tag{C.81}$$

$$\hat{H} = -\frac{\hbar^2}{2m_\mathrm{e}} \nabla^2 + U_{\mathrm{eff}}(\mathbf{r}). \tag{C.82}$$

The eigenvalue of the electron energy for the non-normalized wavefunction $\psi_q(\mathbf{r})$ is defined by the general expression

$$E_q = \frac{\int \psi_q^* \hat{H} \psi_q \mathrm{d}V}{\int \psi_q^* \psi_q \mathrm{d}V}, \tag{C.83}$$

where $\mathrm{d}V = \mathrm{d}x \times \mathrm{d}y \times \mathrm{d}z$. By calculating the integrals in Eq. (C.83) and taking into account orthonormalization of atomic wavefunctions, we obtain

$$E_q = E_{0q} - \left\langle W_q \right\rangle - \sum_n A_{qn} \mathrm{e}^{\mathrm{i}\mathbf{k}\cdot\mathbf{a}_n}. \tag{C.84}$$

Here E_{0q} is the corresponding atomic level, and the average value of the perturbation of an atomic potential calculated using the unperturbed state is defined as

$$\left\langle W_q \right\rangle = \int \psi_{0q}^*(\mathbf{r})[U_{\mathrm{eff}}(\mathbf{r}) - U_0(\mathbf{r})]\psi_{0q}(\mathbf{r})\mathrm{d}V, \tag{C.85}$$

where $U_0(\mathbf{r})$ is the atomic potential. The summation in Eq. (C.84) is done as a rule over the nearest-neighboring atoms which surround the atom which is chosen as the origin. This is because the exchange integral, A_{qn}, responsible for the splitting of the level E_{0q} for a neighboring atom significantly exceeds the exchange integrals for non-neighboring atoms.

Calculation of Eq. (C.84) for a simple cubic lattice gives the following expression for the energy dispersion in the band formed from atomic level E_{0q}:

$$E_q = E_{0q} - \left\langle W_q \right\rangle - 2A_q[\cos(ak_x) + \cos(ak_y) + \cos(ak_z)]. \tag{C.86}$$

When the crystal is formed from individual atoms, the electron energy it had in an isolated atom, E_{0q}, is decreased by $\left\langle W_q \right\rangle$, and is spread out into an energy band as a result of interaction with neighboring atoms. At the limits of the band the electron energy changes periodically with the components of the wavevector $\mathbf{k}$. The same decrease and spreading into a band happen to each stationary energy level of an isolated atom. The width of each allowed energy band is proportional to the exchange integral, $\Delta E_q = 12A_q$, and is mostly defined by the overlap of wavefunctions of neighboring atoms. For valence electrons this overlap is significant and thus the width of the allowed energy band can reach several electron-volts. For electrons of inner shells the splitting is small. Thus, for K-electrons in a crystal of metallic sodium, Na, the splitting is just 2×10^{-19} eV! Therefore, the corresponding band has the form of a narrow energy level.

Let us discuss the dispersion relation for electrons in crystals using expression (C.86) for a cubic lattice. It has very much in common with the equation for elastic waves in a discrete chain of atoms. First of all, let us note that the substitution of wavevector projection

$$k_\alpha \rightarrow k_\alpha + \frac{2\pi}{a} n_\alpha, \quad \alpha = x, y, z, \tag{C.87}$$

where n_α is an integer number, changes neither the dispersion relation nor the wavefunction. Therefore, instead of considering all possible values of the wavevector, it suffices to limit our consideration to the interval

$$-\frac{\pi}{a} \leq k_\alpha \leq \frac{\pi}{a}, \tag{C.88}$$

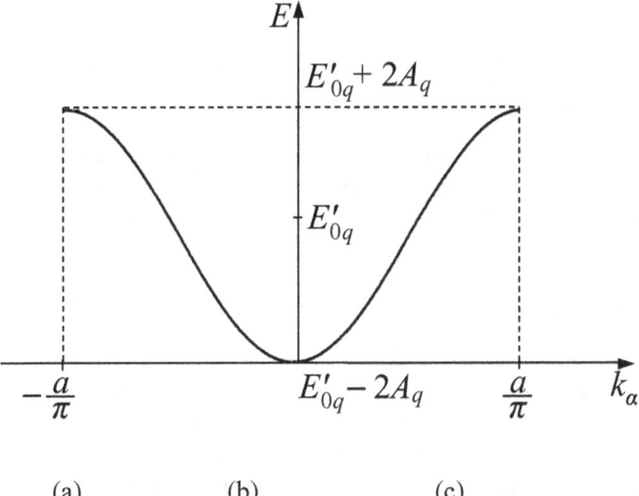

Figure C.10 The first Brillouin zone. Here $E'_{0q} = E_{0q} - \langle W_q \rangle$.

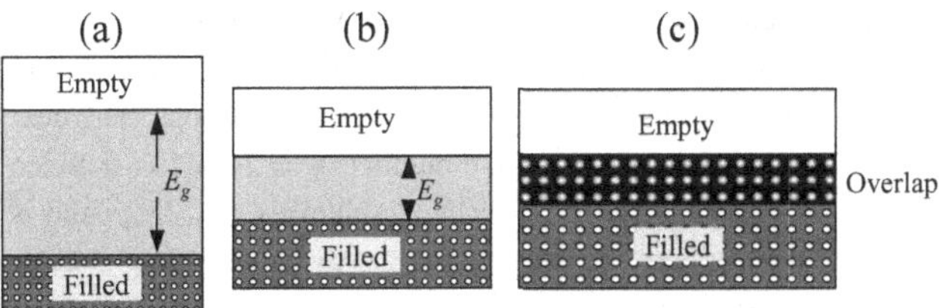

Figure C.11 Energy bands in the cases of (a) a dielectric ($E_g > 5$ eV), (b) a semiconductor ($E_g < 5$ eV), and (c) a metal ($E_g = 0$ eV).

which for electron waves in crystals is called the first *Brillouin zone* (see Fig. C.10). Thus, the electron wavevector, **k**, in a crystal is not defined unambiguously and therefore it is called the quasiwavevector. This vector is related to the quasimomentum by

$$\mathbf{p} = \hbar \mathbf{k}. \tag{C.89}$$

The allowed energy values for stationary states, $E_q(\mathbf{k})$, are limited to a band delimited by

$$E_{0q} - \langle W_q \rangle - 2A_q \le E_q \le E_{0q} - \langle W_q \rangle + 2A_q. \tag{C.90}$$

The energy intervals, E_g, that lie between the allowed energy bands are forbidden for electrons in stationary states (see Fig. C.11). These forbidden energy bands are called *bandgaps*.

Using periodic boundary conditions, we can show that inside of allowed bands the projections of the wavevector k_α have discrete values

$$k_\alpha = \frac{2\pi}{L_\alpha} n_\alpha, \quad \alpha = x, y, z, \tag{C.91}$$

where integer numbers n_α independently take on the values $\pm 1, \pm 2, \ldots, \pm N_\alpha/2$, and the size of the sample is $L_\alpha = N_\alpha a$. Since the number of atoms in the sample is enormous, i.e.,

$$N = \frac{L_x L_y L_z}{a^3} = nV, \tag{C.92}$$

where $n \approx 10^{28}$ m^{-3}, for a macroscopic sample the distance between the adjacent levels even in the widest band is not greater than 10^{-15} eV. For this reason the discreteness of the parameter k_α and the energy $E_q(\mathbf{k})$ in the allowed bands can be neglected.

C.5.3 Filling energy bands

As shown previously, each allowed band with the set of quantum numbers $\{q\}$ corresponds to N stationary states, which can, according to the Pauli principle, contain no more than $2N$ electrons with opposite spins. Therefore, the way of filling bands significantly depends on the valency of the atoms forming the crystal. Let us first consider a crystal of univalent lithium, Li, whose atoms have the electron configuration $1s^2 2s^1$. N sublevels of the lowest allowed band, which is formed from atomic levels with quantum numbers $n = 1$ and $l = 0$, are completely filled by 1s-electrons. N atoms have $2N$ such electrons. The next allowed band is formed from atomic levels with quantum numbers $n = 2$ and $l = 0$. Since there is only one 2s-electron, N electrons can fill only $N/2$ sublevels, i.e., this band is only half-filled. The picture for the filling of the two highest energy bands (completely or partly filled) of univalent metals is similar.

If the valency of an element is equal to two (or is even), then the last band must be completely filled. However, if the adjacent bands overlap, which is widely encountered in real crystals, then incompletely filled bands can have elements with arbitrary valency. Thus, for beryllium, Be, with electron configuration $1s^2 2s^2$, the energy bands which correspond to atomic levels with $n = 2$ and $l = 0$ and with $n = 2$ and $l = 1$ overlap, forming one wide band. Therefore, beryllium, Be, is a metal.

The difference in the form of filling of the highest allowed band is the basis for crystals' classification according to their conducting properties. The highest of the completely filled bands is usually called the *valence band*. The maximum energy in this band is denoted as E_v and is called the *top of the valence band*. The lowest partially filled or completely unfilled band is called the *conduction band*. The minimum energy in this band is denoted as E_c and is called the *bottom of the conduction band*. *Metals* are crystals whose valence band is filled completely and whose conduction band is not completely filled. *Dielectrics* are crystals whose valence band is completely filled and whose conduction band is empty. The region separating the valence and conduction bands, called the *bandgap*, has, for good dielectrics a width, E_g, of several electron-volts (see Fig. C.11).

C.5.4 Effective mass

The solutions of the Schrödinger equation (C.81) describe the behavior of electrons in a crystal in the absence of external fields. However, the response of an electron system to applied external fields is of practical interest. For analysis of electron behavior in a crystal in an external field an approximation called the *effective mass method* is widely used. The effective mass itself can be introduced using the dispersion relation.

Let us expand the energy $E_q(\mathbf{k})$ in a series near one of the extremum points $\mathbf{k}_0$ in the first Brillouin zone:

$$E_q(\mathbf{k}) = E_q(\mathbf{k}_0) + \frac{1}{2}\sum_{\alpha}\left(\frac{\partial^2 E_q}{\partial k_\alpha^2}\right)_{k_\alpha = k_{0\alpha}} (k_\alpha - k_{0\alpha})^2 . \tag{C.93}$$

To describe the motion of an electron in a periodic field of a crystal in the same way as we describe the motion of a free electron, let us introduce the tensor of inverse effective mass, the components of which on the main axes have the form

$$\frac{1}{m^*_{q\alpha}} = \frac{1}{\hbar^2}\left(\frac{\partial^2 E_q}{\partial k_\alpha^2}\right)_{k_\alpha = k_{0\alpha}} . \tag{C.94}$$

Then, the electron dispersion relation in a crystal (Eq. (C.93)) takes the same form as that of a freely moving particle whose inertial properties are different along each of the axes:

$$E_q(\mathbf{k}) = E_q(\mathbf{k}_0) + \sum_{\alpha} \frac{\hbar^2 (k_\alpha - k_{0\alpha})^2}{2m^*_{q\alpha}} . \tag{C.95}$$

If all three components of the tensor $1/m^*_{q\alpha}$ are the same in crystals with a cubic lattice, we can use the scalar effective mass m^*_q. In this case the dispersion relation can be presented in the form

$$E_q(\mathbf{k}) = E_q(\mathbf{k}_0) + \frac{\hbar^2 (k - k_0)^2}{2m^*_q} . \tag{C.96}$$

Near the bottom of the band, where the function $E_q(\mathbf{k})$ has its minimum, the effective mass of the electron m^*_q is positive, since at the point of the minimum

$$\left(\frac{\partial^2 E_q}{\partial k_\alpha^2}\right)_{k = k_0} > 0. \tag{C.97}$$

Near the top of the band the function $E_q(\mathbf{k})$ has its maximum, and the second derivative (C.97) is negative and $m^*_q < 0$. The effective mass, just like an ordinary mass, relates a force applied to the particle to the acceleration produced. If $m^*_q < 0$ then the direction of electron acceleration is opposite to the direction of the applied force. This must not be a surprise, since the effective mass of the electron takes into account the influence of the periodic crystalline potential on the electron. The influence of a periodic field and an external force on the electron, which has wave properties, can lead to such an effect.

Let us write the one-electron Schrödinger equation for an electron in a crystal in the presence of external forces:

$$-\frac{\hbar^2}{2m_{\mathrm{e}}}\nabla^2 \psi_q(\mathbf{r}) + [E_q - U_{\mathrm{eff}}(\mathbf{r}) + U(\mathbf{r})]\psi_q(\mathbf{r}) = 0. \tag{C.98}$$

Here, the inner-crystal potential $U_{\mathrm{eff}}(\mathbf{r})$ is a periodic function, which is usually unknown. The function $U(\mathbf{r})$ describes a non-periodic field, which affects the electron through the defects of crystalline structure and external electric and magnetic fields. Equation (C.98) in the approximation of the effective mass for a cubic lattice can be reduced to the following equation:

$$-\frac{\hbar^2}{2m_q^*}\nabla^2\varphi_q(\mathbf{r}) + [E_q - U(\mathbf{r})]\varphi_q(\mathbf{r}) = 0, \tag{C.99}$$

where the unknown function $U_{\text{eff}}(\mathbf{r})$ is absent and, instead of the mass of a free electron, m_e, we introduce the electron effective mass, m_q^*. The new wavefunction $\varphi_q(\mathbf{r})$ is related to Bloch's modulation factor, $G_q(\mathbf{r})$ which is a property of the electron wavefunction, (C.77) by,

$$\psi_q(\mathbf{r}) = G_q(\mathbf{r})\varphi_q(\mathbf{r}). \tag{C.100}$$

Equation (C.99) has the simplest form in the absence of external fields in an ideal (i.e., without defects) crystalline structure:

$$-\frac{\hbar^2}{2m_q^*}\nabla^2\varphi_q(\mathbf{r}) = E_q\varphi_q(\mathbf{r}). \tag{C.101}$$

Equation (C.101) describes the motion of an electron in a crystal as the motion of a free particle with mass equal to the effective mass, m_q^*, and energy, E_q, which defines a stationary energy state of an electron in one of the allowed bands. The electron wavefunction in this case has the form of a de Broglie wave:

$$\varphi_q(\mathbf{r}, t) = A\mathrm{e}^{\mathrm{i}(\mathbf{k}\cdot\mathbf{r} - E_q(\mathbf{k})t/\hbar)}, \tag{C.102}$$

where the dependence of the electron energy, E_q, on the wavevector, $\mathbf{k}$, is defined by Eq. (C.86). Knowledge of the form of the dispersion relation, $E_q(\mathbf{k})$, and the wavefunction, $\varphi_q(\mathbf{k}, t)$, allows us to find the most important characteristics related to the behavior of the electron subsystem in general.

C.5.5 Quasiparticles – electrons and holes

In the expression for the wavefunction (C.102) the energy $E_q(\mathbf{k})$ is the total electron energy in the energy band defined by the set of quantum numbers $\{q\}$. It contains the electron kinetic energy and energy of interaction with all other electrons and nuclei. Let us expand the expression (C.86) in series over the wavenumbers near the extremum point $\mathbf{k}_0 = 0$:

$$E_q(\mathbf{k}) = E_{0q} - \langle W_q \rangle - 6A_q + A_q a^2 k^2 = E_q^- + \frac{p^2}{2m_q^*}, \tag{C.103}$$

where we have introduced $\mathbf{p} = \hbar\mathbf{k}$ and

$$E_q^- = E_{0q} - \langle W_q \rangle - 6A_q, \tag{C.104}$$

$$m_q^* = \frac{\hbar^2}{2A_q a^2} = \frac{6\hbar^2}{\Delta E_q a^2}, \tag{C.105}$$

where ΔE_q is the width of the allowed band. From the equations obtained above it follows that the magnitude of the electron effective mass, m_q^*, is inversely proportional to the width of the band, ΔE_q. For the inner bands the magnitudes of A_q and ΔE_q are small and therefore the electron effective mass there is large. The consequence of this is larger inertia of electrons of inner bands and their weak response to external influences. Only in the conduction band is the magnitude A_q such that the effective mass of an electron, m_q^*, is similar to the mass of a free electron, m_e. Thus, in the approximation considered here, we can present the total electron energy in a crystal in the form of the kinetic energy of some *pseudo-electron* with mass m_q^*:

$$E_{qn} = E_q(\mathbf{k}) - E_q^- = \frac{p^2}{2m_q^*}. \tag{C.106}$$

The electron in a crystal is called a *pseudo-particle* because its "momentum" is a *quasimomentum*, i.e., it is a periodic quantity in momentum space and all its physically different quantum states can be reduced to the first Brillouin zone:

$$-\frac{\hbar\pi}{a} \leq p_\alpha \leq \frac{\hbar\pi}{a}. \tag{C.107}$$

Moreover, the mass of such an "electron" has not much in common with the mass of a free electron and, as we have already noted, the effective mass of an electron at the top of a band may even be negative. Therefore, we named it a "pseudo-electron," which reflects the unusual properties of this particle. Thus, with the help of the formal method considered here, we can reduce the complex problem of the motion of electrons interacting with each other and with nuclei to the problem of the motion of free pseudo-electrons.

Let us now discuss the question of how to present the dispersion relation near the maximum of the $E_q(\mathbf{k})$ dependence, i.e., near the top of a band near the points $k_{0\alpha} = \pm\pi/a$. In this case the expression for the electron energy can be written as

$$E_q(\mathbf{k}) = E_{0q} - \langle W_q \rangle + 6A_q - A_q a^2(\mathbf{k} - \mathbf{k}_0)^2. \tag{C.108}$$

If we make the substitution $\mathbf{k} - \mathbf{k}_0 = \mathbf{k}'$ and shift the origin of the energy to the top of the band, then the dispersion relation can be presented as

$$E_{qn} = E_q(\mathbf{k}) - E_q^+ = -A_q a^2 k'^2 = \frac{p^2}{2m_q^*}, \tag{C.109}$$

where we have introduced $\mathbf{p} = \hbar\mathbf{k}'$ and

$$E_q^+ = E_{0q} - \langle W_q \rangle + 6A_q, \tag{C.110}$$

$$m_q^* = -\frac{\hbar^2}{2A_q a^2}. \tag{C.111}$$

Therefore, the pseudo-electrons near the top of the band are characterized by a *negative effective mass* and a *negative kinetic energy*, which is not usual for the behavior of free electrons. We can introduce near the top of the band instead of pseudo-electrons with negative effective mass quasiparticles called *holes*, whose mass is positive. Since these quasiparticles are related to vacancies of quantum states near the top of the band, their charge is positive and equal to the charge of the electron. It is commonly accepted that parameters of quasiparticles – electrons and holes – may be written with corresponding indexes "n" and "p" that define their charge (*negative* and *positive*). Further, introducing these indexes will allow us to omit the index "q", which defines the set of quantum numbers of an electron in a band.

In order to describe a hole in terms of its usual physical characteristics, its kinetic energy is counted down from the top of the allowed and not completely filled band (usually this is a *valence band*). The energy of a hole is positive since

$$E_{\mathrm{p}} = -\frac{p_{\mathrm{n}}^2}{2m_{\mathrm{n}}} = -\frac{p_{\mathrm{p}}^2}{2m_{\mathrm{n}}} = \frac{p_{\mathrm{p}}^2}{2m_{\mathrm{p}}} > 0, \tag{C.112}$$

where $\mathbf{p}_{\mathrm{p}} = -\mathbf{p}_{\mathrm{n}}$ and $m_{\mathrm{p}} = -m_{\mathrm{n}} > 0$. Here, $\mathbf{p}_{\mathrm{n}}$ and $\mathbf{p}_{\mathrm{p}}$ are the quasimomenta of the electron and hole, which are related to the group velocities of the corresponding particles:

$$\mathbf{v}_{\mathrm{n}} = \frac{\mathbf{p}_{\mathrm{n}}}{m_{\mathrm{n}}}, \tag{C.113}$$

$$\mathbf{v}_{\mathrm{p}} = \frac{\mathbf{p}_{\mathrm{p}}}{m_{\mathrm{p}}} = \frac{-\mathbf{p}_{\mathrm{n}}}{-m_{\mathrm{n}}} = \mathbf{v}_{\mathrm{n}}. \tag{C.114}$$

Thus, the group velocities of these two quasiparticles are the same, i.e., $\mathbf{v}_{\mathrm{n}} = \mathbf{v}_{\mathrm{p}}$. For quasiwavevectors there are the following relationships:

$$\mathbf{k}_{\mathrm{p}} = \frac{\mathbf{p}_{\mathrm{p}}}{\hbar} = -\frac{\mathbf{p}_{\mathrm{n}}}{\hbar} = -\mathbf{k}_{\mathrm{n}}. \tag{C.115}$$

Thus, near the top of the energy band, we can consider the system of interacting electrons in a crystal from two points of view. First, this system is an array of free quasielectrons with negative charge having negative mass. Second, it is an array of free holes with positive charge and positive mass, which are moving in the opposite direction to the electrons. The description of the physical processes in a crystal which take place with the participation of electrons near the top of the valence band with the help of the notion of holes is more suitable and commonly accepted.

Example C.5. Under the conditions of the Hall effect the magnitude of the magnetic flux is $B = 0.1$ T and the current density along the x-axis in a semiconductor sample of n-type is equal to $j_x = 10^3\,\mathrm{A\,m^{-2}}$. Estimate the Hall voltage, $\Delta\varphi_{\mathrm{H}}$, and Hall constant, R_{H}, if the concentration of carriers is equal to $n = 10^{21}\ \mathrm{m^{-3}}$ and the length of the sample is equal to $L = 5 \times 10^{-3}$ m (see Fig. C.12).
Reasoning. The current density of charge carriers is related to their concentration and velocity by

$$j_x = nqv_x. \tag{C.116}$$

From the last expression we can find the carrier velocity:

$$v_x = \frac{j_x}{nq}. \tag{C.117}$$

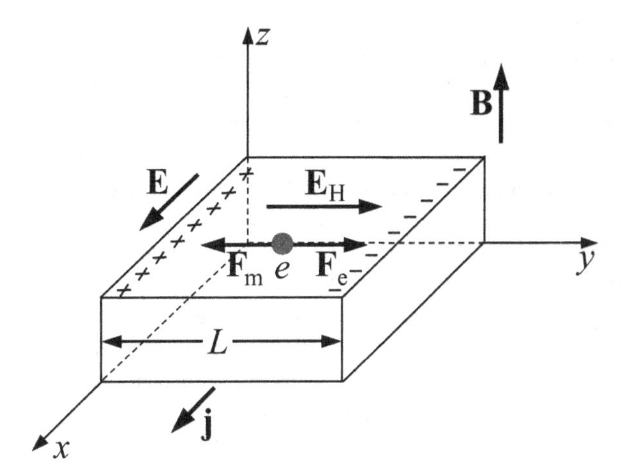

Figure C.12 The Hall effect.

The Hall effect is the occurrence of a transverse (with respect to the current direction) voltage in a sample placed in perpendicular electric and magnetic fields (i.e., $\mathbf{B} \perp \mathbf{E}$). This voltage can be found from the condition that, when the current flows along the x-direction, the electric force, $\mathbf{F}_e = q\mathbf{E}_H$, and magnetic force, $\mathbf{F}_m = q\mathbf{v} \times \mathbf{B} = j_x B/n$, which affect the carriers counterbalance each other, i.e.,

$$qE_H = \frac{j_x B}{n}, \tag{C.118}$$

where $\mathbf{E}_H$ is the electric field intensity of the transverse Hall field. Taking into account that $E_H = \Delta\varphi_H/L$, we obtain

$$E_H = \frac{\Delta\varphi_H}{L} = R_H j_x B, \tag{C.119}$$

where the Hall constant is equal to $R_H = 1/(nq)$. For the semiconductors with their different types of carriers we have

$$R_H = \pm\frac{1}{ne}, \tag{C.120}$$

where e is the modulus of the electron charge. Using the values of n and e, we obtain $R_H = 6.25 \times 10^{-3}$ m^3 C^{-1} and $\Delta\varphi_H = 3.1 \times 10^{-3}$ V. By finding the sign of the Hall constant, i.e., the direction of the Hall voltage, we can find the type of current carrier in semiconductors. Note that Fig. C.12 corresponds to charge of carriers $q = e > 0$.

C.6 Electrons in semiconductors

C.6.1 Intrinsic and extrinsic conductivity

Semiconductors are a special group of crystals. These materials have such a name because their resistivity is intermediate between those of dielectrics and metals. Thus, at room temperature the resistivity of metals is $\rho \approx 10^{-8}$ Ω m; for dielectrics the magnitude of this quantity is in the interval 10^{12}–10^{20} Ω m; whereas for semiconductors this quantity is in the interval 10^{-4}–10^{7} Ω m (pure germanium, Ge, has $\rho \approx 0.5$ Ω m). The materials which have semiconductor properties can be divided into two groups. The first group includes the so-called "elemental" semiconductors, which consist of atoms of only one element. The second group includes semiconductor compounds, which consist of atoms of two or more types.

The main representatives of the first group are semiconductors such as silicon, Si, whose atomic number is $Z = 14$, and germanium, Ge, whose atomic number is $Z = 32$. Both elements have four valence electrons on their outermost shell. The crystalline lattices of these semiconductors consist of atoms surrounded by four neighboring atoms, each of which is bound to a particular atom by covalent two-electron bonds. Therefore, in contrast to metals, whose valence electrons in a crystalline lattice become free, in semiconductors at $T = 0$ K there are no free carriers. This makes a semiconductor have the properties of a dielectric. However, there is a significant difference between semiconductors and dielectrics. The energy which is necessary to break the electron bond is defined by the width of the bandgap, E_g, which for different types of semiconductors ranges from 0.01 eV to 2 eV, is significantly smaller than that in dielectrics. For comparison, the width of the bandgap in Si is equal to $E_g = 1.1$ eV and in Ge $E_g = 0.7$ eV, whereas good dielectrics have E_g of about several electron-volts (for example, diamond has $E_g = 5.4$ eV).

The second group of semiconductor materials is large. We can mention first of all binary compounds of elements of the third and fifth groups of the Periodic Table of the elements (for example, GaAs and InSb). These compounds are usually denoted as A_3B_5 and their properties are very similar to those of "elemental" semiconductors from the first group. Binary semiconductors also include A_2B_6 compounds (for example, ZnS), metal oxides (for example, Cu_2O), and organic compounds.

External influences on such semiconductors (for example, heating or exposure to radiation) may detach some of the electrons from atoms and free them. This means that these electrons from the valence band move to the conduction band, where they occupy the lowest levels. Being in the conduction band, electrons in an external electric field behave as the electrons in a metal would, i.e., they participate in the directed transfer of charge. Simultaneously with the appearance of electrons in the conduction band, vacancies are formed in the valence band, which are called *holes*, i.e., not completely occupied levels. These levels may, because of external influences, be occupied by the electrons residing on the lower energy levels of a valence band. The direction of a displacement of these electrons in a valence band under the influence of an external electric field is equivalent to the direction of a displacement of positively charged holes. Thus, the existence of holes makes additional input to the electrical conductivity of a semiconductor. Its total conductivity is composed from electron and hole conductivities and, taking into account that the concentration of electrons, n_n, in the conduction band and that of holes, n_p, in the valence band are equal to each other, the conductivity can be written as

$$\sigma = q_n n_n \mu_n + q_p n_p \mu_p, \tag{C.121}$$

where q_n, μ_n and q_p, μ_p are the charge and mobility of electrons and holes, respectively. The electron and hole mobilities are similar in magnitude despite the motion of a hole being a peculiar "relay-race" motion of many electrons. Therefore, we can assume that the total conductivity is approximately twice the electron conductivity. This type of conductivity is called *intrinsic conductivity*, and semiconductors with this type of conductivity are called *intrinsic semiconductors*.

If the lattice of a semiconductor has breaks in periodicity because of the introduction of a small concentration of impurities, additional impurity levels appear in the bandgap of the semiconductor. If the impurity is of a *donor type*, i.e., the valence of the impurity atom is higher than the valence of the lattice atom (the outer shell of the impurity atom has one electron more than the lattice atom has), then the donor levels appear in the bandgap close to the bottom of the conduction band. If the valence of an impurity atom is lower than the valence of the lattice atom (the outer shell of the impurity atom has one electron fewer than the lattice atom has), then the impurity is of an *acceptor type* and the acceptor levels appear in the bandgap close to the top of the valence band. Thus, the impurity levels are close to the *bottom of the conduction band* or close to the *top of the valence band*. Such levels are called *shallow*, in contrast to *deep* levels, which are closer to the middle of the bandgap of a semiconductor. As a result of an external disturbance, including under the influence of an electric field, an electron can move either from the occupied donor level to the conduction band or from the valence band to an unoccupied acceptor level. The extrinsic conductivity, which is given by the electrons that move to the conduction band from donor levels, is called *n-type conductivity*. If the charge carriers are holes, which are created in the valence band as a result of electron transitions from this band to acceptor levels, we have *p-type conductivity*. In the case of p-type or n-type conductivity the number of electrons in the conduction band is no longer equal to the number of holes in the valence band. At a sufficiently high concentration of one type of impurities, the extrinsic conductivity is higher than the intrinsic conductivity. By changing the concentration of impurities we can control the conductivity of a semiconductor.

C.6.2 The position of the Fermi energy

As we have already mentioned, at $T = 0$ K there are no free carriers in an intrinsic semiconductor. All energy levels in the conduction band are unoccupied and all energy levels in the valence band are completely filled. In the general case the probability of filling the level with energy E is defined by the Fermi function:

$$f_{\mathrm{n}}(E) = \frac{1}{\mathrm{e}^{(E-E_{\mathrm{F}})/(k_{\mathrm{B}}T)} + 1}. \tag{C.122}$$

The probability that the level with energy E is not occupied by an electron is defined by the expression

$$f_{\mathrm{p}}(E) = 1 - f_{\mathrm{n}}(E) = \frac{1}{\mathrm{e}^{-(E-E_{\mathrm{F}})/(k_{\mathrm{B}}T)} + 1}. \tag{C.123}$$

Note that in these two equations the energy E is measured from an origin (for example, from the top of the valence band). Therefore, the energy E which corresponds to the conduction band is not equal to the energy E which corresponds to the valence band.

Since usually the width of the bandgap in semiconductors is much larger than the electron thermal energy, $E_{\mathrm{g}} \gg k_{\mathrm{B}}T$, the probability of electron transition from valence band to conduction band is very small. Therefore, the values of the function $f_{\mathrm{n}}(E)$, which defines the probability of filling by electrons of energy levels in the conduction band, are small, as are the values of the function $f_p(E)$, which defines the "filling" of energy levels in the valence band by holes. But if $f_n(E) \ll 1$ and $f_p(E) \ll 1$, then in Eqs. (C.122) and (C.123) we can neglect unity in comparison with the exponent and we can present these equations in the following forms:

$$f_n(E) = e^{-(E-E_F)/(k_B T)}, \quad \text{(C.124)}$$

$$f_p(E) = e^{(E-E_F)/(k_B T)}. \quad \text{(C.125)}$$

Each of the equations (C.124) and (C.125) has the form of Boltzmann's function, which defines the distribution of particles in an ideal gas over the energy. This means that, for a small concentration of free carriers, as occurs in intrinsic semiconductors at low temperatures, electrons and holes form non-degenerate electron and hole gases, which have properties very similar to those of the classical ideal gas of particles.

Using Eqs. (C.124) and (C.125), we can find the dependences on temperature of the electron concentration in the conduction band and the hole concentration in the valence band, which can be defined as follows:

$$n_n = 2\int_{E_g}^{E_1} f_n(E) g_n(E) \mathrm{d}E, \quad \text{(C.126)}$$

$$n_p = 2\int_{-E_2}^{0} f_p(E) g_p(E) \mathrm{d}E. \quad \text{(C.127)}$$

Here, the coefficient 2 takes into account filling of each energy level by two electrons (or holes) with opposite spins. The density of states of electrons $g_n(E)$ and that of holes $g_p(E)$ in the corresponding bands for unit crystal volume are defined by expressions like Eq. (7.16):

$$g_n(E) = \frac{(2m_n)^{3/2}}{4\pi^2\hbar^3}\sqrt{E - E_g}, \quad \text{(C.128)}$$

$$g_p(E) = \frac{(2m_p)^{3/2}}{4\pi^2\hbar^3}\sqrt{-E}, \quad \text{(C.129)}$$

where m_n and m_p are the effective masses of electrons and holes, respectively. On substituting Eqs. (C.128) and (C.129) into Eqs. (C.126) and (C.127) and taking into account the form of the distribution functions $f_n(E)$ and $f_p(E)$, we get the following expressions for concentrations:

$$n_n = \frac{(2m_n)^{3/2}}{2\pi^2\hbar^3}\int_{E_g}^{\infty} \sqrt{E - E_g} e^{-(E-E_F)/(k_B T)} \mathrm{d}E, \quad \text{(C.130)}$$

$$n_p = \frac{(2m_p)^{3/2}}{2\pi^2\hbar^3}\int_{-\infty}^{0} \sqrt{-E} e^{(E-E_F)/(k_B T)} \mathrm{d}E. \quad \text{(C.131)}$$

When we wrote the limits of integration in the above expressions we took into account that in the conduction band only the lowest levels are filled and the probability of filling the highest levels is practically equal to zero. In the valence band, holes fill only the highest levels and the probability of filling the lowest levels is equal to zero.

After carrying out the integration (see Example C.6 later) we come to the following expressions for carrier concentrations:

$$n_{\mathrm{n}} = \frac{1}{4\hbar^3}\left(\frac{2m_{\mathrm{n}}k_{\mathrm{B}}T}{\pi}\right)^{3/2} \mathrm{e}^{(E_{\mathrm{F}}-E_{\mathrm{g}})/(k_{\mathrm{B}}T)}, \tag{C.132}$$

$$n_{\mathrm{p}} = \frac{1}{4\hbar^3}\left(\frac{2m_{\mathrm{p}}k_{\mathrm{B}}T}{\pi}\right)^{3/2} \mathrm{e}^{-E_{\mathrm{F}}/(k_{\mathrm{B}}T)}. \tag{C.133}$$

As we have already mentioned, in an intrinsic semiconductor the concentrations of electrons and holes are equal to each other, i.e., $n_{\mathrm{n}} = n_{\mathrm{p}}$. From this condition we can find the following relation:

$$m_{\mathrm{n}}^{3/2}\mathrm{e}^{(E_{\mathrm{F}}-E_{\mathrm{g}})/(k_{\mathrm{B}}T)} = m_{\mathrm{p}}^{3/2}\mathrm{e}^{-E_{\mathrm{F}}/(k_{\mathrm{B}}T)}. \tag{C.134}$$

From the last expression we can find the Fermi energy of a non-degenerate intrinsic semiconductor:

$$E_{\mathrm{F}}(T) = \frac{E_{\mathrm{g}}}{2} + \frac{3}{4}k_{\mathrm{B}}T\ln\left(\frac{m_{\mathrm{p}}}{m_{\mathrm{n}}}\right). \tag{C.135}$$

We see that at $T = 0$ K the Fermi level, E_{F}, is exactly in the middle of the bandgap:

$$E_{\mathrm{F}}(0) = \frac{E_{\mathrm{g}}}{2}. \tag{C.136}$$

If the effective masses of electrons and holes were the same, then, even at $T \neq 0$ K, in the intrinsic semiconductor the Fermi level would stay in the middle of the bandgap. However, the electron mass in a semiconductor is not equal to the hole mass. Usually, $m_{\mathrm{p}} > m_{\mathrm{n}}$, and therefore the Fermi level in an intrinsic semiconductor lies higher than in the middle of the bandgap, i.e., it is closer to the bottom of the conduction band. When the temperature increases, the Fermi energy increases linearly and the Fermi level goes up. Using the expression (C.135) obtained for Fermi energy, the expressions for carrier concentrations (C.132) and (C.133) can be presented in the following form, which does not depend on E_{F}:

$$n_{\mathrm{n}} = n_{\mathrm{p}} = \frac{(m_{\mathrm{n}}m_{\mathrm{p}})^{3/4}}{4\hbar^3}\left(\frac{2k_{\mathrm{B}}T}{\pi}\right)^{3/2} \mathrm{e}^{-E_{\mathrm{g}}/(2k_{\mathrm{B}}T)}. \tag{C.137}$$

This expression can be rewritten as

$$n_{\mathrm{n}} = n_{\mathrm{p}} = n_0\mathrm{e}^{-E_{\mathrm{g}}/(2k_{\mathrm{B}}T)}, \tag{C.138}$$

where we introduced the quantity n_0:

$$n_0 = \frac{(m_n m_p)^{3/4}}{4\hbar^3}\left(\frac{2k_B T}{\pi}\right)^{3/2}. \tag{C.139}$$

Let us make estimates of the obtained quantities for a GaAs intrinsic semiconductor. The width of the bandgap for GaAs is equal to $E_g = 1.424$ eV, the effective masses of carriers are $m_n = 0.067m_e$ and $m_p = 0.47m_e$, and, according to Eq. (C.137), at $T = 300$ K

$$n_n = n_p \approx 2.25 \times 10^6 \text{ cm}^{-3}. \tag{C.140}$$

This is a very small quantity in comparison with the concentration of electrons in metals, which is about 10^{22} cm^{-3}. Nevertheless, an intrinsic semiconductor has a very small but finite conductivity. Using expression (C.121) and, for the carrier concentrations Eq. (C.138), we can find the dependence of the conductivity on temperature. Since the main dependence on temperature is related to the exponential dependence of the carrier concentration, for a pure (i.e., free of defects) intrinsic semiconductor we obtain the following law:

$$\sigma(T) = \sigma_0 e^{-E_g/(2k_B T)}, \tag{C.141}$$

where the quantity

$$\sigma_0 = en_0(\mu_n + \mu_p) \tag{C.142}$$

weakly depends on temperature. We see that, with temperature tending to zero, the conductivity also tends to zero. With increasing temperature the conductivity of an intrinsic semiconductor also increases. Using the experimental data for $\sigma(T)$ and Eq. (C.141), we can find the width of the bandgap, E_g. If on the ordinate axis we put $\ln \sigma$ and along the abscissa T^{-1}, then the slope of

$$\ln \sigma = \ln \sigma_0 - \frac{E_g}{2k_B}\frac{1}{T} \tag{C.143}$$

is equal to $-E_g/(2k_B)$.

C.6.3 Donor and acceptor levels

To find the distribution of impurity levels in the bandgap for a particular extrinsic semiconductor, we will use Bohr's model of the hydrogen atom. This model allows a qualitatively and sometimes quantitatively correct description of the distribution of impurity levels.

It is well known that impurity atoms do not occupy the interstitial space of the main lattice but substitute for individual atoms in the lattice sites. Let us consider first extrinsic silicon, Si, where the role of impurities is played by atoms of the fifth group of the Periodic Table of the elements (for example, atoms of As). The As atoms have five valence electrons. Four of them form covalent bonds with the nearest neighbors and the fifth electron becomes free. If the impurity atom is not ionized, this fifth electron is bound to the impurity atom, i.e., it rotates around the impurity atom. Thus, an atomic system very similar to the hydrogen atom is formed: near a positively charged As^+ ion there rotates one electron. It is easy to remove this electron from this bond, i.e., ionization of the As atom may be done easily. Such an impurity is called a *donor impurity*, since the impurity atom becomes the source (*donor*) of a free electron.

The impurity atom (As^+ plus an electron) has two substantial differences from the hydrogen atom, H. First of all, the electron effective mass in a semiconductor, m_n, is an order of magnitude smaller than the free-electron mass, m_e; second, the dielectric constant of a semiconductor is equal to $\epsilon \approx 10$, while for vacuum this parameter is equal to unity. Taking into account these peculiarities of an impurity atom and using Eq. (6.7), written for quantum states of the hydrogen atom, we can find for the energy states of an electron in an impurity atom

$$E_n^{(d)} = -\frac{k_e^2 m_n e^4}{2\epsilon^2 \hbar^2} \frac{1}{n^2}. \tag{C.144}$$

The ionization energy, $E_i^{(d)}$, of an impurity atom is equal to the modulus of the energy of its ground state ($n = 1$):

$$E_i^{(d)} = |E_1^{(d)}| = -\frac{k_e^2 m_n e^4}{2\epsilon^2 \hbar^2} = \frac{|E_1|}{\epsilon^2} \frac{m_n}{m_e}, \tag{C.145}$$

where $|E_1| = 13.6$ eV is the ionization energy of the hydrogen atom.

Similarly, taking into account Eq. (6.6) for the radius, r_1, of the hydrogen atom for the first Bohr orbit, we can write an expression for the size of an impurity atom, i.e., the average radius of its first Bohr orbit in a crystalline lattice:

$$r_i^{(d)} = \frac{\epsilon \hbar^2}{k_e m_n e^2} = r_1 \frac{m_e}{m_n} \epsilon. \tag{C.146}$$

Silicon, Si, has a dielectric constant equal to $\epsilon = 12$ and an electron effective mass $m_n \approx 0.1 m_e$. Thus, the ionization energy of the impurity is less by a factor of $\epsilon^2 m_e / m_n \approx 1.5 \times 10^3$ than the ionization energy of the hydrogen atom, i.e., $E_i^{(d)} \approx 10^{-2}$ eV. Since the width of the bandgap is about 1 eV, the impurity levels in the bandgap are fairly close to the bottom of the conduction band. Therefore, these levels are considered "shallow." The size of an impurity atom is, according to Eq. (C.146), $\epsilon m_e / m_n$ times larger than the size of the hydrogen atom, i.e., $r_i^{(d)} = 100 r_1 \approx 5$ nm, which is of the order of 10 interatomic distances in a crystalline lattice.

Let us consider an impurity atom from the third group of the Periodic Table of the elements (for example, an indium atom, In) in the crystalline lattice of silicon, Si. The In atom has only three valence electrons, which are not sufficient to form four covalent bonds with the nearest four Si atoms. This unoccupied bond, i.e., positively charged hole, may be filled as a result of transition of an electron from a neighboring filled bond. At the place from which this electron has come, a new hole may be formed, which in its turn can attract another electron, and so on. Such an impurity is called an *acceptor impurity*. As a

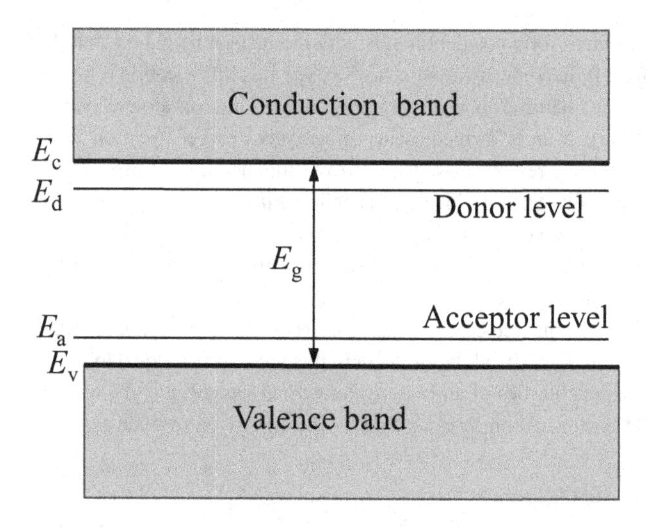

Figure C.13 Impurity levels in an extrinsic semiconductor.

result, near a negatively charged In^- ion (with four occupied covalent bonds) a positively charged hole is rotating. The system (In^- plus hole) is also similar to the hydrogen atom.

The hole may become free, i.e., leave the negatively charged impurity ion. From the energy point of view, the transition of an electron from the saturated bond to the unsaturated bond is the same as the electron transition from the filled valence band of a crystal to the local impurity level, which is in the bandgap close to the top of the valence band. This transition frees one of the levels in the upper part of the valence band, creating in the valence band a *vacant state* – a hole, which can freely move throughout the entire crystal under an applied electric field. For an acceptor impurity the corresponding energy levels (when counting from the top of the valence band) are defined by an expression analogous to Eq. (C.144):

$$E_{\mathrm{n}}^{(\mathrm{a})} = -\frac{k_{\mathrm{e}}^2 m_{\mathrm{p}} e^4}{2\epsilon^2 \hbar^2}\frac{1}{n^2} = \frac{|E_1|}{\epsilon^2}\frac{m_{\mathrm{p}}}{m_{\mathrm{e}}}\frac{1}{n^2} \approx \frac{10^{-2}}{n^2}\ \mathrm{eV}, \tag{C.147}$$

and the size of the corresponding impurity atom can be defined as

$$r_{\mathrm{i}}^{(\mathrm{a})} = \frac{\epsilon \hbar^2}{k_{\mathrm{e}} m_{\mathrm{p}} e^2} = r_1 \frac{m_{\mathrm{e}}}{m_{\mathrm{p}}}\epsilon. \tag{C.148}$$

The energies $E_1^{(\mathrm{d})}$ and $E_1^{(\mathrm{a})}$ of most impurity atoms are of the order of 10–20 meV, and the size of these atoms is two orders of magnitude larger than the size of the hydrogen atom in vacuum. Thus, an impurity atom in a crystal, with an electron or hole bound to it, is a system with very weak bonding but of big size. The impurity levels of an extrinsic semiconductor are shown schematically in Fig. C.13.

Example C.6. Carry out the integration in the expressions for carrier concentrations in an intrinsic semiconductor (C.130) and (C.131).

Reasoning. For shorter notation in the above-mentioned expressions, let us introduce the new variable

$$\beta = \frac{1}{k_B T}. \tag{C.149}$$

First, let us carry out the integration in the expression (C.130). Let us transfer to the new variable

$$u = (E - E_g)\beta. \tag{C.150}$$

Then,

$$dE = \frac{du}{\beta}. \tag{C.151}$$

The argument in the exponent takes the form

$$(E_F - E)\beta = (E_F - E_g)\beta - u. \tag{C.152}$$

Therefore, the integral is equal to

$$I_n = \int_{E_g}^{\infty} (E - E_g)^{1/2} e^{-(E-E_F)/(k_B T)}\, dE = \frac{1}{\beta^{3/2}} e^{(E_F - E_g)\beta} \int_0^{\infty} u^{1/2} e^{-u}\, du. \tag{C.153}$$

In the above integral, let us make the change of variable $u = x^2$. Then $du = 2x\, dx$ and we obtain

$$\int_0^{\infty} u^{1/2} e^{-u}\, du = 2 \int_0^{\infty} x^2 e^{-x^2}\, dx. \tag{C.154}$$

Let us integrate the last integral by parts:

$$2 \int_0^{\infty} x^2 e^{-x^2}\, dx = \int_0^{\infty} x e^{-x^2}\, d(x^2) = -x e^{-x^2}\Big|_0^{\infty} + \int_0^{\infty} e^{-x^2}\, dx = \frac{\sqrt{\pi}}{2}. \tag{C.155}$$

Thus, as a result of integration we obtain

$$I_n = \frac{\sqrt{\pi}}{2} (k_B T)^{3/2} e^{(E_F - E_g)/(k_B T)}. \tag{C.156}$$

On substituting the integral obtained into Eq. (C.130), we get Eq. (C.132) for the electron concentration in the conduction band of an intrinsic semiconductor.

To carry out the integration in Eq. (C.131), let us introduce a new variable $u = -\beta E$. Then, $dE = -du/\beta$ and

$$I_p = \int_{-\infty}^{0} (-E)^{1/2} e^{(E-E_F)/(k_B T)}\, dE = \frac{1}{\beta^{3/2}} e^{-\beta E_F} \int_0^{\infty} u^{1/2} e^{-u}\, du. \tag{C.157}$$

The last integral is, according to Eq. (C.155), equal to $\sqrt{\pi}/2$. As a result of substitution of I_{p} into Eq. (C.131), we obtain the expression (C.133).

C.7 Summary

1. Three-dimensional superlattices whose sites are occupied by quantum dots of various geometries can be considered as nanocrystals. The structural element of such a nanocrystal is a nanoparticle, which consists of a group of a relatively large number (from 10 to 1000) of atoms or ions, which are bound to each other and are organized in some ordered fashion.
2. The variety of nanocrystal structures as well as of conventional crystals is due to the differences in distribution of electrons among quantum states in atoms. The most important role in the formation of an individual nanoparticle and of a crystal is played by the electrons in the outermost atomic shells.
3. The most important property of crystalline structures is their translational symmetry, i.e., the ordered arrangement of particles throughout the entire crystal. In addition to translational symmetry, ideal crystals possess point symmetry. A crystal that has point symmetry coincides with itself under certain rotational operations and reflections.
4. The stability of the crystalline state indicates the existence of attraction forces between atoms (ions) at large distances, which at distances of about 0.1 nm become repulsive forces. In crystals the main, and energetically comparable types of bonding are ionic, metallic, and covalent bonding. Other types of bonding (intermolecular and hydrogen bonding) are weaker by one or two orders of magnitude.
5. Covalent bonding is one of the most important phenomena for the formation of various crystalline configurations. This bonding is realized between neutral atoms and is characterized by the saturation of bonding with a pronounced orientation. The saturation of bonding shows that attraction occurs only between atoms whose valence electrons are able to form pairs with oppositely directed spins. The orientation of bonding is defined by the orientation of the wavefunctions of the valence electrons forming bound states.
6. The repulsion between atoms (ions) at short distances is of purely quantum origin. Under uniform compression of a crystalline sample, the penetration of electron shells of neighboring atoms into each other takes place and therefore an increase of the electron density at each point of the sample occurs. This leads to an increase of the kinetic energy of electrons (which is a consequence of Heisenberg's uncertainty principle) and as a result it leads to an increase of the intrinsic pressure.
7. According to quantum statistics, the valence electrons in a crystal fill the discrete energy levels of the valence band one after another up to the Fermi level, E_{F}. Under conditions of thermodynamical equilibrium the probability of the filling by electrons of the corresponding level is defined by the Fermi function, $f(E)$.
8. A degenerate electron gas obeys the laws of quantum statistics. In metals the electron gas is degenerate at all temperatures acceptable for the crystalline state. The temperature at which an electron gas transforms from a degenerate state to a non-degenerate state is called the *degeneracy temperature*. The electron gas in intrinsic and extrinsic semiconductors with a concentration of free electrons less than 10^{19} cm^{-3} stays non-degenerate, i.e., it obeys the laws of classical statistics.

9. The main indication that an electron gas is degenerate is that the energy of the entire ensemble or of the individual electrons does not depend on temperature. For an individual degenerate electron, the average energy is equal to $\langle E\rangle = 3E_F/5$, whereas for a non-degenerate electron it is equal to $\langle E\rangle = 3k_B T/2$.
10. When a crystal is formed from individual atoms, each energy level of an individual atom, as a result of its interaction with neighboring atoms, is decreased and is split into an energy band. The width of each allowed energy band is proportional to the exchange integral and in many respects is defined by the overlap of the wavefunctions of the neighboring atoms. For valence electrons this overlap is large and therefore the width of the allowed band can be as great as several electron-volts. For electrons from the inner shells this splitting is small. Within the limits of the allowed band, the electron energy has a periodic dependence on the wavevector.
11. Each allowed band corresponds to N stationary states, which, according to the Pauli principle, can contain no more than $2N$ electrons with oppositely directed spins. The difference in character of the filling of the last of the allowed bands is the basis for a classification of crystals according to their conducting properties.
12. The highest of the bands filled by electrons is called the *valence band* and the highest energy in this band is denoted E_v. The lowest of the partially filled or completely empty bands is called the *conduction band* and the lowest energy in this band is denoted E_c. Metals are crystals whose conduction band is partially filled. Dielectrics are crystals whose valence band is completely filled and whose conduction band is empty. The bandgap which divides conduction and valence bands has a width, E_g, of several electron-volts for good dielectrics.
13. Near the bottom of the qth band, where the function $E_q(\mathbf{k})$ has its minimum, the effective mass of an electron, m_q^*, is positive. Near the top of the band, where the function $E_q(\mathbf{k})$ has its maximum, the effective mass of an electron, m_q^*, is negative.
14. Near the top of the valence band, the system of interacting electrons in a crystal usually is considered as a system of non-interacting quasiparticles called *holes*, which have positive charge and mass, but move in the direction opposite to the direction of electron motion.

C.8 Problems

Problem C.8.1. Copper has a face-centered cubic lattice. The density of copper is $\rho = 8.96 \times 10^3$ kg m^{-3} and the molar mass of copper is $\mu = 6.4 \times 10^{-2}$ kg mol^{-1}. Find the lattice constant, a, the atomic radius, r_a, and the number of atoms, n, per unit volume for a copper crystal. (Answer: $a \approx 3.61 \times 10^{-10}$ m, $r_a \approx 1.28 \times 10^{-10}$ m, and $n \approx 8.49 \times 10^{28}$ m^{-3}.)

Problem C.8.2. The energy per atom of a one-dimensional molecular crystal is given by the expression

$$u(x) = A\left[\left(\frac{a}{x}\right)^{12} - 2\left(\frac{a}{x}\right)^{6}\right], \tag{C.158}$$

where x is the distance between neighboring atoms; A and a are constants. Find the equilibrium interatomic distance, the binding energy per atom, and the coefficient of thermal expansion of such a crystal.

Problem C.8.3. Find the relation between the volumes of unit cells of direct, V_a, and inverse, V_b, lattices of crystals.

Problem C.8.4. Crystals of salt, NaCl, have a cubic structure, their molar mass is $\mu = 58.4 \times 10^{-3}$ kg mol^{-1} and their density is $\rho = 2.17 \times 10^3$ kg m^{-3}. Find the lattice constant, a, and the distances between crystallographic planes d_{100}, d_{110}, and d_{111}. (Answer: $a \approx 5.64 \times 10^{-10}$ m, $d_{100} \approx 2.82 \times 10^{-10}$ m, $d_{110} \approx 1.99 \times 10^{-10}$ m, and $d_{111} \approx 1.63 \times 10^{-10}$ m.)

Problem C.8.5. A unit volume of a metal with a simple cubic lattice contains n atoms. Each atom has Z valence electrons. In the free-electron approximation, find the radius of the Fermi sphere in **k**-space and in **p**-space, and the Fermi energy at $T = 0$ K.

Problem C.8.6. The work function for electrons in lithium, Li, is equal to $A_{wf} = 2.36$ eV, the density is $\rho = 5.34 \times 10^2$ kg m^{-3}, and the molar mass is $\mu = 6.94 \times 10^{-3}$ kg mol^{-1}. Find the depth of the potential well U_0, where conduction electrons of metals are confined. (Answer: $U_0 \approx 7.06$ eV.)

Problem C.8.7. The dispersion relation for electrons in a crystal with a simple cubic lattice has the form

$$E_n(\mathbf{k}) = E_n(0) + 2E_n(a)[\cos(k_x a) + \cos(k_y a) + \cos(k_z a)], \tag{C.159}$$

where a is the lattice constant and $E_n(0)$ is the electron energy in the nth state of an isolated atom. Find the electron effective masses and group velocities in the center and close to the apex of the Brillouin zone.

Problem C.8.8. The As atom with five valence electrons is a donor impurity in Ge, which has four valence electrons. Estimate the ionization energy, E_i, and the radius of the first electron orbit, r_1, of the valence electrons in an impurity atom. For germanium, Ge, the dielectric constant $\epsilon = 16$ and the effective mass of the electron is $m^* = m_e/2$. (Answer: $E_i \approx 0.026$ eV and $r_1 \approx 17 \times 10^{-10}$ m.)

Problem C.8.9. Find the electron concentration in the conduction band and the hole concentration in the valence band in an intrinsic non-degenerate semiconductor at temperature T.

Problem C.8.10. Find the position of the Fermi level in an intrinsic non-degenerate semiconductor, whose bandgap width, E_g, depends linearly on temperature, i.e., $E_g = E_g^0 - \gamma T$, where E_g^0 is the bandgap at $T = 0$ K.

Appendix D
Tables of units

Table D.1. *SI base units*

	Unit	
Quantity	Name	Symbol
Length	meter	m
Mass	kilogram	kg
Time	second	s
Electric current	ampere	A
Temperature	kelvin	K
Amount of substance	mole	mol

Table D.2. *SI derived units*

	Unit		
Quantity	Name	Symbol	Equivalent
Plane angle	radian	rad	$m/m = 1$
Solid angle	steradian	sr	$m^2/m^2 = 1$
Speed, velocity			$m\ s^{-1}$
Acceleration			$m\ s^{-2}$
Angular velocity			$rad\ s^{-1}$
Angular acceleration			$rad\ s^{-2}$
Frequency	hertz	Hz	s^{-1}
Force	newton	N	$kg\ m\ s^{-2}$
Pressure, stress	pascal	Pa	$N\ m^{-2}$
Work, energy, heat	joule	J	$N\ m$, $kg\ m^2\ s^{-2}$
Impulse, momentum			$N\ s$, $kg\ m\ s^{-1}$
Power	watt	W	$J\ s^{-1}$
Electric charge	coulomb	C	$A\ s$

(*cont.*)

Table D.2. *(cont.)*

Quantity	Unit: Name	Symbol	Equivalent
Electric potential, emf	volt	V	$\mathrm{J\,C^{-1}}$, $\mathrm{W\,A^{-1}}$
Resistance	ohm	Ω	$\mathrm{V\,A^{-1}}$
Conductance	siemens	S	$\mathrm{A\,V^{-1}}$, Ω^{-1}
Magnetic flux	weber	Wb	V s
Inductance	henry	H	$\mathrm{Wb\,A^{-1}}$
Capacitance	farad	F	$\mathrm{C\,V^{-1}}$
Electric field strength			$\mathrm{V\,m^{-1}}$, $\mathrm{N\,C^{-1}}$
Magnetic flux density	tesla	T	$\mathrm{Wb\,m^{-2}}$, $\mathrm{N\,A^{-1}\,m^{-1}}$
Electric displacement			$\mathrm{C\,m^{-2}}$
Magnetic field strength			$\mathrm{A\,m^{-1}}$
Celsius temperature	degree Celsius	°C	K
Luminous flux	lumen	lm	cd sr
Illuminance	lux	lx	$\mathrm{lm\,m^{-2}}$
Radioactivity	becquerel	Bq	$\mathrm{s^{-1}}$
Catalytic activity	katal	kat	$\mathrm{mol\,s^{-1}}$

Table D.3. *Physical constants*

Constant	Symbol	Value	Units
Speed of light in vacuum	c	2.9979×10^8 $\approx 3 \times 10^8$	$\mathrm{m\,s^{-1}}$
Elementary charge	e	1.602×10^{-19}	C
Electron mass	m_e	9.11×10^{-31}	kg
Electron charge to mass ratio	e/m_e	1.76×10^{11}	$\mathrm{C\,kg^{-1}}$
Proton mass	m_p	1.67×10^{-27}	kg
Boltzmann constant	k_B	1.38×10^{-23}	$\mathrm{J\,K^{-1}}$
Gravitation constant	G	6.67×10^{-11}	$\mathrm{m^3\,kg^{-1}\,s^{-2}}$
Standard acceleration of gravity	g	9.807	$\mathrm{m\,s^{-2}}$
Permittivity of free space	ϵ_0	8.854×10^{-12} $\approx 10^{-19}/(36\pi)$	$\mathrm{F\,m^{-1}}$
Permeability of free space	μ_0	$4\pi \times 10^{-7}$	$\mathrm{H\,m^{-1}}$
Planck's constant	h	6.6256×10^{-34}	J s
Impedance of free space	$\eta_0 = \sqrt{\mu_0/\epsilon_0}$	$376.73 \approx 120\pi$	Ω
Avogadro constant	N_A	6.022×10^{23}	$\mathrm{mol^{-1}}$

Table D.4. *Standard prefixes used with SI units*

Prefix	Abbreviation	Meaning	Prefix	Abbreviation	Meaning
atto-	a-	10^{-18}	deka-	da-	10^{1}
femto-	f-	10^{-15}	hecto-	h-	10^{2}
pico-	p-	10^{-12}	kilo-	k-	10^{3}
nano-	n-	10^{-9}	mega-	M-	10^{6}
micro-	μ-	10^{-6}	giga-	G-	10^{9}
milli-	m-	10^{-3}	tera-	T-	10^{12}
centi-	c-	10^{-2}	peta-	P-	10^{15}
deci-	d-	10^{-1}	exa-	E-	10^{18}

Table D.5. *Conversion of SI units to Gaussian units*

Quantity	SI unit	Gaussian unit
Length	1 m	10^2 cm
Mass	1 kg	10^3 g
Force	1 N	10^5 dyne $= 10^5$ g cm s^{-2}
Energy	1 J	10^7 erg $= 10^7$ g cm^2 s^{-2}
1 eV $= 1.602 \times 10^{-19}$ J $= 1.602 \times 10^{-12}$ erg		

Index

Printed in the United States
By Bookmasters